TRANSPORT SURVEY METHODS: KEEPING UP WITH A CHANGING WORLD

Related Books

HENSHER	Traffic Safety and Human Behavior
HENSHER & BUTTON	Handbooks in Transport — 6 Volume set
STOPHER & STECHER	Travel Survey Methods
HIMANEN, LEE-GOSSELIN & PERRELS	Building Blocks for Sustainable Transport
ELVIK & VAA	The Handbook of Road Safety Measures
BEN-AKIVA, MEERSMAN & VAN DE VOORDE	Recent Developments in Transport Modeling: Lessons for the Freight Sector

TRANSPORT SURVEY METHODS: KEEPING UP WITH A CHANGING WORLD

EDITED BY

PATRICK BONNEL
ENTPE-LET, Lyon, France

MARTIN LEE-GOSSELIN
Université Laval, Québec, Canada

JOHANNA ZMUD
NuStats, USA

JEAN-LOUP MADRE
INRETS, Paris, France

Emerald

United Kingdom • North America • Japan
India • Malaysia • China

Emerald Group Publishing Limited
Howard House, Wagon Lane, Bingley BD16 1WA, UK

First edition 2009

Reprints and permission service
Contact: booksandseries@emeraldinsight.com

British Library Cataloguing in Publication Data
A catalogue record for this book is available from the British Library

ISBN: 978-1-84855-844-1

ISOQAR QUALITY ASSURED ISO 9001 UKAS QUALITY MANAGEMENT 026
Certificate number 1985

Awarded in recognition of Emerald's production department's adherence to quality systems and processes when preparing scholarly journals for print

INVESTOR IN PEOPLE

Contents

PART III: GLOBAL SOCIAL ISSUES

PART IV: FREIGHT AND TRANSIT PLANNING

List of Contributors

Elizabeth S. Ampt	Sinclair Knight Merz, Adelaide, Australia
Carlos H. Arce	Mygistics-NuStats, Austin, TX, USA
Jimmy Armoogum	French National Institute of Research on Transports and Safety (INRETS-DEST), Noisy, France
Kay W. Axhausen	Swiss Federal Institute of Technology Zurich — Institut for Transport Planning and Systems (ETHZ-IVT), Zurich, Switzerland
Earl J. Baker	Florida State University, Tallahassee, FL, USA
Caroline Bayart	LET-ENTPE-CNRS-Université Lyon 2, Lyon, France
Roger Behrens	University of Cape Town, Cape Town, South Africa
Patrick Bonnel	LET-ENTPE-CNRS-Université Lyon 2, Lyon, France
Peter Bonsall	Institute for Transport Studies, University of Leeds, Leeds, UK
Mark Bradley	Bradley Research and Consulting, Santa Barbara, CA, USA
Stacey Bricka	NuStats, Austin, TX, USA
Werner Brög	Socialdata, Munich, Germany
Michael Browne	University of Westminster, London, UK
Linda Cherrington	Texas Transportation Institute, The Texas A&M University System, Houston, TX, USA
Kelly J. Clifton	University of Maryland, College Park, MD, USA
Heather Contrino	Federal Highway Administration, USA
Eric Cornelis	FUNDP, University of Namur, Belgium
Matthieu de Lapparant	INRETS, Noisy-le-Grand Cedex, Paris, France

Sean T. Doherty	Department of Geography & Environmental Studies, Wilfrid Laurier University, Waterloo, Ontario, Canada
Mark Freedman	Westat, Rockville, MD, USA
Regine Gerike	University of Technology, Munich, Germany
Jane Gould	University of California, Los Angeles, CA, USA
E. Hato	Behavior in Networks Studies Unit, University of Tokyo, Tokyo, Japan
Peter Jones	University College London, London, UK
Dominika Kalinowska	German Institute for Economic Research, DIW Berlin, Germany
Ian Ker	CATALYST, WA, Australia
Kara Kockelman	University of Texas at Austin, TX, USA
Stephan Krygsman	University of Stellenbosch, South Africa
Catherine T. Lawson	University at Albany, State University of New York, Albany, NY, USA
Martin Lee-Gosselin	Université Laval, ESAD-CRAD, Québec, Canada
Jacques Leonardi	University of Westminster, London, UK
Bob Leore	Department of Transport, Ottawa, Canada
Jean-Loup Madre	INRETS, Paris, France
Michael Manore	Vispective Management Consulting, USA
Nancy McGuckin	Travel Behavior Associates, South Pasadena, CA, USA
Alan McKinnon	Heriot-Watt University, Edinburgh, UK
Arnim Meyburg	Cornell University, Ithaca, NY, USA
Catherine Morency	Ecole Polytechnique de Montréal, MADITUC-CIRRELT, Canada
Elaine Murakami	Federal Highway Administration, Seattle, WA, USA
Sharon O'Connor	Resource Systems Group, White River Junction, VT, USA
Juan de Dios Ortúzar	Department of Transport Engineering and Logistics, Pontificia Universidad Católica de Chile, Santiago, Chile

Peter Ottmann	Institute for Transport Studies, Karlsruhe, Germany
Danièle Patier	LET, Université de Lyon, France
Alan Pisarski	Consultant, Falls Church, VA, USA
John W. Polak	Centre for Transport Studies, Imperial College London, UK
Martine Quaglia	INED-SFS, Paris, France
Tim Raimond	Transport Data Centre, NSW Transport and Infrastructure, Australia
Benoît Riandey	INED-SFS, Paris, France
Anthony J. Richardson	The Urban Transport Institute (TUTI), Victoria, Australia
Matthew Roorda	University of Toronto, Toronto, Canada
Jean-Louis Routhier	LET, Université de Lyon, France
Gerd Sammer	Institute for Transport Studies, University of Natural Resources and Applied Life Sciences, Vienna, Austria
Giorgia Servente	DITIC, Politecnico di Torino, Torino, Italy
Peter R. Stopher	The Institute of Transport and Logistics Studies, The University of Sydney, Australia
Orlando Strambi	Escola Politécnica da Universidade de São Paulo, Brazil
Christophe Terrier	INSEE, France
H. J. P. Timmermans	Urban Planning Group, Eindhoven University of Technology, The Netherlands
Martin Trépanier	Ecole Polytechnique, University of Montreal, Canada
Klaas van Zyl	SSI Engineers and Environmental Consultants, South Africa
Chester Wilmot	Louisiana State University, Baton Rouge, LA, USA
Jean Wolf	GeoStats, Atlanta, GA, USA
Johanna Zmud	NuStats, USA
Dirk Zumkeller	Institute for Transport Studies, Karlsruhe, Germany

Preface

About every three years the international transport survey community gets together to discuss innovation and quality in transport survey methods and document those discussions in a publication. As this has been happening since 1979, one has to ask: Do we really need another book on survey methods in transport? The answer is "Yes, If..." with the "If" involving whether there have been recent changes that make the capture of transportation data through surveys materially different than they were years ago when other books on the topic were published. One obvious example of a recent change is the need for new data, models, and other analytical tools to support greenhouse gas (GHG) reduction and energy efficiency policies in nations around the globe. Another example is the proliferation of new and affordable information technologies that survey designers can employ to collect and process data, helping them confront increasing barriers to participation in surveys, barriers that in some cases arise from the same technologies, such as developments in telephony. This book focuses on such changes, and on the opportunities and challenges they represent, both for improved survey methods and for the comparability of the data that they provide to different agencies and countries.

As statistical surveys attempt to address GHG issues, and other important transportation policy and planning challenges, they exhibit evidence of success, yet at the same time they frequently come under threat. The evidence of success is that surveys are ubiquitous in the transport world. Almost all countries in the world use them to measure passenger travel, freight movements, or public transit ridership. Leaders in government use survey results to guide policy, and the call from their advisors for more data to address greater challenges is increasing. However, funding for data collection is too often an easy target in difficult economic times, such as the present, and surveys must be credible, transparent, and of assured quality. The papers in this book are thus relevant to government, transport industry practitioners, academic scientists, and commercial researchers.

The book provides a review of the current state of transport survey methods for capturing data in several key areas: freight, personal travel, tourism, evacuations and related travel, and the environmental footprint of transport, among others. It captures the essence of discussions at the 8th International Conference on Survey Methods in Transport that took place in Annecy, France, in 2008. Conference participants from over 25 countries included leading survey researchers and transport professionals representing industry and government policy makers, as well as academic scholars and researchers.

The Annecy Conference succeeded in its main objectives: sharing up-to-date information and experiences on transport survey methods; fostering discussion of mutual problems and issues that affect survey design, data processing, and reporting; proposing and suggesting new initiatives and future approaches for the measurement of critical transportation system indicators; and feeding the results of these discussions into a permanent record in the form of this peer-reviewed book. The book is not a proceedings volume, but a peer-reviewed selection of about one-third of the papers that were presented, as well as a synthesis of 16 workshops.

An editorial committee guided the work that led to this book. It consisted of the four co-editors: Patrick Bonnel, Martin Lee-Gosselin, Jean-Loup Madre, and Johanna Zmud, who also served as co-chairs of the Conference. These four, together with the help of Jimmy Armoogum, divided up the editorial oversight. They built on the considerable efforts of the many people and organizations, recognized in the Acknowledgements that follow. We are indebted to all those who donated their time and energy to review, critique, and add to our body of knowledge about transport survey methods, in order to continuously improve the quality of transport surveys and enhance the value and utility of the data that such surveys provide for transport policy and decision-making.

Finally we, the co-chairs of the ISCTSC, thank all the Annecy authors for their diligence and hard work. We are confident that their continued diligence will lead to new insights for, and new approaches in, transport survey methods.

Martin Lee-Gosselin
Johanna Zmud
ISCTSC Co-Chairs

Acknowledgements

The series of international transport survey conferences and the published record is under the responsibility of the ISCTSC. For the period leading up to the Annecy Conference through to the completion of this book, the members were:

International Steering Committee for Travel Survey Conferences (ISCTSC)		
Tom Adler	Resource Systems Group, Inc.	U.S.A.
Carlos Arce	NuStats, Inc.	U.S.A.
Chandra Bhat[2]	University of Texas	U.S.A.
Werner Brög	Socialdata, GmbH	Germany
Patrick Bonnel	ENTPE	France
Kelly Clifton[2]	University of Maryland	U.S.A.
Peter Jones	University College, London	United Kingdom
Ryuichi Kitamura[1]	Kyoto University	Japan
Stephan Krygsman[2]	Stewart Scott, Inc.	South Africa
Martin Lee-Gosselin (Co-Chair)	Université Laval	Canada
Jean-Loup Madre	INRETS	France
Arnim Meyburg	Cornell University	U.S.A.
Catherine Morency[2]	Polytechnique de Montréal	Canada
Elaine Murakami[1]	FHWA	U.S.A.
Juan de Dios Ortuzar	Pontificia Universidad Católica	Chile
Alan Pisarski	Consultant	U.S.A.
Tony Richardson	The Urban Transport Institute	Australia
Gerd Sammer	Universität für Bodenkultuur	Austria
Cheryl Stecher[1]	IBM	U.S.A.
Orlando Strambi	Universidade de São Paulo	Brazil
Peter Stopher	University of Sydney	Australia
Harry Timmermans	Technical University of Eindhoven	Netherlands
Klaas van Zyl[1]	Stewart Scott, Inc.	South Africa
Chester Wilmot	Louisiana State University	U.S.A.
Johanna Zmud (Co-Chair)	NuStats, Inc.	U.S.A.
Dirk Zumkeller	Universität Karlsruhe	Germany

[1]Member until October 2008.
[2]Member from October 2008.

It was with great sadness that we learnt, in February 2009, of the death of Ryuichi Kitamura. Ryuichi was a source of endless energy and inspiration among the international community of those concerned with data on travel, communication and activities. He saw well beyond transport networks, caring deeply about the data needed to help communities become more responsive to human values. He was one of a kind and we miss him very much.

The success of the Annecy Conference was owed to nearly two years of painstaking preparations by the Local Organising Committee (LOC) in France, co-chaired by Patrick Bonnel and Jean-Loup Madre, both of whom have also been longstanding members of the International Steering Committee (ISCTSC), and who took a very active role in the scientific programme as well. The LOC comprised:

Patrick Bonnel, Laboratoire d'Economie des Transports (LET-ENTPE) (Co-Chair)

Jean-Loup Madre, Institut National de la Recherche sur les Transports et leur Sécurité (INRETS) (Co-Chair)

Jimmy Armoogum, Institut National de la Recherche sur les Transports et leur Sécurité (INRETS)

Gerard Brun, Ministère français de l'Ecologie, de l'Energie du développement durable et de la Mer, direction de la recherche et de l'animation scientifique et technique (DRAST)

Marc Christine, Institut National de la Statistique et des Etudes Economiques (INSEE)

Mary Crass, International Transport Forum (ECMT/OECD)

Marie-Odile Gascon, Centre d'Etudes sur les Réseaux, les Transports, l'Urbanisme et les constructions (CERTU)

Emmanuel Raoul, Ministère français de l'Ecologie, de l'Energie du développement durable et de la Mer

In addition, our sincere thanks to Christophe Rizet of INRETS for all the arrangements needed to make possible holding the final meeting of COST Programme 355 in collaboration with the ISCTSC meeting.

A meeting of the size of the Annecy Conference also depended on the support of many staff. We are particularly indebted to Florence Toilier of ENTPE for her patient resolution of many logistical questions and the production of the conference CDROM, Peter Endemann of Planungsverband Ballungsraum Frankfurt/Rhein-Main and the ETC for so effectively running the Conference secretariat at Annecy (in five languages), to Pierre-Olivier Flavigny and Philippe Marchal of INRETS for organising the most hospitable Internet facilities that any of us have seen heretofore at a conference, and to the staff of Les Balcons du Lac d'Annecy.

The ISCTSC would also like to recognise that the Annecy Conference would not have been possible without the generosity of our sponsors, who provided financial, material and/or staff support. In addition, the sponsors made it possible to continue ISCTSC's commitment to invest in our future by providing scholarships to participants from lower-income countries and some support to new researchers and students. They were:

- La Direction de la recherche et de l'animation scientifique et technique (DRAST), Ministère français de l'Ecologie, de l'Energie du développement durable et de la Mer
- Ecole Nationale des Travaux Publics de l'Etat (ENTPE), France
- Institut National de la Recherche sur les Transports et leur Sécurité (INRETS), France
- Laboratoire d'Economie des Transports (LET), France
- NuStats, United States
- PTV AG, Germany
- Service de l'Observation et des Statistiques (SOeS), Ministère français de l'Ecologie, de l'Energie du développement durable et de la Mer
- The Urban Transport Institute (TUTI), Australia

In addition to these sponsors, we are most grateful for the many individuals in the world community of transport survey researchers and practitioners who donated their time and expertise as authors, reviewers, chairs, discussants and rapporteurs. More than 100 extended abstracts and papers were reviewed, and in the case of papers selected for this book, by three reviewers.

The preparation of the book was greatly aided by the counsels of Zoe Sanders and Claire Ferres of Emerald Group Publishing Limited, and the work of ISCTSC's copy-editor Christopher Parker. We thank them for their attention to detail.

PART I

INTRODUCTION

Chapter 1

Keeping Up with a Changing World: Challenges in the Design of Transport Survey Methods

Patrick Bonnel, Martin Lee-Gosselin, Jean-Loup Madre and Johanna Zmud

Abstract

At the 2008 International Conference on Transport Survey Methods in Annecy, France, transport survey methodologists and practitioners shared their experience with keeping abreast of the data needs of a rapidly changing world. Over the past decade, this has translated into the need for: an expanded travel survey toolkit; methodological innovation for surveys of freight and public transport operations; a growing use of data collection and processing technologies; a need to align surveys with other data streams; and an increased interest in the comparability of international datasets on personal travel and commodity movements in an era of globalisation. We discuss how these guided the choice and scope of the five themes around which both the Annecy Conference and this book were organised.

The International Steering Committee for Travel Survey Conferences (ISCTSC) organises periodic international conferences on the survey methods that support planning, policy development, modelling and evaluation through the observation of person, vehicle and commodity movements at the urban, rural, regional, intercity and international scales. The evolution of the underlying issues, and the methodological response, can be seen in the series of publications that drew on previous conferences, most recently the 1997 Grainau Conference (Stopher & Jones, 2000),

Transport Survey Methods

ISBN: 978-1-84855-844-1

the 2001 Kruger Conference (Stopher & Jones, 2003) and the 2004 Costa Rica Conference (Stopher & Stecher, 2006). That evolution includes:

- a gradual expansion of the travel survey toolkit beyond the needs of 'core business' urban area household travel surveys and national travel surveys;
- a growing recognition of the need for new approaches to collecting data on freight movements and public transport operations;
- an increasing expectation that data from transport user surveys should be aligned with other data streams from administrative and commercial sources;
- a growing application of digital technologies to aid data collection and processing;
- an increased attention to international (and within nation) comparisons of data on personal travel and commodity movements, and to international flows in the context of a progressive globalisation of national economies.

With this in mind, the Transport Survey Methods Conference held in Annecy, France, in May 2008 was designed to continue the emphasis of previous meetings on transport survey quality and on the standards for assessing and maintaining quality (e.g. Stopher, Wilmot, Stecher, & Alsnih 2006), and also to look ahead to transport survey data harmonisation and comparability within and across countries. It was a concerted response to the evolving need to track and compare key policy measures and statistics, and their implications for sustaining mobility, in today's global, interconnected world. For example, how can we track and compare long distance mobility in Belgium versus the United States when the two nations employ different definitions of a long distance trip? (Bonnel, Madre, & Armoogum, 2005) What does it mean that the mobility rate is around three trips per day in a Netherlands metropolitan area and around four trips per day in the Grenoble metropolitan area? Can we attribute this difference to policy measures in respective areas, true travel behaviour differences, different survey methodologies or different spatial boundary definitions? (Bonnel, 2003).

As a community, transport survey researchers, practitioners and planners need advance knowledge of the components of survey and data collection design that are upstream from reliable and accurate intra-national and international comparisons. The topics of the Annecy Conference were intended to facilitate such discussions, and in a few cases to initiate them. It was also hoped that progress would be made towards the development of a framework for harmonising passenger transport survey data and statistics along the lines of what has been done for road freight data at European level (Pasi, 2008). At the same time, it was recognised that some classes of transport surveys are not yet ready for such a framework, and may require some fairly fundamental methodological research in the shorter term. Other classes — for example those that explore hypothetical travel behaviour under a range of possible future environmental pressures — are not intended to generate national statistics, but merit our attention for other sound reasons.

The Annecy Conference thus sought a balance between the themes of data harmonisation and the data quality. With its workshop format, the conference continued the ISCTSC tradition to create an opportunity for networking, collaboration and sharing of knowledge. At the same time, this particular conference was unique in that it was held in conjunction with the final meeting of the European COST Action 355: *Changing Behaviour Towards a More Sustainable Transportation System*, and a review of the activities and recommendations of this COST action pertaining to national surveys is presented in this volume. In Annecy, experts presented papers that provided insight into the many-faceted conceptual dimensions of 16 workshop topics, as well as the practical issues of the impact of data needs, socio-political issues and future policy issues affecting transport surveys. These papers and syntheses of the associated workshops comprise the majority of the chapters of this book.

Conference participants included leading survey researchers and transport professionals from over 25 countries in the private and public sectors, as well as academic scholars and researchers. Three keynote addresses were presented. Simo Pasi provided a snapshot of Eurostats's priorities and activities related to data harmonisation in the freight area. William Dutton, Oxford Internet Institute, provided insight into new research in the areas of communication and information technologies and its implications for data collection in the transport domain. Peter Stopher, University of Sydney, identified his proposal for an innovative, multi-approach transport survey toolkit, and his keynote presentation is included as Chapter 2 of this volume.

The book is organised into five themes that represent the key challenges and opportunity spaces in transport surveys:

I. Sustainability and Traveller Adaptation, Chapters 3–6, providing current thinking on survey protocols related to travel behaviour change;
II. Global Social Issues, Chapters 7–14, describing survey practice on current social challenges — tourist flows, evacuations, physical activity and excluded populations;
III. Freight and Transit Planning, Chapters 15–20, addressing the perennial transportation system challenges of capturing data on freight movements and public transit ridership;
IV. Technology Applications, Chapters 21–28, identifying the new challenges and new opportunities for applying technology solutions to improve survey data quality;
V. Emerging/Persistent Survey Issues and Data Harmonisation, Chapters 29–36, describing best practices for overcoming persistent problems with transport survey data.

The following sections provide more information about each of these five thematic areas by highlighting the pertinent transport survey methods questions, issues, challenges and recommendations that were presented in the resource papers for each workshop, or which came out of the workshop discussions themselves.

1.1. Sustainability and Traveller Adaptation

Sustainability is on the political agenda of every nation. It has spawned numerous policy measures in order to influence travel behaviour towards more environmentally friendly modes of transport. One such policy, *voluntary behavioural change programmes*, has generated great interest. These programmes include a diverse range of initiatives that vary in both who is targeted and the nature of the intervention. While transport survey researchers have long been interested in travellers' responses to external factors such as the price of fuel or provision of new infrastructure, they are on less familiar ground when their traditional dependent variables — the amount of travel and its attributes — are the direct target of an intervention.

Various methods are used to monitor and evaluate these programmes, but all face some common difficulties in applying best practice. For instance, measurement units impact the analysis that can be performed on data. Is it more important to measure the change in behaviour or to measure the impact of the change (e.g. in terms of emissions)? Do we need to analyse the factors that influence change in order to monitor the process of change and improve programme efficiency? What is the best timing for measurement: shortly after programme introduction in order to measure the direct relation between programme and behavioural change or after a longer time in order to observe snowball effect and other changes that take longer time to observe (residence location, for example) but that can have strong effects? How do we measure behavioural change in order to minimise potential bias?

Sustainable initiatives are often part of a more global programme and are introduced in an evolving environment. Thus, their evaluation faces the usual challenges of impact assessments: trying to disentangle outcomes, and determining what is related to the targeted initiatives versus what is related to other changes (e.g. changes in life cycle). The need to adequately and consistently define experimental and control groups to circumvent potential difficulties and alleviate biases was discussed. Difficulty in doing so, however, is increased by the fact that sustainability initiatives might be developed on limited scale and with limited budget. It is therefore not always possible to achieve statistically significant results. Meta studies could be developed to overcome this problem but they may only accentuate the biases of each discrete study and may not always be well documented. It is therefore recommended to triangulate, using various survey methods or gathering data from multiple sources in order to apply triangulation, to deal with or to complement conflicting results, when possible. Transport survey researchers would also benefit from experiences from other disciplines like psychology, health/epidemiology and sociology to glean best practices for effective survey science experiments and programme evaluations.

As noted above, behavioural change does, of course, occur in response to many situations other than programmes that actively seek it. As well as costs and new infrastructure, these can include rising congestion, energy shortages, new vehicle, fuel or communications technologies, and a wide range of policy shifts to address public health, safety or environmental problems. The Annecy discussions focussed in particular on surveys of changing behaviour in the context of an interest in environmental sustainability policy. Importantly, they included stated response surveys to explore potential behaviour under hypothetical situations, and the challenge of obtaining authentic responses about past and future behaviour alike. 'Social desirability bias' is a term used to describe the tendency of respondents to reply in a manner that will be viewed favourably by others, often resulting in respondents' over-reporting of 'good' behaviour or under-reporting of 'bad' behaviour. The potential for social desirability effects in the case of environmentally friendly behaviours, which are generally considered as desirable, is great. Once again, the triangulation of more than one method may be needed to interpret adaptive behaviour.

1.2. Global Social Issues

No discipline can defer addressing contemporary social, political, economic and environmental issues within the course of its scientific discourse. Transport survey research is no different. Surveys are executed in a socio-politico-economic context. Who actually participated in a survey is often more important than what the survey found. Engendering the participation of some population segments is more difficult than others (i.e. illiterate, immigrant, low or high income persons). Often such non-participation is systematic rather than episodic. Hence, a generalised term 'hard-to-reach' population has entered into the transport survey literature. Consequences of this lack of coverage are discussed in this section of the book, especially with regard to the social dimension of sustainable development. Strategies to overcome potential biases are offered and include multi-mode or multi-method surveys, as both under-representation and its impact on data are different depending on population segments concerned.

Public health and fitness are growing 21st century social concerns in both developed and developing countries. Within the transport survey world, physical activity in the built environment is receiving a lot of attention in relation to health concerns, especially obesity. Topics addressed include the potential of travel surveys to inform this social issue in general, as well as methodological issues that affect the utility of surveys to answer questions specific to transport policy and planning. This is an issue that generates opportunities and challenges for interdisciplinary collaboration. Particular attention must be given to the effective capture of short and walking or cycling trips, which are sometimes neglected in travel surveys. We also need to better understand the motivation underlying physical activities within travel, and relate this to the transport system and the built environment.

How to more effectively conduct surveys focussed on transport during exceptional events continues to be of great interest. A major dimension of this is emergency management and evacuations. But this is not just about getting people out of harm's way. There is great variability in the responsiveness of transportation systems, the hazard events, and the key players or detectors of those events. Unfortunately, there are not many, if any, standards to systematically address these factors in a survey context. As a starting point, this section of the book addresses this situation by identifying the variety of exceptional events and proposing a typology related to specific dimensions of an event. Data needs cover both travel behaviour data and its determinants, and also the context of the event. Preference surveys, especially stated choice surveys, are considered important.

Economic sustainability has often been associated with tourism infrastructure, and so effective methods to survey tourists and transients are quite important. Content related to this topic attempts to clarify the use of survey terminology concerning survey accuracy and reliability in tourism research. Definition of tourists and transients is still problematic and depend often on the spatial scale used. The need for standards to assess accuracy is introduced. A unique survey solution appears unrealistic. Therefore data need to be collected from various methods and from various sources — other than transport-related sources. Research is therefore necessary to better integrate these multiple sources of data.

The workshops in this global social issues area appear quite different but some common ground can be emphasised. Interdisciplinary surveys appear promising, and the need to cross traditional boundaries between academic disciplines or schools of thought surfaced in a number of discussions. The need was felt to investigate new paradigms, new approaches and new conceptual frameworks to better understand the determinants of behaviour or the determinants of survey response. The study of global social issues has also emphasised the need for gathering data from various sources (within and outside of the transportation domain), which in turn increases the need to develop methods to integrate and enrich (or merge) these data. In all cases, the potential of technology (i.e. GPS tracking, sensors, RFID) to improve data quality is raised, even if more research is needed to better identify the benefits of and limitations in the information collected using these technologies. Moreover, there are potential roles for several of the survey methods discussed under this theme to contribute to localised alerts in real time, such as to health authorities or incident response agencies. It follows that they would also contribute to collective and cumulative assessments of hazardous conditions.

1.3. Freight and Transit Planning

The interaction between freight flows and transportation infrastructure continues to increase in both importance and complexity. Significant changes in the economy have impacts on the level, type and geographic distribution of freight flows. The variety of uses for freight transportation data is broad and increasing. Estimating the likely impacts of proposed regulations on air quality, greenhouse gas emissions and climate depends greatly on the availability of improved freight flow data. Yet, our traditional sources of freight transportation data struggle to meet new requirements. There are a number of challenges in freight surveys, not the least of which is definitional — exactly what is freight? What do end users want to measure, model, forecast, etc., with freight survey data? Are we interested in surveying what is being moved, how it is moved, or both? What about why it is being moved? What information do modellers need to develop the tools for measuring and predicting freight? These are just some of the questions that need to be answered to allow survey designers to develop and implement surveys that provide not only meaningful information, but also the data for policy makers and planners to answer their questions.

Advancing change and innovation in these types of surveys requires addressing definitional and measurement issues, institutional constraints, private-sector concerns, financial constraints and multi-agency coordination across modes. Intermodality issues are of particular importance: for example, more and more often, a packaged good is not moved by a single vehicle. Thus, the alternative is between shipment surveys (e.g. CFS in the United States, ECHO in France) versus vehicle-based surveys (as mentioned in the keynote presentation by Pasi on truck surveys harmonised by EUROSTAT). There are a number of links to other discussions. In general, it would be fruitful to compare the way freight and travel survey methodologists approach the above fundamental questions. Also, there are new developments in vehicle-based surveys, discussed under the Emerging/Persistent Issues theme, that are relevant here, even though the sampling frame here is mostly heavy-duty vehicles rather than vehicles in general.

Public transit, and transit-supportive land use, is becoming increasingly important to climate change policy as national, regional and local governments develop and implement initiatives to reduce energy use and greenhouse gas emissions. Data needs for public transport planning cover a wide range of domains (e.g. ridership, performance, service quality, customer satisfaction, non-users' expectations). A variety of methods are therefore needed to investigate these domains. Unfortunately, scientific literature about data collection and analysis methods specifically aimed at public transit is not plentiful. Key questions remain, such as where has innovative or best practice been employed? What are the best methods, innovations, hardware and software tools? What administrative data systems have been developed to fulfil the specific requirements of a single study? There has not been a systematic or empirical assessment of methodological improvements in areas such as these. Improvements are needed in areas that include, but are not limited to: questionnaire wording and graphic design to improve comprehension and response quality; selection, training and continuous monitoring of surveyors to assure high performance standards; real-time geocoding and verification to improve accuracy and precision; and a variety of alternative methods for data capture. Of particular interest are new methods for weighting and expanding survey data that account for biases such as short versus long surveyed trips, multi-transfer journeys and multi-modal journeys. The industry is hopeful that various new technologies (i.e. ticketing systems, automatic vehicle positioning, automatic passenger counting, GPS tracking) appear very promising to automatically generate data. But this benefit is currently offset by the challenge faced in processing and organising the huge quantities of data that result.

1.4. Technology Applications

As noted above, technological aids to surveys is a growth area for the transport survey community. There were four workshops on technology topics at the Annecy Conference (see Chapters 21–28 of this book), and technology was actively discussed in several more. This compares to one small session at the 1997 Conference, one workshop at the 2001 Conference and two workshops at the 2004 Conference.

In the case of travel surveys, researchers are facing increasing difficulties, including response rate decreases, sampling frame limitations and privacy concerns at a time when data requirements are getting more complex. New technologies are often seen as a solution to solve or at least mitigate these difficulties. There is no argument that the potential benefits are high. But challenges exist. For example, good user interfaces to help reduce respondent burden, and therefore to increase respondent participation and data reliability, are still not the norm. Research is needed to fully document the strengths and weaknesses of specific technology applications. GPS/GSM helps to produce more accurate data (i.e. fewer trip omissions, more accurate locations and timing, collection of individualised schedule and planning elements) for longer travel periods, but at what cost? Web-based surveys in combination with other modes facilitate response rate increases among certain populations at lower costs, but with what biases? Visualisation techniques are especially promising for generating a better understanding of travel behaviour but require more examination of uses and consequences. Automated data production systems (i.e. fare collection systems, passenger counting systems, mobile phone location systems, credit card transactions, plate number recognition systems) collect huge quantities of data that may have great potential for transport research community. The potential of each technology needs further research to better identify the expected benefits, especially in terms of data quality and level of precision, with implications for the return on investment.

But use of new technologies comes with new problems and new challenges, especially the techniques for processing and enriching the data. New technologies are often seen as a way to reduce non-response in surveys. Benefits are obvious regarding the more complete reporting of trips, or of some of their characteristics. But some individuals are also reluctant to use the technologies, or may not yet have access to them, generating other non-response biases. We need to explore the potential of each technology and each mode to recruit and to motivate respondents, in order to develop strategies to overcome non-response. In this respect, new technologies are like any other methodological innovation — with risks and benefits to be empirically assessed.

Electronic instrument design offers the promise to simplify and customise questionnaires, and therefore to reduce respondent burden and increase participation. But research is still needed to develop interfaces that are congruent with the mental models of how respondents think about the questions embedded in surveys.

New technologies generate huge quantities of data that generally need an *a priori* processing plan in order to be useful for data analysis and planning. This is typically the case for GPS/GSM or cell phone data. Furthermore, such primary data are generally incomplete as survey responses (e.g. only space, time and their derivatives for GPS data). We therefore need to further develop tools to enrich these data, generating mode and purpose from GPS data for example. Research is still necessary to develop more efficient data processing, and where possible, propose some standardisation principles and software tools. A promising avenue seems to be the use of learning algorithms to build essentially passive interpretation of data streams following a period of active involvement of respondents (or at least a sub-sample) in the interpretation or their own data. While improvements have been achieved in the technical performance of survey technology, especially the power management and storage capacities of mobile technologies, or the precision of location and timing, more testing and experimentation is needed. Privacy concerns and ethical rules are expected to assume increasing importance as applications move to larger and larger samples. When considered within the larger context of the issues raised at the Annecy Conference and in this publication, it is clear that technology application will not be a silver bullet. None of the available technologies appears able by itself to cover adequately the growing list of transport data needs. Future work on developing innovative combinations of technologies, survey modes and data fusion methods needs to be conducted, documented and disseminated.

1.5. Emerging/Persistent Survey Issues and Data Harmonisation

Transportation systems are large, complex, multi-component, multi-player collections of interacting elements and subsystems. These systems play critical roles in societies and economies, providing accessibility (and thus value) to places and mobility to people and goods. Decisions about development and operation of the transportation system are of central importance to leadership at all levels, in both the government and the private sectors. Data and the information produced from data are key assets of transportation systems because of the roles they play in support of decision-making: problem identification, design of options and priority setting. These data include not only the measures of travel behaviour information but also its determinants. Yet one of the key sources of these data, transport surveys, tends to be performed at longer intervals due to financial and other institutional constraints. It becomes harder for decision makers, policy makers and planners to have at their disposal up-to-date data and information about the dynamics of travel behaviour that are the necessary underpinnings of good transport policy and planning.

Continuous surveys appear to be a promising solution to address the concomitant problems of both producing up-to-date data and the resulting trend data series. Continuous surveys are a category of survey design that encompass longitudinal and panel methods. With the interest in sustainability and behaviour change programmes, continuous surveys present many advantages, not the least of which is the enabling of a more accurate calculation of time-series for auto-correlated variables (e.g. mobility or car ownership). The growing interest in these types of surveys is illustrated by specific experiences gained from various case study applications around the world. Among the persistent challenges identified, two appear crucial — needing far more research. First, fatigue of staff and interviewers has lead in some countries to declining quality, notably as measured through response rates: permanent monitoring and incentive programmes are necessary to sustain motivation and data quality. Second, and perhaps more important, methods need to be developed to account for the dynamics present in data collected over several years, and resulting from exceptional or unpredictable external factors.

As discussed above, the role of surveys that use vehicles as the sampling frame is not confined to freight and transit. In Canada, for example, vehicle-based surveys of light-duty vehicles have provided valuable national data on vehicle and passenger kilometres of travel in the absence of a national travel survey. At Annecy, there was a wide-ranging discussion of the state of the art of vehicle-based surveys in France, Germany, Finland, the UK, Bangladesh, the USA and Canada. The promise of vehicle-based surveys was assessed, in particular, for the advantages of the parallel collection of data on duty cycles, fuel use, emissions and safety systems, using increasingly sophisticated technological aids that work well on vehicle platforms. The importance of such microdata in a future that may include the more widespread use of automatic systems for charging individual users did not escape the discussions of this workshop.

Most of the workshops have identified the need to combine methodologies because single methods or surveys generally failed to sufficiently cover the problem that motivated the surveys. We also have more and more data that are automatically produced, especially from new technology systems, and which need to be processed and enriched with other sources. Combining survey modes or methods within survey appears to be a good strategy, both to reduce the non-response problem and to improve the representativeness of the population, especially for hard-to-reach groups. This problem is not new, and several methods are available in a number of survey fields besides transport. In addition, progress is being made on the alignment or fusion of data from surveys of users and other sources, notably those from traffic and transit monitoring, administrative data and even financial transactions. Nevertheless, there is a need to better assess the potential of each method on real case studies. There is a particular need to identify methodology to validate fused data, and to assess the implications for data analysis and modelling, especially regarding level of confidence.

The holding of the final COST 355 meeting in conjunction with the Annecy Conference gave participants from around the world the opportunity to appreciate the breadth of European cooperation on the role of behaviour in a more sustainable transport system. While the first two of the COST 355 themes, 'Freight transport and energy consumption' and 'Automobile: panel data analysis' were of great interest to the participants, the third 'Overview of national transport surveys' was of direct relevance to the conference theme, the more so because of its focus on the comparability of survey data. Our final chapter provides a fitting conclusion to this book as it summarises the lessons from more than three years of analysis of the European experience.

1.6. Conclusion

A number of ideas and proposals emerged during the conference, and participants reached some conclusions about how to improve the quality of transport survey data.

The emphasis in most of the workshop discussions was on developing practical, achievable and affordable strategies for collection of essential travel information that would be less contingent upon shifting political and funding priorities. While such points are addressed in more detail in the chapters, some of the priorities that cut across the different workshop topics are presented below:

- continuous longitudinal surveys to improve the frequency, timeliness and responsiveness of personal or freight data to policy and planning issues;
- creation of national mobility panels to obtain better measures of changes in travel demand relative to economic, social and environmental factors;
- consideration of digital methods of data capture and organisation to enhance quality and cost-effectiveness of data capture;
- automated vehicle movement data collection (through ITS, probe vehicles, cell phone probes or other telematic techniques);
- supplementation or replacement of primary micro-level data collection with simulation and modelling approaches to improve the accessibility of information for decision making;
- greater application of computer-assisted techniques for visualising data to facilitate communication and decision making.

It was not difficult to draw an overarching conclusion from the Annecy Conference that international collaboration is fruitful, but that a *strategy* for that collaboration would be even better. Harmonisation of data across our countries would be of great value, but so would be a fuller appreciation of the *variety of experimentation*. The ISCTSC Conference Series is committed to continue the sharing of our experiences: of serving better those who pay for surveys; of working with the tools and technologies that can help surmount the new challenges to our art; and of helping identify the opportunities to make the transport system more socially and environmentally responsible.

References

Bonnel, P. (2003). Postal, telephone, and face-to-face surveys: How comparable are they? In: P. R. Stopher & P. M. Jones (Eds), *Transport survey quality and innovation* (pp. 215–237). Oxford: Pergamon Press.

Bonnel, P., Madre, J. -L., & Armoogum, J. (2005). National transport surveys: What can we learn from international comparisons? *European Transport Conference*, Strasbourg.

Pasi, S. (2008). Harmonisation and comparability of European Union road freight statistics, Eurostat G-5-Transport Statistics, Luxembourg, available at: http://www.isctsc.let.fr/2008Conf/keynote.html (consulted May 2009).

Stopher, P. R., & Jones, P. M. (Eds). (2000). *Transport surveys: Raising the standard.* Washington, DC: Transportation Research Circular E-C008, Transportation Research Board.

Stopher, P. R., & Jones, P. M. (Eds). (2003). *Transport survey quality and innovation* (p. 646). Oxford: Pergamon Press.

Stopher, P. R., & Stecher, CC. (Eds). (2006). *Transport survey methods: Quality and future directions* (p. 685). Oxford: Elsevier.

Stopher, P. R., Wilmot, C. G., Stecher, C. C., & Alsnih, R. (2006). Household travel surveys: Proposed standards and guidelines. In: P. R. Stopher & C. C. Stecher (Eds), *Transport survey methods: Quality and future directions* (pp. 19–74). Oxford: Elsevier.

Chapter 2

The Travel Survey Toolkit: Where to From Here?[☆]

Peter R. Stopher

Abstract

The purpose of this chapter is to provoke thinking about the directions in which the travel survey toolkit should move in the near future based on the author's personal experience and as an outcome of the Travel Survey Methods conference. The chapter begins with a brief historical review that attempts to show some of the major elements of change that have occurred in travel survey methods over the past 40–50 years. A more detailed review is provided about developments over the past 10–15 years. The chapter then explores a number of emerging challenges, including telephone contact of potential respondents, computer-assisted surveys, Internet surveys, mixed-mode surveys, the impacts of language and literacy and the potentials of mobile technologies. Based on this, the chapter then considers future directions that should be pursued. The chapter suggests that it has been changes in survey methodology that have, in the past, sometimes enabled and at other times led to changes in modelling paradigms, and that this may be an appropriate time for travel survey methodology again to enable changes in modelling paradigms. A speculative specification of a new household travel survey that makes use of a number of these developments is then offered. The chapter ends with some concluding remarks that issue a challenge to the travel survey community to think 'outside the box' and foster change and improvement in the accuracy and representativeness of travel surveys.

[☆] Keynote Paper prepared for the 8th International Conference on Transport Survey Methods, Annecy, France, May 2008.

Transport Survey Methods

ISBN: 978-1-84855-844-1

2.1. Introduction

To ask where we are going from here requires, first, that we know where 'here' is. To that end, at the outset of this paper, it seemed worthwhile to reflect on where we have come from in the area of travel survey methods, before setting out to try to see where we are going in the early 21st century. Following that, this paper discusses some of the emerging challenges and also opportunities that appear to be available to those intent upon undertaking travel surveys in the early 21st century. For the most part, this paper focuses on travel surveys of human populations and the main focus within travel surveys of human populations is on the household travel survey. However, some mention is made in passing of other travel survey needs, especially in relation to on-board transit surveys and roadside or intercept surveys.

The purpose of this paper is not to be exhaustive in the treatment of the travel survey toolkit, but rather to open up some ideas for consideration during the course of the 8th International Conference on Travel Survey Methods in Annecy, so that delegates to the conference may be inspired to think outside the box in looking at where we are going and where we need to go in the future. There is little doubt that there are many threats to conducting travel surveys at this time, and that some of these threats are likely to become more serious in the future. There are also many new opportunities to do things better than they have been done in the past, although there are potential pitfalls that could lead towards negative impacts on travel survey quality. It is hoped that this paper will stimulate thinking and also help to ensure that the paths followed will bring about improvements in data quality.

2.2. A Historical Review

Although it was before probably any of those currently active in travel survey methods were actually working in the field, the beginnings of travel survey methods were firmly founded in the early stages of urban transport planning, especially as it was practiced in the United States in the 1950s and 1960s. At that time, urban transport planning was primarily a concern of the largest urban areas in the country, and also of major cities in a few other countries around the world, where US consultants were largely employed to bring this new field of transport planning to those cities, for example, the London Traffic Study of the early 1960s (London County Council, 1964). At the core of the emerging urban transport planning field was the home interview survey, with its associated cordon and screenline surveys, often involving roadside interview surveys. The principal method used to obtain data about the travel movements of people was, therefore, an interview.

Roadside interviews were conducted by having the help of local police to flag down a systematic sample of cars, which were required to drive into a marked-off bay, where a short interview, usually lasting no more than about 2 min, took place. The interview determined vehicle occupancy, origin location, destination location and the purpose of the trip. The frequency with which the trip was made was also often asked. The roadside interview also involved counts being made of the vehicles passing the interview location, so that the total volume on the roadway could be determined. Samples were usually of the order of one vehicle in ten. The roadside interview is still alive and well, judging by a brief search on the Internet (e.g. Cicerone, Sassaman, & Swinney, 2007; Northamptonshire County Council, 2003; Scottish Executive, 2004), although there are numerous cautionary comments about its use on motorways and freeways (where it should not be used), on major arterial roads and in major cities. In the United States, the roadside interview is rarely done for larger urban areas, because of concerns with road rage and adverse publicity resulting from the inevitable delays to drivers when such an interview is conducted.

On-board transit surveys were something of an afterthought in the survey toolkit for transport planning. This resulted from the strong focus on highway-based solutions that emerged from the US-led transport planning field, which only slowly acknowledged an ongoing role for public transport in the urban context, partly as a result of growing interest in transport planning in Europe and partly as a result of a growing recognition in the United States that highways could not solve all of the urban transport issues. Again, in the wake of the home interview and roadside interview, early on-board transit surveys focused on an interview survey either on board the public transport vehicle, or occasionally at the stops and stations where people waited to board. Subsequently, and by the late 1970s, the on-board interview became a self-administered survey, often to be accomplished whilst passengers rode on the public transport vehicle (Stopher, 1985). Household travel surveys and national surveys often result in apparent undercounts of public transport ridership (Transmanagement, 2004), but on-board transit surveys are also subject to very low response rates in many instances. Conducting an on-board transit survey involves facing a number of challenges. These include creating a survey form that can be filled out whilst the vehicle is moving and by passengers who may have nothing on which to rest the form; keeping the survey short enough that most passengers can complete the form during their transit trip; providing forms in more than one language where necessary and the logistics of handing out forms and retrieving them, especially on overcrowded vehicles. On-board surveys, however, continue to be an important part of the travel survey toolkit.

The home interview (BPR, 1954; CATS, 1956; FHWA, 1973) was conducted by sending interviewers out to households, where, unannounced, they knocked on doors of selected sample households and then proceeded to ask a battery of questions about the household, its members and recollections of travel that took place on the previous day. The focus of these surveys was weekday travel, so interviewing was usually conducted throughout the week, on Tuesday through Friday evenings and on Saturdays, so that the data on the previous day was collected for Monday through Friday. In some surveys, interviewing took place only on weekday nights, and the Monday evening surveys asked people to recall their travel on the previous Friday. The sample sizes for these home interview surveys were generally defined as a percentage of the metropolitan region's population, and often ranged from 1% to 3% of the population. Thus, it was not uncommon for the sample to comprise as many as 20,000 or 30,000 households in large metropolitan areas. The Chicago Area

Transportation Study Home Interview Manual (CATS, 1956) specified the design of a survey to be conducted on a sample of 56,000 households. It was also usually claimed that the response rate was upwards of 90%. However, it is unlikely that this was calculated using the CASRO (1982) or AAPOR (2000) formulae, but was most probably simply estimated as the proportion of households that were successfully approached and had completed the interview.

By the end of the 1970s in the United States, home interview surveys were being replaced by other types of surveys. It was becoming evident in the late 1970s that the home interview survey could not be carried out safely in certain parts of most urban areas, and that alternative methods of surveying were required. One of the last large home interview surveys to be conducted in the United States was probably the one undertaken in Detroit in 1981 (Parvataneni, Brown, & Stopher, 1982). In the next few years, the home interview survey began to be replaced by telephone recruitment, using random-digit dialling (RDD); this was followed by a mail-out, mail-back survey. However, response rates to the mail-back portion of the survey were often disappointingly low, so that the mail-back option began to be replaced by a telephone retrieval of the diary data. The telephone retrieval method largely replaced mail-back options by the early 1990s. A few regions continued to use either a mail survey or the telephone recruitment with mail-out and mail-back, but the home interview in the United States had been almost totally replaced by a telephone recruitment, mail-out, telephone retrieval method of surveying by the mid-1990s.

At the same time that the standard methodology of conducting travel surveys in the United States changed, the sample sizes also dropped and were no longer referred to in terms of the percentages of the population. Rather, sample sizes were more generally described in terms of a range from around two or three thousand up to as much as ten or fifteen thousand in a few cases. Samples of the size used in the early days of the home interview survey were no longer to be seen. During the 1970s, even before the emergence of disaggregate models and their incorporation into the standard transport planning package, there began a process of moving away from the large samples of the early home interview surveys (Foster, Benson, & Stover, 1976). The increasing costs of such surveys, coupled with improved understanding of sampling statistics, produced pressure on the survey work to reduce the sample sizes.

The methodology for household travel surveys has undergone somewhat less change outside North America. The home interview survey remains the methodology of the Sydney Continuous Household Travel Survey, for example (Battelino & Peachman, 2003). In Brazil and Chile (Ortúzar, 2006), face-to-face interviewing is also still being used. In other cities of Australia, one of the methods currently in use is a face-to-face recruitment interview with survey instruments left with the household and either picked up in person by the interviewer subsequently or sent back by mail. In New Zealand, the method is quite similar, except that on the return of the interviewer, a computer-assisted personal interview (CAPI) is done, and the materials left with the household consist only of a memory jogger that covers the designated days of the survey. In Canada, most household travel surveys appear either to follow the U.S. procedure or to be pure telephone surveys. However, the sample sizes are very large comparatively, with the recent Toronto survey covering more than 100,000 households. In the United Kingdom, the face-to-face interview is still the preferred method, using CAPI, and following a pre-notification letter and diary provided to each household. Different methods are recommended for smaller area surveys, namely, a mail survey with CATI retrieval for smaller urban areas and regions. In France, a face-to-face survey is compulsory (to obtain the CERTU label and state financing) in large conurbations, but a telephone survey (CATI) is recommended for medium-sized conurbations surveying only one person in small households and two persons in larger ones (CERTU, 2008). In the Netherlands, surveys are predominantly mail surveys, with telephone contact to remind households to complete and return the surveys, and also to correct problematic or missing data (van Evert, Brög, & Erl, 2006). This survey has evolved from a predominantly face-to-face survey up to 1983, and a telephone survey from 1983 to 1998. In Germany, recruitment by telephone using RDD is the method used in the current German mobility panel (Kuhnimhof, Chlond, & Zumkeller, 2006) and the German long-distance panel (Chlond, Last, Manz, & Zumkeller, 2006), with some data collected in the recruitment stage by CATI, followed by a mail survey for the transport information, whilst the last National Travel Survey in Germany sampled from an address register, but conducted the survey by a mix of CATI and self-administered postal survey. In South Africa, surveys are still largely conducted as face-to-face interviews at the household. These surveys are generally done using paper and pencil, although some are conducted using CAPI.

Irrespective of the method used, household travel surveys, in common with many other types of surveys, have been experiencing declining response rates over the past two or three decades. Some of the decline is clearly associated with changing methods, but there is a growing reluctance from the public to get involved in surveys of any type, and reassurance that the survey is not a marketing device has only modest impact on response rate declines. The Sydney Continuous Household Travel Survey provides a good example of how even the face-to-face interview is experiencing a declining response rate (Battelino & Peachman, 2003). While the statistics in that paper show results only through 1999, the trend towards declining response rates has continued into the 21st century. Another example is provided in the case of the Dutch Travel Survey (van Evert et al., 2006), although this survey shows a significant jump in response rate, due to adoption of an improved survey method in 1999. However, even the improved rate has showed a decline in the four years documented in the paper by van Evert et al. (2006).

During the past 50 or more years, the household travel survey instrument has undergone a significant and substantial evolution. Figure 2.1 shows an example of the type of trip data interview form that was used from the 1950s to early 1970s. This example is similar to many that were used in this period and which were designed specifically for use in a face-to-face interview survey. The survey forms were, to a large extent, designed by traffic engineers and transport planners, who had little or no formal training in survey techniques. Fortunately, because the surveys were conducted as interview surveys, it was possible to train interviewers in the use of the forms

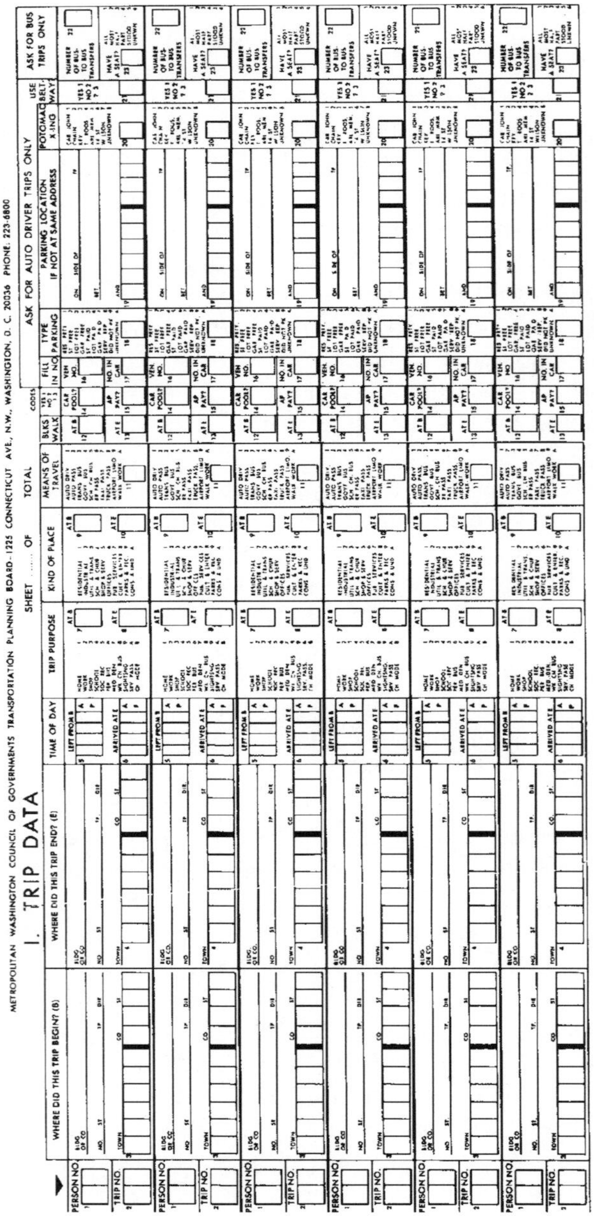

Figure 2.1: Example of a trip data form from FHWA (1973).

The earliest designs of self-administered surveys used forms that were little different from the interviewer forms that were developed in the early days of transport planning. However, during the 1970s, especially as disaggregate models began to move into mainstream modelling, changes began to occur in the layout and structure of the survey form. Probably, the most significant of these changes was made by Brög, Fallast, Katteler, Sammer, and Schwertner (1985a) through the introduction in Germany of the 'Kontiv design' of diary survey. A similar diary design was then introduced in the United States (Brög, Meyburg, Stopher, & Wermuth, 1985b) in the Detroit survey in 1981 and then in a number of subsequent surveys in the United States. This began the development of travel diary booklets in the United States, which have now become more or less the standard for most household travel surveys. Figure 2.2 shows an example of a recent booklet used in the Michigan Household Travel Survey. Differences between Figures 2.1 and 2.2 are indicative of the extent of the changes that have occurred in the design of survey instruments. In the early 1970s, the standard survey instrument was a set of leaflet style sheets, usually of foolscap or legal size in landscape mode that was used by an interviewer to enter the responses to questions asked of a householder. By the middle of the first decade of the 21st century, the standard survey instrument is a booklet with questions to be answered by the respondent.

Recruitment methods have certainly undergone significant change, but this often remains as one of the more difficult aspects of a household-based survey. In many countries around the world, there is no listing of household addresses, which is either publicly available or even for purchase. This creates various problems for undertaking good random sampling. In the United States, this problem has given rise to the almost complete dependence on random-digit dialling as the method to draw the sample, although this method is unpopular in many other countries.

Another major change that has taken place in the method of undertaking household travel surveys is the definition of the day of interest for the survey. In the traditional home interview survey, conducted in the early days of household travel surveys, the interviewer invariably asked questions about the preceding weekday to the day of interview, and relied on the ability of the respondent to recall behaviour from that day. It was recognised early on that this recall survey under-reported travel, as evidenced by discrepancies between screenline counts and expanded home interview survey data. Nevertheless, in the early days of self-administered surveys in the United States, as well as in many surveys elsewhere, this retrospective collection of travel data continued. By 1980, especially with the introduction of the diary survey, prospective diary completion was beginning to be used (Stopher, 1996). In addition, as Stopher (1996) notes, a secondary issue that was also coupled with this change from retrospective to prospective was an overt attempt to reduce the level of proxy reporting. In the conventional retrospective surveys of the 1970s and earlier, it was customary to ask the person who answered the door to the interviewer to report on travel not only for themselves, but also for everyone else in the household. In introducing the prospective diary, each person in the household (often over a certain age, such as 5 years old) was provided with a diary to be filled out by the person themselves, if able to read and understand the questions, and by proxy only if the

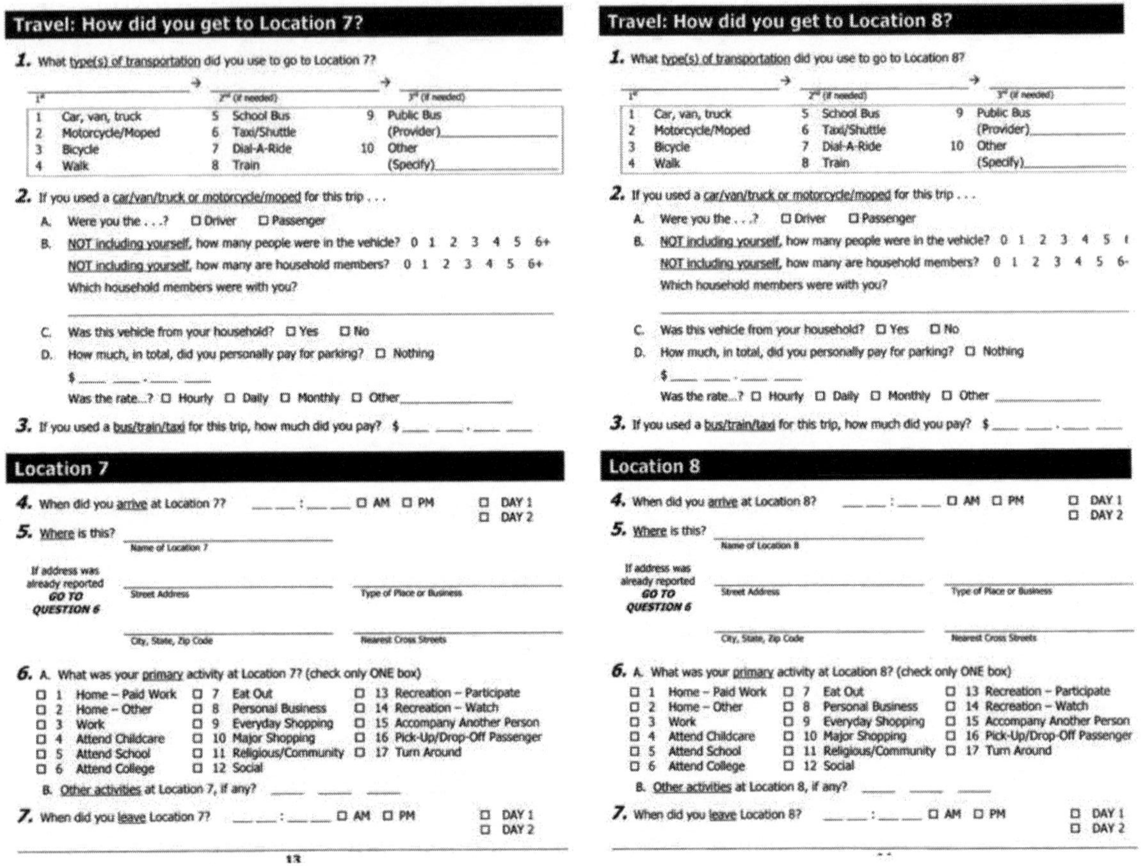

Travel: How did you get to Location 7?

1. What type(s) of transportation did you use to go to Location 7?

1st → 2nd (if needed) → 3rd (if needed)

1 Car, van, truck	5 School Bus	9 Public Bus (Provider)___
2 Motorcycle/Moped	6 Taxi/Shuttle	
3 Bicycle	7 Dial-A-Ride	10 Other (Specify)___
4 Walk	8 Train	

2. If you used a car/van/truck or motorcycle/moped for this trip . . .
   A. Were you the . . .? ☐ Driver ☐ Passenger
   B. NOT including yourself, how many people were in the vehicle? 0 1 2 3 4 5 6+
   NOT including yourself, how many are household members? 0 1 2 3 4 5 6+
   Which household members were with you?
   C. Was this vehicle from your household? ☐ Yes ☐ No
   D. How much, in total, did you personally pay for parking? ☐ Nothing
   $__ __ . __ __
   Was the rate...? ☐ Hourly ☐ Daily ☐ Monthly ☐ Other___

3. If you used a bus/train/taxi for this trip, how much did you pay? $__ __ . __ __

Location 7

4. When did you arrive at Location 7? __ __ : __ __ ☐ AM ☐ PM ☐ DAY 1 ☐ DAY 2

5. Where is this? Name of Location 7

If address was already reported GO TO QUESTION 6 — Street Address; Type of Place or Business; City, State, Zip Code; Nearest Cross Streets

6. A. What was your primary activity at Location 7? (check only ONE box)

☐ 1 Home – Paid Work	☐ 7 Eat Out	☐ 13 Recreation – Participate
☐ 2 Home – Other	☐ 8 Personal Business	☐ 14 Recreation – Watch
☐ 3 Work	☐ 9 Everyday Shopping	☐ 15 Accompany Another Person
☐ 4 Attend Childcare	☐ 10 Major Shopping	☐ 16 Pick-Up/Drop-Off Passenger
☐ 5 Attend School	☐ 11 Religious/Community	☐ 17 Turn Around
☐ 6 Attend College	☐ 12 Social	

B. Other activities at Location 7, if any? __ __ __

7. When did you leave Location 7? __ __ : __ __ ☐ AM ☐ PM ☐ DAY 1 ☐ DAY 2

13

Travel: How did you get to Location 8?

1. What type(s) of transportation did you use to go to Location 8?

1st → 2nd (if needed) → 3rd (if needed)

1 Car, van, truck	5 School Bus	9 Public Bus (Provider)___
2 Motorcycle/Moped	6 Taxi/Shuttle	
3 Bicycle	7 Dial-A-Ride	10 Other (Specify)___
4 Walk	8 Train	

2. If you used a car/van/truck or motorcycle/moped for this trip . . .
   A. Were you the . . .? ☐ Driver ☐ Passenger
   B. NOT including yourself, how many people were in the vehicle? 0 1 2 3 4 5 6
   NOT including yourself, how many are household members? 0 1 2 3 4 5 6
   Which household members were with you?
   C. Was this vehicle from your household? ☐ Yes ☐ No
   D. How much, in total, did you personally pay for parking? ☐ Nothing
   $__ __ . __ __
   Was the rate...? ☐ Hourly ☐ Daily ☐ Monthly ☐ Other___

3. If you used a bus/train/taxi for this trip, how much did you pay? $__ __ . __ __

Location 8

4. When did you arrive at Location 8? __ __ : __ __ ☐ AM ☐ PM ☐ DAY 1 ☐ DAY 2

5. Where is this? Name of Location 8

If address was already reported GO TO QUESTION 6 — Street Address; Type of Place or Business; City, State, Zip Code; Nearest Cross Streets

6. A. What was your primary activity at Location 8? (check only ONE box)

☐ 1 Home – Paid Work	☐ 7 Eat Out	☐ 13 Recreation – Participate
☐ 2 Home – Other	☐ 8 Personal Business	☐ 14 Recreation – Watch
☐ 3 Work	☐ 9 Everyday Shopping	☐ 15 Accompany Another Person
☐ 4 Attend Childcare	☐ 10 Major Shopping	☐ 16 Pick-Up/Drop-Off Passenger
☐ 5 Attend School	☐ 11 Religious/Community	☐ 17 Turn Around
☐ 6 Attend College	☐ 12 Social	

B. Other activities at Location 8, if any? __ __ __

7. When did you leave Location 8? __ __ : __ __ ☐ AM ☐ PM ☐ DAY 1 ☐ DAY 2

Figure 2.2: Pages from Michigan statewide place-based diary (reprinted with permission of Michigan DOT).

person was unable to read and understand the questions (which would be predominantly children under the age of about 12).

Many other changes have taken place, both subtly and more obviously in the design and conduct of household travel surveys. These are likely too many in number to be of interest to catalogue in this paper. However, the changes already noted here are certainly among the significant changes that have taken place and that are leading to improvements in the quality of household travel surveys, and other transport-related surveys in general.

Perhaps, however, the most significant change that has occurred in the history of the household travel survey has been the emerging awareness by those involved in household travel surveys of a vast literature on conducting surveys of various types from other fields. In the early days of the home interview survey, the surveys were designed almost solely by traffic engineers and transport planners who had little or no training in survey design, and were unaware of the efforts of others in survey design. It is, perhaps, rather amazing to see how well these engineers and planners did in a field that was unknown to them and to which they could bring only their engineering and planning training. It was only in the 1970s that the field of household travel surveys began to be a little more informed of the efforts of others outside the field of transport that could provide pointers to better survey design, and it is almost certain that some of the changes that began to appear in the survey designs were as a result of such influences, albeit influences that were largely unrecognised at that time. However, such changes as the move to booklet-style diaries, prospective diary completion and reduction in proxy reporting all mirror information well known in the wider field of survey design. Household travel survey designers have only become aware of some of this work in the last decade or so, and references to such works in papers and articles dealing with household travel surveys have appeared only relatively recently. Nevertheless, the field could clearly have benefited by consulting works such as Sudman and Bradburn (1974), Cialdini et al. (1975), Dillman (1978), Bradburn and Sudman (1979), Sudman and Bradburn (1982), and Groves et al. (1988) among others, most of which have been absent from the references of papers in travel survey methods until quite recently. It is, however, now evident that the field is becoming increasingly well informed about research in other fields on survey design and execution, and that current designs are often now improved through bringing such knowledge to the table.

2.2.1. The Recent Past

Changes in the last 10 or 15 years have, perhaps, accelerated along with so many other things, such as technology and the sense of the passage of time. During that time, we have seen increasing willingness to experiment with technological solutions, such as Internet surveys, GPS devices and computer-based surveys, among others. A review of the books produced from the past three conferences (TRB, 2000; Stopher & Jones, 2003; Stopher & Stecher, 2006) and a comparison to the papers in a much earlier conference (Ampt, Richardson, & Brög, 1985) gives something of a flavour for how far the profession has moved in this respect. Among specific areas worth mentioning is the increasing interest in behavioural processes, as evidenced by the work of Doherty, Papinski, and Lee-Gosselin (2006). In the CHASE methodology developed by Doherty, Nemeth, Roorda, and Miller (2004), the way in which people plan for and then reschedule and change particular activities and associated travel has been explored. There has also been an emerging focus on process data, rather than outcome data (e.g. Bradley, 2006; Pendyala & Bricka, 2006). The scheduling and process directions are concerned with both developing a better understanding about how travel decisions are made (as a foundation for a better understanding of how behaviour can potentially be changed) and improving the ability to model travel behaviour, and not just measuring the outputs of the decision process in the form of the travel that people undertake. Also, while transport data has almost always been a significant input to modelling, it has also always been a significant input to policy formulation and policy testing. In this respect, policy issues have changed over time, and these changes also influence the changes in travel survey methodology. Among these one can note particularly concerns with greenhouse gas emissions that are driving the need to look at more than just a typical weekday in the spring or autumn, and the increasing concern with bicycling and walking trips that are driving changes in how accurately we collect such non-motorised travel.

Also, in the last 15 years, we have seen a shift from concentrating on the trip as the unit of measurement to activities (Stopher, 1992), and then to time use (Harvey, 2003). It is perhaps interesting to note here that the shift to activity diaries was made initially in the belief and expectation that this would improve the accuracy of data reporting by respondents, and was not made with any modelling implications in mind. Indeed, consulting the paper by Stopher (1992), one can see that some pains were taken to stress that the resulting survey would still be able to produce the conventional trip-based data. The development of activity-based models largely followed this shift in the data collection mechanism, rather than driving the change in data collection methods. However, the shifts in the modelling paradigms have been instrumental in pushing the data collection methodology more towards an interest in both activities and tours. Again, however, the introduction in some locations of time-use surveys was also motivated by efforts to improve the accuracy and completeness of travel reporting, rather than being driven by a modelling paradigm. Perhaps, however, because modelling paradigms that specifically require time-use data have not emerged, the time-use version of the diary has not moved so clearly into the mainstream of data collection as has the activity-based diary and a close relative of it, namely the place-based diary.

Another development over the past 15 years has been the re-emergence and substantial improvement in the use of Stated Response measurement as an input to both policy evaluation and modelling of travel demand (Louviere, Hensher, & Swait, 2001). Again, the key element here was the development of improving methods for undertaking the surveys that would provide the stated response data and particularly the tying together of revealed choice and stated choice measurement into a coherent package in which the stated choice questions clearly relate to the experience of the respondent, as measured by revealed choice data (Stopher, 1997). Efforts to continue to improve the methods of Stated Response measurement continue (Jones & Bradley, 2006). In the 6th International Conference, Lee-Gosselin (2003) put forward a useful taxonomy of stated response methods, including stated preference/choice, stated tolerance, stated adaptation and stated prospect. At this point in time, only the stated preference/choice subset has really seen attention in the recent past.

Two further interesting developments in the past 15 years or so have been concerned with an aspect of survey methodology. The first of these was the initiation of several transport panel surveys, most notably the Dutch Mobility Panel which began in 1984 and ended in 1989 (van Wissen & Meurs, 1989), the Puget Sound Transportation Panel which began in 1989 and ended in 2002 (Murakami & Watterson, 1989) and the German Mobility Panel which began in 1994 and is continuing (Zumkeller, Madre, Chlond, & Armoogum, 2006). For some reason, transport professionals have been reluctant to embrace panel surveys, despite the fact that panels have many advantages. In 1993, there was a conference held in Lake Arrowhead, California, which was called the First US Conference on Panels for Transportation Planning (Golob, Kitamura, & Long, 1997), but after 15 years there is yet to be a second such conference. Stopher, Swann, and FitzGerald (2007) have been using panels for the past three years for evaluating travel behaviour changes in Australia. However, apart from these and a few other instances, there have, to date, been few panels in transport applications. Perhaps one of the reasons that more panel surveys have not been undertaken is again because modelling has yet to take advantage of the richness of data and the dynamics of behaviour offered by such surveys. As is discussed later in this paper, there are clear reasons for using panels more extensively, but it is doubtful that a strong case will be made for this until the modelling catches up with the potential use of the data.

The other methodological development that has appeared in transport surveys is the continuous survey. This is not to be confused with a panel. In continuous surveys, a fresh sample is drawn each year, but in such a way that the sample members drawn in each year should not include respondents who were included in the recent past years of the continuous survey. The survey is continuous, in that it takes place usually throughout the year and on all days of the week and weeks of the year, and continues from one year to the next. At present, apart from National Travel Surveys in the United Kingdom, which is a continuous survey that has been running since 1988; Sweden, which was continuous from 1994 to 2001; Denmark from 1992 to 2003; the Netherlands since 1978 and the Czech Republic since 2000 (Bonnel & Armoogum, 2005); the only examples of regional continuous household travel surveys appear to be in the Antipodes. Possibly the first of these Antipodean continuous surveys was the Victoria Activity and Travel Survey (VATS) which ran from 1994 to 2002, and which involved collection of data from about 5000 households each year during that period. VATS has now been replaced by the Victoria Integrated Survey of Travel and Activity (VISTA), but it is not yet known if this will be a continuous survey like its predecessor. From 1997, the Transport Data Centre in New South Wales has collected a continuous household travel survey in the Sydney region (Battelino & Peachman, 2003). This survey collects data from about

5000 households each year, throughout the entire year, and provides data based on a three-year rolling average, which therefore covers about 15,000 households on average. The survey is ongoing. The third continuous household travel survey is carried out by the Land Transport Safety Authority in New Zealand and covers the entire country with a sample of about 2200 households each year. It began in 2002 and is continuing at this time, and aims to have a four-year rolling average to represent the New Zealand population. Although not a household survey, there is also a Canadian Vehicle Survey that has been running as a continuous survey since 1999. Again, continuous travel surveys have not yet made it into the mainstream, although there are many good arguments for them, especially the constancy of the budget allocation from year to year, as opposed to having to seek a major budget item for the periodic household travel survey as well as the issue of measuring CO_2, which requires year-round measurement and measurement from year to year to determine the direction of changes in greenhouse gas emissions. One of the major arguments against continuous surveys is the use of rolling averages in modelling, a practice that some transport planners and modellers regard with discomfort, if not outright opposition.

2.3. Emerging Challenges and Directions

In many parts of the world, the paper and pencil interview survey is still the mainstay of the household travel survey. However, in some parts of the world, other survey methods are emerging at this time that are hoped to bring improved accuracy and completeness to the data that are acquired. It is probably important to reiterate at the outset that the main purpose of a household travel survey, and indeed most other types of survey in the field of transport, is to collect data that may be regarded as representing accurately the various travel behaviours of the subject population, be it a city, a region, a state, a nation or whatever. There are two key words in that purpose that should be kept very clearly in mind as we pursue other methods of undertaking surveys — 'accurately' and 'representative'.

We undertake sample surveys always because it is cheaper, quicker and more expedient to conduct a sample survey than a census. Funds to obtain a census of travel behaviour have never been available and are unlikely ever to become so. However, the desire would be to measure the entire population (i.e. undertake a census) if it could be afforded and could be done within the time frame required. Therefore, the requirement of the survey is that it be as representative as possible of the population from which it is drawn. A useful working definition of representativeness is that a representative sample will have means and variances of each item measured in the sample that are identical to those of the population as a whole. Of course, it is unlikely that a small sample will produce means and variances that are identical, but they should be within an acceptable range from the true values of the population. This would mean that the sample could then be regarded as being reasonably representative of the population. Also, the idea of the values being within an acceptable range from the true values brings into the equation the notion of accuracy. A secondary aspect to accuracy is that the survey instrument should be able to measure the attributes and characteristics of interest with sufficient accuracy that the results can be considered to be reliable. Any new direction in survey methodology should, therefore, have as its primary aim the improvement of both accuracy and representativeness. One of the biggest threats to these goals of the survey is non-response, which is widely documented as increasing for surveys of almost any purpose.

One area that this author has not attempted to address is the potential for new technologies to provide data passively for use in policy, modelling and other aspects of transport planning. These may include data that can be obtained from smart fare cards, smart toll systems, mobile telephones and various other technologies, all of which have some ability to track where people are and where and when they travel. This may become an increasingly important area, but one that will almost certainly also face challenges on whether or not such data should be released and used in this way.

2.3.1. Telephone Contact

In some countries, as noted earlier, the main method that has been used for some time in household travel surveys to contact people, if not to undertake most of the survey, is the telephone. Increasing non-response rates are clearly evident in such surveys. The increase in telemarketing that has occurred in the last decade or so in many parts of the world is clearly a major issue for conducting any type of survey or recruitment to a survey over the telephone. As a result of telemarketing, we see the establishment of laws relating to 'Do Not Call' registers, and the establishment of such registers, while individual households use caller identification and answering machines as means to screen calls. Further, in an effort to reduce annoyance calls, more and more households are having their telephone numbers omitted from published telephone directories (unlisted or 'silent' numbers). In the United States, the proportion of unlisted numbers ranged from a low of 7% in Sarasota-Bradenton, Florida, to a high of 71.6% in Sacramento, California, according to Blankenship, Breen and Dutka (1998), and has probably increased significantly since publication of their book, as suggested by Czaja and Blair (2005).

Telephone surveys are appealing, in that in many countries the proportion of the population with telephones is high. Czaja and Blair (2005) suggest that it was as high as almost 96% in 2003 in the United States. In Scotland, telephone coverage of the population is even higher according to Taylor (2003), who estimates that it is as high as 99 5%, while the estimated proportion of households with unlisted (ex-directory) numbers in Scotland was 26% in 1998 (Beerten & Martin, 1999). In South Australia, the proportion of households with silent numbers increased from 17.3% in 1994 to 20.2% in 2002 (Dal Grande, Taylor, & Wilson, 2005). Exclusion of unlisted telephone number households is potentially a threat to the representativeness of the sample, because those households that opt to have silent numbers are not a random cross-section of the population. As noted by Dal Grande et al. (2005), such households are more likely to be single adult households of younger people, which would make them a biased subset of the total population. While RDD is a method that can include both unlisted and listed telephone numbers, it can be an expensive option. In Australia, it has become an increasingly expensive option as a result of lawsuits that currently prevent the publication and use of reverse directories, therefore eliminating much of the ability to filter random digit numbers of those numbers likely to be unproductive for a given survey.

Another aspect of telephone recruitment and surveying that represents a threat to representativeness is the increasing use of and reliance on mobile telephones. This reduces the potential usefulness of telephone-based samples, especially where portability of the mobile telephone number allows people to take their mobile phone number with them wherever they go. In Australia, there is a separate area code (04) that is used for mobile telephones, so that there is no geographic information present in the mobile phone number. In the United States, mobile numbers are included within most area codes but are not necessarily required to be changed when a person moves into a new area code than the one where the cell phone number was acquired. When the population is identified as being less than the coverage of an area code, the cell phone number is clearly of little use to screen the geographic location. In some countries, charging regimes for mobile phones result in the person being called paying part of the cost of the call in addition to the person placing the call. This results in a violation of the survey ethic that participation in the survey should not cost the respondent anything. The issue here is the substitution of wireless telephones (mobile telephones or cell phones) for landline telephones. Blumberg and Luke (2007) estimate that almost 13% of US homes in the latter half of 2006 had wireless telephones only, while other researchers have forecast that within a few years, the proportion of households in some nations with wireless telephones only will rise to more than 50%. Clearly, such trends will make the use of telephone as the first contact method increasingly suspect and subject to increasing bias.

2.3.2. Computer-Assisted Surveys

There are a variety of methods of using the computer to assist in the conduct of the survey. To a large extent, these methods do not have a significant impact on the representativeness of the sample, but may have a substantial impact on the accuracy of the data collected. There are three main types of computer-assisted survey: computer-assisted telephone interviews (CATI), computer-assisted personal interviews (CAPI) and computer-assisted self-administered interviews (CASI). The CATI survey has been around for a long time in the history of household travel surveys and was rapidly adopted in the 1980s and 1990s in the United States as the preferred method to use for both recruitment telephone calls and for retrieving data from self-administered travel diaries and household and vehicle forms. CAPI has been introduced increasingly as an aid in conducting face-to-face surveys. As noted earlier, CAPI is frequently used in the door-knocking interviews in the United Kingdom and

South Africa, among others. There are fewer reports of the use of CASI, although one might include the Internet survey, dealt with in the following section of this paper, as being one instance of a CASI survey.

The main value of computer-assisted surveys is that they can allow internal error checking to take place as the survey unfolds, and that they can also simplify the execution of the survey, when there are many skip patterns and other similar complexities in the survey. The computer-assisted survey also removes one of the potential steps that can result in error in survey data, by combining the data entry with the interview itself. In travel surveys in particular, when a respondent is asked for information about places visited, a built-in gazetteer of locations for different types of activities, along with their full addresses, can help enormously in pinning down the exact locations that respondents visit. The lack of ability for most people to provide street addresses of many of the places they visit in the course of the day has always been a serious issue in travel surveys. Hence, the inclusion of a gazetteer in the software can provide significant improvements in identifying origins and destinations of travel.

2.3.3. Internet Surveys

One of the emerging directions for surveys is the Internet survey. At present, however, the Internet is even more subject to bias than the telephone contact. Even in developed countries, the penetration of Internet into homes is not much above 70%. World region statistics show that the low is 4.7% in Africa to a high of 71.1% in North America (MiniWatt Marketing Group, 2008). Between these extremes, Oceania/Australia has about 57% penetration, while Europe has about 43%. Asia, with nearly 40% of all Internet users, has a penetration rate of only 13.7%. While Internet penetration, especially in India and China, is growing at extremely high rates, it will yet be some time before the Internet is as widely accessed as the telephone. Surveying by drawing a sample of Internet users is, therefore, likely to remain a very biased source for sampling for some time to come. In addition, Internet sampling has the same issue of lacking geographic specificity as discussed for mobile telephones. Hence, for travel surveys in particular, sampling by Internet is likely to be problematic. The role of the Internet survey is more that of an alternative mode for a broad population survey, or as the sole mechanism for a survey where either the sample biases are of no importance, or the target population is Internet users. There are also concerns, when the Internet is used as an alternative mode for some respondents, as to whether the medium of the Internet will produce different answers than those that would be provided in an interview or a self-administered paper survey (Dillman, 2000).

2.3.4. Mixed-Mode Surveys

This is another area that has been visited repeatedly in this conference series (Goulias, 2000; Viegas, 2000; Bonnel, 2003; Morris & Adler, 2003; *inter alia*) and in other work (Ampt & Stopher, 2005). Bonnel (2003) raised the issue of comparability among different survey methods and found some evidence that there are significant differences between alternative methods of administering surveys, although he had few cases in transport where more than one method was used for the same survey. Ampt and Stopher (2005) also researched this issue and concluded that there was evidence that mixed-mode surveys require careful consideration of the design effects, and that the profession needs to conduct surveys in parallel by multiple methods to determine how significant these effects are and how the design may be modified to mitigate these effects. Adding this evidence to that of Dillman (2000) raises issues about the design of multi-mode surveys, that is, surveys in which respondents are offered a choice of modes for response.

Nevertheless, the mixed-mode survey continues to be of substantial interest to travel survey designers. Offering alternative modes by which respondents can respond places some level of control back in the hands of the respondent over when, where and how a response is provided to a survey. In recent work at ITLS (University of Sydney), alternatives of the Internet, telephone and mail have been offered for responses to an odometer survey (Stopher & Swann, 2008). This proved very effective, but it was also in a situation where differences in reporting the survey measurements were not likely to be a factor. Respondents were primarily being asked to report the odometer readings of household vehicles. A survey mode effect was, therefore, unlikely.

2.3.5. Language and Literacy

In many countries around the world, there are increasing challenges from both language and literacy. In terms of language, the past two or three decades have seen increasing in-migration to many countries of people whose native language is different from the primary national language of their new homes. Organisation for Economic Co-operation and Development (OECD) statistics show that the United States, Canada, the European Economic Area, Australia and Japan all have experienced significant in-migration over the decade of the 1990s. The annual rate of in-migration has ranged from a low of 0.1 per 1000 inhabitants in Japan to a high of 5.2 in Canada over the period from 1996 to 2001 (OECD, 2004). This high rate of in-migration, much of which comes from Asian and Hispanic countries, has resulted in a rapid escalation in the number of different languages spoken at home. For example, according to Australian Bureau of Statistics (ABS) (2008), in Australia in 2006, the percent of the population that did not speak English well or possibly not at all ranged by state and territory from a little over 1% to as much as 5% in the more populous states. Furthermore, in Australia in 2006, approximately 21.5% of the population spoke a language other than English at home.

The impact of these figures is that increasingly there is a need to be able to provide written survey instruments in multiple languages, and to have interviewers who are multi-lingual. This is an enormous challenge to any type of survey, whether a written or a spoken survey. In particular, the costs of translating a survey into other languages can be very high, especially to reflect all of the nuances necessary in writing a good survey (it is not just simply a matter of translating phrase for phrase and sentence for sentence, because of issues of idiomatic usage in different languages, ambiguities that will vary from language to language, etc.). It is often the case that the funds available to undertake the survey are limited, so that translation will come at the expense of a smaller sample size, or other trade-offs in the design and execution of the survey. Studies on the effectiveness of surveys in multiple languages do not appear to have been made, but it is likely that the versions in languages other than the language of the principal designers will have various deficiencies that may compromise the results of the survey. In addition, it is often more difficult to appeal to the altruism of those who are recent migrants to help in voluntarily completing the survey. Indeed, some migrants may have come from countries where governments are more repressive; these subjects may see a survey as a threat.

There are also anecdotal reports of increasing illiteracy in developed nations, most of which do not currently compile statistics on literacy. However, the CIA (2008) estimates the literacy rate of the world to be about 82%, currently, with most developed countries achieving literacy rates above 95%. The definition of literate used in compiling these statistics is that a person of age 15 or over can read and write. However, these statistics may be misleading, as documented in a recent study on literacy in the United States (Kirsch, Jungeblut, Jenkins, & Kolstad, 2002). In their study, they found that 21–23% of the population demonstrated literacy skills at the lowest level, which was defined as being able to perform simple routine tasks associated with reading and writing. However, it is quite evident that the majority of such people would be unable to read and respond to a survey written in the English that would be expected of a child of about 10 or 11 years of age. Kirsch et al. (2002) found a further 25% of the population functioned at the next level of literacy. This level was characterised by being '…able to locate information in text, to make low-level inferences using printed materials, and to integrate easily identifiable pieces of information'. Again, this sounds to be a level well below that required to complete a typical travel survey. At the same time, Kirsch et al. (2002) report that a large majority of these two groups, comprising some 90 million adults, described themselves as being literate. Clearly, this casts substantial doubt on the world literacy statistics, which may often be based on self-report surveys. These statistics suggest that about 45–50% of the adult population in the United States would be unable to handle a typical travel survey, if presented as a written document, requiring them to respond. These statistics may well be replicated in many other countries around the world, even though official statistics may suggest that the literacy rate is in the high ninety percents.

Clearly, these literacy statistics must suggest that household travel surveys that are undertaken by recruiting people through a telephone call, then sending a package of survey materials to the household and retrieving the data by telephone, would be compromised in terms of the literacy abilities of almost half the population. It is highly likely that many of those who lack adequate literacy skills to deal with the survey will become non-respondents, whilst those who do respond will do so without the benefit of having been able to record data in the written instruments provided.

Clearly, there are two reactions that are likely to be generated in this respect. On the one hand, one group of people who fall into this 45–50% of the population will simply throw the survey away, as being something that they are not willing to try to read and understand. They may also feel threatened by the survey, in that it may require them to admit that they cannot read well enough to understand the task and accomplish what is being asked. On the other hand, there is also likely to be a group who will believe that they can understand and respond to the survey, but will so completely misunderstand its purposes as to provide a written set of responses that are of little use.

In the author's experience with postal surveys in both the United States and Australia, it is evident that many of those who attempt to complete the diaries lack sufficient literacy to understand what is being asked of them. This is the strength, in one respect, of an interview retrieval, in that those who cannot deal adequately with the written instrument may still be able to report their travel with some confidence in a conversation with an interviewer. However, given that, even in prospective diary surveys with subsequent telephone retrieval, the retrieval call will take place days after the diary day, the survey is probably again actually being undertaken as a retrospective survey, but often now after a lapse of two or more days.

With increasing dependence on visual and aural media for information and entertainment (TV and iPods, in particular), one has to speculate that these levels of functional illiteracy for adults in even developed nations are likely to grow. In another study in the United States (Jackson, 2003), it was found that one third of high school graduates never read another book for the rest of their lives after graduation, and that 42% of college graduates never read another book after graduation. Other, similarly startling statistics are provided by the same author. All of this tends to suggest that our views of literacy are quite at odds with the reality of the ability of people to read and understand the surveys we write. It clearly calls into question any survey methodology that relies on self-reporting in response to a written survey instrument.

2.3.6. *Mobile Technology Surveys*

The increasing sophistication and technological development of mobile technologies, especially mobile telephones and GPS devices, provides a new possibility for undertaking surveys about travel. Most work with mobile technologies has focused on GPS devices (Wolf, Loechl, Thompson, & Arce, 2003; Wolf, 2006) and on the use of these devices to validate conventional diary surveys. In this role, most GPS samples have ranged from less than 100 to around 800 or more. The current French National Travel Survey has a GPS sample of about 1500 individuals, each selected from a different household, while the Washington/Baltimore survey is projected to have 1600 households undertaking GPS, and a survey in Halifax is projected to collect GPS data from 2200 households (Bricka, 2008). To date, mobile technologies have not been used as a complete substitute for a conventional diary survey, largely because it was perceived that the technology had not quite advanced to the point where this was fully practical. Nevertheless, it has been suggested for some time that this is likely to be the eventual role that such technology can play (Wolf, Guensler, & Bachman, 2001; Stopher et al., 2007). Recently, the Ohio Department of Transportation issued a request for proposals for what may well be the first GPS survey as a replacement for a diary survey. This survey is likely to be undertaken in 2009, so information from this experiment may well be available by the time of the next International Survey Methods Conference. To date, however, in respect of concerns raised in the preceding section of this paper, the GPS survey has not been able to eliminate the need for a paper and pencil part of the survey, in that we still require socioeconomic data on the household and persons and vehicles in the household, and we also need information on the addresses of places that are visited by household members routinely (Stopher et al., 2007). In many applications of mobile technology, the recruitment has still relied on the telephone, also.

There are, however, some clear potentials of the mobile technology route to overcome some of the challenges mentioned in this paper. Recruitment and collection of socioeconomic data could be accomplished by door-knocking and face-to-face interviewing. Such a procedure would eliminate the reliance on being able to find the sample through telephoning, which is perhaps becoming more and more difficult, and would also move away from reliance on written materials for collection of data outside of the travel itself. Further developments in mobile technologies, in terms of miniaturisation, improved power management and efficiency, sensitivity of the devices and potential to add a back-up method to cover the problem areas of the main technology all promise a potentially viable future for such technologies as a means to overcome other difficulties in conducting travel surveys. Nevertheless, in the few instances in which GPS has been used as a stand-alone method for travel measurement (Stopher et al., 2007), the response rates on recruitment have been little different than those for other conventional survey methods, so that it does not appear that mobile technologies will provide the answer on response rate decline.

On a broader note, the potentials for application of what may also be termed 'location-aware devices' were explored in a paper by Lee-Gosselin and Harvey (2006). They included a wide range of different technologies and devices, which included GPS, GSM, RFID and other technologies as well as GIS layers and displays, licence plate tracking systems and electronic toll collection systems and explored potentials for using data from these in present and future surveys. They made a number of recommendations for future research in these technologies, many of which have yet to be acted on.

2.4. Some Thoughts on the Way Ahead

While this paper does not attempt to discuss all of the challenges and emerging trends in travel surveys, sufficient have, perhaps, been elucidated for us to think now about where we need to go in the next decades. It seems reasonably clear that 'business as usual' is not an option because the trends are unmistakable in their impact on the future of the conventional travel survey. Neither representativeness nor accuracy is likely to be increased with minor refinements in the survey procedures. As part of the preparation of this paper, the author reviewed an earlier keynote paper (Stopher, 1996) and noted that, while some of the suggestions put forward in that paper had been acted on in the past decade, others equally clearly had not and seem to remain an imperative to consider for the future. In this section are summarised some of the thoughts about directions in which we need to take surveys in the coming years.

2.4.1. Panels

Because of the very small number of panels that have been established in transport, the modelling of travel behaviour has largely not evolved to take advantage of the information provided by a panel. As a result, we have little awareness of the advantages that can be offered to modelling by using panels. Panel surveys offer several important advantages, given the issues raised earlier in this paper. Zumkeller et al. (2006) and Murakami, Greaves and Ruiz (2006) both outline many of the advantages of panels, and also suggest steps that need to be taken to bring panels into greater use in transport. Rather than reiterate those advantages and steps, the reader is referred to the papers cited.

A potential direction in which the household travel survey might evolve is through the recruitment of relatively modest-sized panels, with an occasional larger cross-sectional survey that is undertaken to check the representativeness of the panel from time to time. This design is known technically as a split panel. The panels would need to be surveyed on an annual basis because less-frequent surveying is likely to result in unacceptably high attrition from wave to wave, thereby undermining the advantages of a panel survey. In a large country such as the United States, individual metropolitan areas might maintain their own split panels, with an annual wave of the panel survey, and a cross-sectional survey undertaken every ten years or so. For smaller countries, and for national travel surveys, a somewhat larger panel might be recruited from across the country, again with a cross-sectional survey possibly every 10 or 15 years. In all cases, the panels could be designed as rotating panels to prevent respondent fatigue, with households or persons remaining in the panel for three or four years. While the sample size for such a panel will depend in part on the survey methodology to be applied, the gains from using a panel will include the possibility of using a much smaller sample size, and also reduce even the periodic cross-sectional survey to a smaller sample size, especially if it is conducted close to the time of a national census. A major gain from using a panel is the ability to determine the dynamics of changes in travel behaviour, especially as they relate to both external and internal stimuli to the household. A further potential advantage of a panel is that it offers the opportunity to empanel a representative sample, using a reasonably large incentive, if needed, and even to train the panel members in how to complete the survey, although one must be careful that any training does not produce behaviour changes through a Hawthorne effect (Landsberger, 1958). Given concerns raised in this paper about language and literacy, as well as the issues of how to locate a representative sample and declining response rates, the idea of empanelling a representative sample should be given serious consideration.

2.4.2. *Household versus Person*

Almost all household travel surveys conducted to date have considered the household to be not only the sampling unit, but also the analysis unit. As a result, the issue arises of defining what comprises a complete household for the purposes of determining response rates and such things as sampling errors, etc. As response rates decline, this issue becomes one of increasing concern. In the use of travel survey data for modelling, traditional approaches to modelling have only used the household as a unit of modelling for trip generation, while the remaining models in the four-step process are actually based on the trip as the unit of analysis. If there is a car ownership model, then it also usually operates at the level of the household. With the emergence of tour-based and activity-based modelling, the unit of analysis has changed to a tour or a day of activities. While intra-household interactions are of some concern, they are not always present in the models. The profession needs to address the issue of whether or not a complete household is required for subsequent modelling purposes. Given the increasing sophistication of micro-simulation procedures, it is possible that households could be micro-simulated from good data on individual persons, provided that the survey and secondary data sources contain adequate data on the socio-demographics of those who make up specific households.

It is clear that there must be serious questions raised about the representativeness of samples collected with a definition of a complete household being the receipt of completed survey materials (travel or activity diaries, in particular) from all or most of the members of a household. As argued by Richardson and Meyburg (2003), it is often preferable to accept as valid data that come from only a few members of the household, rather than introducing the biases of rejecting such data as incomplete. There is also likely to be an increasing bias against larger households, from which it will be more difficult to obtain data from every household member. Within households, it is increasingly difficult to get satisfactory responses from teenage members of the family, and also sometimes from the busier one of two adults in two-adult households. Sampling on the basis of the household leaves a dilemma in how to deal with such losses of complete data. Sampling on the basis of persons could lead to selecting only one or two persons from a household with whom to conduct the survey, and much higher completion rates, especially of hard-to-get respondent groups.

This is not necessarily an argument for restricting sampling to only one individual per household; however, definition of the sampling unit and the analysis unit as a person, rather than a household, opens up considerable potential to improve response rates and also improve completion rates of the surveys. Coupled with such improvements as the use of panels and mobile tracking technology devices, this could result in significant improvements in the data collected and the representativeness and accuracy of the data.

What is required to resolve this issue is research into the use of person data only for modelling purposes, albeit with information provided about the household from which the person comes, and the role the person has in the household. This could be accomplished quite readily with existing data sets by decoupling the individuals from other household members, while retaining the household and person data. Furthermore, investigation could be undertaken to determine whether or not it is feasible to simulate the travel of other household members from knowledge of the household characteristics, person characteristics and the travel of a subset of the members of the household. Finally, the issue of representativeness needs to be addressed in this respect, especially in terms of whether or not the sampling of individuals from within a household, as well as the sampling of the households themselves, can be done in such a way as to permit accurate and representative data to be developed for descriptive, modelling and policy purposes.

2.4.3. *Use of Telephones and Mail for Surveys*

The increasing incidence of telemarketing in many countries, the shift from land lines to mobile telephones and the increasing use of screening devices by owners of telephones all tend to suggest that the telephone may already have outlived its usefulness as a means of drawing the sample and recruiting people or households into a survey. If the telephone has not already done so, it is likely that it will become an unsuitable method of drawing a sample and recruiting respondents within the next few years. In countries where no listing of residential addresses exists or where none is available for use, this raises difficult questions as to how to draw a sample and recruit people into a survey. It may result in the need to apply multi-stage sampling or other sampling methods to avoid having to develop a full enumeration of the population from which to draw a random sample. It may also lead to the requirement to re-visit the face-to-face interview, where this has been abandoned in favour of telephone contact. In some societies, it may work well to contact people first through the mail and ask them to respond to a request to provide further contact details for a survey, although this technique is often not very productive.

In the first instance, in countries that rely on the telephone currently as a sampling and recruitment tool, it would be useful to examine the make-up of samples to see to what extent the telephone recruitment has resulted in recruiting households from areas of the metropolitan region where it would not be considered safe to send in interviewers for face-to-face surveys. It is quite possible that the telephone survey has not succeeded in penetrating such areas, in which case face-to-face recruiting may be equally as effective. It may also be worthwhile to compare the completion rates and correctness of surveys where the respondents are introduced to the survey by an interviewer who is physically present against ones where the interviewer is at the other end of a phone line.

The issues raised about literacy and language also tend to direct one away from written surveys and towards surveys that are conducted entirely by oral interview. Indeed, one could perhaps suggest that coupling mobile technology for tracking people's travel, together with a face-to-face recruitment and collection of the household and vehicle data usually required for such a survey, may be the appropriate direction in which to move. The extent to which mobile technology is ready yet for a full-scale substitution for the conventional travel diary is an issue that it is hoped will be addressed by this conference.

2.4.4. *Mixed-Mode Surveys*

The travel survey profession needs a controlled experiment in which a test can be made of the effects on survey measurements of using different survey modes, such as mail, Internet and telephone. The potential increase in response rates from multi-mode surveys also needs to be identified better. Theoretically, the increase should be significant, but this has not been proven as yet. There seems little question that providing respondents with more control over when, how and where they respond to a survey is likely to represent a reduction in respondent burden (Ampt, 2003) and produce a concomitant improvement in response rates. Therefore, while recruitment may need to move away from the telephone, the telephone, along with mail, Internet and face-to-face interview, may continue to provide viable options for how a person or household provides their completed response to the survey. For example, while the Internet, as discussed earlier, is not a good survey mechanism as the sole method because of its lack of penetration into the population, it often offers an excellent mechanism for response for those persons or households inclined to use it.

2.4.5. *Proxy Reporting*

Reducing the level of proxy reporting is another issue that has been discussed at length, and for which attempts have been made to improve the survey methodology. Proxy reports are known to have serious problems, especially for adults when made by other adults in the household and for teenagers when made by their parents (Stopher et al., 2007). Proxy reporting for younger children is probably likely to remain a necessity, because of both ethical concerns about surveying young children and the ability of children to respond to and answer appropriately the questions in a typical travel survey. If the survey is conducted by using a portable mobile technology device, there will remain issues about whether or not young children could or should use the devices, notwithstanding instances where children have used GPS devices (Mackett, Brown, Gong, Kitazawa, & Paskins, 2007). Among the issues that relate to children using mobile devices are whether or not they would lend them to friends to use, and whether or not they would take appropriate care of the devices and not be tempted to tamper with them.

Along with the investigation of the extent to which we need household data as opposed to a random sampling of people could be a more detailed examination of the question of the use made and importance of travel data for children. If it is found that such data represent an important component of the data to estimate travel behaviour models, then it becomes worthwhile to consider how such data may best be collected. There may be feasible methods to have children carry mobile technology devices, in which case this would seem to be the preferable method to collect data on their travel, rather than relying on proxy reporting by a parent or guardian.

The potential of moving to a person-based sample and to the use of mobile technology devices for measuring travel would both assist in reducing the need for and occurrence of proxy reports in travel surveys.

2.4.6. New Directions

The travel survey profession needs to take to heart some of the observations made earlier in this paper, where the evolution of travel surveys was seen as leading, rather than following the development of modelling paradigms. It is clear that, in the recent past, efforts to improve the accuracy and completeness of travel survey data have led to improvements and even significant changes in the ways in which data are collected, which have led, in turn, to the evolution of new modelling paradigms, such as activity modelling. Already, the introduction of GPS devices as a means of measurement is opening up new possibilities in examining travel behaviour, including providing an unprecedented level of detail on route choices and the potential to extend greatly the length of feasible observation periods, not only to multiple days, but to weeks, or (in principle) even months, whereas only rarely have travel surveys observed more than seven days. Modelling paradigms to use these capabilities in the data have not been developed as yet, but their development would probably make certain that such measurement procedures would move rapidly into the mainstream of the household travel survey toolkit.

In the same way, there is a need to explore further and apply methods of measuring the decision process through surveys, so that behavioural process models can be developed. This is likely to be instrumental in allowing exploration into the behavioural foundations of what we have, in the past, simply observed and modelled as travel choices. There may be yet other directions in which travel survey methods might change and thereby build the base for improving the understanding of behaviour, the ways in which to change behaviour and the basis of travel demand modelling. As in the past, the direction to pursue may be more a focus on how to measure more accurately and more completely, and the modelling and policy applications will then be able to take advantage of the new and richer information being provided from the surveys. One of the directions, for example, that may be fruitful to consider is that of the interaction between travel and other methods of communication. Thinking back to the origins of household travel surveys, such things as the Internet, e-mail, mobile phones, mp3 players and various other communication devices that are so prevalent in the world today were all unheard of and unthought of at that time. The increasing availability of communication through Wi-Fi, Bluetooth and similar capabilities is likely to change the interaction between communication and travel yet further. Yet, most of our travel surveys inquire little, if at all, about these other means of communication. The technological changes of the next few years are likely, if the past decade or so is any guide, to be of enormous significance, and it is likely that we will not be able to understand the true behavioural basis of the travel processes we observe without understanding how these various communication capabilities interface and interact with travel.

2.4.7. *Specification of a New Household Travel Survey*

From the thoughts elaborated in this paper, the following might be the specification of a household travel survey in the future. Sampling would be done from an address register, or some form of multi-stage sampling could be used, in which addresses for a subset of the population are enumerated for the purposes of choosing a random sample. Recruitment would be undertaken by first posting a pre-notification letter to all households in the sample, and then having multi-lingual interviewers visit the households with the intent to recruit one or more members of the household, but not necessarily the entire household. In fact, household members may be recruited using a grid procedure to determine which member or members of a household will be asked to take part. Recruitment would be to a panel, with recruited respondents being asked to remain in the panel for a period of three or four years.

At the time of recruitment, the interviewer would use a CAPI procedure to collect certain pertinent data about the household, the persons recruited to the survey and the vehicles available to the household. The interviewer would then provide a personal mobile technology device for each of the recruited household members, and would explain how it is to be used or might have the household members view a DVD, which would have been produced in multiple languages. This DVD would be left with the household members, where the DVD would have spoken instructions, along with visual demonstrations. Lest this idea of a DVD be interpreted as implying that the use of the GPS device is complicated, it is rather suggested as an alternative to written instructions, which, while they would probably be brief (since most mobile technology devices are very simple to use) would preclude active involvement of those with poor language and literacy skills. At an appointed time, the interviewer would return to the household and collect the mobile technology devices, and also ask a few additional questions, again using CAPI. However, if household members found a second visit to be difficult, the interviewer would leave a box and labels for a courier to pick up the devices, and the household members would be offered a choice of the Internet, a telephone call at a pre-arranged time or post/courier for returning the additional information required at the time of completion of the use of the mobile technology device. As noted earlier, this survey design would also reduce the need for proxy reporting, and substantially the occurrence of such reports, especially if it were found possible to have children carry mobile technology devices, or it was found that the travel data of children was not that important to travel behaviour modelling.

Such a survey design would minimise the requirements for respondents to be able to read and respond, thereby offering a greater chance that a representative sample could be obtained. Travel information would be collected unambiguously by the mobile technology devices, which, together with the other data collected by interview or by a method chosen by the respondents, would permit derivation of the complete travel and activity pattern of the respondents for a period of a week or two. (There are also substantial reductions in sample size possible (Stopher, Kockelman, Greaves, & Clifford, 2008) from collecting multi-day data, as well as added richness in the data for determining repetitive patterns, and differences among days of the week.) Repeating this survey as an annual panel, with rotation out of the panel of about one quarter to one third of respondents on each wave, would provide a wealth of information on the dynamics of behaviour, as well as responses to short-run changes such as petrol prices, changes to public transport fares and service, etc.

Perhaps of even greater significance is the potential that such a panel would offer for learning about and eventually understanding much more about the behavioural processes that underlie travel decisions and the interactions between communication devices and travel. With potential training, as suggested for the mobile technology devices themselves, panel members may also be able to be trained to interpret and understand better their own travel and activity decisions, and also be trained to self-report these better than is done at present. While such training may well change behaviour, the opportunity to observe how improved understanding and learning by the respondents themselves leads to behavioural change is likely to lead to significant illumination of the entire behavioural process itself. This opens up some exciting possibilities for new directions of enquiry and research. A variety of other in-depth techniques could be introduced with panel members, these being rooted in knowledge of the current travel and activity patterns as revealed through the more conventional survey elements. This would certainly include, but not be limited to, use, from time to time, of Stated Preference surveys and a variety of other specialised surveys that would both enrich the modelling data and inform policy and investment options. Indeed, the possibilities that could be offered by this move to ongoing panels bear considerable thought and offer exciting opportunities for future directions.

2.5. Concluding Remarks

It is, perhaps, rather too bold to suggest that the telephone survey for household travel surveys is likely to disappear as a useful and useable option in the near future. In 1973, Lee (1973) predicted the imminent demise of large-scale planning models, only to find that these models are still being developed and used 30 years later. However, there are clearly a number of issues that are compromising the telephone-based survey in many countries around the world. Some countries, as noted in the early part of this paper, have never moved to the use of such surveys, and continue to use a combination of postal and face-to-face surveys. However, there are other threats to continuing to do business as usual, no matter which method is currently being used. Declining response rates, increasing globalisation of population — leading to multiple languages being spoken — and declining literacy are just a few of the challenges that face the household travel survey today. In many nations, also, people perceive that they have less time available for such activities as completing a voluntary survey, so that such survey mechanisms as a roadside interview or, a self-administered or personal interview to be done at home are seen by many people as wasting valuable time. While these challenges are clear, the transport profession is demanding more data and more accurate data for the purposes of planning and policy development.

Along with these challenges, there are also emerging new potentials for how to conduct surveys and gain accurate measurement of how people travel. Many of these potential opportunities are to be discussed during this 8th International Conference on Survey Methods. Issues relating, for example, to mobile technologies and Web-based surveys will continue to occupy a prominent place in the discussions in this conference. Other aspects of what is put forward in this paper may need to be considered across a number of workshops, for example, households versus persons as the analysis unit, the minimum age for which a mobile technology device can be used, the importance of children's travel in travel models, the potential use of panels, avoidance of proxy reports, methods to obtain representative samples (especially as the potential to use RDD declines) and methods to measure the underlying behavioural processes and the interactions between travel and activity choices on the one hand and new communication capabilities on the other.

As has been noted in previous conferences, the health science professions, among others, have long used panels as a means to trace the effects of different treatments and to monitor the health of populations. The transport profession needs to address more clearly the issue of whether or not panels could generate the data needed to trace the effects of different policies and investments in transport and to monitor and understand the dynamics of travel behaviour and the processes that lead to behaviour patterns in general. It may be time to question the long-held notion that data must be collected to populate an origin–destination matrix in order to be able to undertake transport planning with the results of surveys.

It seems likely that travel surveys as they have been practised in the past 50 years or so will need to undergo significant and substantial change in the near future. To a large degree, a review of past changes in the methods and approaches to travel surveys suggests that the profession has tended to be reactionary, rather than visionary. Perhaps it is time to set out clear goals as to what travel surveys should achieve and then to seek methods that will allow these goals to be met, rather than continually reacting to constraints that are imposed upon the travel survey as a result of societal and other changes. In this mode, it is well to take seriously the fact that often in the past it is the survey methods that have taken the lead in improvements in the uses to which the data are put, such as the development of activity diaries, which have since helped to open up a field of activity modelling. The challenge to this conference is to see where the travel survey toolkit might go in the near future to open up new possibilities and new frontiers in our understanding of the mechanisms that produce the travel that we are so anxious to measure.

Acknowledgements

The author acknowledges particularly many helpful suggestions that were provided by Martin Lee-Gosselin, who reviewed two earlier versions of this paper. The added insights he provided, and even actual phrases in some cases, have added much to this paper. However, any remaining errors and omissions in the paper are entirely those of the author.

References

AAPOR. (2000). *Standard definitions: Final call dispositions of case codes and outcome rates for surveys*. American Association for Public Opinion Research, available at: http://www.aapor.org/uploads/standarddefs_4.pdf, accessed 14th April 2008.

Australian Bureau of Statistics (ABS). (2008). *2006 Census tables: Australia*, available from the ABS Web site www.abs.gov.au, accessed 26th April, 2008.

Ampt, E. S. (2003). Respondent burden. In: P. Stopher & P. Jones (Eds), *Transport survey quality and innovation* (pp. 507–521). Oxford: Pergamon.

Ampt, E. S., Richardson, A. J., & Brog, W. (1985). *New survey methods in transport*. Utrecht, The Netherlands: NU Science Press.

Ampt, E. S., & Stopher, P. R. (2005). Mixed method data collection in travel surveys: Challenges and opportunities. Paper presented to the 28th Australasian Transport Research Forum, Sydney, Australia, September, available at: http://www.patrec.org/atrf/papers/2005/Ampt%20&%20Stopher%20(2005).pdf, accessed 25th April 2008.

Battelino, H., & Peachman, J. (2003). The joys and tribulations of a continuous survey. In: P. Stopher & P. Jones (Eds), *Transport survey quality and innovation* (pp. 49–68). Oxford: Pergamon.

Beerten, R., & Martin, J. (1999). Household ownership of telephones and other communication links: Implications for telephone surveys. *Survey Methodology Bulletin*, *44*, 1–7.

Blankenship, A. B., Breen, G. E., & Dutka, A. F. (1998). *State of the art marketing research* (2nd ed., pp. 155–157). New York: McGraw-Hill.

Blumberg, S. J., & Luke, S. V. (2007). Wireless substitution: Early release of estimates based on data from the national health interview survey, July–December 2006, National Center for Health Statistics, available at: http://www.cdc.gov/nchs/nhis.htm, accessed May 14, 2007.

Bonnel, P. (2003). Postal, telephone, and face-to-face surveys: How comparable are they? In: P. Stopher & P. Jones (Eds), *Transport survey quality and innovation* (pp. 215–237). Oxford: Pergamon Press.

Bonnel, P., & Armoogum, J. (2005). *National transport surveys – What can we learn from international comparisons?* European Transport Conference, 3–5 October 2005, Strasbourg, Actes sur cd rom, 24pp.

BPR (1954). *Manual of procedures for home interview traffic studies*. Washington, DC: US Department of Commerce, Bureau of Public Roads.

Bradburn, N., & Sudman, S. (1979). *Improving interview method and questionnaire design*. San Francisco, CA: Jossey-Bass Publishers.

Bradley, M. (2006). Process data for understanding and modelling travel behaviour. In: P. Stopher & C. Stecher (Eds), *Travel survey methods: Quality and future directions* (pp. 491–510). Oxford: Elsevier.

Bricka, S. (2008). Non-response challenges in GPS-based surveys. Paper to be presented at the 8th International Conference on Travel Survey Methods, Annecy, France, May (in this volume).

Brög, W., Fallast, K., Katteler, H., Sammer, G., & Schwertner, B. (1985a). Selected results of a standardised survey instrument for large-scale travel surveys in several European countries. In: E. S. Ampt, A. J. Richardson & W. Brög (Eds), *New survey methods in transport* (pp. 173–191). Utrecht: VNU Science Press.

Brög, W., Meyburg, A. H., Stopher, P. R., & Wermuth, M. J. (1985b). Collection of household travel and activity data: Development of a survey instrument. In: E. S. Ampt, A. J. Richardson & W. Brög (Eds), *New survey methods in transport* (pp. 151–172). Utrecht: VNU Science Press.

CASRO. (1982). *On the Definition of Response Rates: A Special Report of the CASRO Task Force on Completion Rates*, Council of American Survey Research Organizations, available at: www.casro.org, accessed 14 April 2008.

CATS (1956). *Home interview manual 1956*. Chicago, IL: Chicago Area Transportation Study.

CERTU (2008). *L'enquête ménages déplacements "standard standard", guide méthodologique*. Collections du CERTU, éditions du CERTU, Lyon, 204p.

Chlond, B., Last, J., Manz, W., & Zumkeller, D. (2006). Long-distance travel in a longitudinal perspective — The INVERMO approach in Germany. Conference paper presented at the 85th Annual Meeting of the Transportation Research Board, Washington, DC, January.

CIA (2008). *The world fact book*, available at: https://www.cia.gov/library/publications/the-world-factbook/geos/xx.html, accessed on 26th April, 2008.

Cialdini, R. B., Vincent, J. E., Lewis, S. K., Catalan, J., Wheeler, D., & Darby, B. (1975). Reciprocal concessions procedure for inducing compliance: The door-in-the-face technique. *Journal of Personality and Social Psychology*, *31*, 206–215.

Cicerone, R. A., Sassaman, R., & Swinney, J. (2007). Controlling motor vehicle driver behavior at roadside interview sites. *ITE Journal*, *27*(8), 18–20.

Czaja, R., & Blair, J. (2005). *Designing surveys: A guide to decisions and procedures* (p. 45). Thousand Oaks, CA: Pine Forge Press.

Dal Grande, E., Taylor, A., & Wilson, D. (2005). Is there a difference in health estimates between people with listed and unlisted telephone numbers?. *Australian and New Zealand Journal of Public Health*, *29*(5), 448–456.

Dillman, D. A. (1978). *Mail and telephone surveys: The total design method* (344 pp). New York: Wiley.

Dillman, D. A. (2000). *Mail and internet surveys: The tailored design method*. New York: Wiley.

Doherty, S. T., Nemeth, E., Roorda, M., & Miller, E. J. (2004). Design and assessment of the Toronto area computerized household activity scheduling survey, *Transportation Research Record 1894*.Washington, DC: Transportation Research Board, pp. 140–149.

Doherty, S. T., Papinski, D., & Lee-Gosselin, M. (2006). An internet-based prompted recall diary with automated GPS activity-trip detection: System design. Paper presented to the 85th Annual Meeting of the Transportation Research Board, January 22–26.

FHWA. (1973). *Urban origin–destination surveys*. Washington, DC: Federal Highway Administration, US Department of Transportation.

Foster, R. E., Benson, J. D., & Stover, V. (1976). *Consequences of small sample O–D data collection in the transportation planning process*. Final Report to US Department of Transportation, Federal Highway Administration, Offices of Research and Development, Report No. FHWA-RD-78-56, Washington, DC.

Golob, T. F., Kitamura, R., & Long, L. (Eds). (1997). *Panels for transportation planning: Methods and applications*. Dordrecht, The Netherlands: Kluwer Academic Publishers.

Goulias, K. G. (2000). Surveys using multiple approaches. *Transportation research circular E-C008, transport surveys: Raising the standard*. Washington, DC: Transportation Research Board, pp. II-A/1-16.

Groves, R. M., Biemer, P. P., Lyberg, L. E., Massey, J. T., Nicholls, W. L., II, & Waksberg, J. (Eds). (1988). *Telephone survey methodology*. New York: Wiley.

Harvey, A. S. (2003). Time–space diaries: Merging traditions. In: P. Stopher & P. Jones (Eds), *Transport survey quality and innovation* (pp. 151–180). Oxford: Pergamon.

Jackson, R. (2003). *Some startling statistics*, University of Dayton, Erma Bombeck Writers' Workshop, available at: http://www.humorwriters.org/startlingstats.html, accessed on 26th April 2008.

Jones, P., & Bradley, M. (2006). Stated preference surveys: An assessment. In: P. Stopher & C. Stecher (Eds), *Travel survey methods: Quality and future directions* (pp. 347–361). Oxford: Elsevier.

Kirsch, I. S., Jungeblut, A., Jenkins, L., & Kolstad, A. (2002). *Adult literacy in America*, National Center for Education Statistics, US Department of Education, Report NCES 1995-275.

Kuhnimhof, T., Chlond, B., & Zumkeller, D. (2006). Nonresponse, selectivity and data quality in travel surveys — Experiences from analyzing the recruitment for the German mobility panel. *Transportation Research Record, No. 1972*, pp. 29–37.

Landsberger, H. A. (1958). *Hawthorne revisited*. Ithaca, NY: Cornell University.

Lee, D. B. (1973). Requiem for large-scale models. *Journal of the American Institute of Planners*, *39*(3), 163–178.

Lee-Gosselin, M. E. H. (2003). Can you get there from here? A viewpoint on stated response survey innovation and quality. In: P. Stopher & P. Jones (Eds), *Transport survey quality and innovation* (pp. 331–345). Oxford: Pergamon Press.

Lee-Gosselin, M. E. H., & Harvey, A. S. (2006). Non-web technologies. In: P. Stopher & C. Stecher (Eds), *Travel survey methods: Quality and future directions* (pp. 561–568). Oxford: Elsevier.

London County Council. (1964). *The London traffic survey*. Report in two volumes, London County Council, The County Hall, London, England.

Louviere, J. J., Hensher, D. A., & Swait, J. D. (2001). *Stated choice methods: Analysis and application*. Cambridge: Cambridge University Press.

Mackett, R., Brown, B., Gong, Y., Kitazawa, K., & Paskins, J. (2007). Children's independent movement in the local environment. *Built Environment*, *33*(4), 454–468.

Miniwatt Marketing Group. (2008). Internet world stats: Usage and population statistics, available at: http://www.internetworldstats.com/stats.htm, accessed 25th April, 2008.

Morris, J., & Adler, T. (2003). Mixed-mode surveys. In: P. Stopher & P. Jones (Eds), *Transport survey quality and innovation* (pp. 239–252). Oxford: Pergamon Press.

Murakami, E., Greaves, S., & Ruiz, T. (2006). Moving panel surveys from concept to implementation. In: P. Stopher & C. Stecher (Eds), *Travel survey methods: Quality and future directions* (pp. 399–412). Oxford: Elsevier.

Murakami, E., & Watterson, W. T. (1989). Developing a household travel panel survey for the Puget sound region. *Transportation Research Record No. 1285*, pp. 40–46.

Northamptonshire County Council (2003). A509 roadside interview observed origin–destination matrix. *Technical Note 1, A509 Isham Bypass*, Northamptonshire County Council, December.

Organisation for Economic Co-operation and Development (OECD) (2004). *Trends in international migration*. OECD Publishing, p. 19.

Ortúzar, J. de D. (2006). Travel survey methods in Latin America. In: P. Stopher & C. Stecher (Eds), *Travel survey methods: Quality and future directions* (pp. 1–18). Oxford: Elsevier.

Parvataneni, R., Stopher, P. R., & Brown, C. (1982). Origin–destination travel survey for Southeast Michigan. *Transportation Research Record No. 886*, pp. 1–8.

Pendyala, R., & Bricka, S. (2006). Collection and analysis of behavioural process data: Challenges and opportunities. In: P. Stopher & C. Stecher (Eds), *Travel survey methods: Quality and future directions* (pp. 511–530). Oxford: Elsevier.

Richardson, A. J., & Meyburg, A. H. (2003). Definitions of unit nonresponse in travel surveys. In: P. Stopher & P. Jones (Eds), *Transport survey quality and innovation* (pp. 587–604). Oxford: Pergamon Press.

Scottish Executive. (2004). *Roadside interview survey guidance in Scotland*, available at: http://www.scotland.gov.uk/Publications/2004/12/20455/49117, accessed 24th April 2008.

Stopher, P., & Jones, P. (Eds). (2003). *Transport survey quality and innovation*. Oxford: Pergamon Press.

Stopher, P., & Stecher, C. (Eds). (2006). *Travel survey methods: Quality and future directions*. Oxford: Elsevier.

Stopher, P. R. (1985). The design and execution of on-board bus surveys: Some case studies. In: E. S. Ampt, A. J. Richardson & W. Brög (Eds), *New survey methods in transport* (pp. 209–225). Utrecht, The Netherlands: VNU Science Press.

Stopher, P. R. (1992). Use of an activity-based diary to collect household travel data. *Transportation*, *19*, 159–176.

Stopher, P. R. (1996). Household travel surveys: Cutting-edge concepts for the next century. *Conference on Household Travel Surveys: New Concepts and Research Needs, Conference Proceedings 10*, Transportation Research Board, Washington, DC, pp. 11–23.

Stopher, P. R. (1997). A review of separate and joint strategies for the use of data on revealed and stated choices. In: P. Bonnel, R. Chapleau, M. Lee-Gosselin & C. Raux (Eds), *Urban travel survey methods: Measuring the present, simulating the future*(pp. 15–32).

Stopher, P. R., Kockelman, K., Greaves, S. P., & Clifford, E. (2008). Sample size requirements for multi-day surveys: Some findings. Paper to be presented to the 8th International Conference on Travel Survey Methods, Annecy, France, May.

Stopher, P. R., & Swann, N. (2008). A 6-wave odometer panel for the evaluation of voluntary travel behaviour change programs. Paper presented to the 87th Annual Meeting of the Transportation Research Board, Washington, DC, January.

Stopher, P. R., Swann, N., & FitzGerald, C. (2007). Using an odometer and a GPS panel to evaluate travel behaviour changes. Paper presented to the 11th TRB National Planning Applications Conference, Daytona Beach, FL, May.

Sudman, S., & Bradburn, N. (1974). *Response effects in surveys*. Chicago, IL: Aldine Press.

Sudman, S., & Bradburn, N. (1982). *Asking questions: A practical guide to questionnaire design*. San Francisco, CA: Jossey-Bass Publishers.

Taylor, S. (2003). Telephone surveying for household social surveys: The good, the bad and the ugly. *Survey Methodology Bulletin*, *52*, 10–21.

Transmanagement. (2004). *Counting transit so that transit counts*. Report prepared for the American Public Transportation Association, Washington, DC, 64pp.

TRB (2000). *Transportation Research Circular E-C008, Transport surveys: Raising the standard*. Transportation Research Board, Washington, DC.

Van Evert, H., Brög, W., & Erl, E. (2006). Survey design: The past, the present, and the future. In: P. Stopher & C. Stecher (Eds), *Travel survey methods: Quality and future directions* (pp. 75–93). Oxford: Elsevier.

Van Wissen, L. J. G., & Meurs, H. J. (1989). The Dutch mobility panel: Experiences and evaluation. *Transportation*, *16*(2), 99–119.

Viegas, J. (2000). Workshop Summary, *Transportation Research Circular E-C008*, *Transport Surveys: Raising the Standard*, Transportation Research Board, Washington, DC, pp. II-A/17–21.

Wolf, J. (2006). Applications of new technologies in travel surveys. In: P. Stopher & C. Stecher (Eds), *Travel survey methods: Quality and future directions* (pp. 531–544). Oxford: Elsevier.

Wolf, J., Guensler, R., & Bachman, W. (2001). Elimination of the travel diary: An experiment to derive trip purpose from GPS travel data. *Transportation Research Record No. 1768*, Transportation Research Board, Washington, DC, pp. 125–134.

Wolf, J., Loechl, M., Thompson, M., & Arce, C. (2003). Trip rate analysis in GPS-enhanced personal travel surveys. In: P. Stopher & P. Jones (Eds), *Transport survey quality and innovation* (pp. 483–498). Oxford: Pergamon Press.

Zumkeller, D., Madre, J.-L., Chlond, B., & Armoogum, J. (2006). Panel surveys. In: P. Stopher & C. Stecher (Eds), *Travel survey methods: Quality and future directions* (pp. 363–398). Oxford: Elsevier.

PART II

SUSTAINABILITY AND TRAVELLER ADAPTATION

Chapter 3

What is so Special about Surveys Designed to Investigate the Environmental Sustainability of Travel Behaviour?

Peter Bonsall

Abstract

The increasing prominence of environmental sustainability as an objective of transport policy, and in people's thinking about their travel-related decisions, brings new challenges for data collection because, uniquely in the transport sector, it brings three issues to the fore: firstly, the difficulties associated with gathering data and opinions on a topic which is socially and morally charged; secondly, the fact that the resulting adjustments to behaviour may be in dimensions (such as choice of vehicle characteristics and mode choice for infrequent trips) which are not traditionally covered very fully in travel surveys; and thirdly, the likelihood that any change in behaviour may emerge only slowly over an extended period of time (as and when the opportunity arises for the individuals' behaviour to be adjusted to fit their aspirations).

It is argued that, because the topic is socially charged, data on sustainable travel behaviour is particularly prone to social desirability bias and other related biases. The nature and implications of these biases are addressed and it is concluded that individuals will tend to exaggerate the likelihood of behavioural change in response to sustainability concerns and policy initiatives. Methods, which might be used to study the emergence of sustainable patterns of behaviour in general, and responses to sustainability orientated policy initiatives in particular, are discussed and attention is given to ways of minimising bias in the data.

Transport Survey Methods

ISBN: 978-1-84855-844-1

3.1. Introduction: The Emerging Phenomenon

Environmental sustainability is an increasingly important objective of transport policy — as witnessed by the increased prominence of investment in public transport, walking and cycling facilities and of the use of regulations, fiscal measures and 'soft' measures to promote the use of such modes in preference to those deemed less sustainable. These policies are being pursued against a background of changes in lifestyle facilitated by developments in telecommunications, the availability of 'greener' technologies and, according to several recent polls, the emergence of sustainability as a socially desirable goal for an increasing proportion of the population.

The increasing role of environmental sustainability as a consideration in individual decision-making is evidenced by the fact that commercial organisations are competing to align themselves with the green agenda. Their desire to stress their products' green credentials is already well established but the next stage appears to be offering to assuage their customers' sense of guilt about purchasing unsustainable products or services by subscribing to a carbon emission offset scheme. Such actions by commercial organisations tend to reinforce the prevailing climate in which it has become de-rigour to espouse environmental values.

Adoption of more environmentally sustainable lifestyles and travel patterns, whether in response to deliberate policy initiatives by public authorities, in response to social pressures or as a result of individual motivation, is becoming an increasingly important phenomenon among certain groups. Meanwhile other trends in society, notably the emergence of more dispersed activity patterns in response to increased mobility, changing patterns of land use and more extensive social networks (see, e.g. Button & Nijkamp, 1997) are pulling in the opposite direction.

All this is already happening and if the transport analyst wishes to understand and monitor developments and, more ambitiously, to predict trends and influence the course of future developments, there is an urgent need to obtain reliable data on changes in attitudes and behaviour.

This paper considers what the particular data needs are and outlines some particular challenges brought about by the nature of the subject matter and the prevailing social climate. Potential sources of data are reviewed and some examples from recent literature are discussed before conclusions are drawn on the best way forward.

3.2. Issues

3.2.1. *How does the Data Requirement Differ?*

Table 3.1 summarises the data, which is required to help understand, monitor, predict or promote the emergence of environmentally sustainable travel behaviour. The most obvious point to make here is that, for the most part, this list is remarkably

Table 3.1: Data required for investigating the environmental sustainability of travel.

Activities engaged in	Nature of activity Locations **Whether physical presence is required?**
Travel undertaken	Mode Timing Origin and destination Frequency Cost incurred Time taken
Usage of communications media	Medium Timing Frequency Cost incurred Time
Consequences of engagement via chosen mode/medium	Environmental (**greenhouse emissions, local air quality, energy consumption, environmental degradation**) Other (social, economic, employment generation, etc.)
Characteristics of individual actors	Socioeconomics (age, gender, income, education etc.) Access to facilities, modes and media (home and work location, car ownership, computer access, physical fitness etc.) **Awareness of relevant facilities, modes and media** **Attitudes towards relevant facilities, modes and media**
Characteristics of available travel modes, communications media and other relevant facilities	Cost incurred Time taken Comfort/ease of use Hours of operation (including departure time, frequency etc.) **Environmental consequences of use** Other consequences of use (social, economic etc.) Legislative restrictions on use Physical restrictions on use **Visibility ('known-ness') and image**

similar to that required for almost any analysis of travel behaviour. The difference lies in the increased prominence accorded to items (highlighted in bold), which relate to the environmental consequences of the behaviour, to the achievement of activity goals without engaging in travel, and to the motivations and attitudes affecting the decision to adopt an environmentally sustainable pattern of behaviour.

Our interest in the environmental consequences (greenhouses gasses and local pollutants emitted, energy consumed, bio-hazards and ecological damage) of a given pattern of behaviour reflects the fact that these are the most direct indicators of environmental sustainability.

Our interest in the achievement of activity goals without needing to travel reflects the fact that, given the burgeoning opportunities offered by ICT, adoption of telecommuting, tele-shopping and other travel substitutes may be seen as an obvious means of reducing ones travel-related environmental footprint — even though the actual result may again be far from straightforward.

Our interest in motivational and attitudinal factors reflects the fact that they are thought to play a key role in the adoption of environmentally sustainable lifestyles, whether as a result of a direct policy intervention or as a response to a social or moral climate in which environmental sustainability is seen as a desirable trait.

One dimension which is not apparent from Table 3.1, but which is nevertheless important, is that, because adaptation of lifestyles is not instantaneous, we need to be interested in the dynamic evolution of attitudes, aspirations and behaviours over time. This suggests the use of panels rather than relying on repeated cross-section surveys. However, it may still be difficult to attribute an observed change, even if it can be detected, to a specific cause and, as we will see, the very fact of participation in a panel may affect the individual's attitudes and behaviours.

The collection of information affecting the adoption of patterns of travel behaviour is clearly important when the analyst is seeking to understand behavioural outcomes. But it is also important when the aim is to monitor the effectiveness of initiatives designed to promote sustainable behaviour and, as is the case whenever data is being collected to assess the effect of a specific intervention, it is important to design the study such that it can shed light on the counterfactual (what would have happened if the intervention had not occurred).

3.2.2. Some Potential Biases

All data collection is potentially subject to bias and this is not the place to seek to provide a complete review of this phenomenon.[1] However, the collection of data on sustainable travel behaviour is perhaps particularly prone to response bias and reporting bias. The implications of these potential biases, and discussion of how they

1. Some indication of the range of such biases can be obtained from Groves (1989), Richardson, Ampt, and Meyburg (1995) and from Appendix A of Birol, Karousakis, and Koundouri (2006).

might be reduced, are addressed at various points in the paper but it is appropriate to draw attention to them at this early stage because they pervade much of what is to follow.

Environmental sustainability is now widely seen as a socially desirable phenomenon and this, taken alongside widespread publicity for the sustainability agenda and its strong association with transport and travel, variously engenders feelings of pride, guilt or bravado which may affect people's willingness to reveal their behaviour patterns. Psychologists use the term '*social desirability bias*' to refer to the tendency of people, when addressing issues characterised by a moral or societal overtone, to deceive themselves, and others, into thinking that they subscribe to the prevailing social norm (see, e.g. Ganster, Hennessey, & Luthans, 1983; Paulhus, 1984, 1991; Nederhof, 1985; King & Brunner, 2000). The presence of this bias can threaten the validity of research by obscuring the true nature of relationships between variables and can lead to spurious correlation, or to the suppression or moderation of relationships. To date, much of the evidence on social desirability bias has come from research into voting intentions and responses to health-related promotion of healthy lifestyles but, against a background of increasing environmental concern, and given that different transport modes are associated with very different scores on social desirability scales (e.g. Kaiser, 1998) it seems that data on sustainable travel patterns are likely to be at increasing risk of such bias.

People are showing a growing interest in environmental sustainability and share a generalised belief that individuals have some duty to do something about it. Since participation in a survey on the subject might be regarded as part of one's contribution to the cause, some people will be particularly keen to participate in a survey related to the topic. Survey organisers have been tempted to tap into this enthusiasm by publicising the fact that the survey is concerned with environmental sustainability (e.g. by using tag lines such as '... a study of how we can make travel more environmentally friendly'). But in doing so they risk creating a *response bias* towards respondents who put a high value on environmental issues and, given the socially charged nature of environmental sustainability, are likely to attract those who are proud of their record more than those who think their record is embarrassingly poor. Discussing the results of her own survey of constraints on pro-environmental behaviour, Tanner (1999) admits that her sample was probably biased towards people 'who were more likely to engage in (pro) environmental behaviour' but then disarmingly states that 'most environmental studies' are biased in this way.

The problem is not confined to the achievement of a representative sample of respondents; it can also result in *biased reporting* of behaviours and attitudes. It is well known that, when asked about something that is socially charged, people are tempted to try and project a socially desirable picture of themselves and that this can lead to them suppress facts or opinions which might attract opprobrium and exaggerate those that are likely to attract approval (see Leary & Kowalski, 1990, for a description of the phenomenon known as '*impression management*'). In the current context this could lead to biased reporting of travel and activity patterns and of attitudes to environmental issues. Such misreporting is perhaps more likely to involve omission of negatives than invention of positives (because the consequences of discovery are less serious) and in the current context would probably be manifest in some respondents toning down any beliefs which they thought might be unpopular and 'forgetting' to report trips that might be regarded as environmentally unsound (some respondents, seeking to avoid cognitive dissonance, might use a range of excuses to persuade themselves that their record was not *really* incorrect; perhaps arguing that the omitted journey was very short or atypical).

It should be noted that these potential biases are likely to be particularly serious where there has been a campaign designed to raise peoples' awareness of the link between transport and environmental sustainability issues and that this creates a particular problem if one is seeking to monitor the effectiveness of such a campaign. The problem may be reduced by distancing the survey from the campaign but, particularly if the campaign has been successful, people will be aware that some behaviours are more environmentally sustainable (thus socially desirable) than others and so, from the standpoint of the surveyor hoping for honest answers to her questions, the damage will already have been done.

Campaigns to raise environmental awareness or promote adoption of more sustainable travel patterns often tap into a sense of community spirit and collective responsibility (even when the overt message is that it makes sense *for the individual* to revise their behaviour, the idea that it is also better for the community is not far below the surface). The sense that the individual has a stake in the success of the campaign may be enhanced through regular communication and feedback on campaign developments. This approach, which is in line with the marketing textbooks, has had considerable success. Unfortunately, from a monitoring point of view, it raises the possibility of *stakeholder bias* whereby individuals associate themselves with the objectives of the campaign and may be affected by a desire, or assumed obligation, to show the programme in a good light — particularly if they feel that they have benefited from it.

Stakeholder bias is clearly a danger if active participants in a scheme are relied on to report changes in their own behaviour; the participant list offers easy access to a source of enthusiastic survey respondents but they cannot be regarded as disinterested. A less obvious, and therefore potentially more dangerous, problem is that stakeholder bias may not be restricted to active participants. It can affect any members of the wider community who feel that they share the objectives of the campaign — and may indeed exist even in the absence of any specific campaign, simply as a result of a generalised sense of common purpose caused by public debate. This kind of generalised stakeholder bias might be expected to be particularly marked among people with a strong sense of community obligation (the very people most likely to respond to questionnaires) and may be reinforced by membership of a panel because repeated consideration of one's travel and activity patterns might lead to increased awareness and introspection.

Affirmation bias (whereby respondents, unconsciously or otherwise, seek to give consonant responses) is a related phenomenon, which can affect efforts to monitor the success of environmentally positive initiatives if the respondents assume that the monitors *want* to find evidence of success. The risk of this kind of bias is reduced by maintaining a distance between the survey exercise and any ongoing initiatives; however, it may not be realistic to avoid a perceived link between the programme and the monitoring. It is unlikely that the sponsoring authority would want to hide the fact that it was commissioning an evaluation, and a member of the target population would have to be particularly slow-witted to avoid drawing any conclusion from the coincidence of being asked about their travel patterns within a few months of being invited to consider how they might make their travel patterns more sustainable.

3.3. Potential Data Sources

Table 3.2 summarises, in no particular order, the range of potential data sources and the types of investigation to which they might contribute. The remainder of this section discusses each of these sources in more detail.

3.3.1. Travel and Activity Surveys

This section deals with surveys, which involve asking individuals to report details of their travel and/or activity patterns. The data thus collected can be used to establish the status quo, derive trends or monitor the success of a particular initiative. It can be presented as a description of travel and/or activity patterns (via trip rates, activity

Table 3.2: Data sources and roles.

Data sources	**Indicative[a] assessment of the importance of the data source for...**			
	...general monitoring	**...evaluation or assessment of specific initiatives or developments**	**...prediction of future outcomes**	**...behavioural investigation**
Travel and activity surveys	XXXX	XXX	XX	X
Stated choice experiments etc.			XX	XXX
Attitudinal questionnaires	XXXX	XX	XX	XXX
Depth interviews	X	XX	X	XXX
Focus groups	X	X	X	XXX
Objective monitoring	XXXX	XXX	XX	

[a]The actual importance will clearly depend on the approach used (e.g. a prediction based simply on aggregate trends could rely exclusively on data from aggregate modelling, and some forms of behavioural investigations would make no use of data on actual behaviour or on stated choices).

participation rates, mode shares etc.) or processed to provide an indication of the sustainability of the patterns (e.g. via calculation of fuel consumption, greenhouse gasses emitted etc.).

The direct involvement of the individual traveller/actor in the survey process means that their willingness to participate and to provide an objective record become important issues which will feature in several of the following paragraphs.

The issues involved in the design of travel surveys are well known and have been revisited many times in previous conferences in this series. They include:

- whether to seek diaries (and if so over what period?) or to concentrate on examples of a particular type of journey or activity (e.g. the journey to work, the most recent trip, the most recent journey of over 100 miles or the most recent purchase of groceries);
- how to obtain data about changes in behaviour over time (e.g. via panel surveys, repeated cross-section surveys or retrospective questions);
- what level of detail to seek on the selected travel or activities;
- whether to seek information from selected individuals or from all household members, and whether to accept proxy data;
- what mode of data collection to use: self-completion questionnaire, interview or some combination;
- whether to offer incentives or inducements; and
- how to check or verify the data collected.

There is a range of opinions on most of these issues. For example, one can contrast the KONTIV method (Brög, 2000) with its one-day self-completion diary subject to subsequent verification via a phone call, with the method used in the UK National Travel Survey (DfT, 2007a), which involves seven-day records collated and checked by a visiting interviewer. Similarly one can contrast the use of interactive software such as CHASE (Doherty & Miller, 2000) with the continued dominance of hard copy questionnaires.

Different solutions will clearly be appropriate in different circumstances and so the question to be addressed in the current paper is whether, given that we are considering the environmental sustainability of travel and activity patterns, there is any reason to favour one position rather than another on any of these design issues. The following paragraphs focus on those issues where this may be the case.

3.3.1.1. Whether to employ diaries or to seek information about specific categories of journey or event The assessment of environmental sustainability of travel patterns requires a representative picture of overall travel and it is generally considered appropriate to use diaries to obtain complete pictures rather than to focus on a particular type of activity or trip. However, given that some categories of activity or trip have a disproportionate impact on overall sustainability (e.g. long-distance journeys because they are likely to have a disproportionate environmental impact, or walk/cycle trips because they may indicate the feasibility of low-impact behaviour or provide a barometer on lifestyle choices), it may seem logical to focus on them (it being statistically indefensible simply to extend the diary period with the aim of increasing the likelihood of capturing rare events). This focus might be achieved by asking about, for example, the most recent trip of greater than 100 km, or the most recent cycle trip, or by asking the frequency of walk trips. However, a note of caution is necessary here because, if the interviewee is simply asked to state the frequency of a class of trip which might be seen as having high (or low) social desirability, he might, in the absence of the constraint of reporting a plausible diary sequence, tend to over- (or under-) state its true frequency. The fact that data collected as part of the NTS on the frequency of cycling trips (DfT, 2007b) implies a higher occurrence of such trips than is recorded in the NTS diary data may be evidence of this kind of distortion.

3.3.1.2. How to obtain information about changes over time Retrospective questions (e.g. 'How many walk trips per week were you making this time last year?' or 'Compared to this time last year, do you think you are now making about the same number of walk trips, more walk trips or fewer walk trips?') appear to be an efficient way of finding out about change in behaviour over time. However, the usual caveats about relying on retrospective recall probably apply with increased vigour because of the risk of distorted recall of socially charged phenomena and so this technique can probably be ruled out as a means of gaining an accurate, as opposed to a perceived, estimate of the degree of change in behaviour.

So, it becomes necessary to use pairs or sequences of contemporaneous records of behaviour and to consider whether it is better to use a panel or repeated cross-sections. Again, the arguments for and against each of these options have been well rehearsed in the literature (see, e.g. Zumkeller, Madre, Chlond, & Armoogum, 2006) and the general conclusion remains that panels bring additional complication but have particular advantages if the analyst wishes to study the process of change (e.g. trajectories of behaviour over time). Three issues might be of extra relevance in the current context. One is that, since the changes may take place only slowly, the panel will need to be maintained for an extended period, and would thereby be more exposed to problems of attrition. The second is the likelihood that the panellists' repeated exposure to the questions would engender increased awareness of their own travel behaviour and the possibility that introspection on a socially charged issue might itself affect behaviour. The third is that involvement in a panel might engender the sense of being a stakeholder in the monitoring exercise with some kind of duty to report 'improvements'.

3.3.1.3. Whether to allow proxy reporting Given our concern about the risk of biased reporting, the case for seeking information directly from all household members is strengthened, in the context of diary surveys, by the fact that the involvement of the whole household is likely to make it more difficult for individuals to provide incomplete or inaccurate records. But even if it were not so, the importance of intra-household linkages in the detailed planning of activities means that a full picture of a household's travel and activities is likely to be particularly important to an analyst seeking to understand the constraints affecting the adoption of sustainable patterns of behaviour.

3.3.1.4. How to decide on the best mode of data collection The achievement of a representative sample within an affordable survey budget is clearly important but no more so than in other survey contexts and so, although we must be concerned about the increasing problems confronting methods which rely on contact being made by phone,[2] this concern is not specific to investigations into sustainability.

Two issues which are of particular concern in the context of investigations into sustainability are, firstly, the need to minimise non-response (because of the risk of under-representation by people who are not motivated to respond to a survey on sustainability issues), and secondly, the need to minimise incomplete or biased reporting (because of the risk of social desirability bias).

Reliance on self-completion questionnaires (pen and paper or computer-based) with no provision for chasing up non-respondents or verifying the information provided is clearly undesirable in the light of our fears about the likelihood of non-response bias and biased or incomplete reporting.

Provided that resources are available to fund the repeat visits sometimes necessary to catch those respondents whose busy schedules mean that they are rarely at home, the use of interview staff to visit home addresses, to conduct interviews or collect and check previously completed records can achieve high response rates and can overcome any selection bias inherent in reliance on a telephone-accessible (or computer-accessible) sample. It might also be assumed that visiting interviewers would be more able than telephone interviewers to detect missing or misleading data. For example, they might be better able to detect that records from late-responding households have been completed rapidly and with little regard to voracity simply to avoid receiving yet another chase-up call. However, there is some debate as to whether, when dealing with socially charged phenomena, the physical presence of an interviewer will necessarily reduce the extent of misreporting. Some authorities agree with Goudy and Potter (1975) that the presence of an interviewer will lead to increased accuracy because of the increased perceived risk and embarrassment associated with committing a falsehood, and that development of a good rapport between interviewer and subject is the soundest basis for an honest exchange (e.g. research by O'Sullivan, 2000 suggests that people find it easier to give an inflated impression of themselves and their actions in writing or on the telephone than in a face-to-face interview).

However, other authorities (e.g. Wiseman, 1972; Nederhof, 1984) suggest that, given that the misreporting is a means of avoiding social censure, it is less likely to occur when reporting anonymously than when reporting directly to an individual.

2. The increased prevalence of call-filtering via voicemail or answering machines, and the growing proportion of people whose phone number is not directory-listed, are making it increasingly difficult to recruit representative samples by phone or to use phone as the medium for chasing up non-respondents.

This expectation was confirmed by Booth-Kewley, Edwards, and Rosenfeld (1992) who found less evidence of deception and impression management among anonymous respondents. Of course, survey companies who use interviewers will insist that well-trained staff (e.g. their own) would remain impartial and would have the skills to obtain complete and unbiased records, but it is clearly important, when using interviewers, to ensure that they remain detached from the question of the desirability of sustainable behaviour and do not use environmental arguments to encourage co-operation.

3.3.1.5. Whether to use incentives and inducements There is a significant literature on the use of incentives (my own modest contribution, Bonsall, 2002, has a particular focus on their use in transport surveys) and the only additional point to make here is to ensure that any incentives for survey respondents, as opposed to individual participants in a voluntary behaviour change programme, should not be related to any particular mode of transport nor reflect any particular environmental stance.

3.3.1.6. Whether biases may be introduced when raw data is subject to a process of verification Data verification, whether conducted face-to-face within an interview, over the phone or electronically, is clearly an important phase in the collection of diary data. Inappropriate or inadequate procedures during this phase can have a significant effect on the data and the conclusions drawn from it. One issue which needs to be considered is whether bias may be introduced during this phase of data collection due to a difference in the rigour or skill with which staff seek to identify gaps or errors in the raw record. For example, one would expect that, with increasing experience, staff would get better at their job, so how does one avoid them identifying and filling more gaps in a second phase survey than they did in a first phase survey? Also, if staff is aware whether they are dealing with records from a target group or control group, might their loyalty to the project (another manifestation of stakeholder bias) lead them to apply different levels of rigour to records from one or other group?

3.3.2. Stated Choice Experiments

Experiments involving carefully designed questionnaires to elicit stated preferences, transfer prices, stated adaptations or stated intentions have a particular role in helping to understand the rationale for individual behaviour and so improve forecasts of behaviour or assist in the design of products and services. Their particular advantage over field observation is that they allow the experimenter to manipulate the attributes of choice options and thereby, with great speed and statistical efficiency, explore the effect of changes in attributes, which could not otherwise be observed. Their drawback, of course, is that, since it is difficult or impossible to replicate all aspects of the real-world choice scenario, the preferences and intentions gathered through this kind of experiment may not necessarily be a reliable guide to real-world behaviour.

Nevertheless, if the results can be trusted (see Sammer & Hoessinger, 2008), this kind of technique might be used to explore the value which people place on modes or activities with different sustainability credentials. For example, respondents might be asked to choose between a series of option pairs in which one alternative is always 'greener' while the other is more attractive in some other respect (e.g. cheaper/faster/more convenient) and, with careful manipulation of the relative 'greenness', cost, speed and convenience of each option, the trade-offs between greenness and other attributes might be deduced. Alternatively, using a more direct approach, respondents might be asked how they would behave under specified future scenarios (if fuel prices were doubled, if fuel were rationed, if they were charged according to their carbon emissions etc.).

Although such experiments would be easy to construct and administer via a questionnaire (these days usually computer-based but generally still involving an interviewer), and the results would be interesting, the socially charged nature of the subject matter could lead to some significant biases. The natural (socially compliant) tendency would be for people to exaggerate their preparedness to sacrifice money, time or convenience for additional greenness. This tendency might be overcome by drawing respondents' attention to the way that extra daily costs of an environmentally sound option could mount up but it is not clear how to avoid drawing so much attention to the costs that the 'true' influence of environmental credentials is lost. Recent work (Bonsall & Shires, 2009) has demonstrated the ability of briefing material to influence stated expectations and attribute valuations and led to the suggestion that the robustness of respondents' expectations and opinions could be tested by deliberately exposing different sub-samples to very different briefing material and, although this approach would not identify all bias, it might at least quantify its potential size.

A rather different type of stated choice experiment is the use of stated adaptation questions within an intensive, interviewer moderated exploration of possible adaptations to future scenarios. Using this technique, the respondent is encouraged to consider, and reflect on, the wider consequences of potential future patterns of behaviour — a process that may be aided by one of the specialist tools that have been developed for that purpose (see below).

3.3.3. Specialist Software for Exploratory Investigations

One of the most interesting developments in travel behaviour research in recent decades has been the emergence of specialist tools to assist in the exploration of potential responses to changed constraints and opportunities. Examples include HATS (the Household Activity Travel Simulator developed by Jones, 1979), CUPIG (the Car Use Pattern Interview Game devised by Lee-Gosselin, 1989), CHASE (the Computerised Household Activity Scheduling Elicitor developed by

Doherty & Miller, 2000) and TACAsim (an immersive travel-demand investigation tool described by Watcharasukarn, Krumdieck, Dantas, & Green, 2009).

The idea behind these tools is that, by helping people to appreciate the consequences that a change in one element of their travel or activity pattern might have on other elements and on other household members, the feasibility of potential changes and adaptations to an existing pattern can be explored and assessed in greater depth and with greater confidence than would otherwise be possible, and more realistic conclusions can be drawn on the likelihood of different behavioural outcomes. The particular relevance of such tools to the investigation of potential responses to policy initiatives, and to other developments which might promote environmentally sustainable travel patterns, is that the individuals' attention is drawn to the knock-on effects of what might otherwise have been an unrealistic, purely aspirational response. Recent work (Doherty, Lee-Gosselin, Burns, & Audrey, 2003) suggests that the discipline provided by CHASE does indeed help to ensure that the anticipated responses to policy are more realistic but that the process is time-consuming and even so, does not stimulate consideration of all the consequences of a change in behaviour. Nonetheless, it does seem reasonable to assume that the process facilitated by such tools will lead to the identification of realistic responses which are relatively unaffected by social desirability bias because respondents who feel the need to express solidarity with the prevailing social norms are able to do so in general terms before finding/showing that, given their personal circumstances and constraints, a more modest change in travel patterns is all that they can realistically achieve.

An alternative use of these techniques is, of course, to challenge the respondent to justify adherence to one activity-travel pattern rather than another that, according to the simulation, appears equally feasible. This might seem a useful way of teasing out underlying attitudes and perceptions but would clearly be open to social desirability bias.

3.3.4. Attitudinal Questionnaires, Individual Interviews and Focus Groups

Data on attitudes and values can provide snapshots of the distribution of opinions and attitudes within a target population. Such data can play an important role in understanding behaviour and can contribute to more accurate forecasts and more effective policy. Questionnaires can be used to explore attitudes in various ways but the most common are batteries of multiple-choice questions or of statements with which respondents are asked to indicate their degree of agreement. In the current context, such data may sometimes be designed to reveal a general picture and sometimes to test a particular theory (e.g. Schwartz and Howard's (1981) Norm-activation model of Altruism or Azjen's (1988) Theory of Planned Behaviour).

Such data is, of course, potentially very susceptible to social desirability bias but experienced practitioners are well aware of the potential problems and, while recognising that they can never be fully overcome, have developed rules of thumb to help minimise them. For example, several authors have noted the particular danger of bias in situations where the respondent has to make explicit statements to an interviewer or where questions could be seen as intimidating or loaded with respect to a particular value system. Nederhof (1985) reports that the best strategy is to use self-completion wherever possible and, where the involvement of an interviewer is essential, to maintain a social distance between the interviewer and respondent.

The anonymity and lack of human intervention which characterises online self-completion would suggest that it is an ideal medium for these kinds of survey but the possibility of a response bias (e.g. a tendency to get a higher response from people who espouse the values assumed to underlie the questions) must be a worry because there is no obvious way to quantify, or correct for, it. This argues for the incorporation of such questions as modules within general omnibus surveys deemed to espouse no particular cause, rather than for their use in one-off, specialist, surveys.

There is a very large literature on the link between (pro-) environmental attitudes and behaviour. Most studies have concluded that environmental concern and attitudinal variables fail to correlate with actual behaviour (e.g. Hines, Hungerford, & Tomera, 1986–1987; Scott and Willis, 1994; Kraus, 1995; Diekmann & Preisendorfer, 1998). Some theorists have suggested that this lack of correlation is due to a failure to consider behavioural intent and perceived agency (e.g. Azjen & Fishbein, 1980). However, more recent work (e.g. King & Brunner, 2000; Ewert & Galloway, 2004) suggests that it is due, at least in part, to the presence of 'environmentally desirable' rather than truthful, responses to attitudinal questions. In either case it is clearly important to ensure that very careful attention is paid to the phrasing of attitudinal questions and that, where comparisons are to be made over time or space, no changes are made to the wording or medium of delivery.

In-depth interviewing provides an extremely valuable means of obtaining insights into an individual's motivations and attitudes and can play a particularly important role during the early stages of an investigation when the researcher needs to be open to new models and constructs. However, given the socially charged nature of the subject matter, there must inevitably be some concern about the potential influence of the interviewer on the outcome. A skilled interviewer may be able to remain detached from the process and avoid imposing their own opinions but cannot avoid the possibility that the interviewee is seeking to create a socially desirable image of himself or herself.

Focus groups can similarly fulfil an important role during the early stages of an investigation and the interaction between participants can produce insights, which might not be forthcoming from a one-to-one interview. Focus group moderators always need to possess a quite extraordinary degree of skill and tact to avoid the exchange being unduly influenced by the expressed attitudes of dominant participants (or the perceived values of the moderator!) but, when the subject matter is socially charged, it is perhaps impossible to avoid the exchanges being influenced by social desirability.

It would seem wise to assume that the findings from in-depth interviews and focus groups will inevitably have been affected by social desirability bias and that allowance should be made for this in interpreting the results. One way of exploring this issue might be to focus on the arguments people use to legitimise their actions and decisions — so called 'structural stories' (Freudendal-Pedersen, 2005). Further study might involve focussing on the perception of constraints, which obstruct the adoption of environmentally sustainable travel patterns and on the extent to which efforts are (or are not) made to overcome them when a change in behaviour is forced by a change of house, job or personal circumstances.

3.3.5. Objective Monitoring of Travel Behaviour and its Consequences

Data that can be collected without the active involvement of the individual traveller will be less affected by the problems associated with the socially charged nature of sustainable travel.

As was indicated in Table 3.2, this type of data is particularly useful when the aim is to monitor general trends or to evaluate policy initiatives or commercial/technological developments. Interest is likely to focus on the amount of use being made of different modes of transport and the environmental consequences of that use but, depending on the purpose of the investigation, other indicators may be relevant, or even central, to the investigation. For example, it might be important to monitor the sales of hybrid-fuel vehicles, the membership of car-sharing schemes, the percentage of retail sales occurring via the Internet, the number of people making a voluntary carbon offset payment for their air trips and so on. It not appropriate to seek to enumerate or discuss the full list here, but it is appropriate to say a little more about monitoring the use of the relevant modes of transport and the consequential environmental impacts.

3.3.5.1. Data on mode use Although it might be desirable to monitor the use of all modes, it may in practice be necessary to concentrate on those which are particularly relevant to the specific investigation; for example, cycle use if monitoring the effectiveness of a campaign to encourage cycling, car use if monitoring a campaign to reduce car use or plane use if concerned with the effect of transport on gasses in the upper atmosphere.

An indication of overall usage of the various modes of transport may be gained by monitoring flows of traffic, public transport passengers, cyclists or pedestrians at strategic locations in an area and by monitoring public transport ticket sales, airport passenger numbers etc. Unfortunately, however, it is often difficult or impossible to deduce what part of the total usage is associated with a particular population and, if this information is needed, it becomes necessary to monitor the behaviour of a representative sample of people within that population.

For vehicle use, this measurement is probably best done via periodic odometer readings taken by survey staff or obtained second-hand from other sources such as annual returns to vehicle licensing authorities (although cheaper, there may be reason to doubt the objectivity or accuracy of readings taken by the motorists themselves). The use of satellite or mobile phone tracking (Stopher, 2009) may offer an alternative approach but the additional expense and practical problems may be difficult to justify unless one needs to know where and when the vehicles were being used rather than simply the total distance travelled.

Turning to other modes, it will sometimes be possible to log an individual's use of public transport via smart card records, while pedometers or satellite tracking may offer a means of monitoring their cycling and walking behaviour. However, a problem with all these approaches is that they rely on the assent and, in some cases, the active participation of the individuals and, for reasons discussed earlier in this paper, it may be impossible to achieve an unbiased sample of volunteers.

3.3.5.2. Data on environmental impact Monitoring the environmental consequences of a given activity-travel pattern is far from straightforward and is certainly not achieved by simply monitoring local air quality. Account has also to be taken of environmental consequences, the occurrence of which may be quite remote from the activity or travel in question, but which are nevertheless attributable to it. For example, if a vehicle is used, a pro-rata account should be taken of the environmental cost of its construction and eventual recycling and of the production of the fuel to power it. If someone works from home rather than travelling to an office then the travel-related environmental costs may be avoided but those associated with domestic energy consumption may be increased.

The measurement of pollutants in the local atmosphere will obviously reveal only a small part of the overall picture and even here the vagaries of local atmospheric conditions and presence of other local sources of emissions will make it impossible to deduce the proportion of overall emissions attributable to transport (given sufficient monitoring stations, appropriately located, it may be possible to produce a reliable estimate of change in local air quality over time but not the cause of that change). The preferred approach is therefore a desk study to estimate the emissions likely to be associated with a given activity-travel pattern. Such a study would ideally have access to a range of data items including traffic flow by each mode, car ownership, public transport fleet characteristics and domestic energy consumption. This, together with appropriate formulae and assumptions, would be used to calculate overall changes in emissions.

The estimation of fuel and energy consumed by a given population, or in a given area, might contribute to the above calculation but is also is important in its own right. Where records of the amount and type of fuel consumed by public transport and sold to private motorists are available, this data can be very useful for largely self-contained areas. However, it is normally difficult to attribute fuel purchases to a specific population and it will usually again be better to use a desk-study approach (applying known rates of fuel consumption to estimated traffic flow via formulae or, if greater accuracy is required, models which allow for different rates of fuel consumption in different traffic conditions). The distinction between different types of fuel each of which may have different environmental credentials (e.g. different %s renewable) might be derived quite simply if the basic data source is fuel sales but would have to be implied in some pro-rata fashion if the fuel consumption has been estimated from a formula or models.

3.4. Monitoring the Effectiveness of Initiatives Designed to Promote Environmentally Sustainable Behaviour

A number of the issues and problems raised above can be illustrated by reference studies where the aim was to monitor the effectiveness of initiatives designed to promote environmentally sustainable behaviour.

The University of Leeds conducts periodic surveys among its staff and students to monitor progress towards its well-publicised aim of increasing the proportion of journeys to the campus being on foot, by bicycle or by public transport. The main component of the periodic surveys is a Web-based questionnaire that was adapted from an original developed by HEEPI[3] for the specific purpose of monitoring travel plans at universities and colleges. In line with HEEPI procedures, e-mail invitations are sent to all (38,000) staff and students just prior to the census date inviting them to complete the Web survey, notices are displayed on campus urging people to complete the survey and e-mail reminders are sent out on the day. The Web-survey response rate is typically about 35% for staff and 12% for students. The Web-based survey is accompanied, when resources permit, by counts of vehicles and cycles parked on campus and of people arriving on campus by each mode of transport. Recent results from the Web survey suggest a reduction in the use of cars, an increase in walking and cycling and an increased prevalence of environmentally aware attitudes (support for measures to discourage travel and for a more restrictive parking policy, opposition to making car parking easier and cheaper). However, closer examination shows that the Web survey always reports much higher levels of walking and cycling (and lower levels of car use) than are apparent from the on-site monitoring and that the Web-reported share of walking and cycling, and of pro-environment opinion, has been increasing against a background of falling response rates. It may therefore be that the Web survey is increasingly subject to a response bias (with an under-representation of people who travel by car and whose attitudes are less 'green') and that, as the response rate falls, the share of green behaviour and attitudes is increasingly exaggerated. Of course, the presence of a response bias should not be unexpected because the annual survey has been billed as constituting part of the university's commitment to its green travel plan (thus appealing most strongly to those who support the plan). However, it does highlight the fact that, when seeking to monitor changes in behaviour or attitudes in this sensitive domain, it is important to have some 'objective' data and that it is dangerous to rely on voluntary self-reporting.

The University of Leeds example is instructive but far from unique. In my recent review of the results of studies designed to establish the effectiveness of individual travel marketing programmes in Australia, the United Kingdom and elsewhere (Bonsall, 2007, 2009), I noted a number of problems with the methods used. I drew

3. Higher Education Environmental Performance Index — an initiative funded by the Higher Education Funding Council for England which seeks to promote and monitor environmental standards in higher education institutions.

attention to problems in the definition of control areas, the use of inconsistent procedures to correct for sampling bias and the frequent absence of full documentation. I noted that most of the well-known studies had relied on self-completion diaries as the main source of evidence on changes in individual behaviour among the target population and that several studies had not sought to establish the counterfactual or to monitor public transport ridership, the amount of vehicular traffic, cycling or walking. I concluded that, despite significant expenditure on the monitoring programmes, some understandable scepticism remained about the significance and duration of any impacts on behaviour; this conclusion echoed earlier work by Stopher and Bullock (2003) and others in regretting the paucity of evidence from independent sources.

It is sometimes suggested (e.g. Brög & Ker, 2009) that, while the results of individual studies will inevitably contain some error, one can rely on the result obtained by averaging over numerous studies. I am not convinced and suggest that, even when studies are conducted to the highest standards of objectivity, there is a natural tendency for news of successes to be broadcast more widely than news of failures. This reflects editors' preference for positive results and the reluctance of scheme sponsors and their consultants to publicise failures (although determined doom-mongers and dispensers of cold water may be keen to point to failures, they tend not to be so well resourced or to have access to the relevant data). A trawl of the literature is therefore unlikely to produce a full picture of the spectrum of outcomes and thus the tendency of numerous published studies to point in the same direction may simply reflect the fact that they all suffer from the same biases. Meta-analysts beware!

3.5. Overall Conclusion

The overall conclusion from the above is that the collection of information on the emergence of environmentally sustainable travel and activity patterns is fraught with difficulties. It seems that no single technique can be relied on to give a complete and unbiased picture. The obvious implication is that analysts should proceed with care and, where possible, try to use data from a variety of sources (preferably ones which are unlikely to suffer from the same biases!). The choice of which sources to use will, of course, depend on the purpose of the exercise but the following may provide a good starting point.

For monitoring trends: Personal travel records (e.g. from national travel surveys), car ownership data, omnibus attitude surveys, traffic monitoring (all modes), odometer records, fuel sales and telework statistics.

For assessing the effect of local initiatives: Local travel diary surveys (with control studies) conducted by disinterested parties, traffic monitoring (all modes), odometer readings and GPS records (if detailed information is required).

In order to increase understanding of factors affecting behaviour: Exploratory interviews, stated adaptation surveys, and stated preference surveys (with robustness testing à la Bonsall-Shires). There is doubtless room for more studies of the link between behaviour and environmental attitudes/aspirations, but I believe that this area is likely to be fraught with difficulties because of the problem of social desirability bias. A less heavily researched, and possibly more fruitful, area for further study might involve focussing on the perception of constraints that obstruct the adoption of environmentally sustainable travel patterns and on the extent to which efforts are (or are not) made to overcome them when a change in behaviour is forced by a change of house, job or personal circumstances.

As a basis for aggregate modelling: Trend data on behaviour and attitudes (see above), together with equivalent trend data on the price and level of service offered by different modes (including tele-modes).

As a basis for behavioural modelling: Exploratory interviews and stated adaptation surveys, stated preference surveys (with robustness testing à la Bonsall-Shires), detailed travel activity diaries from panels and cross-sections and attitudinal questionnaires.

References

Azjen, I. (1988). *Attitudes, personality and behaviour*. Milton Keynes: Open University Press.

Azjen, I., & Fishbein, M. (1980). *Understanding attitudes and predicting social behaviour*. Englewood Cliffs, NJ: Prentice Hall.

Birol, E., Karousakis, K., & Koundouri, P. (2006). Using economic valuation techniques to inform water resources management: A survey and critical appraisal of available techniques and an application. *Science of the Total Environment*, *365*(1–3), 105–122.

Bonsall, P. W. (2002). Motivating the respondent — How far should you go? In: H. S. Mahmassani (Ed.), *Perpetual motion: Travel behaviour research opportunities and applications challenges*. Oxford: Pergamon.

Bonsall, P. W. (2007). Does individualised travel marketing really work? *Proceedings of the 35th European Transport Conference*, Amsterdam, Oct. 2007. PTRC.

Bonsall, P. W. (2009). Do we know whether personal travel planning really works? *Journal of Transport Policy* (in press).

Bonsall, P. W., & Shires, J. D. (2009). Estimating the robustness of questionnaire results: Lessons from a mixed-mode survey of expectations for tele-working and business-related travel. *Transportation*, *36*, 47–64.

Booth-Kewley, S., Edwards, J. E., & Rosenfeld, P. (1992). Impression management, social desirability, and computer administration of attitude questionnaires: Does the computer make a difference?. *Journal of Applied Psychology*, *77*(4), 562–566.

Brög, W. (2000). The New KONTIV design: A total design for surveys on mobility behavior. Presented at the International Conference on Establishment Surveys (II), Buffalo, NY, July 2000.

Brög, W., & Ker, I. (2009). Myths, (mis)perceptions and reality in measuring voluntary behavioural changes. In: P. Bonnel, M. Lee-Gosselin, J. Zmud & J.-L. Madre (Eds), *Transport survey methods: Keeping up with a changing world*. Bingley, UK: Emerald.

Button, K., & Nijkamp, P. (1997). Social change and sustainable transport. *Journal of Transport Geography*, *5*(3), 215–218.

Department for Transport. (2007a). National Travel Survey 2006, Technical Report, available at: http://www.dft.gov.uk/pgr/statistics/datatablespublications/personal/methodology/ntstechreports/pdfntstechrep06

Department for Transport. (2007b). National Travel Survey 2006, Table 5.6a of Annual Bulletin.

Diekmann, A., & Preisendorfer, P. (1998). Environmental behaviour: Discrepancies between aspirations and reality. *Rationality and Society*, *10*(1), 79–102.

Doherty, S.T., Lee-Gosselin, M., Burns, K., & Audrey, J. (2003). Household activity rescheduling in response to automobile reduction scenarios. *Transportation Research Record*, 1807, Washington, DC.

Doherty, S. T., & Miller, E. J. (2000). A computerised household activity scheduling survey. *Transportation*, *27*(1), 75–97.

Ewert A., & Galloway, G. (2004). Expressed environmental attitudes and actual behaviour: Exploring the concept of environmentally desirable response. *Proceedings of the International Conference on Outdoor Education*, Latrobe University, Australia.

Freudendal-Pedersen, M. (2005). Structural stories, mobilities and (un)freedom. In: T. U. Thomsen, L. D. Nielsen & H. Gudmundsson (Eds), *Social perspectives on mobility*. Aldershot: Ashgate.

Ganster, D. C., Hennessey, H. W., & Luthans, F. (1983). Social desirability response effects: Three alternative models. *Academy of Management Journal*, *26*, 321–331.

Goudy, W. J., & Potter, H. R. (1975). Interview rapport: Demise of a concept. *Public Opinion Quarterly*, *39*, 529–543.

Groves, R. (1989). *Survey errors and survey costs*. New York: Wiley.

Hines, J. M., Hungerford, H. R., & Tomera, A. N. (1986–87). Analysis and synthesis of research on environmental behaviour: A meta analysis. *Journal of Environmental Education*, *18*, 1–8.

Jones, P. M. (1979). HATS: A technique for investigating household decisions. *Environment & Planning*, *11A*, 59–70.

Kaiser, F. G. (1998). A general measure of ecological behavior. *Journal of Applied Social Psychology*, *28*(5), 395–422.

King, M., & Brunner, G. (2000). Social desirability bias: A neglected aspect of validity testing. *Psychology and Marketing*, *17*(2), 79–103.

Kraus, S. J. (1995). Attitudes and prediction of behaviour: A meta-analysis of empirical literature. *Personality and Social Psychology Bulletin*, *21*(1), 58–75.

Leary, M. R., & Kowalski, R. M. (1990). Impression management: A literature review and two-component model. *Psychological Bulletin*, *107*(1), 34–47.

Lee-Gosselin, M. (1989). In-depth research on lifestyle and household car use under future conditions in Canada. In: *Travel behaviour research*. 5th International Conference on Travel Behaviour. International Association for Travel Behaviour, Avebury, Aldershot. UK, pp. 102–118.

Nederhof, A. J. (1984). Visibility of response, a mediating factor in equity research. *Journal of Social Psychology*, *122*, 211–215.

Nederhof, A. J. (1985). Methods of coping with social desirability bias. *European Journal of Social Psychology*, *15*, 263–280.

O'Sullivan, B. (2000). What you don't know won't hurt *me*: Impression management functions of communication channels in relationships. *Human Communication Research*, *26*(3), 403–431.

Paulhus, D. L. (1984). Two-component models of socially desirable responding. *Journal of Personality and Social Psychology*, *43*, 838–852.

Paulhus, D. L. (1991). Measurement and control of response bias. In: J. Robinson, P. Shaver & L. Wrightsman (Eds), *Measures of personality and social psychological attitudes* (pp. 17–59). New York: Academic Press.

Richardson, A. J., Ampt, E. S., & Meyburg, A. (1995). *Survey methods for transport planning*. Melbourne: Eucalyptus Press.

Sammer, G., & Hoessinger, R. (2008). Level of knowledge and awareness about choice alternatives — A missing link of stated response surveys? A hypothesis. Presented at the 8th International Conference on Survey Methods in Transport, Annecy, France, May 2008.

Schwartz, S. H., & Howard, J. A. (1981). A normative decision making model of altruism. In: J. P. Rushton & R. M. Sorrentono (Eds), *Altruism and helping behavior* (pp. 189–211). Hillsdale, NJ: Lawrence Erbaum.

Scott, D., & Willis, F. K. (1994). Environmental attitudes and behaviour: A Pennsylvania survey. *Environment and Behavior*, *26*, 239–260.

Stopher, P., & Bullock, P. (2003). Travel behaviour modification: A critical appraisal. Paper 13 of 26th Australasian Transportation Research Forum, 1–3 October 2003, Wellington, New Zealand.

Stopher, P. R. (2009). Collecting and processing data from mobile technologies. In: P. Bonnel, M. Lee-Gosselin, J. Zmud & J.-L. Madre (Eds), *Transport survey methods: Keeping up with a changing world*. Bingley, UK: Emerald.

Tanner, C. (1999). Constraints on environmental behaviour. *Journal of Environmental Psychology*, *19*, 145–157.

Watcharasukarn, M., Krumdieck, S., Dantas, A., & Green, R. (2009). A game-based survey for core travel demand and fuel price adaptability assessment. Paper to be presented at TRB January 2009 (paper 09-2389).

Wiseman, F. (1972). Methodological bias in public opinion surveys. *Public Opinion Quarterly*, *36*, 105–108.

Zumkeller, D., Madre, J.-L., Chlond, B., & Armoogum, J. (2006). Panel surveys. In: P. Stopher & C. Stecher (Eds), *Travel survey methods: Quality and future directions*. Oxford: Elsevier.

Chapter 4

Evolving Behaviour in the Context of Interest in Environmental Sustainability: Synthesis of a Workshop

Mark Bradley

4.1. Introduction

The workshop was focused on issues in using surveys to measure behaviour in the context of respondents' growing awareness of and sensitivity to the environmental consequences of their travel choices and habits. An empirical research focus is how peoples' travel behaviour is changing over time, and how such changes interact with awareness and attitudes regarding climate change and other environmental issues. For survey practitioners, a crucial, related focus is how those same factors can introduce possible biases in completing travel surveys. In other words, will growing environmental concern also affect the accuracy with which people tell us about their travel decisions?

The workshop topic attracted a large number of participants: 25 in total, including 18 Europeans from 12 different countries, 5 participants from North America and 2 participants from Australia. Although this was a new workshop topic for the conference series, there is a long history behind it. In the 1970s, in response to fuel supply disruptions and emerging air pollution problems, many began to question the sustainability of growing automobile use. In fact, some people were convinced that by the time the 21st century arrived, the gasoline automobile would be a thing of the past. Improved emission control technologies and increased oil production soon allayed some of the concerns, however, and fossil fuel use for private travel has continued to increase around the globe.

The 2008 situation is different from what we have faced before. Most people now believe that climate change and declining production of fossil fuels will be global

Transport Survey Methods

ISBN: 978-1-84855-844-1

problems that will not diminish until people around the world change their lifestyles, particularly the way they travel. The threat of global warming is quickly becoming one of the most emotionally fraught social, political and economic issues of our time. The resulting impacts on travel behaviour can arise through conscious lifestyle decisions, and also through less conscious reactions to everyday choice situations.

4.2. Types of Behaviour Considered

To set the context for our discussion, we considered a range of behaviour that is associated with reducing the use of fossil fuels towards more sustainable alternatives, as listed below. The first four changes involve using a different type of vehicle or using a vehicle more efficiently. The next four changes involve reducing one's travel by shortening trips or avoiding them altogether. Only the last two involve changing the mode of travel away from the automobile. Although mode choice is the primary focus of much of travel demand research, it is important to consider this wider range of behaviours, particularly in suburban or rural areas where opportunities for walking, biking and using transit are often quite limited:

- buy smaller, more fuel-efficient vehicles;
- buy alternative fuel vehicles;
- drive more slowly and efficiently;
- increase car occupancy/share rides more often;
- shorten trips by chaining trips together and choosing new destinations;
- shorten trips by moving home and/or work to a more compact area;
- reduce work trips by telecommuting;
- reduce other trips by substituting in-home activities;
- use public transportation more often;
- walk and bike more often.

4.3. Types of Surveys Considered

The workshop discussion covered a wide range of survey methods, including qualitative interviews; revealed preference (RP) surveys that record actual behaviour using travel/activity diaries, GPS devices or retrospective methods; stated response (SR) surveys, including stated preference (SP) methods that ask respondents to predict their own behaviour under hypothetical scenarios; and attitudinal surveys. To lead off the discussion, the authors of four different contributed papers provided brief introductions to the specific survey methods described in their papers.

Janssens, Bellemans, Moons, and Wets (2008) describe methods used in a recent Belgian study to capture RP data on detailed activity and travel patterns using a combination of paper-and-pencil and GPS/PDA handheld data capture devices. This method allows capturing data for a week or more to analyse weekly behaviour patterns as well as daily patterns. This type of data will be useful in creating activity-based travel demand simulation models capable of responding to a wide range of future scenarios and policies.

Behrens and Del Mistro (2008) report on a retrospective survey method used in South Africa to capture past shifts that have occurred in typical travel patterns. The paper illustrates both the strengths of the retrospective method — the ability to capture changes over time rather than a single snapshot — and the weaknesses — possible inaccuracies due to difficulty or distortion in recall of past events.

Van Bladel et al. (2008) provide examples of stated adaptation experiments that have been carried out in Belgium and the Netherlands. Stated adaptation is a particular type of SR method in which the possible behavioural responses are not artificially limited to estimate a particular type of model, but are broad enough to capture the true range of possible responses. For example, the authors present an example choice scenario with eight possible responses, including various combinations of changing mode, changing destination, rescheduling or cancelling the activity altogether.

Sammer, Gruber, Hoessinger, and Roeschel (2008) also provide examples of SR surveys, this time carried out in Austria. The authors focus on the role of information given to respondents during the survey, and how the level of information provided can significantly influence the responses. This property of hypothetical survey methods may make them particularly vulnerable to instrument effects when information about the environmental consequences of travel alternatives is provided as part of the survey.

4.4. Potential Survey Biases

The workshop resource paper by Prof. Peter Bonsall, also included in this volume, provides a very useful and comprehensive discussion of specific types of biases and inaccuracies that can arise in surveys designed to investigate the sustainability of travel behaviour. Although the biases that he identifies may also arise in other research contexts, the environmental sustainability context is one that is likely to exacerbate them. In particular, Bonsall focuses on:

- *Social desirability bias*: Wishing to project a positive image of themselves, respondents tend to exaggerate positive aspects of their behaviour and downplay the negatives (‘*impression management*’).
- *(Non) Response bias*: People are more likely to respond to a survey if they share the (typically pro-environmental) values of its sponsors.
- *Affirmation bias*: People tend to report what they think the interviewer wants to hear.
- *Stakeholder bias*: People tend to report favourably on something to which they feel an affinity, particularly if they feel their responses may influence some policy decision (strategic bias).
- *Selective reporting bias*: There is a tendency for policy or intervention successes to be reported more widely than failures.

These biases can be closely related to one another, and may work in combination. Clearly, the more that surveys use objective methods (GPS) or mechanistic methods (travel diaries) to record behaviour, the fewer opportunities there are for such biases to occur. On the other hand, methods that are very retrospective, or else very prospective such as SP methods, provide more scope for respondents to consciously or unconsciously distort their answers. Bonsall (2009) concludes that no single technique can be relied upon to give a complete and unbiased picture, and that, where possible, we should use data from a variety of sources, preferably the ones that are not prone to the same types of biases.

As a brief exercise to illustrate the potential for biases, participants completed a short SP questionnaire during the workshop. First, participants were asked if they would have used a video conference hook-up to attend the conference via their own personal computer rather than travelling to Annecy, if such a facility had been available at a reduced registration fee of €300. Participants gave their responses anonymously on paper. Subsequent analysis revealed that not a single respondent chose the video conference option in this first scenario. This result may have been strongly influenced by information availability — participants had already experienced the conference setting in Annecy, and had no experience with attending such a gathering via video conference. The same SP question was then repeated, but this time giving people specific information on the environmental impacts of travelling by air. This time, 20% of the respondents said they would choose the video conference option, at the same price of €300. The substantial change in responses indicates that the extra information may have brought affirmation bias into play — some respondents now gave the answer that they imagined the survey designer was looking for. (Because the responses were anonymous, social desirability bias was less of an issue.) The same SP question was then asked one final time, with the respondents told that their aggregate choices would be reported to everyone at the conference during the workshop summary presentation. Now, 33% of respondents said they would choose the video conference option. This shift illustrates potential social desirability bias and stakeholder bias, with some participants changing their responses so that our workshop group might appear more environmentally conscious to our colleagues. (Note that the small sample size and obvious illustrational context of this exercise would make it inappropriate to draw any conclusions about the actual potential for video conferencing for future research conferences.)

4.5. Survey Methods for Specific Research Purposes

Workshop participants discussed the issues introduced above for specific research contexts, in the interest of providing some general recommendations. The first context was that of monitoring general trends in travel behaviour. For example, there has been a great deal of research interest in how vehicle-kilometres travelled per household has changed recently in response to volatile fuel prices and growing awareness of reducing one's 'carbon footprint'. Survey methods mentioned in this context included:

- personal travel diary data (e.g., national or regional household travel surveys);
- traffic monitoring/passenger counts on different modes;
- auto ownership/sales data;
- odometer records and fuel sales records;
- omnibus attitude surveys;
- time use surveys;
- employer statistics on tele-working, employer-based ridesharing, etc.;
- commercial vehicle/business travel surveys.

In order to measure *change*, these surveys must be repeated cross-sections, continuous/rotating cross-sections or else longitudinal panel surveys. In practice, large-scale panel surveys of travel behaviour are quite rare, although they have been carried out on a national scale in the past in the Netherlands and currently in Germany. More typical are national-level cross-sectional household travel surveys carried out at various time intervals. The NHTS survey in the United States has been done only at 6-year intervals. Other large-scale surveys done at more frequent intervals, such as the UK household travel survey (a continuous cross-sectional national survey), will tend to give more accurate and timely information to measure general behavioural changes. As mentioned above, there is also the option to add longer term retrospective and/or prospective (behavioural intentions) questions to these surveys, but such questions are subject to greater inaccuracy and potential bias. The increasing use of passive data-collection methods such as wearable and vehicle-based GPS devices, on the other hand, will tend to enhance the accuracy of the data and avoid some sources of potential bias, although any trend towards less survey interaction with respondents may also reduce the availability of data on situational and personal variables which could be useful in explaining observed behavioural shifts.

A more focused variation on the general research context above is where the researcher wishes to assess the behavioural effects of local policy interventions. The types of survey methods to be used are generally the same as those listed above. In this case, however, the sample design may be more crucial. It is often important to include a control sample for comparison purposes, to isolate any policy-specific effects. It is also a natural context for the use of before-and-after study designs, with either a panel approach or repeated cross-sections. In that case, the study should go on long enough after the policy change to measure longer-term behavioural shifts as well. For example, some people may not be able to substantially change their travel habits until they change their work location or residence location. It is also very important to conduct the survey research in a way that does not focus too much attention to the policy initiative being studied, which can lead to the types of biases discussed earlier. This is especially true for panel survey approaches, where participation in the study could influence peoples' travel decisions. Use of GPS data-collection methods could be very attractive in this longitudinal context to avoid subtle response biases.

Some travel surveys are done, either primarily or secondarily, to provide data for estimating predictive models of travel demand. In this context, a critical question is whether or not peoples' preferences or sensitivity to specific situational variables are changing over time. For example, because of awareness and concern for environmental problems, does a person today have a higher probability of using transit compared to a person who was in 'exactly' the same situation a few years ago? If these types of behavioural shifts are indeed taking place, then our typical forecasting assumption of extrapolating past behaviour into future situations becomes more questionable. We will need to create more complex models that can anticipate 'threshold' levels or other phenomena where people may respond to situations differently than they have in the past. This, in turn, will require a deeper understanding of the behavioural and attitudinal processes that lead to travel choices. In that case, collecting data only on the observed outcomes will not be sufficient.

The last research context discussed, and one of great interest to the workshop participants, is the collection of data to improve our understanding of the processes through which awareness and attitudes about the environmental sustainability of travel options influence the actual use of those options. It was agreed that this will require combining the types of quantitative survey methods described above with more qualitative research methods. The latter could include focus groups, in-depth personal interviews and various types of more structured attitudinal surveys. Following the example from the Sammer et al. (2008) paper, there may be potential for imaginative SR surveys that use computer-based 'virtual' methods to put respondents in possible future choice situations, and then offer them various types of information from different angles in order to gauge how it influences attitudes and intentions. It was noted that exploratory research could be greatly informed by work in related fields such as psychology, advertising, sociology, political science and even criminology. In the field of psychology, for example, Winter and Kroger (2004) explain how attitudes and behaviour related to environmental problems are viewed under several competing strands of psychological theory.

4.6. Summary and Priorities for Future Research

There was a consensus among participants that existing survey practices can tell us a great deal about general trends in sustainable travel behaviour, and can be used to assess the success of particular policy interventions. However, these methods may be prone to serious biases that are related to environmental attitudes, so a great deal of care should be taken in survey design to avoid such biases to the extent possible. Furthermore, if we wish to not just observe, but also understand and possibly predict changes in travel behaviour related to sustainability, we need to undertake more exploratory research and enhance the range of survey methods that are used.

As the workshop concluded, participants created a list of research priorities in these directions:

Research into the propensity for biases in different survey methods:

- Carry out controlled studies comparing the results of different survey methods used in the same choice contexts related to travel behaviour and sustainability, so that biases of each method relative to the others can be identified and quantified.
- Conduct controlled studies to measure the effects on survey results related to the amount and type of information given to respondents concerning the environmental effects of different travel choices.
- Focus research on the relationships between respondents' attitudes about the environment, sustainability and social/community participation and their propensity to participate in and complete travel surveys, including typical large-scale household travel surveys.
- Perform carefully designed studies using SR methods in this policy context, ideally with ex-post validation against observed behavioural shifts.

Research into behavioural processes:

- Recommend best practices to integrate quantitative and qualitative survey techniques.
- Compile a guide to best survey practice for measuring and interpreting attitudes and their relationship to behaviour (survey protocols, references, caveats/warnings), paying particular attention to lessons learned in related fields such as sociology, psychology, public health and criminology.
- Carry out historical research into travel marketing themes, political and activist campaign themes, advertising of consumer goods (i.e., cars) to study how attitudes towards sustainability have been shaped in the past.
- Look for opportunities to field targeted exploratory surveys as 'add-ons' to existing large-scale surveys.
- Devise and test survey methods to identify and understand how travellers perceive constraints on their choices and their preparedness and ability to change those constraints.
- Explicitly investigate how intentions become behaviour, and the differences between spontaneous and planned behaviours.
- Test methods of visualising complex data and relationships, in order to help respondents understand the linkages between individual behaviour and the environment.
- Undertake studies to apply existing theoretical frameworks, such as the theory of planned behaviour, social network theory and social learning theory, to the context of environmentally sustainable travel.
- Focus on the substitution between telecommunications and travel, and between in-home and out-of-home activities.
- Study the relationship between sustainable travel behaviour with other types of sustainable behaviour — do they mainly substitute for or complement one another in respect to expenditure of money and effort?

- Focus on the role of information search and provision in forming attitudes and choices.
- Study the ways in which people (could) get feedback on the environmental consequences of their own travel decisions, and how such feedback affects subsequent decisions.

To conclude, given the large number of participants in the workshop, the growing interest in the topic and the realisation that we still have much to learn, it seems safe to say that additional workshops on this topic in the future would be very worthwhile.

Acknowledgements

In addition to the workshop authors whose papers are cited in this summary, I wish to acknowledge the contributions of Dr. Randi Hjorthol, who was the discussant in the workshop, and Dr. Stacey Bricka, who served as scribe. I also thank Dr. Eric Molin, who was originally designated as workshop Chair and did much of the organising work, but could not attend the conference.

The workshop participants, in alphabetical order, were: Peter Bonsall, United Kingdom; Mark Bradley, United States; Stacey Bricka, United States; Beat Brunner, Switzerland; Olivia Christophersen, United Kingdom; Andrew Collins, Australia; Blake Emery, United Kingdom; Peter Endemann, Germany; Pavlina Filemonova Rocakova, Czech Republic; Randi Hjorthol, Norway; Reinhard Hoessinger, Austria; Ian Ker, Australia; Solveig Meland, Norway; Andres Monzon de Caceres, Spain; Elke Moons, Belgium; Linda Nijland, Netherlands; Francis Papon, France; Dimitris Potoglou, Greece/United Kingdom; Anais Rocci, France; Sophie Roux, France; Tomas Ruiz, Spain; Karel Schmeidler, Czech Republic; Felix Schiffner, Germany; Giorgia Servente, Italy; and Justyna Sczcepanska, Poland.

References

Behrens, R., & Del Mistro, R. (2008). Analysing changing personal travel behaviour over time: Methodological lessons from the application of retrospective surveys in Cape Town. Presented at the 8th International Conference on Transport Survey Methods, Annecy, France, May 2008.

Bonsall, P. (2009). What is so special about surveys designed to investigate the environmental sustainability of travel behaviour? In: P. Bonnel, M. Lee-Gosselin, J. Zmud & J.-L. Madre (Eds), *Transport survey methods: Keeping up with a changing world*. Bingley, UK: Emerald.

Janssens, D., Bellemans, T., Moons, E., & Wets, G. (2008). Simulating emergent behavior and evolution of activity-travel patterns: Data collection challenges. Presented at the 8th International Conference on Transport Survey Methods, Annecy, France, May 2008.

Sammer, G., Gruber, C., Hoessinger, R., & Roeschel, G. (2008). Level of knowledge and awareness about choice alternatives — a missing link of stated response surveys?

A hypothesis. Presented at the 8th International Conference on Transport Survey Methods, Annecy, France, May 2008.

Van Bladel, K., Bellemans, T., Janssens, D., Wets, G., Nijland, L., Arentze, T. A., & Timmermans, H. J. P. (2008). Design of stated adaptation experiments: Discussion of some issues and experiences. Presented at the 8th International Conference on Transport Survey Methods, Annecy, France, May 2008.

Winter, D. D., & Kroger, S. M. (2004). *The psychology of environmental problems*. New York: Lawrence Erlbaum.

Chapter 5

Myths, (Mis)Perceptions and Reality in Measuring Voluntary Behavioural Changes

Werner Brög and Ian Ker

Abstract

Increasing interest in 'soft' policy approaches to travel demand management, poses the question of how to measure the effectiveness of interventions. Much of the focus has been on statistical reliability of measured change where sample surveys are the primary means of estimating change. Sample surveys also pose issues of non-sampling errors, especially when the 'measure' is the difference between 'before' and 'after'. This paper outlines the principles and pitfalls in measuring behaviour change. It draws on voluntary travel behaviour change (VTBC), using a number of approaches, including but not limited to Individualised Marketing (a method developed by the authors). A key issue in VTBC is the extent to which repeated experience can validate the effectiveness of voluntary behaviour change interventions in general, despite statistical errors of individual measurements. Measurement is fundamental to evaluation of outcomes. It can also aid the selection of locations with high potential to achieve change through identification of key success factors. In the specific case of travel behaviour change, there is now a substantial body of research that potentially allows outcomes to be related to other factors. To date, no strong relationships have been identified, but this would be a useful area for further research. Experience does demonstrate, however, that the scale of the intervention is important. Interventions with more than 5000 households are consistently more successful than small ones, even allowing for the greater statistical variability of measurement for smaller projects. Large scale also offers opportunities for intervention design to benefit from the potential for diffusion beyond those directly involved in the project.

Transport Survey Methods

ISBN: 978-1-84855-844-1

5.1. Introduction

The focus of urban transport policy has shifted from roads to public transport and non-motorised transport, from infrastructure to services and from demand satisfaction to demand management. This is a reflection of changing values about what represents viable and sustainable cities and communities as much as it is of the financial, physical and social realities of attempting to provide major road infrastructure to satisfy continually growing demand for car travel.

Demand management has been given further impetus by the twin pressures of global climate change (private motorised transport is a large and a growing source of greenhouse gas emissions) and, more recently, the so-called 'Peak Oil' phenomenon. Peak Oil contends that conventional oil production has either peaked or will do so within the next decade; this would be a major challenge in its own right but is exacerbated by the continuing rapid growth in oil demand from developing economies, particularly China. This paper is not the place to debate the extent or consequences of Peak Oil (see, e.g. http://www.peakoil.net), but there is widespread agreement that energy for transport (whether from conventional oil, unconventional oil [oil shale, tar sands] or alternative sources) will be more costly.

At the same time, we are more aware of and concerned by the adverse health effects of lower levels of physical activity, partly resulting from increasing car use.

There is a large amount of research over more than half a century to support the analysis and estimation of the impacts of 'demand satisfaction', although much of this is problematic when there are substantial impacts on land use. Demand management does not have the same history of application, monitoring and analysis. Voluntary travel behaviour change (VTBC) is relatively new in the context of demand management.

The primary issue with any new transport initiative is identifying and estimating the impacts on the amount and type (including mode) of travel. This paper outlines the state of current practice, the issues arising and some ways of resolving them for VTBC through household-based programmes, primarily in transport but also in the emerging areas of water and electricity.

5.2. Voluntary Behaviour Change

5.2.1. What is Voluntary Travel Behaviour Change?

VTBC is one part of the travel demand management toolkit, not a single 'silver bullet'. It is the face of demand management that operates largely without 'carrots' (direct incentives, including system improvements) or 'sticks' (disincentives, such as additional charges or regulation). In most cases, VTBC raises awareness, improves availability of information and provides support for people to try alternatives to driving their cars, working through empowerment and motivation.

VTBC itself has a range of tools that operate in the household, school and workplace contexts, each of which is almost a separate area of development. Household-based programmes are of two principal types:

- Those that deal with the whole identified population (usually of an area), but with a structured programme that focuses on those with identified sub-populations that have the greatest potential for change. The major programme of this type is *Individualised Marketing* (*IndiMark*®). See, for example: http://www.travelsmart.gov.au/training/packaging_comm_indi.html.
- Those that deal with a sub-population, often self-selected. These are largely based on *Travel Blending*® and its derivative *Living Neighbourhoods*® (see, e.g. http://www.travelsmart.gov.au/training/packaging_comm_blend.html). *Travel Blending*®/*Living Neighbourhoods*® may also include aspects of community development, but we are concerned only with the travel behaviour outcomes in this paper.

A third group (such as the 'Households on the Move' project in the Australian Capital Territory) is aimed at people or households making changes such as change of residential location that will inevitably have an impact on travel behaviour. Such an approach is potentially very effective for individuals, but because the participating people or households would change their travel behaviour anyway (because they are moving house or changing jobs, for example) there is no way of estimating the extent of travel behaviour change or even demonstrating that it has occurred.

Typically, there are no changes to transport infrastructure or services, other than the provision of route and timetable information at bus stops (in the case of IndiMark, new stands with route maps and stop-specific timetable information). Whilst there is some evidence that service improvements implemented in conjunction with *IndiMark*® can have a compounding (positive) effect (UITP, 1998), service changes *during* an initiative can have a detrimental impact by invalidating information and confusing participants.[1]

5.2.2. What is Voluntary Behaviour Change Trying to Achieve?

From a public policy perspective, it is important to be clear about the objectives that are being sought by an intervention, before we can establish how best to measure the extent to which they have been achieved by the intervention. This sounds trite or

1. The change in bus boardings in the City of Subiaco, Perth, Western Australia, as measured through the Transperth ticketing system, was much lower in those areas where bus services were re-organised during the Indimark® period (personal communication from Department for Planning and Infrastructure, Perth, Western Australia).

self-evident, but is often not clearly articulated in the professional debates on such topics.

In the case of water and power consumption, the objective is usually expressed in terms of actual consumption of water or electricity. In the case of transport, there is less convergence on a single outcome and this has sometimes led to confusion about the nature and validity of measurements.

There are many potential outcomes of travel behaviour change that are of interest, across the full range of economic, environmental and social outcomes, including:

- *Economic*
  - The economic cost of transport
  - Traffic congestion
- *Environmental*
  - Greenhouse gas emissions
  - Local air pollution
  - Noise
  - Water pollution
- *Social*
  - Accessibility
  - Social inclusion

In transport terms, these objectives are often reduced to the shorthand of mode share targets, although these are usually aspirational targets rather than achievable outcomes. There are good reasons for the aspirational nature of targets, including the fact that the primary strategic objective is to change from increasing personal car use. It is also important that, since most of the policy measures that could contribute to this objective are new or have not previously been subjected to rigorous analysis, we do not know how effective they are likely to be, individually or in combination.

At the strategic level, the extent of car driver travel is a direct measure of the amount of private motor vehicle travel, which is the principal determinant of most of the higher-level outcomes. There is much less certainty about the achievability of targets of the non-car modes than for the car driver mode and measurement of change is also correspondingly more difficult as they represent a smaller initial quantum.

5.3. Indicators of Change

It is difficult to measure the impacts of behavioural changes as a result of marketing actions. There are several methods available and each has advantages and disadvantages. So when measuring the success of a household-based travel behaviour change project, it is preferable to combine these methods and if they all point in the same direction and are consistent with the same size of change, then it is most likely that there was real success.

There are three main measures that should be considered:

- *Marketing Indicators.* These are determined by the amount and type of information requests compared with the total target group, and the quantitative feedback from residents throughout the project. For example, one of the desired changes is to increase the use of public transport, and the volume of stop-related timetables by people would be an indicator of impact. It is extremely unlikely that thousands of households would order specific, address-based timetables that they are not interested in and will never use.

  In traditional direct marketing, these types of indicators are the only success factors used. They are reliable, precise and easy-to-measure indicators, which should not be ignored.

  Further, it has to be remembered that behaviour change tries to effect a mind or culture change, and it achieves this quite often as documented by personal comments. Unprompted comments from real people can be as important as changes measured in counts or surveys: 'Your programme was a great success, because it was the first time in 25 years that my husband used public transport and left his beloved Jaguar in the garage' (Nürnberg, 2001).

  And, when it comes to changes in the mindset which affect longer term planning, this type of indicator should not be dismissed: 'We are moving house soon and have now disregarded the option with bad public transport connections' (Vancouver, 2006). After moving house, these people may no longer live in the TravelSmart area, but they are still part of the outcome.
- *External Indicators* include measuring public transport patronage. In the TravelSmart programme in Western Australia, bus boardings are collected and analysed independently. The ability to do so will be enhanced by the recent introduction of a comprehensive Smart Card ticketing system (SmartRider). ITP (2007) supports the value of such 'robust corroborative data'.

  External indicators are primarily of value to corroborate the measured outcomes of travel behaviour change. They are, inevitably, partial (e.g. dealing only with one mode) and care is needed to separate out the influence of factors other than VTBC (e.g. road traffic in the intervention area is a mix of locally generated and non-local car trips). The opportunities for using external indicators vary between interventions, as setting up measurement systems purely for VTBC is difficult and may be prohibitively costly.
- *Behavioural Indicators.* The effectiveness of travel behaviour change projects can also be evaluated by measuring changes in the mobility characteristics of residents, by conducting extensive 'before-and-after' travel surveys. The analysis is detailed, based on mode share, activities and travel time, and shows whether there has been a mode shift from car-as-driver trips to environmentally friendly modes. Of key importance, these variables can then be used to estimate any reduction in greenhouse gas emissions.

  The most frequently reported behavioural indicator is the change in car driver trips. This can be estimated more robustly than the change in any of the other modes, because car driver trips are the largest single component of personal travel across the population. However, *car travel distance* is the critical parameter for many of the objectives of VTBC and this has not been consistently estimated across interventions. Where it has been estimated, the percentage reduction in car-kilometres of travel generally exceeds the reduction in car trips. It is in this area that alternative methods of measurement, such as odometer readings, may be able to play a useful supplementary role, but this has its own problems (see 'Measuring Behaviour Change', below).

Most attention has been focussed on behavioural indicators, often to the exclusion of the marketing and external indicators. For external and behavioural indicators, it is important to measure changes against a control group (ITP, 2007, para 9.48). Marketing indicators pose some problems of comparative assessment as no common metric or scale has been developed for them.

Whatever methods are used, acceptance of the results will be highly dependent upon comprehensive and consistent documentation of processes and outcomes.

5.4. Issues Raised by Voluntary Travel Behaviour Change

5.4.1. Engagement of the Community

A key success factor for any programme of VTBC is the engagement of the community. According to AGO (2006, p. 23):

There appears to be a correlation between changes in travel behaviour and:

- personal engagement
- individualising materials to people's particular circumstances
- scale of the intervention — securing community support, as well as individual participation
- public visibility of the project (this is less important than the other points).

The first two of these are difficult to measure, but there is clear evidence that the scale of travel behaviour change intervention has an impact on the outcome measured in terms of the change in car driver trips (Figure 5.1).

Public visibility will have a more substantial impact where diffusion is designed into the project (see Section 5.4.3 below).

5.4.2. Induced Demand?

Stopher (2003) argues that reduction in car driver trips from one part of the community will improve traffic conditions and induce additional car use. This is not a valid argument for car travel generated by households in the target area, as any such effect will already be included in the survey estimates. It might, however, have

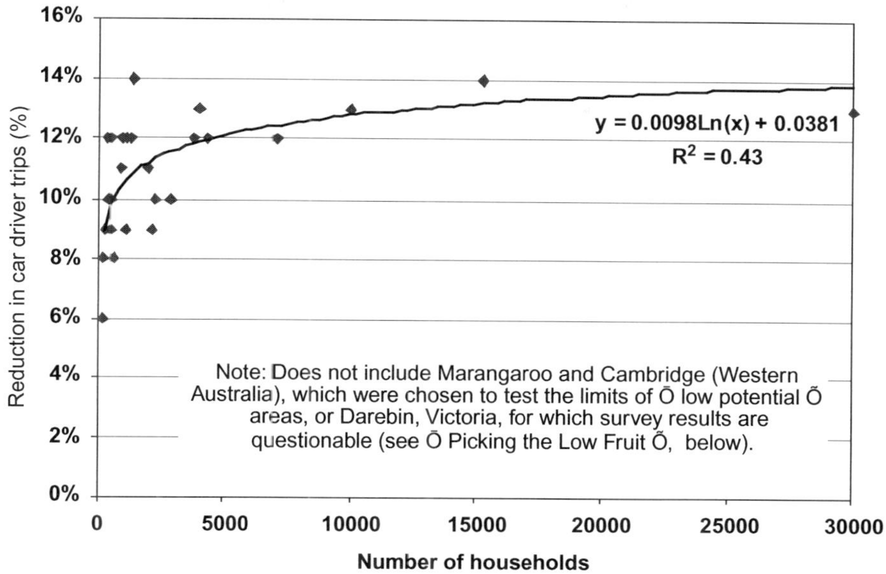

Figure 5.1: Travel behaviour change outcomes and scale of intervention (IndiMark®).

some validity for car trips generated beyond that area, primarily for trips undertaken in the most congested times and places in peak periods.

No evaluations have attempted to estimate any such effect largely because it is likely to be diffuse and could only be estimated through a detailed network-based traffic model. The authors are not aware of any successful attempt to incorporate travel behaviour change into a network-based travel demand and traffic model, although the requirements for doing so have been established (Ker, Lloyd, Ringvall, & Sidebottom, 2005). It should be noted, however, that any such analysis would have to deal appropriately with statistical and modelling errors inherent in all traffic models.

The only reported measurement of VTBC impacts on actual road traffic volumes was for Darebin, Victoria, Australia. This appeared to show a reduction, relative to the metropolitan Melbourne average, of between 1% and 2% on roads outside the project area and between 2% and 3% on both main and local roads within the project area (Richardson, 2005). These data are consistent with diminishing impact of travel behaviour change by Darebin residents with distance from Darebin, but it is not possible to disentangle the extent of induced traffic from outside Darebin resulting from the lower traffic volumes. The same point is made by ITP (2007, p. 110).

The practical value of the Darebin data is significantly reduced by apparent problems with the other measurements of travel behaviour change for Darebin, including low response rates and sample losses from the panel.

5.4.3. Diffusion into the Broader Community

Diffusion is not a theoretical statistical concept, but a very real contributor to behaviour change across the whole population, including those outside the identified project population. It is well-known in many areas of commercial marketing. Diffusion relates to the extent to which awareness, attitudes and behaviours are transmitted within a population without external intervention (see box).

In the most common type of travel behaviour change applications, large-scale diffusion could take place only outside the area of application (as the total population was targeted). Since before-and-after surveys are only carried out in the identified target population, there has been no measure of any diffusion effect into the wider community. Indeed, to the extent that the diffusion effect might influence the control group (where one was established), diffusion would have led to underestimation of car driver trip reduction in the identified target population.

Diffusion

There are three main models of innovation diffusion models, each arising from a different account of how innovations spread (Young, 2007):

- *Contagion*. People adopt an innovation when they come in contact with someone who has already adopted.
- *Social threshold*. People adopt when enough other people in the group have adopted.
- *Social learning*. People adopt once they see enough evidence among prior adopters to convince them that the innovation is worth adopting.

Contagion was deliberately avoided in the original South Perth pilot project, in order to test the effectiveness of the IndiMark® methodology itself. Subsequent larger-scale applications have not attempted to limit contagion, but have effectively kept its influence low by targeting a very large proportion of the population of an area; many of each person's contacts will already be part of the initiative.

Social threshold effects are most likely to occur where the innovation is clearly visible to the casual observer, as in the case of clothing fashion or the use of mobile phones. Travel behaviour will have some elements of this (e.g. if you see more people cycling), but in any area such visibility is limited by the extent to which travel in an area is undertaken by people from outside that area.

Social learning provides an explanation of why a person would adopt an innovation given that others have adopted it: the adoption decision flows directly from expected utility maximisation. Specifically, the decision is based on reason to believe the innovation is better than what the person is doing now, where the evidence comes from directly observing the outcomes among prior adopters. For example, when a new product becomes available, many people will want to see how it works for others over time before trying it themselves.

Working against diffusion is *inertia*, the fact that people delay in acting on new information. In the case of TravelSmart/IndiMark®, inertia is mitigated by its being known that the initiative is of a finite duration, even though the outcomes are expected to be durable and long term.

In North Brisbane, Queensland, only 66% of households were contacted as part of the Household TravelSmart project. This approach simplifies and reduces the costs of making initial contact to establish a target population. At the same time, it leaves 'space' within total population of the area for diffusion to occur.

The measured travel behaviour change for the target population was consistent with expectation from previous IndiMark applications (13.4% reduction in car trips). There was no measurement of travel behaviour change in the rest of the area population.

The North Brisbane project created space in the intervention communities in which *contagion, social threshold effects and social learning* could be manifest; only 66% of households were directly contacted. The very large scale of the North Brisbane project will have enhanced the likelihood of *social threshold* effects. There is also likely to have been a diffusion impact from several component (sub-area) interventions being under way at any point in time.

The key indicators of the extent to which the social learning model is likely to apply are those relating to the satisfaction of participants first with the process and second the outcomes (i.e., their actual experience with the non-car alternatives they use).

The evidence regarding the process is largely anecdotal and positive. It is extremely rare for people to express dissatisfaction with the TravelSmart/IndiMark® process, whereas many positive responses are received.

The evidence regarding the experience of the non-car alternatives is, however, more tangible. In the North Brisbane project, the 'after' survey of travel behaviour was undertaken in March/April 2007, between 3 and 12 months after the TravelSmart/IndiMark® intervention (the intervention was progressively rolled out across the area over a nine-month period, April–December 2006). It is unlikely that those who had a negative experience with the non-car alternatives would still be using them months afterwards.

In turn, this supports the contention that the impacts of VTBC will continue as long as the people who make the change continue to receive the same quality of travel experience as they do initially. This will require continued resourcing of public transport and walking/cycling infrastructure to ensure that additional demands that factors such as population growth and higher fuel prices put on other transport systems do not reduce the level of service they offer.

There are sound reasons to believe that diffusion effects will work better with large-scale rather than smaller applications:

- Materials for supporting travel behaviour change are available for the target area only;

- The larger the target area, the greater is the proportion of participants' family, friends, work colleagues and others (to whom information and attitudes might be transferred though such means as general publicity and 'word of mouth') who are likely to live in the area covered by the intervention;
- A large-scale project takes place over a longer timeframe, creating additional opportunities for diffusion while the project is still in progress and
- A large-scale application increases the likelihood that when a household moves to a new location it will be able to access comparable location-specific information such as local access maps and public-transport timetables in a familiar form.

5.4.4. Picking the 'Low Fruit'?

Figure 5.1 does not include three interventions that produced very low car trip reduction. These are:

- Marangaroo (4%) and Cambridge (7%) in Western Australia, which were specifically chosen to test the lower bounds of achievable outcomes. Both had poor public transport, high car use and poor provision of local facilities (important for walking and cycling). Including these would have introduced substantial variation in a range of variables that were not represented in the regression.
- Darebin, Victoria, which had a low survey response rate (50% or 1346 responses out of the 2772 sample) for the 'before' survey and substantial further losses for the 'after' survey — 881 responses out of 1346 in the 'before' survey (Richardson, 2005). Differences in response rates have been shown to affect the reported travel activity (Brög & Erl, 1999) and the sample losses between the 'before' and 'after' surveys are likely to be statistically biased.

This prompts the question of whether these outcomes shown in Figure 5.1 are the result of 'picking the low fruit' or selecting target areas with high potential (Bonsall, 2007). The Marangaroo and Cambridge examples indicate that there is potential for this to occur, but the wide range of situations (e.g. the pre-existing car driver mode share in VTBC applications in the United Kingdom and Australia varies from less than 50% to at least 60%) in which comparable results have been achieved indicates wide applicability rather than 'low fruit'.

Limited analysis of Australian IndiMark interventions has not demonstrated any strong relationships between outcomes and socio-demographic parameters (Ker, 2007). However, this analysis was undertaken for the purposes of providing benchmarks for evaluation of a specific project. A more rigorous multi-variate analysis would potentially allow interventions from other countries to be included and issues of multi-colinearity to be dealt with.

The recent North Brisbane Household TravelSmart project has demonstrated that a wide range of values for key socio-demographic parameters does not preclude the achievement of car driver trip reduction at least comparable to those achieved in smaller, more homogenous and apparently favourable locations. For example, the North Brisbane area included areas with:

- 1.8–3.1 (whole area 2.5) persons per household;
- 4–27% (whole area 12%) of households having no car;
- 42–77% (whole area 54%) of population in employment;
- 21–62% (whole area 38%) managerial/professional employment status and
- 40–63% (whole area 53%) of employed people driving their car to work (ABS, 2007).

5.4.5. Durability

There is limited evidence of the durability of travel behaviour change outcomes, although what there is tends to show sustained reductions in car use for 4–7 years (Roth et al., 2003). Bus ticketing data in Cambridge, Western Australia, has shown a consistent (and, if anything, growing) increase in public transport use (see Figure 5.2, below).

The most likely time for 'disillusionment' to be experienced and reversion to previous behaviour to occur is within the first few months of the changed behaviour.

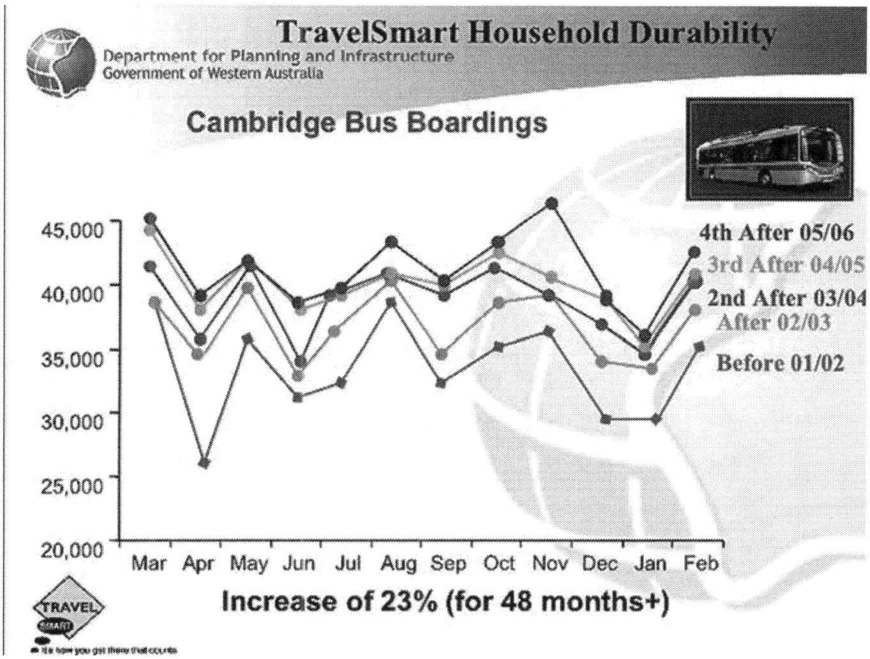

Figure 5.2: Public transport boardings and IndiMark®-Cambridge, WA. *Source*: Government of Western Australia (2009). Reprinted with permission.

Since 'after' surveys are typically undertaken 6–9 months after the intervention, as noted above it is unlikely that those who had a negative experience with the non-car alternatives would still be using them months afterwards. According to Maunsell Australia (2004a, p. 57), *previous experience indicates that for household/community initiatives there appears to be some reversion to previous travel choices over the first nine months following the TBhC project but that people who have not reverted by this time tend to stay with their new travel choice.* On this basis, the 'after' survey will already incorporate the most substantial part of any 'reversion' to previous behaviour.

There are strong arguments to support at least medium-term durability, provided those who change continue to experience the same level of service in their newly adopted modes. If there are doubts about the durability of behaviour change, it would be appropriate to use a shorter period for evaluation: for example, New Zealand uses a 10-year evaluation period for travel behaviour change and other 'soft' policies (Maunsell Australia, 2004a). At typical public sector discount rates, this would reduce the present value of benefits by 40% compared to a 25-year evaluation period. Whilst this is a substantial reduction, the impact on benefit–cost outcomes is unlikely to be critical, with benefit–cost ratios in excess of 30:1 typically being recorded for *IndiMark®* projects with 25-year evaluation periods.

5.5. Measuring Behaviour

5.5.1. Direct Measurement?

The behaviours in which we try to measure change are often not directly measurable. Water, electricity and gas, in Australia, are directly metered at the consumer household level. However, we still do not know what the water or electricity is used for and the effect of exogenous factors can be difficult to identify. Some of the complexity of doing so, even at a highly aggregated level, is outlined in http://www.daa.com.au/case-studies/water-usage.

A common problem with direct metering is ensuring accurate and timely readings on a regular basis. These problems can be eliminated by using self-reporting meters, which are already available on a walk-by basis for water, electricity and gas (see, e.g. http://www.neptunetg.com) with no need for access to premises or manual recording of meter reading and could be adapted to automatically send readings by phone at specified times. Alternatively, data loggers could be attached to meters for periodic automated walk-by download.

For the basic parameter of concern in VTBC, car use as driver, the extent of experience now available to support the effectiveness of the main travel behaviour change tools, has been interpreted (AGO, 2006) to support the use of simpler single measures such as odometer readings. However, this is not without its own problems, particularly in the case of households with multiple motor vehicles, where vehicles are used both for personal and business travel — a common issue in Australia with its Fringe Benefits Tax providing concessional taxation of cars as part of remuneration packages — and where vehicles have been bought or sold during the monitoring period.

Some Issues with Meters

In the recent behaviour change literature, it is often said that the use of meters (in transport: odometers) would be a promising tool to get better results. The main arguments to support the use of meters are that respondents in surveys are self-reporting and might want to report project-friendly results; meters are more precise and statistically valid. However:

- Even in water projects (where meter readings provide the only source for evaluation), the readings are reported readings. In the case of water, reporting is done by an 'independent' person. But painful experience from five water projects has shown that meter readers make mistakes, are sloppy or at times do not even read the meter and make the results up.
- In an odometer-reading project, these problems get greater. It is self-reporting again, in this case by a household member. And might they, as is sometimes argued in survey-based evaluations, want to give project-friendly results? This risk alone would cut out odometer readings as a viable option, because there is nothing easier than to simply report 1000 km less, to pretend great success.
- Additionally, the common odometer reading projects achieve very low participation rates. Stopher, Swann, and FitzGerald (2007a, Table 1) report that rates for newly recruited households vary between 12% and 25% and need a strict regime to be kept by the households. Adding to the problems of low participation, *in any given wave, about 25% of households will fail to provide odometer readings* (Stopher et al., 2007a). At best this increases the required sample size, but in conjunction with low recruitment rates, there is concern for systematic bias that cannot be addressed by sample size. In a recent validation survey on odometer readings conducted by Socialdata Australia, about two thirds of the readings were reported in doubt (collected at the wrong time, wrong day, wrong car, reading invented, later readings lower than earlier ones, distances travelled of several thousands of kilometres a day, etc.).

Even if the meter reading is correct, it tells us nothing about the type of travel, its frequency (number of trips) or trip purpose.

For travel generally, and car use in particular, we are forced to use less direct means of estimation, unless the outcome is so large that it can feasibly be measured through road traffic volumes. Even then, we face the problem of differentiating the traffic generated by the target population from that generated by the rest of the city.

It has been argued that geographical positioning systems (GPS) technology can provide a means of directly measuring travel and, by inferential means, mode use (Stopher et al., 2007a). However, this is, as yet, unproven in at least two key areas:

- The robustness of the algorithms for inferring mode use; and
- The potential for the measurement actually to influence travel behaviour — for example, a person 'equipped' with a GPS device might be more aware of their travel behaviour (as distinct from its being habitual) and use transport more efficiently while carrying the device.

Unfortunately, there do not appear to have been any comparisons between GPS and surveys in respect of either error or data capture (AGO, 2006, p. 57).

GPS might have potential value in validating a travel diary or survey (e.g. in terms of places visited and times of travel), subject to the caveat of measurement influencing behaviour, but the reported difficulties of recruiting people to take part in trials (Stopher et al., 2007a) and retaining them in subsequent waves of a GPS panel (Table 5.1) suggests that it might be difficult to achieve a sampling basis that is sufficiently representative of the community overall for useful results to be derived.

Alternatively, GPS might have a more robust role if attached to the vehicle rather than the person, subject to the caveats already noted about multi-vehicle households, vehicle changes and identifying business use of vehicles.

For public transport, ticketing systems may allow direct measurement of trips originating in an area. Where this was possible, ticketing data has been consistent with the change estimated from travel surveys (see Figure 5.2).

Implementation of smartcard ticketing systems will facilitate use of ticketing information to estimate changes in public transport patronage related to VTBC.

5.5.2. *Surveys*

VTBC initiatives have most often been assessed by surveys of various types (see Section 5.5.3 below). We generally measure change as the difference between 'before-and-after' surveys, with 'corrections' indicated by a control group (see Section 5.7.1 below). It follows that it is critical that the 'before', 'after' and 'control group'

Table 5.1: Recruitment and retention in GPS trial (South Australian panel).

Status	**South Australian panel (50 households nominally)**		
	Wave 1	**Wave 2**	**Wave 3**
Recruited (new to wave)	57	17	0
Completed	50 (88%)	14 (82%)	0
Continuing (recruited)	–	35 (61% of 57)	44 (59% of (57 + 17)
Completed (continuing)	–	32 (56% of 57)	36 (49% of (57 + 17)

Source: Stopher et al. (2007a, Table 3).

surveys are undertaken using the same survey design and methodology. Differences in design or methodology will affect response rates (see Section 5.6.6 below) or have more subtle impacts on the accuracy of responses that make comparisons problematic.

5.5.3. Cross-Sections or Panels for Surveys

Many travel behaviour change interventions have adopted a cross-sectional survey approach to estimating 'before-and-after' travel behaviour. For small projects, this can create issues of statistical reliability of the difference between the two surveys, but these are less severe for larger applications.

It has been suggested that a panel survey approach, using the same people for both surveys, would be more suitable. However, panel surveys have their own problems that can be more systematic and less amenable to treatment by statistical analysis.

For example, an aging population ages more slowly than the individuals that comprise it. At a more aggregated level, it ages more slowly than the households or other groups that comprise it. A panel that remains intact over a period of time will not be the same at the end of the period as it was initially. Whilst this might not make very much difference for the typical 'before–after' time period of 12 months, it has an increasing impact over time and could adversely affect the comparability over longer time periods and, hence, the estimation of the durability of behaviour change. Also, in practice, there is attrition of panels. If no action is taken to replace, the integrity of the panel is compromised by the possibility of self-selection bias among those opting out. It also introduces the difficult problem of identifying and recruiting panel replacements that have the same characteristics as those who leave. People also suffer from 'survey fatigue' and may drop out or become less reliable for that reason. Whilst this can be addressed through 'refreshment' (i.e., replacing a proportion of the panel on a periodic basis), this also raises the issue of how to ensure that replacements have the same characteristics, including travel, as those who are discarded.

It has been argued that a benefit of panels is that they require lower samples sizes, but at the same time, some of the parameters enabling smaller sample sizes also give rise to survey designs that are more difficult to undertake. For example, panel survey data is more difficult to obtain (with full control of other biases) than repeated cross-sectional data (Richardson, Seethaler, & Harbutt, 2003). Acquiring repeated data from the same respondents is challenging and a reduced response rate in the 'after' survey can lead to sampling bias (ITP, 2007, para 9.24).

One of the examples of a panel survey to estimate the effect of a VTBC initiative had a low initial response rate (49%, or 1346 out of a sample of 2772 households) for the 'before' survey and substantial further loss between the 'before' and 'after' surveys. Furthermore, in the 'after' survey only 682 of the 881 households that responded to the 'after' survey had the same composition as in the 'before' survey (Richardson, 2005). At the individual level, people may exhibit substantially different travel behaviours at different times for reasons that are purely idiosyncratic and not related to the intervention being investigated.

In two cases where both panel and cross-section surveys were carried out (North Brisbane, Queensland and Victoria Park, Western Australia), the panel survey recorded slightly smaller, but still very substantial, reductions in car driver trips (11% vs. 13% and 12% vs. 14%, respectively). This might be a result of people who are less likely to change residential location, and hence more likely to remain in the panel, also being less likely to change other behaviours, including travel (see box).

Integral measurement of travel behaviour, as part of the intervention, is a form of panel survey that highlights some of these problems, most notably those of attrition and measurement actually influencing the outcome (i.e. travel behaviour).

Comparing Panels and Cross-Section Surveys: Two Case Studies

Two unpublished studies in Australia have evaluated a TravelSmart project (Brisbane North and Victoria Park, Perth) with a panel *and* repeated cross-sectional (random) surveys.

Sample sizes (respondents–persons)

	Before	After panel	After cross-section
Brisbane north	1309	825	1381
Victorial Park	905	780	766

The response rates were in the seventies and eighties, the combined response rate for the panels in the mid-sixties. If we compare the results of the panels with the cross-section survey, we can see that the behaviour changes picked up in cross-section were greater than in the panel.

Measured travel behaviour change — panel and cross-section (random) surveys

	Brisbane north		Victoria park	
	Relative change		Relative change	
	Panel	Random	Panel	Random
Walking	+ 32%	+ 49%	+ 12%	+ 12%
Bicycle	+ 33%	+ 58%	+ 60%	+ 48%
Motorcycle	n/a	n/a	n/a	n/a
Car driver	– 11%	– 13%	– 12%	– 14%
Car passenger	+ 8%	+ 8%	– 1%	0%
PT	+ 16%	+ 22%	+ 14%	+ 17%

This supports the anecdotal evidence accumulated over many projects that people who are flexible, move more, change more easily and are under-represented in panels, which tend to focus on stability, less change and the status-quo. Behaviour changes measured in panels therefore seem to be smaller than in reality.

This conclusion is supported by current research in travel behaviour change projects for which panel evaluation has been selected as the method of evaluation. In all these surveys, the response rate of the 'before' survey was at least in the mid-seventies and the response rate of the 'after' always in the eighties. Thus the combined panel response was always over 60%.

Using the speed of response technique, it was possible to simulate a traditional before-and-after panel with a response rate in the low forties (before) and the low seventies (after) — a common combined response rate of about 30%. In terms of the analysis above, the selection worked even more in favour of stability and no change. This is clearly reflected in the results. Behaviour changes in a panel with a combined response rate of about 30% are smaller than in a panel with a combined response rate of 60 + %.

5.6. Using Surveys

5.6.1. Sample Size

With a sample survey, the reliability of a sample estimate (in this case, the mean proportion of trips by car as driver) is inversely related to the sample size, both in absolute terms and as a proportion of the population being sampled.

A great deal of attention has been paid to the theoretical requirements for sample sizes to identify relatively small changes at a reasonable level of confidence. Stopher & Greaves (2006) suggests that to identify a change of 2% at the 95% confidence level it might be necessary to gather data from around 8000 people in a before-and-after study. Such a requirement would be beyond the resourcing likely to be available for monitoring any such project and would, in any case, substantially diminish the resources available for actually delivering the VTBC initiative.

The focus on establishing statistical confidence requirements for small levels of behaviour change appears to stem from an erroneous assumption that the behaviour change was measured only over the actively participating population (Stopher, 2003), which is not always the case. The Australian Greenhouse Office has noted that: *In some of the reports evaluated, there appears to have been misunderstanding of statistical methods used to interpret results, and hence problems in the sample sizes, sampling methods and evaluation periods employed* (AGO, 2006, p. 58).

Much of the contention about sample sizes is predicated on a desire for accurate estimation of the level of travel behaviour change, whereas the public policy imperative is for assurance that the level is greater than the level that would justify the initiative in socio-economic terms. Given the high benefit levels from travel behaviour change, this level is a very low bar: the prospective benefit–cost ratio for completion of the Perth TravelSmart Household programme estimated a benefit–cost ratio of 67:1 on the basis of a 10.7% reduction in car driver trips (which was the weighted average of projects to date in Perth, including two project areas (Marangaroo and Cambridge) specifically chosen for their 'low potential') (Ker, 2004). Typically, urban transport projects are regarded as 'justified' if they achieve a BCR of 3:1 or 4:1. On this basis, VTBC would be justified in benefit–cost terms with a car trip reduction as low as 1%.

There has also been a strong tendency to focus on the probability of the change being *less than* that estimated through the sample surveys, with little or no recognition of the corresponding probability of its being greater. If the population mean is greater than 2% (as *all* the interventions except Darebin indicate), by a factor of up to seven times and an average of around four times, the sample size requirements become much less onerous.

Most of the debate on sample size has been on the basis of random sampling error estimation. In practice, the error due to systematic (design-related) factors (e.g. response rate, qv, below) can be many times larger than statistical error (Table 5.2). This is a serious problem, even if we do only focus on significance, because both random and systematic errors must be considered in empirical studies.

Since there are always systematic errors, the calculated random errors are always based on a wrong assumption (that there are no systematic errors). But the real problem of these two types of errors is their nature:

- Random errors can be calculated exactly, but not corrected, whereas
- Systematic errors can be corrected but not (precisely) calculated.

In addition, whilst random (sampling) errors can be reduced by using larger sample sizes, systematic errors cannot be reduced in this way. In travel behaviour, the systematic errors often significantly outweigh the random errors in size.

One of the most useful factors to understand, calculate and correct systematic errors is the response rate of a survey, and one of the best variables is the number of trips per person per day. In a mail-back survey, the response can be analysed by the speed of response. This has two advantages: lower response rates can be easily simulated, and the whole process can be analysed to estimate the final (unknown) result for the total. Both things have been done and this function has been used to

Table 5.2: Random and systematic errors.

Net responses	66,500 respondents (net)	
Response rate	40%	72%
Gross sample required	166,250	92,350
Trips per person per day	3.4	3.1
Random statistical error	±0.02 trips per person per day	
Systematic (response rate) error	±0.30 trips per person per day	

Note: Note that the comparison between random and systematic errors shown here is illustrative rather than definitive, as random errors are dependent on sample size, whereas systematic errors are not.

analyse the likely shortcomings of a parallel survey that achieved only 25% response (Figure 5.3).

The technique developed and used has been further improved and applied in the Netherlands' National Travel Survey, the only long-running continuous travel survey in the world (MVWRAVV, 2007). As a consequence, a specific self-validating design has been developed and applied. This design can correct for non-reported items, non-reported trips and non-response and is not dependent on external data.

If we look at the respective response curve, we see that detailed analysis of reasons and effects with improved design of survey methodology has reduced the non-response effect significantly, yet it is still there: with an estimate of the total trip rate to be 3.1 (100), the trip rate at a 40% response rate would still be 10% higher (Figure 5.4).

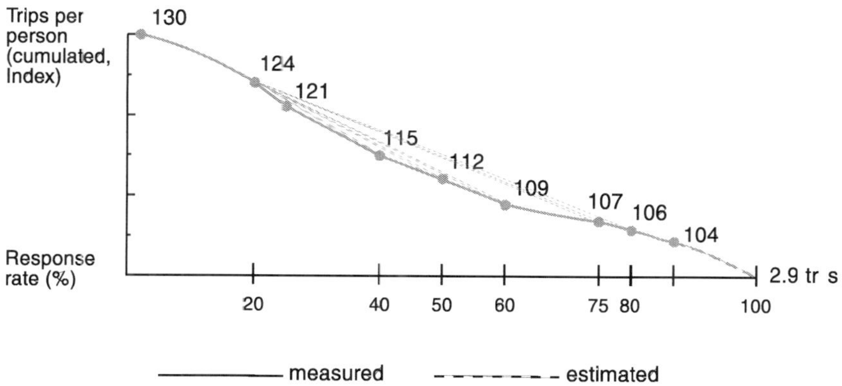

Figure 5.3: Trip rate by response rate: Vienna 1993 (response rate 85%). *Source*: Brög and Erl (1999).

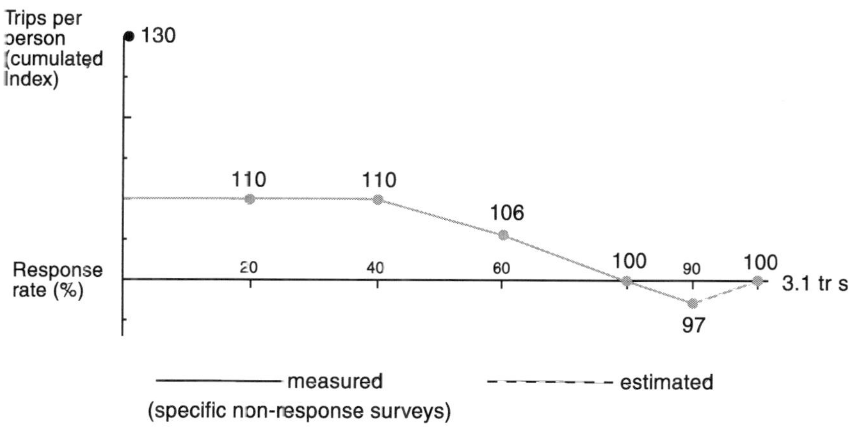

Figure 5.4: Trip Rate by Response Rate: Netherlands (MON-Mobiliteitsonderzoek Nederland).

Many current designs applied to evaluate behavioural change projects have a much higher influence on systematic errors and the difference in the trip rate at a 40% level (which is normally regarded as good) would be considerably greater. However, if we use the MON data and simulate a situation of a 40% response rate we would have needed a gross sample of 166,250 people and achieved a trip rate of 3.4. The 'real' MON achieved an overall response rate of 72% and a trip rate of 3.1.[2]

In the first case, the systematic error would be 0.3 trips per person and day, and the random error only 0.02. This one type of systematic error, on its own, using only one variable shows already that the systematic error is fifteen times the (widely and commonly used) random error (Table 5.2).

It has been suggested that the use of GPS for measuring travel behaviour would permit a smaller sample size than for a conventional travel survey (Stopher, FitzGerald, Greaves, & Khang, 2007b). For example, we understand that the sample size proposed for long-term monitoring of travel behaviour change in North Brisbane, using GPS, is 300 households. It is not clear why this is any more than would be the case for multi-day diaries, as the variability due to random sampling would be the same in both cases. The potential for non-sampling and response errors exists, in different ways, for both methods, but is not related to sample size. Indeed, the difficulties of recruiting and retaining households in GPS panel surveys (Stopher et al., 2007a) suggest that non-sampling errors are likely to be a severe problem.

5.6.2. Daily or Weekly 'Diaries'?

Most travel surveys are based on a single day for each household, rather than a full week, despite the fact that travel behaviour for any individual will vary by day of week. This is most notable, on a systematic basis, between weekdays and weekends, with most trip purposes, other than leisure and recreation and some types of shopping trip, being focussed on weekdays. The rationale behind a daily diary is precisely that this differentiation *is* systematic and therefore can be picked up by appropriate sampling on each day of the week.

Both daily and weekly diaries have some advantages and some limits. Daily diaries, when used, should represent each day of the week equally to overcome limitations. A weekly travel diary will pick up the variation between days for individuals and individual households, but does so at the expense of:

- A lower response rate, as fewer people are willing to commit to the extended commitment;
- A higher attrition rate, as people cease to record their travel during the week and/or
- Lower quality data towards the end of the survey week, as the novelty wears off.

2. Non-reported trips are very similar at all response rates (c5%), so the inverse relationship between trip rate and survey response rate is not a function of non-reporting.

5.6.3. Repetition

Experience with repeated application of VTBC (Figure 5.1, for *IndiMark®*) shows a high degree of consistency with:

- greater variability of outcomes for small applications (consistent with statistical expectations resulting from smaller survey sample sizes) and
- larger reductions in car driver trips (and lower variability) for larger applications.

The probability of *all* of these outcomes being over-estimates of the population value (let alone sufficiently large over-estimates to invalidate the principal evaluation conclusions for VTBC) is very small and gets smaller with every repeated application.

Alternatively, if the individual projects were to be aggregated into a single notional 'programme' with a number of different spatial locations for evaluation purposes, the total sample size would now exceed the theoretical sample size required for measuring a level of change that is less than a quarter of that actually achieved. If these had been undertaken as one project, even this requirement would have been achieved. The consolidated result for Perth TravelSmart projects, including locations chosen for their likely poor performance (Marangaroo and Cambridge), is a 10% reduction in car driver trips (http://www.dpi.wa.gov.au/travelsmart/14960.asp).

This conclusion is supported by the Australian Greenhouse Office (AGO, 2006, p. 53): *For households, predicting the effectiveness of a Travelsmart project is now essentially a solved problem. The eleven Australian projects evaluated here join over 50 other formal evaluations conducted internationally and many other informal assessments (See* Ker, 2003; Maunsell Australia, 2004; UK Department of Transport, 2004 *for a list of some other evaluations.) While individual outcomes vary with geographic location, what can be said broadly is that community-based household projects will achieve a reduction in car travel of 5–15%, and this change appears to be sustained for several years without further intervention. Methods for achieving these results are now well understood, and further evaluations are unlikely to add much to existing knowledge.*

The practical value of repetition is evident from the City of Nürnberg, Germany (500,000 people), which uses Individualised Marketing (IndiMark®) to promote public transport (i.e., not in multi-modal mode). It started in 1996 with a small project of nearly 5000 people. From then on, every year several projects have been conducted. Since 2006, early areas are worked on for the second time and by the end of 2007 a total of 600,000 people have been part of the programme (including repetition). All projects were successful and attitudes in the entire city have been changed.

Each project has been evaluated separately. In 2004, all evaluation surveys up to a certain point were amalgamated and analysed. The result is an increase in public transport by 13% (fully supported by counts and calculated against control groups)

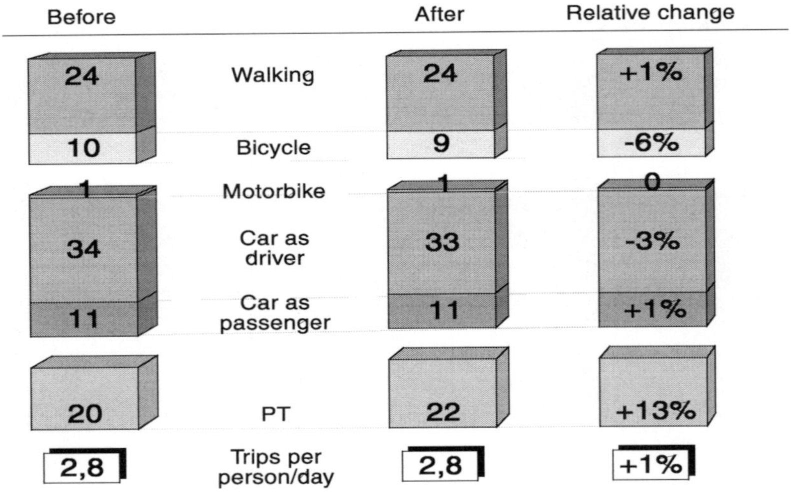

Figure 5.5: Consolidated behaviour change 1996–2004: Nürnberg, Germany. *Source*: Nürnberg (2006).

and — although not targeted — a reduction in car driver trips of 3% (Figure 5.5). This is still continuing.

5.6.4. *Response Bias*

There is sometimes a danger that the very fact that a person is part of a programme will influence that individual's actual or reported behaviour. In the case of VTBC, the influence on *actual* behaviour is precisely what is sought by the programme, provided that this is an ongoing response and not a temporary one simply to be able to present in a better light for the programme itself.

The greater danger lies with 'programme-compliant' *reported* behaviour, although this is more likely to be found in panel surveys than in repeated cross-section surveys. Participants in a panel survey will know what the base travel behaviour was (the 'before' survey) and could, theoretically, report appropriately changed behaviour in the 'after' survey, even if actual behaviour had not changed. Participants in a cross-sectional 'after' survey will not have a 'base' to compare themselves against and will be less familiar with the detailed programme objectives.

Misreporting travel behaviour in surveys happens for a variety of reasons, irrespective of any potential desire of the 'observed' to please the 'observer'. The best way to guard against the latter is exactly the same as for the former. Since even apparently simple activity and travel patterns can be quite complex, involving issues of location, duration, speed and mode, it is more difficult than might be expected to falsify a survey response in a realistic way. Detailed inspection of survey responses will show up unusual, inconsistent or infeasible travel behaviour that can then be followed up for clarification and, where necessary, correction.[3]

Bonsall (2007) draws attention to the possibility that the repeated cross-section surveys will *by happenstance, result in some households being contacted more than once*, which he says could have negative (e.g. households feeling hassled by repeated surveys) or positive (e.g. the ability to explore changes in individual behaviour) consequences. In practice, unless sample sizes are larger than necessary or surveys are continually repeated to assess durability, multiple-contact households will be too few for any such issues or opportunities to be substantial.

5.6.5. Control Groups

Behaviour change initiatives do not operate in isolation. Almost by definition, such initiatives are funded because there is a sympathetic government policy environment. In recent times, the increasing price of petrol will also be having a systematic impact on travel behaviour. This requires that, wherever possible, we identify 'control groups' of people or households similar to those subject to the intervention, so that the impact of factors external to the intervention can be identified and appropriate adjustments made to the measured outcomes in the intervention group.

This poses a problem for interventions that do not attempt to measure change over the whole target population, as there is a potentially high degree of self-selection in the intervention group that makes them different from any potential control group.

The key criteria for a control group include:

- Spatial location
- Demographics (age, gender, household size/structure, income, education, employment type and status)
- Opportunities in the relevant area (e.g. similar transport systems and activities)
- Exposure to the same exogenous factors such as petrol price, public transport fares and transport system development.

The only group that, in principle, meets all of these criteria (subject to sampling variability) would actually be a random sample from the population of the intervention area itself. This is particularly so with a very large-scale application that covers a spatially large and demographically diverse area. In practice, however, even those who are not part of the target population are highly likely to be influenced by the intervention itself (see Section 5.4.3 above).

3. The second author has direct experience of this in an *IndiMark*® project, having been part of the 'after' survey and contacted for clarification of his household's 'unusual' travel pattern — only one car and one licensed driver in a household of four adults; circular trips involving delivery-on-foot of local newspapers; low level of travel activity; very low car driver trips.

Control groups, where used, are usually external to the intervention community and any measured travel behaviour changes in them have been small, as would be expected over a short period of 12 months. In recent years, the major systematic influence has been from increasing fuel prices and it is highly likely that the same impact would be found in both the control group and the identified target population. The effects of this can be unexpected, however — for example, in the recent North Brisbane project, the petrol price in March/April 2007 ('after') was very similar to the price in March/April 2006 ('before'), but in between there was an increase (followed by a decrease back to earlier price levels). The reduction in price leading up to March/April 2007, coming after a 2-year period of sustained price increases, could account for a measured increase in car use in the control group (Ker, 2008).

5.6.6. Response Rates

The response rate to a survey is an important measure of its reliability. It also affects the value of key parameter estimates from the survey. To obtain accurate data, it is important to interview a representative sample. When a survey's response rates are low, it is unlikely that the sample interviewed is completely representative. The data collected from a survey with low response rates can suffer from non-response bias and the decisions made using those data may be flawed. Response rates in all forms of travel surveys have been declining and telephone survey response rates are rarely above 60% and are often 20–30% (http://rns.trb.org/dproject.asp?n = 14124).

With telephone surveys, non-respondents can include high-mobility people (i.e., who are often not at home) as well as those who do not consider themselves to have a sufficiently strong 'stake' in travel issues (e.g. because they travel very little).

With mail-back surveys, it has been shown that the average number of trips per person per day varies inversely with the response rate (Brög & Erl, 1999). In other words, people who are less active in terms of travel are less likely to respond to surveys. This has double importance when we are seeking to estimate the *difference* between two estimates:

- For any percentage change, the number of trips will depend upon the absolute values, which in turn is inversely related to the response rate. It follows that a low response rate is likely to overestimate not only the before-and-after trip-making but also the difference between them.
- A difference in response rate between the 'before' and 'after' surveys will also affect the estimate of the difference between them. If the 'before' response rate is higher than the 'after' rate, the difference is likely to be under-estimated, and vice versa.

The best estimate of difference between two cross-sectional surveys of given sample size will come from surveys that achieve a similarly high response rate. VTBC evaluations with high response rates (in excess of 70%) for both 'before' and 'after' surveys also show a consistency of frequency of out-of-home activities between the 'before' and 'after' surveys, indicating that the difference between them is unlikely to be a major influence on trip rates or mode usage. Low and varying (between 'before' and 'after' surveys) response rates might be a partial explanation for differences in measured effectiveness between types of intervention.

5.7. Measuring and Reporting Behaviour Change

5.7.1. Measuring Differences in Behaviour

It is intrinsically more difficult to measure changes in behaviour than behaviour itself, largely because changes can only be measured as the difference between behaviour before and after. Estimation of differences is, therefore, affected by all of the issues associated with the measurement of both the before and after states.

In principle, this problem can be eliminated in the limited context of measuring travel behaviour across the whole participant population (as, e.g. with early *Travel Blending* applications), with the 'before' and 'after' travel diaries being an integral part of the overall intervention. Unfortunately, this provides no information about behaviour change among those who do not participate right through the process and no generalisation to the whole population is possible.

For those interventions that do measure behaviour change across the whole identified target population, irrespective of their level of involvement (including zero), it is still possible to make some useful assessment of the reliability of estimates. As a specific example, the North Brisbane Household TravelSmart project achieved an estimated 13.4% reduction in car driver trips (eight percentage points, from 58.4% to 50.4%), which is significantly different from zero with a 99%-plus probability[4] (Socialdata, 2007). However, the tests do not estimate the potential range of the magnitude of the difference. This result can be used inferentially to estimate the upper and lower confidence limits for the value of the difference between the means:

- The 'before' and 'after' surveys each have a sampling distribution for the estimated mean value of car driver trips;
- The sample size for each of the surveys is sufficiently close so that we can assume, for practical purposes, that the sampling distribution for each is the same;
- The sampling distribution can be described in terms of the standard deviation;
- A 99% probability that the proportion of car driver trips was lower after IndiMark® means that there is less than 1% overlap between the two sampling distributions;
- With a normal distribution, 99% of the sampling distribution is within 2.5758 standard deviations of the mean;

4. Target Group net sample size 1309 ('before') and 1381 ('after'), with 76 and 79% response rates, respectively. Total population: 277,000 people or 113,000 households.

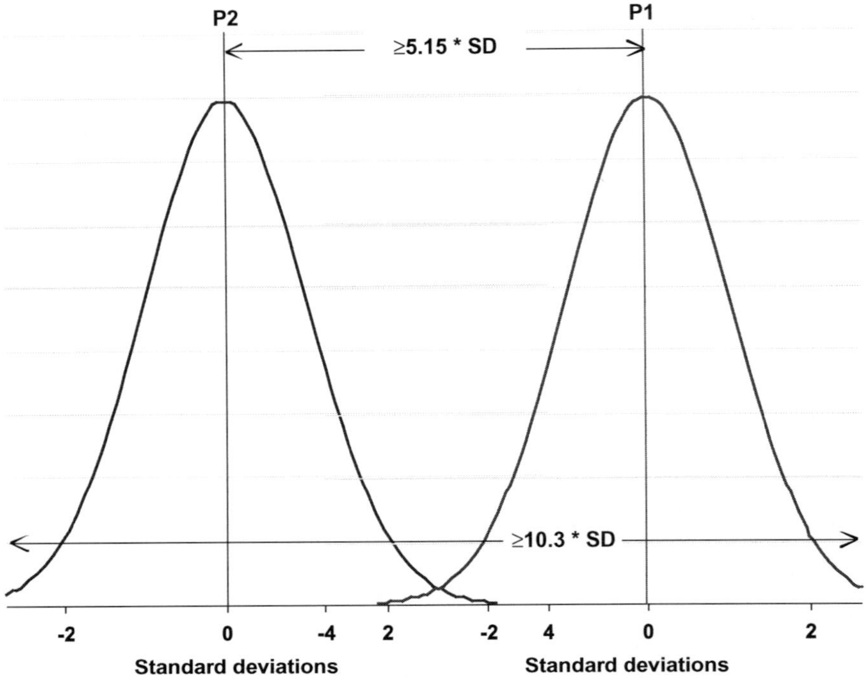

Figure 5.6: Sampling distribution of the mean (before (P1) and after (P2) IndiMark®).

- It follows that the difference between the estimated means is at least 5.15 standard deviations with 99% probability (Figure 5.6);
- Whilst the distributions of the 'before' and 'after' sample means indicates that the true value (99% probability) could be anywhere between zero and 15.6 percentage points (26.8% of previous car driver trips), the distribution is heavily weighted towards the mid-range;
- The difference between any two normal distributions is itself normally distributed. A normal distribution with a 99% probability range of 0–15.6 has a standard deviation of 3.0.

Since Figure 5.7 shows an 80% probability of the reduction in car driver trips being *between* 3.8% and 12.2% of all trips and the range outside that evenly distributed above and below that range, the probability of the reduction being greater than 3.8% is 90%. There is a 10% probability that the reduction exceeded 12.2% of trips.

In terms of the more commonly stated percentage reduction in car driver trips, these figures are equivalent to:

- An 80% probability of the reduction being between 6.4% and 20.4%;
- A 90% probability of the reduction being greater than 6.4% and
- A 10% probability of the reduction being greater than 20.4%.

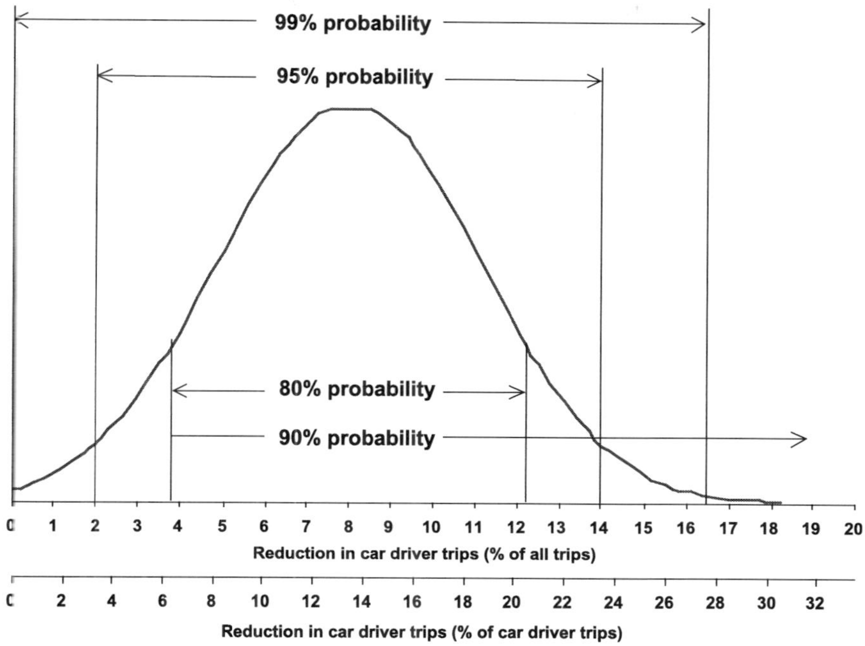

Figure 5.7: Sampling distribution of the difference between means.

If we redo this for a 95% probability that the difference exceeds zero and for a more 'typical' measured reduction of six percentage points or 10% (from, say, 60% car driver trips to 54%), we get the following:

- The range of estimates will be between zero and 12 percentage points (20% of car driver trips);
- There is an 80% probability of the reduction being between 2.1 and 9.9 percentage points (3.5–16.5%);
- There is a 90% probability of the reduction being greater than 2.1 percentage points (3.5% of car driver trips).

There is a 10% probability of the reduction being greater than 9.9 percentage points (16.5% of car driver trips).

5.7.2. Population or Participants?

A specific issue that has received little attention is that intensive interventions may have substantial impacts on particular individuals or households but, because of their intensive or intrusive nature, may achieve lower participation rates. It is essential that comparative evaluations are on a like-for-like basis and, given that the strategic objectives for travel behaviour change are population-based, reflect the outcomes across the identified target population as whole, not just for the participating individuals or households.

Reporting travel behaviour change across the whole identified target population, including those who do not actively participate, such as is done by *IndiMark*®, facilitates the use of control groups (qv) not affected by the intervention. Although one would not normally expect external factors to have a major impact on travel behaviour over the typical 12-month period between the 'before' and 'after' surveys, the rapidity of petrol price changes in recent years is an exception to this.

Travel Blending® and other approaches that use travel diaries as an integral part of the process are only able to report outcomes across the participating population — and specifically only among those who complete both 'before' and 'after' travel diaries. Because of the risk that participants will self-select through their choice to be part of the programme, there is a further risk of self-selection in the 'loss' of participants between the Stage 1 and Stage 2 diaries. Perkins (2002) estimated that the reported results of Travel Blending in South Australian pilot projects would have to be factored down by 0.6 to estimate population-wide outcomes on a basis comparable to *IndiMark*®.[5]

5.8. Conclusion

VTBC aims to reduce car driver trips and car-km of travel without specific investments in physical infrastructure or transport services, and without the regulation of transport activity by price or other means. The principal techniques that are applied at the household level differ in their targets, the means of identifying those who actively participate, the methods of participation and the way the extent of travel behaviour change is assessed.

Some interventions work only with self-selected participants, and the proponents of these programmes tend to report measures of behaviour change for those participants alone. The proponents of some other interventions report behaviour change that is measured across the whole identified target population. This difference has led to some confusion in the literature, including the assumption — in our view erroneous — that results measured across the whole target population ought to be 'scaled back' as if they had been obtained from a self-selected subset.

VTBC is almost unique among transport initiatives in that it has been developed from a strong theoretical and observational basis, developed through a series of interventions with proper experimental design (including control groups) and has been subject to comprehensive monitoring and evaluation of outcomes with the

5. This is not a criticism of *Travel Blending*® alone. It is an observation of the difficulty of identifying policy-relevant outcomes from interventions that have the measurement tool as an integral part of the process. Behaviour change can only be estimated for those who participate right through the process and cannot be extrapolated to any larger group.

process and outcomes being widely documented in the public domain. Despite (or perhaps because of) that, it has been subject to more intensive scrutiny than any comparable development. Much of this has focussed on the sample size requirements for measuring quite small changes with a 'satisfactory' degree of statistical reliability (usually 95% probability).

In practice, measured estimates of travel behaviour change have consistently been in the range of 5–15% reduction in car driver trips. This consistency of measured impact, along with repetition of results from successive applications and the cumulative sample size now achieved, successfully counters any doubts about effectiveness based on the method of measurement.

The most straightforward way of overcoming any remaining issues of statistical reliability in measuring behaviour change for an individual VTBC initiative is to undertake very large-scale interventions. This reduces the relative sample sizes required to achieve a specified level of statistical reliability and, hence, the cost of monitoring relative to delivery. A recent application in North Brisbane, to 74,500 households out of an area population of 113,000 households, has demonstrated that the change in car driver trips is significantly different from zero with 99% confidence. This is equivalent to a 90% probability that the reduction in car driver trips exceeds 6.4% (or 3.8 percentage points).

Very large-scale interventions, appropriately designed, also create substantial opportunities for achieving travel behaviour change through diffusion at little or no cost to the project. Further research into the extent of such diffusion would be useful.

Confidence in the estimates of outcomes is further enhanced by consistent high survey response rates, which minimises the effects of non-response bias.

Further development is required of the potential for direct or indirect measurement of car driver trips/travel as an alternative to surveys, to provide robust estimates of the primary travel behaviour outcomes, including car-kilometres of travel as well as car driver trips. This might be achieved through measurement related to household vehicles rather than to individual members of households. However, such approaches do not automatically overcome issues related to sampling and non-sampling errors associated with surveys.

Further research would also be desirable to know the extent of induced car travel resulting from the reduction in car traffic generated by the population of the intervention area. In particular, this should focus on the circumstances in which such induced traffic might be significant.

References

ABS. (2007). *Census 2006: Community profiles*. Australian Bureau of Statistics: Canberra, ACT. Available at: http://www.abs.gov.au/websitedbs/d3310114.nsf/Home/census

AGO. (2006). *Evaluation of Australian TravelSmart Projects in the ACT, South Australia, Queensland, Victoria and Western Australia 2001–2005*. Australian Greenhouse Office, Department of Environment and Heritage: Canberra, ACT. Available at: http://www.travelsmart.gov.au/publications/pubs/evaluation-2005.pdf

Bonsall, P. (2007). *Does individualised travel marketing really work?*. Leeds, UK: Institute for Transport Studies.
Brög, W., & Erl, E. (1999). Systematic errors in mobility surveys. *Proceedings of the 23rd Australasian Transport Research Forum*, Perth, Western Australia. Available at: http://www.patrec.org/atrf/index.php?forum = 12&showResults = Search&hdnShowResults = Yes&session = 0&surname = brog&title = Systematic
Government of Western Australia. (2009). Department of Planning and Infrastructure. Available at: http://www.dpi.wa.gov.au/travelsmart/14974.asp, accessed 26 June 2009.
ITP. (2007). *Making personal travel planning work*. London, England: Integrated Transport Planning Ltd FOR Department for Transport.
Ker, I. R. (2003). *Travel demand management: Public transport business case*. Melbourne, Victoria: ARRB Transport Research FOR Department of Infrastructure.
Ker, I. R. (2004). *TravelSmart household (individualised marketing): Evaluation*. Perth, Western Australia: ARRB Transport Research FOR Department for Planning and Infrastructure.
Ker, I. R. (2007). *North Brisbane household TravelSmart: Peer review and evaluation — Context, methodology and values*. Brisbane, Queensland: CATALYST FOR Brisbane City Council, Queensland Transport and Australian Greenhouse Office.
Ker, I. R. (2008). *North Brisbane household TravelSmart: Peer review and evaluation — Appraisal of outcomes*. Brisbane, Queensland: CATALYST FOR Brisbane City Council, Queensland Transport and Australian Greenhouse Office.
Ker, I. R., Lloyd, R., Ringvall, K., & Sidebottom, A. (2005). *Effects of TravelSmart on urban congestion: Discussion paper*. Melbourne, Victoria: ARRB Consulting FOR Department of Infrastructure.
Maunsell Australia. (2004). *Travel behaviour change evaluation procedures and guidelines: Literature review*. Melbourne, Victoria: Maunsell.
Maunsell Australia. (2004a). *Travel behaviour change evaluation procedures*. Maunsell Australia Pty Ltd, Pinnacle research and Booz Allen Hamilton FOR Transfund New Zealand/EECA. Available at: http://www.ltsa.govt.nz/funding/travel-behaviour/technical-report.pdf
MVWRAVV. (2007). *Mobiliteitsonderzoek Nederland 2004 – 2007*. Rotterdam, Netherlands: Ministerie van Verkeer en Waterstaat Rijkswaterstaat Adviesdienst Verkeer en Vervoer.
Nürnberg. (2001). Individualisiertes marketing in Tullnau. Laufamholz und Mögeldorf: Social-data for VAG Nürnberg.
Nürnberg. (2006). Die Bewertung eines funktionierenden ÖPNV für eine funktionierende Kommune im internationalen Vergleich, Nürnberg und seine VAG: Nürnberg 26.10.2006.
Perkins, A. (2002). *The greenhouseabatement potential of travel behaviour change initiatives*. Transport SA and transport WA FOR national taskforce on national greenhouse strategy measure 5.3 — Best practice in transport and land use planning. Available from Transport Queensland, Brisbane, Queensland.
Richardson, A. J. (2005). *TravelSmart Communities Project 2004 — Evaluation Report*. Melbourne, Victoria: The Urban Transport Institute FOR Department of Infrastructure.
Richardson, A. J., Seethaler, R. K., & Harbutt, P. L. (2003). Design issues for before and after surveys of travel behaviour change. *Proceedings of the 26th Australasian Transport Research Forum*, Wellington, New Zealand. Available at: http://www.patrec.org/atrf/index.php?forum = 35&showResults = Search&hdnShowResults = Yes&session = 0&surname = richardson&title = Design
Roth, M., Ker, I., James, B., Brög, W., Ashton-Graham, C., Ryle, J., Goulias, K., & Richardson, E. (2003). A dialogue on individualised marketing: Addressing misperceptions. *Proceedings of the 26th Australasian Transport Research Forum*, Wellington, New Zealand.

Socialdata. (2007). *Brisbane North TravelSmart Communities Project— Final Report*. Brisbane, Queensland: FOR Queensland Transport.

Stopher, P. (2003). Travel behaviour modification: A critical appraisal. *Proceedings of the. 26th Australasian Transport Research Forum*: Wellington, New Zealand. See also Roth et al. (2003) for rebuttal of many of the issues raised by Stopher.

Stopher, P., FitzGerald, C., Greaves, S., & Khang, J. (2007b). Can GPS measurement replace travel diaries? *Proceedings of Transport — the next 50 Years Conference*.

Stopher, P., & Greaves, S. (2006). Guidelines for samplers measuring a change in behaviour from before and after surveys. *Transportation, 34*, 1–16.

Stopher, P., Swann, N., & FitzGerald, C. (2007a). Using an odometer and a GPS panel to evaluate travel behaviour changes. Paper presented to 12th TRB National Transportation Planning Applications Conference. Transportation Research Board. Available at: http://www.trb-appcon.org/2007conf/program.html#s4

UITP (1998). *Switching to public transport*. Brussels, Belgium: Socialdata FOR International Association of Public Transport.

UK Department for Transport. (2004). *A review of the effectiveness of personalised journey planning techniques*. London, UK: Department for Transport.

Vancouver. (2006). Vancouver Showcase Project: Social data for TransLink Greater Vancouver Transportation Authority.

Young, H. P. (2007). *Innovation diffusion in heterogeneous populations: Contagion, social influence, and social learning*. Center on Social and Economic Dynamics Working Paper No 51, Center on Social and Economic Dynamics at Brookings, Brookings Institution: Washington, DC, USA. Available at: http://www.brookings.edu/papers/2007/10_diffusion_young.aspx#_ftn1, accessed 8 November 2007.

Chapter 6

Surveys for Behavioural Experiments: Synthesis of a Workshop

Peter Jones, Regine Gerike and Giorgia Servente

6.1. Introduction

National and local governments and other public sector bodies are increasingly showing interest in adopting voluntary behaviour change (VBC) initiatives, as a means of encouraging people to modify their behaviour patterns in order to contribute to meeting a range of policy objectives (e.g. improving health or reducing CO_2 emissions). In the case of transport, the focus is usually on reducing car use and encouraging the use of more environmentally friendly modes of transport.

A diverse range of initiatives is included under this broad heading, varying both in who they are designed to target (e.g. workers, students, shoppers, households) and the nature of the intervention (e.g. provision of general maps and timetables, marketing materials, customised information). As an important part of such initiatives, there is a need to monitor their impacts in order to establish their effectiveness in changing behaviour, and to identify which kinds of instruments are most effective among different target groups and in particular contexts. However, monitoring the impacts of VBC initiatives is non-trivial and has been quite contentious among researchers and practitioners.

The purpose of this workshop was to identify and assess recent experiences with, and the special challenges involved in, designing and executing surveys to support the development and evaluation of VBC initiatives in different countries. The workshop concentrated particularly on initiatives aimed at influencing household travel behaviour, using customised information. The primary focus, however, was methodological — on examining the survey and analysis procedures used to identify impacts and their causes — rather than on findings concerning the success of different VBC initiatives. The workshop benefited considerably from the preparation of a comprehensive resource paper, drawing on Australian and European experience

Transport Survey Methods

ISBN: 978-1-84855-844-1

(Brög & Ker, 2009), and detailed papers describing particular studies in Switzerland (Brunne & Haefeli, 2008) and in France (Papon, Armoogum, & Diana, 2008). Very valuable insights were also provided by our discussant, Karl Sieber.

This workshop report summarises two days of discussions concerning the best ways of monitoring and evaluating VBC initiatives, looking in detail at a number of methodological issues. It sought to identify a series of best practice recommendations and to establish key remaining research challenges. The paper begins with an overview of the survey design process and then looks in more detail at six key issues. It concludes by identifying a number of remaining research challenges.

6.2. Survey Design Process

The context in which surveys are carried out to monitor and assess the impacts of VBC initiatives is shown in Figure 6.1, below. A particular population group or study area displays a set of travel and consumption patterns which give rise to certain types of concern (e.g. high levels of air pollution) and trigger the need to encourage changes in some aspects of that behaviour. A set of policy objectives is identified and a VBC intervention is designed to achieve these objectives. Following the intervention, there is likely to be a different set of travel and consumption patterns. From a methodological viewpoint, the challenges lie in first accurately measuring these 'before' and 'after' travel and consumption patterns, establishing whether they are 'different' in a statistically significant sense, and second, whether such differences can be attributed to the VBC initiative or are due to other factors that have changed concurrently. The study may also go on to try and identify the processes by which the VBC intervention influenced behaviour.

Figure 6.2 shows how a programme concerned with monitoring and evaluating the VBC intervention would tie in with this process. The primary emphasis is on comparing the behaviour patterns 'before' and 'after' the VBC intervention, in order to establish what impacts it might have had on those who were directly targeted by

Figure 6.1: The VBC procedure.

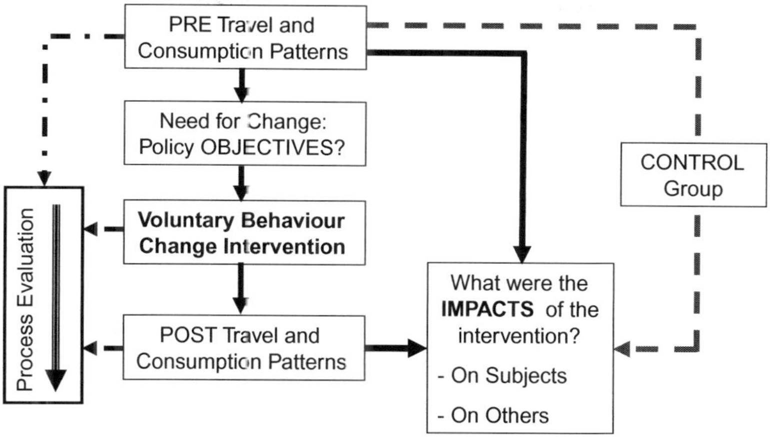

Figure 6.2: Procedure for monitoring the impacts of a VBC intervention.

the initiative (the 'subjects'), and on others. The exercise of comparing two data sets in itself requires that great care be taken in ensuring consistency of methods, and in matching the two samples. However, no matter how carefully this is done, the question remains as to whether any statistically significant differences in behaviour that are observed are due to the VBC intervention or to other exogenous changes that occurred during the same time period (e.g. a sharp increase in fuel prices). This difficulty can be overcome, to a considerable extent, through the use of a 'control group' consisting of an equivalent sample of people who have not been exposed to the VBC intervention, but who would have been influenced by any other concurrent changes. The main challenge here is to identify such a group.

Finally, some of the monitoring and evaluation studies seek to go beyond simply identifying the changes in behaviour resulting from a VBC initiative, to understanding the processes by which the initiative influenced behaviour, and in which ways. This requires a more in-depth analysis of the subjects' perceptions, attitudes and motivations, and there are further methodological challenges involved in exploring these issues.

6.3. Key Methodological Issues

The workshop discussed a wide range of methodological issues, which have been grouped under six headings, both identifying concerns and recommending best practice.

6.3.1. Deciding on What We Are Trying to Measure

The starting point in developing a VBC monitoring and evaluation research programme is to be clear on what the intervention is trying to achieve in the way of impacts on behaviour, and how these changes are expected to contribute to meeting primary policy objectives. What are these objectives: are they to reduce car use, save energy, increase physical activity, or what? Answering this question affects both what is measured, before and after, and how 'success' is judged in the evaluation process.

However, it is important not just to measure the direct, intended outcomes, but also to look for broader and unintended outcomes, in order to obtain a balanced view of the consequences of the intervention. Here it is useful to look at the literature, as well as using a degree of professional judgement.

Examples of factors to take into account include:

- Where an intervention is aimed at individuals in households (e.g. the car commuter, the parent driving the child to school), any changes in car use might have impacts on other household members; for example, if the car commuter switches to public transport, then the car might be used by another person in the household.
- Changes in behaviour resulting from a VBC intervention might have wider impacts on the economy; for example, switching trips from car to walking and cycling might encourage a greater use of local shops and services, thereby encouraging local businesses.

6.3.2. *Alternative Ways of Measuring Impacts*

The workshop identified a number of ways in which impacts might be measured, with consequences for the appropriate definition of a 'successful' intervention. In particular:

- The take-up of marketing information. The marketing profession commonly uses take-up and recall measures as an indication of success, such as the number of requests for brochures, or the percentage of a population who recall seeing a particular advertisement. This is of very limited value from the point of view of measuring impacts on travel behaviour, but might be useful information when carrying out a process evaluation.
- Psychological and sociological analysis. Changes in knowledge, perceptions and attitudes relating to transport and the availability of goods and services: improved awareness of public transport services or cycle lanes, changes in attitudes to alternative modes, changes in support for particular policy measures.
- Reported travel behaviour. This type of data is the most familiar to the transport analyst, and includes data that might typically be obtained from a travel diary, such as trip rates by travel mode, or trip and activity durations. While this kind of data is well suited to assessing impacts on travel patterns (allowing for certain deficiencies and biases in reporting and recall — see below), it does not directly address the question of whether certain higher-level policy objectives have been met (e.g. reductions in local congestion).
- Direct measures of individual consumption. There are various ways in which patterns and levels of individual consumption can be directly measured (e.g. using vehicle odometer readings, GPS or mobile phone tracking, public transport

smartcards, gas or electricity meter readings etc). Depending on the technology and local circumstances, the measurements might be reported by respondents (and hence be subject to misreporting error), or relayed directly to the analyst.

- Aggregate area measures, both of travel behaviour (e.g. bus passenger numbers, car traffic levels) and of the wider economic and social consequences of that behaviour (e.g. area accident rates, changes in local retail turnover). By definition, these figures are reported by agencies rather than the subjects of the VBC intervention; this generally provides a much greater degree of objectivity, but there may be technical problems with the data that are unknown to the analyst and it may be more difficult to attribute causation.

6.3.3. How to Identify 'Real' Changes in Behaviour

If the analyst finds that travel behaviour is 'different' before and after a VBC intervention, this does not, of itself, indicate that the intervention has contributed significantly to the observed changes in behaviour.

Several factors need to be considered here:

- The 'difference' may not be large or consistent enough to be statistically significant.
- It may reflect a cyclical variation in behaviour in the study area. In particular, in many countries there are major seasonal differences in travel behaviour and patterns of mode use.
- It may be due to changes in conditions that coincided with the VBC intervention, such as increases in fuel prices, or reductions in employment rates.

The recommended means of overcoming many of these problems is to include a 'control group' sample of people who are similar in socio-demographic characteristics to the subject group, who live in the same type of area (with similar levels of access etc.) and are subject to the same set of exogenous factors, but have not been subject to, or influenced by, the VBC intervention. In practice, the workshop recognised that this is very difficult to achieve. Firstly, in finding a good match between control and subject groups; and secondly, in ensuring that the former have not been exposed to or have been indirectly influenced by the intervention. The ideal control group would be drawn from the same community as the subject group, as is often the case in medical trials, but this is likely to result in a highly 'contaminated' control group since the effects of most VBC initiatives cannot be confined to particular individuals.

One issue that the workshop was particularly concerned about was the dangers of respondent bias in subjects reporting their behaviour, before and after the VBC intervention. Issues discussed included:

- A concern that respondents might (consciously or subconsciously) report 'after' patterns of behaviour that show them as having responded favourably to the initiative; this problem can arise with any form of intervention, but is particularly likely with the kind of intervention that encourages attitude change in a direction believed to make the desired behaviour change more likely.

- The added dangers of conducting only one interview per respondent, by using a retrospective survey to obtain 'before' data at the same time as the 'after' data, where the intervention being examined is subjective rather than objective in nature.
- The key importance of question sequencing and framing, particularly when trying to establish the factors affecting behaviour and the reasons for any behavioural change.
- The risks associated with using the same company to carry out the VBC intervention and to conduct the monitoring, both in terms of potentially raising respondents' awareness of the purpose of the monitoring exercise, thereby increasing response biases; and also due to suspicions among professionals of the impartiality of the analysis.

Some VBC monitoring studies use panel surveys, in which the same individuals are asked to report their behaviour in successive interviews, before and after the intervention. From a statistical point of view, this is more efficient and enables greater precision for a given sample size, but most workshop members had reservations as they felt that this increases the likelihood of a policy response bias among respondents in the reporting of their behaviour when dealing with VBC interventions. The use of statistically less efficient, but independent, repeated cross-section surveys would largely avoid this problem.

6.3.4. How to Assess Statistical Significance

Statistical procedures are particularly effective at taking account of random errors in the observations among a sample of respondents, but in this type of VBC monitoring exercise it is the systematic errors that tend to dominate, for reasons noted above.

The workshop addressed the question of what level of statistical significance is *really* required. Large samples, that give tighter confidence intervals, are expensive to collect, and as VBC initiatives are often relatively low-cost interventions, there is a danger that the monitoring exercise might consume a substantial proportion of the available budget.

Several conclusions were drawn:

- It might be appropriate to accept lower significance levels (i.e. using smaller sample sizes) where interventions are relatively low cost.
- If a process evaluation is included (i.e. looking at how the VBC intervention is affecting subjects' attitudes and behaviour), then this kind of information could complement the statistical analysis of impacts, and compensate for lower significance levels.
- There is a strong case for providing a monitoring research budget, so that a small number of VBC interventions can be monitored in detail, in order to provide a

sound basis for future business cases, and to help identify the most cost-effective kinds of interventions.
- There is scope for carrying out a meta analysis, by pooling data sets from different VBC monitoring studies, as discussed in the paper by Brög and Ker.

One cautionary note emerges from the analysis reported by Brög and Ker, who show that mean sample trip rates change in accordance with the response rate. More precisely, they found that respondents who return their self-completed travel diaries after several reminders report lower trip rates than those who are early completers, although it is sometimes difficult to distinguish between truly lower trip rates and incomplete reporting by those less motivated to respond. Moreover, some highly mobile, busy people may not respond at all, despite the reminders.

6.3.5. *Timing Issues Relating to Impact Assessment*

A number of different issues were discussed under the general heading of 'timing' issues:

- It is important to allow time for stable behaviour patterns to develop after the VBC intervention. This is particularly necessary where limited period incentives have been provided to subjects (e.g. free bus tickets), but in other cases too there may be a temporary period of experimentation, perhaps before pre-existing conditions are reestablished.
- Where possible, match before and after measurements seasonally, to avoid this known source of variation.
- The period of time over which impacts are likely to last. Very few studies have looked at the longer-term impacts of VBC initiatives. These temporal factors might be hypothesised to work in either direction: a diminution with time, or a snowballing effect, as behaviour change spreads through the local community. They might also vary by type of person (e.g. younger vs. older people).
- Respondents might also be more susceptible to influence at certain points in their lives; for example, when first moving to a new residential area, or on changing their life cycle stage (e.g. having a child or retiring) — when habitual behaviour patterns have not yet been established.

6.3.6. *Understanding Causes of Change and Willingness to Change*

There is considerable value in carrying out a 'process' evaluation, alongside the monitoring of changes in behaviour, in order to understand more about the ways in which VBC initiatives can influence attitudes and behaviour, and how best to develop and target them in future applications. Much of this work is likely to be qualitative in nature, although there may be scope for a quantitative measurement and analysis of attitudes, for example.

Points discussed included:

- Trips are more than simply descriptions of movement in time and space. Different modes have different images and associations and allow different kinds of supplementary activities to be conducted on route. Trips can also be closely associated with the destination activities they serve, and so there is a need to look at trips as multi-dimensional packages.
- There are many theories of behaviour change, mostly grounded in psychological theory (e.g. the Theory of Planned Behaviour) that can provide insights into the factors that trigger or inhibit behaviour change, and what might be done to enhance the positives and minimise the negatives.

6.4. Key Research Areas

Eight areas were identified for further research.

6.4.1. Accuracy and Role of 'Meter' Readings

The greater availability of meter readings of various kinds provides new ways of measuring and analysing behaviour, subject to the analyst being allowed to obtain the information under data-protection legislation, and being able to cross match it to other data sources. Examples include odometer readings, mobile phone and GPS tracking, and public transport smartcards, as well as the measurement of in-home energy consumption. However, while these offer valuable sources of additional information, research is needed to validate these types of readings, and to establish to what extent they can enhance or replace traditional measurement techniques. Where the analyst relies on self-reporting of these measurements, there are further issues associated with respondent error and bias.

6.4.2. How Best to Integrate Data Sources

The analyst now potentially has available a wide range of data types, covering reported and observed behaviour at the individual level, and measures of attitudes and awareness, through to measures of aggregate changes in behaviour (e.g. changes in motor vehicle or cycle flows, air pollution or congestion or in retail turnover in an area). These pose two kinds of challenge. First, how to ensure consistency and compatibility among the data sources; and second, how they can be brought together and used collectively to determine the impacts and causation of VBC and other policy initiatives. There is a growing literature on 'triangulation' that can assist in drawing on diverse data sources to reach a balanced judgement.

6.4.3. Defining Effective Control Groups

It is often very difficult to define and recruit effective control groups, particularly for programmes such as VBC, where the information that is provided may not be tightly targeted (e.g. using local advertising) and where there is likely to be 'leakage' through social networks and word of mouth. Screening procedures might be developed, both for the before survey (e.g. to exclude people with friends or relatives in the targeted area) and the after survey (e.g. excluding those with an awareness of the VBC initiative). It might also be necessary to look further afield for control groups than is commonly the case, by choosing a different sector of a city, or a nearby town with similar characteristics.

6.4.4. Cultural Effects (Survey and Behaviour)

The researcher needs to be sensitive to cultural factors that may influence both the effects of the VBC programme and its monitoring. Different cultural groups may be responsive to different kinds of information and marketing materials, and may have different scope to modify behaviour patterns due to cultural traditions or taboos. Response rates may differ, and so might patterns of item non-response. In more qualitative and subjective surveys, some groups may feel less comfortable exposing their personal views to strangers, or may rationalise behaviours in different ways.

6.4.5. Deeper Understanding of Change Processes

Previous research has provided some indication of the variations in the overall, aggregate effectiveness of a VBC programme in an area, but there is much less understanding of the ways in which the initiative has led to the observed changes in behaviour, and what makes some initiatives appear to be more successful than others. To what extent are some communities more amenable to change than others (e.g. due to existing latent pressures for change, or a more comprehensive set of alternatives), and are some forms of intervention or materials more effective than others? Guidance is needed on the techniques available for carrying out an effective process evaluation.

6.4.6. Importance of 'Interviewer' Effects

The responses in traditional travel surveys can be subject to interviewer effects, both in terms of response rates and the nature of the answers given. This is more of an issue where perceptual/attitudinal data is being collected, and has proved to be significant in interviewer-administered stated preference surveys. The additional concern that VBC initiatives raise is that interviewer effects may not only relate to the reporting of attitudes and behaviour, but may also possibility influence the behaviour itself. These potential effects need investigating and then isolating — or, alternatively, could be fed back into the design of more effective VBC programmes.

6.4.7. Greater Use of Experimentation

The VBC initiatives that are introduced and evaluated are usually multi-faceted in nature and involve a range of actions and materials. In addition, the collection of monitoring survey data usually includes a mix of objective, quantitative data on travel patterns and more subjective and qualitative data on perceptions, attitudes and motivations. Given this rich set of interventions and measurement instruments, it is difficult to obtain a clear understanding of how particular components of the VBC intervention have influenced attitudes and behaviour: could some of the measured effects be artefacts of the survey instrument? Greater use of experimental designs, both in terms of the structuring of the VBC interventions and the ways in which responses are measured, would provide greater clarity and precision, enabling reduced sample sizes. It might be possible to go further and test some kinds of potential interventions in experimental laboratory situations.

6.4.8. Developing a 'Tool Kit' for VBC Evaluation — Particularly Low-Cost Options for Developing Countries

By pooling knowledge on the monitoring and evaluation of VBC initiatives, it should be possible to provide guidance on how to cost-effectively evaluate such initiatives. This would cover sample selection (including a control group), type of survey (panel vs. repeated cross-section, self-completion vs. interviewer-administered etc.), questionnaire framing and content and key evaluation performance indicators. This might be linked to advice on which kinds of VBC measures are suitable under different conditions for achieving certain goals.

Acknowledgements

We gratefully acknowledge the contributions of the paper authors, the discussant and the other workshop participants for sharing their knowledge and contributing to a stimulating debate. The workshop participants were Roger Behrens, South Africa; Peter Bonsall, United Kingdom; Beat Brunner, Switzerland; Marco Diana, Italy; Blake Emery, United States; Pavlina Filemonova Rocakova, Czech Republic; Christine Fox, United States; Regine Gerike, Germany; Thomas Haupt, Germany; Reinhard Hoessinger, Austria; Peter Jones, United Kingdom; Stephen Jones, United States; Ian Ker, Australia; Martin Lee-Gosselin, Canada; Matthias Lenz, France; Michael Meschik, Austria; Linda Nijland, The Netherlands; Barbara Noble, United

Kingdom; Francis Papon, France; David Perroud, Switzerland; Valentina Rappazzo, Italy; Anaïs Rocci, INRETS, France; Karel Schmeidler, Czech Republic; Giorgia Servente, Italy; Karl Sieber, United States [Discussant]; Anu Siren, Denmark; Juliane Stark, Austria; Justyna Szczepanska, Poland; Gordon Taylor, Canada; Maisa Tobias, Brazil; Klaas Van Zyl, South Africa; Ramakrishna Yennamani, United States; and Johanna Zmud, United States.

References

Brög, W., & Ker, I. (2009). Myths, (mis)perceptions and reality in measuring voluntary behavioural changes. In: P. Bonnel, M. Lee-Gosselin, J. Zmud & J.-L. Madre (Eds), *Transport survey methods: Keeping up with a changing world.* Bingley, UK: Emerald.

Brunne, B., & Haefeli, U. (2008). Moving toward sustainability? The consequences of residential relocation for mobility and the built environment — Methodological aspects of our experimental intervention study. Presented at the 8th International Conference on Transport Survey Methods, Annecy, France.

Papon, F., Armoogum, J., & Diana, M. (2008). Specific experimental trials versus large-scale mobility surveys insets to investigate transport-related behavioural issues: The case of the primary utility of travel. Presented at the 8th International Conference on Transport Survey Methods, Annecy, France.

PART III

GLOBAL SOCIAL ISSUES

Chapter 7

Surveying Hard-to-Reach Groups

Benoît Riandey and Martine Quaglia

Abstract

Surveying hard-to-reach groups is difficult but necessary to prevent selection effects and biased sampling. Their diversity makes it difficult to recommend efficient solutions because they bring challenges that are specific to each group. Among these are limited ability in official languages, literacy problems, physical or mental disabilities or the particularities of subgroups such as ethnic, religious and cultural minorities, adolescents and the elderly. Drawing notably on lessons from migration research, this paper reviews the contemporary issues associated with five sets of circumstances that may result in groups being unreached by transport surveys.

7.1. Introduction

Migration research is a field that is particularly sensitive to the problems of reaching all parts of sampling frames for mobile populations, and is thus a useful source of experience for those designing transport surveys.

There are five major sets of circumstances that contribute to chronic states of being "hard to reach":

- people missing from the sampling frames,
- people out of home during hours or days of data collection,
- people refusing contact,
- people refusing to answer the questionnaire, and
- people who cannot answer to the interviewer.

We shall examine each of these in turn.

Transport Survey Methods

ISBN: 978-1-84855-844-1

7.2. People Absent from the Sampling Frame

This coverage error exists in surveys and in censuses in two ways:

a. *The subpopulation is completely absent from the sampling frame*: the population of interest has to be distinguished from the population of the studied population.
b. *Only some units of the population are missing from the sampling frame*: this is a sampling bias that could be corrected by weighting, but only if the researcher can obtain unbiased statistics of the population.

There are many obvious classes of subpopulations missing from general surveys. Some include people living in institutions such as hospitals, rehabilitation centres, chronic care facilities, retirement colonies, universities, foreign workers' hostels or prisons. Other classes include those who are homeless.

Students living on campus are not different from those living in town, but their clustering has implications for travel surveys. However, weighting by sex and age usually corrects the resulting "distortion." At the opposite extreme, the other examples represent subpopulations with a way of life completely different from ordinary households. When measuring activity or mobility, whether or not these subpopulations are included in the survey frame will have a non-trivial impact on the resulting estimates. But they cannot just be excluded given the importance of some topics, for example the decline in physical capability that comes with aging; surveys focussing on handicaps cannot but include hospitals and communities of retired people. If the respondent is not able to answer, the interviewer chooses a proxy: a nurse or a member of the family.

Surveys in prison are difficult to carry out: entering a prison is generally involuntary. But specific surveys may be conducted there. Recently, INSEE[1] has carried out a family survey and a physical disability (HID) survey (Desesquelles, 2005a, 2005b) among prisoners. For the family survey, a large majority of interviewers consisted of persons with prison experience such as prison visitors, former lawyers and researchers working on the prison population. The test of the HID survey showed that disabilities were far more prevalent among inmates than in the total population and that INSEE interviewers would be perfectly suited to the task: it is easier to interview inmates about their disabilities than about their families. Prison is an institution "almost" like any other, with the obvious difference of restriction of movement. The questionnaire of the survey HID prison was as similar as possible to those of the general HID survey, except for the wording of questions on travel; other questions on purchases and vacations were eliminated. In subpopulation surveys, an optimal balance between specificity and comparability is difficult to reach. A weak comparability is often observed in analysis.

In surveys on immigrants' integration, immigrant communities are obviously of relevance. In France, the 58,000 immigrants who live in these communities are the

1. The French office national of statistics.

least integrated, even though these foreign workers have lived in the country for a very long time. Of course, language is a difficulty, discussed later (Scott, Vaessen, Coulibaly, & Verral, 1988). In the INED's MGIS survey,[2] these communities were a stratum of the sample. Data collection was very difficult, despite the employment of interpreters as interviewers in this stratum, and these immigrants will be out of the coverage of the next survey. Similarly, the 15,000 women living in women's shelters are out of the coverage of the ENVEFF survey, the first French survey on violence against women (Jaspard & The ENVEFF Group, 2001), although they are probably the main group of these victims. Surveys in communities are often difficult to sample in a general survey; data collection can be difficult as well, especially in phone surveys such as ENVEFF. For this first reason, in the French HID survey on physical disabilities (Mormiche, 2000), the essential sample of communities was interviewed one year later, as an independent survey, but this was not a perfect approach in this longitudinal survey because (fortunately) some disabled persons recover health and leave the community for a private dwelling.

Most surveys are carried out at the home of the respondents. Homeless and mobile home people cannot be directly included in the sample, such as gypsies, workers living on-site, sailors or barge drivers. Statistics are often lacking for enumerating those subpopulations. Generally, their number is small and can be ignored, but in some surveys, it would be unacceptable to exclude them. In social dwelling or poverty surveys, homeless people cannot be excluded. Specific survey methodologies have been invented by the RTI[3] team (Dennis & Iachan, 1993), and then applied in France (Marpsat & Firdion, 1996) and Europe (Brousse, de la Rochère, & Massé, 2003). In these surveys, statisticians draw an indirect sample of individual services (nights in a shelter or meals, etc.) offered to homeless people. The consumer of this service is then interviewed and his or her questionnaire weighted by inverse of the number of consumptions of this person during a day or a week. The methodological challenge is the construction of the sampling frame by enumerating all services offered to homeless people as well as those used by the respondent for estimating his probability of selection. Indirect sampling is a common practice, but in contrast to this example, it is generally applied top–down, drawing a sample of dwellings for observing individual behaviour, but also sampling persons for observing households (consumption, dwellings, etc.)

Home surveys exclude some people who are difficult to reach because they live in a sensitive situation; for instance, women victims of violence, drug users and undocumented migrants cannot be contacted in the normal way. It is therefore indispensable that associations or institutions in contact with such people act as intermediaries (Quaglia & Vivier, 2006). Building a sampling frame with the lists of associations or institutions is the first step of the methodological process. If not,

2. Institut National d'Etudes Démographiques (France). MGIS: survey on integration of immigrants and their children.
3. Research Triangle Institute.

random sampling has to be abandoned and replaced by a snowball sample with the risks of omitting a part of the population, unknown unequal probabilities and cluster effects.

When surveys are conducted outside, for instance by intercepting people while travelling, statisticians must be attentive to the probabilities of inclusion, notably probabilities proportional to the duration of the opportunity to intercept, such as the duration of a walking or a bus trip, or the length of waiting time at a transit stop or terminus. Inexperienced statisticians must be aware of classical biases that have to be avoided in these surveys. For example, delayed buses are more crowded than on-time buses, and therefore drawing the next bus at a bus stop is a biased sampling method.

Of concern to transport surveys, emigrants are a subpopulation always lacking in samples. Generally, they are out of the design, but not always. For instance, they cannot be excluded from international transport surveys or household income surveys in countries with a large emigrant worker population. In these countries, household revenues cannot be estimated without the estimation of financial and material transfers made by migrants related to the household (Fergany, 1987). Building activities and transformations in the villages cannot be understood without taking into account the role of migrants' savings. The commercial transfers use the familial network spread out on each side of the border. No surveys on emigration can be done without samples of migrants.

In such research, indirect sampling has been carried out very efficiently, mainly in islands. The historical model of this methodology was developed experimentally in Barbados by a team of demographers, mainly Kenneth Hill (1984) and Jorge Somoza (1984). They used the respondent's familial network with a doubly indirect method: it was not a sample of immigrants but a sample of residents. Then as a proxy, the respondent described the migrations of his children or siblings living abroad and not his own migrations. Because everyone knows the number of his own children or siblings, the selection probabilities of an emigrant can be estimated. It is the number of familial witnesses of his emigration. Of course, statisticians have to correct a bias: all the siblings of an emigrant may have emigrated as well and his parents may be dead; no familial witness is then resident in the country. With fertility and mortality tables and limited hypotheses, demographers can estimate the bias, but with a small underestimation of emigration. This is the methodology used by Fergany (1985) for estimating the amount of labour emigration from Egypt.

Some current studies of international migration try to connect surveys carried on households living in their country and surveys carried out abroad among households of emigrated members of their family. For example, Hill's survey on Mexican emigration into the United States (Hill & Wong, 2005) or the NIDI's survey on Moroccan migration into the Netherlands (Esveldt, de Valk, Henkens, van Solinge, & Idema, 2000), and more recently the Mexican Migration Project (Massey, Fisher, & Capoferro, 2006) in Mexico and the United States or the MAFE project in Senegal, France, Italy and Spain. These push-and-pull surveys are carried in a southern country where emigrants are described by their familial proxies living in the country, and then interviewed in their European country of immigration (Eurostat, 2000).

The quality of indirect estimation is conditioned by the accuracy of the number of consumed services or familial witnesses. On the other hand, the social network is not relevant because the friendly relationship is not necessarily symmetrical and we cannot enumerate the network of a friend. In the same way, we can easily use services specifically offered to homeless people (beds in shelters or meals in a soup kitchen), but it would be very difficult to organize a random selection of individuals in services where homeless people are mixed with the general population, as with some medical services. A similar problem occurs with specific services for homeless where they are not enumerated, such as in day centres. Selection probabilities could not be estimated for these services.

In face-to-face surveys, deceased people are out of the frame. But this does not completely solve the challenges of demographic or epidemiologic surveys that concern illness and death. The main challenges of an epidemiologist are his or her right of access to the civil register and knowing the patient's vital status and cause of death. On the other hand, surveys on end of life use civil registers as sampling frame. The respondent is the physician that certified the death and observed its conditions and the final medical care (Bilsen, Cohen, & Deliens, 2007). Such a survey needs the agreement of the medical profession and a guarantee of statistical anonymity, particularly in the numerous countries where all practices of euthanasia are prohibited.

In the context of these subpopulations, there is a fundamental methodological difference between retrospective surveys and panels with regard to some sampling biases. The death or emigration bias is a common difficulty of sampling in retrospective surveys. We may add exits from inclusion criteria: an example in surveys collected among foreigners is a change of nationality. Consider the MGIS survey on immigrant integration (Tribalat, Riandey, & Simon, 1995). In this retrospective survey, immigrants were asked to tell their whole life history in France. Each life history is complete, but statistics on the immigrant population living in France 10 years ago are biased because data concerning dead or returned immigrants at the time of data collection are missing. At the opposite extreme, bias caused by exit from the frame has been avoided by the INED's choice to survey immigrants and not foreigners. Deceased and returned immigrants cannot be selected, although they were present during the observation period. So the retrospective estimation concerns only the population still present at the time of data collection. On the other hand, in panels,[4] they are present and available in the sample before their attrition. The main problem of panels is determining the cause of attrition. In an epidemiological cohort, the research will succeed only if lost, dead or emigrated respondents can be distinguished. If this is not possible, data are rightly censored, but the selection bias inevitable in retrospective surveys will have been avoided in the cohort.

4. For demographers, retrospective surveys and panels are two kinds of longitudinal surveys. In the economics domain, retrospective surveys are generally not used because they do not lend themselves to econometric measurement tools, so economic statisticians tend not to distinguish between panels and longitudinal surveys.

A frequent source of sampling bias is not updating the sampling frame, and particularly the addresses, but a more fundamental source is the delay before including recent cohorts into the sampling frame. In the retrospective MGIS survey, the sample was drawn from the French census before introduction of efficient optical data capture methods. As a consequence, two and a half years were needed to produce the sampling frame. Therefore, the population surveyed was not the immigrant population at the date of the survey, but the immigrant population at the date of the census and still resident at the date of the survey. However, to the extent that the object of study was the fulfilment of the slow process of integration, the new waves of immigration would not have added much to the findings. It would have been different had the reception of immigrants been central to the study. Of course, publications must clearly specify the population of interest.

Tourism surveys raise related questions about sampling bias. They may have two different aims. If it is a social one, resident households will be interviewed about their holidays spent in the country or abroad. If it is economic, the topic would concern the entirety of tourist consumption in the country, and particularly foreigners' consumption during their holidays in the country. However, foreign tourists cannot be interviewed at home, but only during their time abroad or while travelling. Cordon surveys were often conducted at custom stops at borders. In the European Union, the Schengen Treaty has made this methodology impossible for use at land borders. An indirect sampling of tourist services offers an alternative methodology (Deville & Maumy-Bertrand, 2006). This methodology resembles the one described for surveys of the homeless, but with a difference: the sampling units are not the consumptions of necessity, but more the consumptions of affluence. Interviews are done in hotels, bakeries, camps, tourist sites, airports and so forth.

Some very rare groups — for instance very wealthy people — cannot be efficiently reached through usual sampling frames in which these groups are present but not specified. However, it may be efficient to increase the size of this subsample because this subpopulation is very mobile (and, for example, probably pollutes a lot relative to the per capita criteria of the Kyoto Protocol). As people in this group are frequently away, they are very likely to be absent from the classical frames, and there is no general method for resolving this. In 2006, the top 1% of the US population held 17% of the national revenue (Piketty, 2005). What is their contribution to the CO_2 emission in the world? Without limiting this question to one country, statisticians may be expected to answer this question, but random surveys are probably not the best approach. Access panels and snowball sampling are probably the only methods for improving this knowledge.

7.3. People Out of the Home During Hour or Day of Data Collection

In a general sense, the temporary absence of an expected respondent generates a sensitive effect of selection, strongly correlated with transport behaviour. Their absence can be due to travel during this specific day or for several days. The longer this absence, most likely, the greater will be the distance travelled. Estimates by gender, age, occupational status, mobility and distance from place of employment would be biased.

Some strategies can minimize the bias induced by an unsuitable hour or day of the interviewer's visit: extending the data collection period, proposing a multi-mode data collection, surveying a proxy and so forth. The benefits or defaults of these methods may be compared in different cases.

Extending the data collection period and the number of attempts of contact is an expensive caution. It is irrelevant to some topics (polls, media audience measurement, time use and transport) because the sample design includes the day of collection. No survey can ask detailed questions about broadcast programmes heard a week ago. There are two solutions, possibly used together: considering the absent person as a non-respondent and surveying in a later sample with a non-respondent questionnaire concerning the circumstances of the missed response. This would improve the knowledge of non-response factors and perhaps enable them to be modelled, thus averting a bias from factors that can intervene every day. In a transport survey, the inability to respond on a specific day can be related to the phenomena being measured (see the example of wealthy people above). Delaying the date of interview does not allow the respondent to change the observed day.[5] A delayed interview is disturbed by memory errors (omission of trips, imprecise diaries or activity purposes). The modelling of the information forgotten because the interview is delayed could correct a large part of the measurement error.

Surveying a proxy is a second solution, but some answers will be lacking, and it is a new cause of error. However, these partial data may be improved by an imputation model.

A respondent missing an interview can also be invited to complete a mailed or an Internet-based self-administered questionnaire. Of course, heterogeneity caused by multi-modal data collection has to be measured and corrected by relevant experimentation (Goulia, 2001; de Leeuw, 2005; Cobben, Janssen, van Berkel, & van den Brakel, 2007; Jäckle & Roberts, 2007; Dauplait, de Clédat, & Vichnevskaia, 2008).

These three solutions face similar difficulties: the overlapping effects of non-response bias and of the alternative mode of data collection cannot be directly estimated. The cause and effect of initial non-response are mixed. An experimental design is needed to separate the two effects. At the outset of a study, random subsamples could be drawn. Each subsample would be completely surveyed with one of the alternative modes of data collection to measure its specific effects. This theoretical principle would thus be submitted to a realistic test. But how much does the correction of this potential bias cost? The frequency of the non-response and the

5. In transport surveys, while the assignment of a specific date or dates may be part of the sample design, this may be relaxed for surveys of urban travel, in which an assigned randomized day or days of the week (e.g. a contiguous Friday and Saturday) may be re-assigned to a later week.

permanence of the survey must be weighed before making a decision. In a continuous labour force survey or a continuous transport survey, it does matter.

At the same time, it is important to keep in mind that the causes of absence or the characteristics of absent people are often very specific: holidays, students rarely coming back to their parents' home, employment far from the familial dwelling, trade travellers moving through the region or abroad and so forth.

7.4. People Impossible to Contact or Avoiding the Contact

Many buildings are equipped with technical means (digicodes or recorders) to prevent strangers from entering the building. Carrying out surveys is therefore becoming more and more difficult, but the induced bias is not as problematic as that discussed in the previous section. This equipment, which used to be found only in certain districts and among certain social groups, now concerns a large proportion of the population. Nevertheless, its presence, and the selection bias it introduces, are usually independent of the theme of the survey. An efficient weighting could correct it.

In an area sample, a cluster of interviews may be lost, but even one successful interview may open all the doors, as important official surveys such as labour force surveys or the census frequently have procedures for this situation.

A common reason making contact impossible is the change in home address of the person. Residential mobility is very selective, and the non-response bias it introduces is evidently of major importance to surveys of daily mobility.

If dwellings are the sampling units, the new inhabitants at the addresses will measure the migration to the arrival place, not from the departure one. There is no bias in the description of internal migration, but information is lacking on international emigration.

In longitudinal surveys, if the sampling unit is a person and not a dwelling, all possible efforts must be made to discover the new address of any respondents expected to move, to prevent a broken relationship with them and to determine whether they are still within the territory covered by the survey.

7.4.1. Prevention of Attrition in Panels

How can we avoid panel attrition? Can we trust weighting or imputation for correcting its bias? Many European countries have population registers, but not France. In Northern Europe, their quality enables the recovery of new addresses for people who moved, preventing attrition. The access to administrative registers by official statisticians varies between countries. In some countries their access is quite free. French law is stricter, and the phone register used to be the only available file for tracking mobile persons. It is more efficient for males than for females and, particularly, for divorced women who may change at the same time their address and their surname. However, the proliferation of mobile phones limits the efficiency of such registers. Therefore, prevention of attrition must be done before the loss of addresses. Some regular contact with the panel respondents is an elementary precaution. A person about to move would give his future address to the post office, which will then be able to redirect mail. Previous neighbours may keep contact with them But these solutions do not eliminate the need for statisticians to be careful before data collection. The best way is the collection of the names and addresses of intermediary contact people during the first interview. Parents generally keep contact with their children and are less mobile than younger people, but the agreement of several contact persons is more prudent.

If interviewers cannot use administrative files to get back to the contact, they will try to contact a piece of the social network of the mobile person: family, job, school, neighbourhood, leisure and associations, among others. This strategy was optimized in Philippe Collomb's survey (1979) of a sample of emigrants from the Lauragais, a declining rural zone. Nineteen years after their emigration, Collomb succeeded in "surveying" 97% of this sample (including persons living and dead). In the departure region, he could recover two-thirds of the sample by contacting the social network still living in the region. For the last third, he invented an original strategy by which he was able to track 90% of the remaining sample because he knew that emigrants from a rural village keep contact among themselves. The surveyed people transmitted the addresses of almost all the lost sample. People in southern France are very faithful to their village. This qualitative insight led to the quantitative success of this survey. Other key factors were Collomb's willpower and the perseverance of four or five exceptional interviewers. Such a dense social network cannot always be found in towns, but is often present among migrants. However, it has to be successfully contacted.

7.5. People Who Do Not Answer or Refuse to Answer

Interviewers often cannot contact certain people because they don't wish to be reached. This failure to contact is a hidden refusal. For this reason, this paper concerns refusals too. The same strategy is often used for no contact and for refusal.

At first we have to distinguish passive and active refusals of answering. Passive non-answering is observed in self-administered surveys, mainly mail or CAWI data collection.

Self-administered surveys without interviewers lose the benefit of an interpersonal incitation. Interviewed people have five reasons for responding: obligation and fear of penalization, interest in the topic, common benefit, kindness to the interviewer or payment. Kindness to the interviewer and the pleasure of the encounter are often the main reasons of acceptance. Their absence is an important reason of non-response to mail surveys. A second reason is specific to self-administered surveys without interviewers. It is the downside of giving a free choice of when to respond. This freedom facilitates answering by absent respondents, but the answers can be always delayed without any refusal. This is why, when an interviewer hands a paper self-administered questionnaire over to a respondent, it would be better to wait at the respondent's home for the questionnaire to be filled up, or to come back later to collect the questionnaire, than to hope for its return through the mail. Mail, phone or Internet follow-up is the best antidote against non-response in a survey without interviewers. During a phone follow-up, it is wiser to ask to retrieve the questionnaire responses immediately, or at a later appointment, than to wait for a return by mail. The multi-modal bias is certainly less problematic than unit non-response. When using a self-administered method, an attractive questionnaire is the best caution against non-response. CAWI surveys often benefit from this attractiveness as well as from the surprise effect: the CAWI can be programmed so that questions cannot be read before responding to the previous one. Similarly, it is possible to prevent a response to a previous question from being changed after reading the next question (for instance in the case of a closed question following an open one of the same topic). This appears to present a significant advantage of CAWI or CASI over paper survey. Nevertheless, researchers have to be careful with using this approach as experience has shown that it can also be a disadvantage. In the French INED Survey EFE among establishments and their employees (Dauplait et al., 2008), the establishment's paper questionnaire proved to be more efficient than the Internet in posing a complex question for which the answer could be facilitated by previewing the following question.

These non-response factors are nonexistent in the presence of an interviewer. On the other hand, interviewers must deal with a potential refusal because a respondent may be uncomfortable not knowing what the interview will cover, or is unavailable. Phone appointments are less expensive than face-to-face interviews and limit the chances of refusal. Mailing an introductory letter is a good practice. A mail provides authenticity to the survey; such authenticity is lacking in most phone surveys. In a political poll, how can the recipient be sure that it is an effective poll and not an invasive call? The European Directive (8th article) of October 24, 1995, concerning privacy protection stipulates that collection of political opinion requires a clear agreement (in French law a signed agreement). How can we respect this prescription without an introductory letter? A well-written letter improves the response rate by 10–15% (ACSF group, 1992).

Refusals can be classified into three categories: systematic, occasional and thematic. The first category causes little bias because they are non-informative and effectively corrected by weighting. As an exception, right-wing French voters refuse more surveys than left-wing voters. This systematic refusal is only informative in political surveys. In the same way, in surveys on revenues or capital, very frequent refusals from people who are self-employed or who are not wage earners are not easily corrected by weighting.

Occasional refusals could be prevented by an appointment at a better hour or by a proposal of another mode of answering, for instance a paper or CAWI self-reported interview. A small team of super-interviewers for resolving these non-responses has been an efficient method, but this may not be enough in many fields where public spiritedness has decreased.

Thematic refusals cause the most difficult biases because they are very informative. Introductory information and the contact of a very qualified and well-paid interviewer are the best formula. Preserving confidentiality and privacy is an ethical obligation and a condition of success for a survey, particularly if it seems likely to encounter thematic refusal. There are very efficient, inexpensive techniques, such as CASI (Rogers, Miller, Forsyth, Smith, & Turner, 1996; Turner et al., 1998). Paper self-administered surveys generally penalize illiterate and semiliterate respondents. Audio CASI data collection is an excellent solution, with a good formula for response being: "Press the green button to say 'Yes'; press the red button to say 'No.'"

Giving confidence to the respondent is more difficult than protecting his or her privacy. Random questions clearly illustrate this challenge in surveys of sensitive topics such as drug use. The respondent draws a question at random from two possible questions without saying the result of this sampling. For instance, the respondent chooses blindly one card from a pack of cards with the text of one of two possible questions. One question is: "Have you used drugs during the past year?" (response yes or no). The other one is: "Choose one person of your family without telling me your choice. Is this person born before the fifteenth day of a month?" (response yes or no). The global result of the question, and the knowledge of the probability of drawing each question and of answering "Yes" or "No" at the second question, give the frequency of the answer "Yes" to the first question without ever knowing what question anybody has been asked. So the rate of drug users is estimated without the interviewer ever hearing any respondent tell him he is a drug user. This methodology is so inspired that many respondents think it to be a subterfuge and don't answer with the sincerity allowed by the strict confidentiality of this methodology.

Confidence is given to a person more than to a technique. Therefore, the interviewers' motivation is the best tool for fighting non-response and is fed by their supervisor's energy in training the interviewers. Delegating the training rather than travelling to all the field offices to meet the interviewers at the time they start their training creates very different dynamics. Supervising each interviewer's first questionnaires, correcting them without delay and encouraging them improve the team's motivation. It is much easier in a phone survey because interviewers work generally in one or two rooms. As an example, two institutes collected the 2000 ACSF questionnaires on a very sensitive topic: sexual behaviour. Training the interviewers was not a difficult task for the research team. Being present each day during the five months of data collection in order to encourage the interviewers was much more demanding. However, as a consequence, the young interviewers' motivation was very strong in this survey about AIDS prevention at a time when no medical treatment was yet available, and they called the respondents up to 12 times, if necessary even the following month, rather than abandoning the phone numbers that had not been reached. Moreover, the researchers' checks revealed that many gays had been reached only during the last calls. In a similar way, INED and INSEE researchers kept almost daily contact with interviewers during their national surveys of homeless people and drug users.

In order to minimize non-responses, the best interviewers of each region are selected, and they try to contact each respondent who has refused, who was never reached or for whom a forwarding address has not been obtained. A good team resolves 50% of non-responses. The interviewer's personality is the key, as shown by this veteran interviewer's experience recalled at a survey institute: when he arrived at an upstairs apartment, he looked like a very tired and reliable person and, as soon as the door opened, the respondent offered him a chair to take some rest. The experience of a female interviewer is another example: after one of her rare refusals, she began to inquire immediately as to the cause of the refusal, making sure she was not becoming too old for that kind of work. Assembling such a team is a good solution to the problem of hard-to-reach respondents. However, its cost is high, and the interviewers' performance is correlated with their wage rate, which has been declining for years.

Some husbands or parents refuse to let their wife or adolescent child answer the survey. In the ACSF survey, the first person to answer the phone was not necessarily the survey respondent. The respondent was selected according to the anniversary method: starting from the date of the contact one chooses the household member whose anniversary would be first celebrated. A new call might benefit from a kinder respondent and draw an acceptance from a household member. But the French privacy authority, the CNIL, prevents this solution, limiting the liberty of the unknown respondent to answer the questionnaire. Ten years later, this limitation has been avoided in the ENVEFF survey and the response rate has been improved, thanks to this change (Jaspard & The ENVEFF Group, 2001).

In most countries, interviewers are likely not allowed to interview minor children without their parents' permission. Then the adolescent has to be interviewed with his or her parent's supervision. Audio CASI is the most relevant way of collection. It is sometimes easier to collect data at school, even if the school head and the parents' associations have to be convinced (Lagrange & Lhomond, 1998). For practical purposes, there is a choice between two methods: interviews and a simultaneous paper self-report of the whole class (Baccaïni & Gani, 2002). This latter method of cluster collection is very efficient, but it did not work well for a survey on adolescents' early sexual experiences, carried out in high schools. Everyone can imagine the classroom atmosphere during such a collection, but overall a self-reported questionnaire could not be contemplated. This sensitive survey had to resolve a lot of ethical challenges, particularly that the interview could not initiate the respondents into new knowledge of sexual practices. So the questionnaire could not be put in the respondent's hands, and a face-to-face interview was chosen.

Incentives (in the form of gifts or payments) could be another way to induce the non-respondent. Peter Stopher (1997) greatly improved the transport survey in Boston in this way (from 28% to 40%). It was a large success and demanded a lengthy investment of effort by the respondent, so offering nothing in return seemed inappropriate. This practice is more and more usual from private institutes, notably with access panels. Of course, interest in the gift could be a source of selection bias. In France, it has been assumed conventionally that official surveys, whether compulsory or not, can rely on citizenship. However, gifts (lottery tickets) have been given in long consumption surveys. In surveys of homeless people or drug users, researchers have chosen not to use the gift as an incentive, but as a real gift/compensation of the time spent answering the questionnaire. It was made clear during the training of interviewers that the gift should never be mentioned before the end of the interview. Only a few cases of people having heard about the gift and asking to be interviewed were reported from the field. The interviewers had also been trained to explain that people could not volunteer to be interviewed (unless they also happen to be sampled), and no real problem was reported. Different kinds of gifts were presented; phone cards or radios were very much appreciated by homeless people as a social link. Some people refused the gifts, explaining that they had made the choice to participate because the aim of the survey was to report on their life conditions. Therefore, they did not need any compensation for the time spent answering the questionnaire. Where incentives are used, they are now most often in the form of gift vouchers, allowing the respondent to keep the final choice.

7.6. People Who Cannot Answer

Researchers should not forget people who cannot answer because of a health or language handicap. Interviewing a proxy is the most common solution (Scott et al., 1988). Often children of immigrant households speak the local language better than their parents and are therefore their parents' interpreters. They often enjoy this role, but the translation of the questionnaire by children is not the most professional solution. Furthermore, the child may be the real respondent instead of their parent because the child can easily use the computer in a CASI survey or answer a pencil and paper questionnaire. This does not respect the statistical confidentiality when the questionnaire data are confidential or sensitive. It is always a very approximate solution, and it is certainly not a correct methodology for opinion surveys.

In some surveys, interpreters are employed. The methodology of multilingual surveys is based on a double translation with two translators. Going from a language and back to it via the translations of another language introduces impreciseness and conceptual errors. Difficulties in international surveys are known, particularly their possible ethnocentrism. The methodological challenges of translation, interpretation and cultural context of international and multicultural surveys have been very well studied by Lord Kendall's team for the World Fertility Survey, for which documentation is available (Scott et al., 1988). More recently, the CSDI[6] has been leading a workshop on this topic. See its 2005 session (Edwards, 2005; Yuling, 2005; Harkness & Joye, 2005) and its 2008 conference.

Everyone can imagine the cost of translation and its limited effectiveness. In a survey of new immigrants to France (Bèque, 2008), interpreters were employed as an interviewer team. They administrated the questionnaire in 13 foreign languages, but only 36% of the necessary translation was done. Of the interviewed people, 3.5% did not speak any of the 14 languages of the survey and could not be interviewed.

6. International workshop on Comparative Survey Design and Implementation (http://www.gesis.org/en/research/eccs/csdi).

Nevertheless, the survey was facilitated by the administrative knowledge of the interviewed people's origin and an indication of their level of competency in French at the time that their residency permit was issued. Of course, illegal immigrants are beyond coverage because the sample frame is the annual file of issued permits. The large regularizations of the status of immigrants, as has recently been done in Italy and Spain, are rare opportunities to obtain statistics on illegal immigrants. Nevertheless, some categories of illegal immigrants are not allowed to regularize their situation, and we can suspect systematic measurement errors.

The MGIS survey had to solve the same linguistic challenge, but with less diversity because the sample design was limited to the most common immigrants' nationalities and done according to the time spent in France (those who had stayed longer). If the interviewer could not be fully understood in French, he called on an interpreter interviewer. The regional and urban concentration of immigrants has made this organization easier.

This expensive team could not work across the whole country because Inter Service Migrants, the interviewers' association, could employ them only in the main immigration regions. Therefore, the rare immigrants who live in rural areas could not be interviewed by interpreters. Thus, this solution was expensive and incomplete. Evidently, mobile telephony now presents a solution: the interpreter translates at home, possibly after making an appointment. This solution needs to be optimized, particularly to improve the confidentiality of the interview. We suppose these experiments have not yet been done in France. The coverage survey carried out by INED among non-French-speaking homeless people showed the methodological and economical limits of employing and training interpreters as interviewers. The practices of these two jobs can somehow be very similar, and interpreters could not always distinguish the difference. They sometimes modified the formulation hoping to help the respondent to understand the questions, not realizing, because they were not aware of the context of homeless life conditions in France, how this change could also modify the whole meaning of the question. Neutrality was also difficult to maintain when facing a compatriot homeless person in distress. The less difficult experiments of the MGIS or the Bèque surveys are much positive. These three surveys point out the importance of the training of interpreters for becoming qualified interviewers.

Of course, an audio CASI data collection would be an efficient practice, the best one when facing many languages. Audio CATI and CAWI surveys help the respondent to answer in his or her own language without having to ask for an interpreter to assist. The cost depends on several factors, such as the number of foreign languages. This is a key challenge in immigrant surveys but less important than in general surveys because the local language is rarely unknown. In this topic, a Canadian experiment is probably the best example, first because of the English–French bilingual situation, but also because they have to carry specific surveys among Indian or Inuit populations (Rochette, 2006). A recent health survey has been entirely centred on the aboriginal community. This was a realistic decision, not only because of the linguistic challenges of a survey on the sensitive topic of health, but also because of the difficulties of access in the far north of Canada.

Ill and handicapped people present multiple difficulties in surveys (Cowham, Dye, & Crowther, 2008). They often live in institutions, have been absent from home for a long time or are unavailable at home. The bias is generally small because not too many people are concerned. It would be different when specifically surveying elderly or handicapped persons. These surveys require data collection in institutions and often a two-phase sampling. For a long time, the Canadian long census questionnaire has screened handicapped people, but the screening question has been so imprecise that a post-screening of the general population had to be done. Developing the Canadian experiment, the French two-phase survey included a 13-question screening questionnaire in addition to the census bulletin in a sample of districts (Ravaud, Letourmy, & Ville, 2002). A two-phase survey is always a large and expensive collection. The best solution is not to pay for the first phase, for instance by inserting it into the census!

Despite its label, the French Vespa survey is neither a transport nor a road security survey but a survey of AIDS patients. There is no French AIDS register, and there never will be. The sampling and data collection had to be carried out in concerned hospitals. Hospital departments were drawn, and sample size determined, according to the hospital patient file. The interviewers stayed two weeks in the department, and the patients were selected only when they came to the hospital (Peretti-Watel et al., 2005). Interviewers encountered the difficulty of screening AIDS patients, before having to face the difficulty of interviewing weak people in a moment of little availability. As the hospital physician chose the patients to be interviewed, depressive or very ill patients were less selected. Nevertheless, to avoid conclusions such as "medical surveys are always successful but with those who are ill," some biological measures extracted from medical files, such as the DC4 rate, were used for weighing data. The strict control of medical confidentiality requires a double team for dissociating the two operations. Surveying sick people is always a delicate task.

7.7. Conclusion

Many survey difficulties have been detailed above. Our main conclusion is the recall of a fundamental principle: data collection is an experimental task. Qualitative experimentation before testing is essential, but it requires time. Some knowledge may be transferred to transport surveys although they too vary in their objectives, their target populations and the territories involved. Our experience is mainly with face-to-face surveys in developed countries. In developing countries, methodologies are necessarily different, particularly in how space is represented (Godard, Diaz-Olvera, Dieng, & Kane, 2001). The ISCTSC conferences are quite attentive to the specificity of surveys in developing countries.

Three main recommendations may be proposed:

- To further experiment with multi-modal and Internet modes of data collection. It seems to be the best way to capture data from highly mobile persons who are often

hard to reach. Attention must be paid to the unequal selection caused by Internet data collection. A reference source is needed for weighting. Attention must also be paid to modal effect errors. An experimental design has to be carried out to separate these two biases.
- Immigrants must be presented with an interview spoken in an adequate language. Skilled immigrants are often very mobile. Unskilled ones may not speak the survey language fluently. Methodology has to be improved, for instance by phone interpreting.
- Developed countries are paying more attention to offering a normal way of life to physically handicapped persons. Mobility is a condition of success of this social priority. The transport system must accommodate this, based on first-hand knowledge of handicapped persons' difficulties. There is, therefore, a real demand for transport surveys to be conducted on this topic.

Acknowledgements

The authors are grateful to Anne-Marie Dussaix and Martin Lee-Gosselin for their suggestions. This version of the paper was adapted by Martin Lee-Gosselin from a Resource Paper presented at the Annecy Conference, with reference to an original draft in French, and with the editing assistance of Christopher Parker.

References

ACSF Group. (1992). Analysis of sexual behaviour in France (ACSF): What kind of advance letter increases the acceptance rate in a telephone survey behaviour. *BMS*, 35.

Baccaïni, B., & Gani, L. (2002). La population en questions. *Cahier de l'INED*. 146. Paris: INED-PUF.

Bèque, M. (2008). Enquête longitudinale sur le parcours et le profil des migrants récemment arrivés ou régularisés en France. In: Ph. Guilbert, D. Haziza, A. Ruiz-Gazen, & Y. Tille (Eds.), *Enquêtes et sondages. Cinquième colloque francophone sur les sondages. Marseille 2007*. Paris: Dunod.

Bilsen, J., Cohen, J., & Deliens, L. (2007). End of life in Europe: An overview of medical practices. *Population and Societies*. 430. Paris: INED.

Brousse, C., de la Rochère, B., & Massé, E. (2003). The INSEE survey on users of shelters and soup kitchens: An original methodology for studying the homeless. *Courrier des statistiques*, English series, 9. Available at: www.insee.fr/publications

Cobben, F., Janssen, B., van Berkel, K., & van den Brakel, J. (2007). Statistical inference in a mixed-mode data collection setting. *ISI Session*, *56*.

Collomb, P. (1979). Les émigrants de l'ouest audois dix-neuf ans après. *Population*, *1*.

Cowham, M., Dye, J., & Crowther, B. (2008). Surveys for disabled persons. ISCTSC Conference, Annecy.

Dauplait, C., de Clédat, B., & Vichnevskaia, T. (2008). Enquête famille et employeurs: Retour méthodologique sur la collecte (papier et Internet). In: Ph. Guilbert, D. Haziza,

A. Ruiz-Gazen, & Y. Tille (Eds.), *Enquêtes et sondages. Cinquième colloque francophone sur les sondages. Marseille 2007*. Paris: Dunod.

Dennis, M. L., & Iachan, R. (1993). A multiple frame approach to sampling the homeless and transient population. *Journal of Official Statistics, 4*.

Desesquelles, A. (2005a). Handicaps, disabilities, and dependency in French prisons. Lessons of a special survey. *Le Courrier des statistiques*. English series 11. Available at: www.insee.fr/publications

Desesquelles, A. (2005b). Disability in French prison: How does the situation differ from that of the general population. *Population English series, 1*.

Deville, J. C., & Maumy-Bertrand, M. (2006). Extension of the indirect sampling method and its application to tourism. *Survey Methodology, 2*.

Edwards, B. (2005). *Interpreting in surveys*. CSDI 2005 Workshop session, Madrid.

Esveldt, I., de Valk, H., Henkens, C. J. I. M., van Solinge, H., & Idema, M. (2000). Push and pull factors of international migration: country report: The Netherlands. Eurostat XVI, NIDI.

Eurostat. (2000). Push and pull factors of international migration: A comparative report. *Eurostat, XXI*.

Fergany, N. (1987). Une enquête nationale égyptienne sur la migration internationale de travail (1974–1985). *Revue Européenne des Migrations Internationales, 1/2*.

Fergany, N. (1985). *Egyptian survey on labour emigration*. Paris: National Population Council.

Godard, X., Diaz-Olvera, L., Dieng, A., & Kane, C. (2001). *Guide méthodologique d'enquête-ménages de mobilité dans les villes en développement*. Paris: SITRASS.

Goulia, K. (2001). *Surveys using multiple approaches*. Resource paper. Workshop on multi-method and multi-instrument surveys. ISCTSC conference.

Harkness, J. & Joye, D. (2005). Oral translation in telephone surveys. CSDI 2005 Workshop session, Madrid.

Hill, K. (1984). A note on the use of information of residence of surviving siblings to estimate emigration. *IUSSP Newsletter, 20*.

Hill, K., & Wong, R. (2005). Mexico-US migration: Views from both sides of the border. *Population and Development Review, 1*.

Jäckle, A., & Roberts, C. (2007). Assessing the effect of data collection mode on measurement. *ISI Session, 56*.

Jaspard, M., & The ENVEFF Group. (2001). Violence against women. The first French national survey. *Population and Societies,* 364, Available at: www.ined.fr

Lagrange, H., & Lhomond, B. (dir.) (1998). L'entrée dans la sexualité. Le comportement des jeunes dans le contexte sida. Paris: La Découverte.

de Leeuw, E. (2005). To mix or not to mix data collection modes in survey. *Journal of Official Statistics, 2*.

Massey, D. S., Fisher, M. J., & Capoferro, C. (2006). International migration and gender in Latin America: A comparative analysis. *International Migration, 44*(5).

Marpsat, M., & Firdion, J.M. (1996). Devenir sans domicile: Ni hasard, ni fatalité. *Population et Sociétés*, n° 313.

Mormiche, P. (2000). The INSEE survey on handicaps, disabilities and dependency: Aims and organization. *Courrier des Statistiques,* English Series, 6. Available at: www.insee.fr: publications

Peretti-Watel, P., Riandey, B., Dray-Spira, R., Bouhnik, A.-D., Sitta, R., & Obadia, Y. (2005). Surveying the HIV-positive population in France. The VESPA 2003 survey. *Population, English Edition, 4*.

Piketty, T. (2005). Top income shares in the long run: An overview. *Journal of the European Economic Association*, *3*(2–3).

Quaglia, M., & Vivier, G. (2006). Enquêter des populations difficiles à joindre: Principes méthodologiques et adaptations pratiques. In: P. Lavallée & L. P. Rivest (Eds), *Méthodes d'enquêtes et sondages. Pratiques européenne et nord-américaine*. Paris: Dunod, Collection Sciences sup.

Ravaud, J. F., Letourmy, A., & Ville, I. (2002). Identifying the population with disability: The approach of an INSEE survey on daily life and health. *Population, English Edition*, *3*.

Rochette, L. (2006). Réalisation d'une enquête de santé dans le nord québécois: Défis méthodologiques. In: P. Lavallée & L. P. Rivest (Eds), *Méthodes d'enquêtes et sondages. Pratiques européenne et nord-américaine*. Paris: Dunod, Collection Sciences sup.

Rogers, S. M., Miller, H. G., Forsyth, B. H., Smith, T. K., & Turner, C. F. (1996). Audio-CASI: The impact of operational characteristics on data quality. In: *Joint proceedings of AAPOR/ASA (Survey method Research Section)*. Alexandria, VA.

Scott, C., Vaessen, M., Coulibaly, S., & Verral, J. (1988). Verbatim questionnaires versus field translation or schedules: An experimental study. *International Statistical Review*, *56*(3).

Somoza, J. (1984). Barbados experimental migration survey: Method and results. *IUSSP Newsletter*, *20*.

Stopher, P. R. (1997). A review of separate and joint strategies of data on revealed and stated choices. In: P. Bonnel, R. Chapleau, M. Lee Gosselin & C. Raux (Eds), *Les enquêtes de déplacements urbains: Mesurer le présent, simuler le futur* (*Urban travel survey methods*: *Measuring the present*, *simulating the future*). Programme Rhône-Alpes, Recherches en Sciences Humaines. Lyon: Centre Jacques Cartier.

Tribalat, M., Riandey, B., & Simon, P. (1995). *De l'immigration à l'assimilation. Enquête sur les populations d'origine étrangère en France*. Paris: La Découverte.

Turner, C. F., Ku, L., Rogers, M., Lindberg, L. D., Pleck, J. H., & Sonenstein, F. L. (1998 May). Adolescent sexual behaviour, drug use and violence: Increased reporting with computer survey technology. *Science*, *280*.

Yuling, P. (2005). The use of interpreters in the conduct of households surveys. *CSDI 2005 Workshop session*, Madrid.

Chapter 8

The Challenges of Surveying 'Hard to Reach' Groups: Synthesis of a Workshop

Roger Behrens, Mark Freedman and Nancy McGuckin

8.1. Introduction

The basic premise of a probability sample is that the sample will represent the universe or target population. To achieve that goal, pertinent information about each sample unit should be obtained. When some groups are not included, representation of the universe within the sample is brought into question. So, one goal of survey research is to maximise the probability of obtaining information about each sample unit.

In every survey, there are groups that may be underrepresented because they are difficult to include. This chapter discusses challenges facing the inclusion of so-called 'hard to reach' groups in travel surveys, and explores methods that might be applied by travel survey practitioners to ensure representation of these groups. The content of the chapter is based upon deliberations within a workshop session on this topic, at the 8th International Conference on Transport Survey Methods.

The chapter is divided into five sections. The following section identifies examples of 'hard to reach' groups, and speculates on how the relative extent and composition of these groups could change with time. The third section defines the problem, by identifying the bias that can result from inadequate 'hard to reach' group participation in surveys. The fourth section discusses measures that can be taken to ensure adequate participation of these groups, with particular reference to improving the coverage of samples, reducing non-response and improving survey instrument cognition. The final section concludes with a discussion on the transferability of solutions to non-coverage and non-response problems, and an identification of key research gaps.

Transport Survey Methods

ISBN: 978-1-84855-844-1

8.2. 'Hard to Reach' Groups

Who, then, are the 'hard to reach' groups in travel surveys? Two, at times overlapping, categories can be identified. The first category includes groups that are hard to reach because of their frequent omission from sampling frames. Such non-coverage 'hard to reach' groups include:

- residents of informal housing that is not connected to utility servicing networks, who are not easily identified because of the absence of a numbered and named address;
- people who have recently changed residence, changed type of telephone service or lost telephone service and are not currently associated with a common sample frame;
- sub-letters of sections of formal dwellings (e.g. garage apartments, backyard shacks) who are indirectly connected to utility servicing networks and do not have separate street addresses;
- residents of group quarters (army barracks, student residences, migrant worker hostels, hospices, etc.);
- illegal immigrants who typically avoid or resist involvement in activities that they perceive might lead to their identification by government agencies;
- transients who remain in a place only for a relatively brief period of time (e.g. the occupants of mobile homes, casual seasonal workers, workers resident on construction sites, etc.) and
- homeless people who, without a fixed abode, cannot be easily identified or contacted.

The second category includes groups that are hard to reach because they are disproportionately persistent non-responders in surveys. Such non-response 'hard to reach' groups include:

- people who are unable to speak the language, or languages, in which the survey is conducted, and are therefore unable to respond;
- (in the case of self-completion questionnaires) people who are functionally illiterate;
- people who are, to varying degrees, disengaged from broader society and do not wish to be contacted (e.g. residents of 'gated communities', religious retreats, etc.);
- adolescents and members of very high and very low income households: categories of people who are often unwilling, for a variety of reasons, to make their time available;
- (in societies experiencing high levels of crime) people who are afraid of allowing strangers into their homes, or of divulging information about their lives that they perceive would make them more vulnerable to criminals;
- 'gate-kept' individuals who are difficult to access without the permission of a third party (e.g. women within strictly patriarchal households, farm workers residing on commercial farms, etc.) and

- people with various forms of physical or cognitive disabilities (e.g. aurally impaired people in the case of telephone surveys, visually impaired people in the case of mail-back questionnaires, etc.).

Current events and trends in many parts of the world suggest that the proportion of these 'hard to reach' groups within target populations is likely to grow. The key drivers of this expected increase include changing population demographics, socio-economies, technologies, and geo-political stability.[1] Clearly exposure to, and the impact of, these drivers of change will be different and uneven in various parts of the world. Consequently, trends with respect to the growth or decline in the relative extent and composition of 'hard to reach' groups is likely to vary considerably across different contexts.

8.3. Problems Resulting from Inadequate Inclusion of 'Hard to Reach' Groups

The main problems of data unreliability and bias that are introduced by the absence or underrepresentation of certain groups in travel surveys are well documented. Decreasing participation in surveys, generally, leads to reduced data reliability because of reduced sample size, and, more specifically, may increase coverage errors and non-response bias in the likely event that the travel behaviour profile of non-contacts and refusals systematically differ from those of respondents. It is possible of course to address the problem of the underrepresentation of certain groups in sample respondents through weighting corrections and non-response adjustments. In some cases, however, these techniques can raise problems of their own (e.g. unreliable or outdated census data introduce weighting errors, reduced 'hard to reach' respondent sub-samples may not be representative of the whole 'hard to reach' group, etc.) and offer a partial solution at best.

The extent of the bias and error that emerge from non-coverage of, and non-response by, 'hard to reach' groups will clearly vary, depending on context, method and purpose.

The relative size of the 'hard to reach' group population, relative to the whole population, will vary from country to country and from city to city. In some cases,

1. Examples of population demographic changes include: an increase in the number of child-headed households in countries with high rates of HIV infection; and aging societies in wealthier countries. Examples of socio-economic changes include: in stagnant or 'jobless growth' economies, increasing levels of poverty and associated increases in informal and sublet housing and homelessness; the casualisation of labour forces and increased disengagement amongst wealthier households in response to perceptions of greater security risks. Examples of technological changes include: increases in the proportion of cell phone-only households, and of households with unlisted landline telephone numbers. Examples of geo-political changes include: increased immigration and emigration in response to political conflicts, natural disasters and the projected impacts of climate change and resource depletion.

the errors resulting from non-coverage and non-response might be slight and of relatively minor consequence for data reliability, while in other cases they might be significant.

The various 'hard to reach' groups are impacted differently by alternative survey methods, and depending on the relative composition of the 'hard to reach' population, the extent of bias and error will be determined by the choice of method. For instance, in populations with high levels of no-phone, unlisted or cell phone-only households, coverage error would be compounded by the use of telephone interviews. Similarly, in populations with high rates of functional illiteracy and low Internet penetration, self-completion questionnaires and web-based surveys, respectively, would compound non-response and coverage errors.

The purpose of the survey is also important. For instance, in the case of surveys to calibrate a city-wide travel demand forecasting model for the purpose of long-term transport infrastructure planning, the omission of relatively small sections of the population, with typically lower trip-generation rates, is of relatively low significance. By comparison, in the case of surveys intended to inform the development of public policies relating to social equity issues (e.g. public transport subsidy policies, social inclusion policies, etc.), the omission or underrepresentation of even relatively small 'hard to reach' groups is of critical concern because these are the very groups the policy is wishing to target.

8.4. Potential Measures to Ensure the Participation of 'Hard to Reach' Groups

Having established that 'hard to reach' groups are heterogeneous and therefore require different approaches, what then are some of the measures that have the best chance of reducing the barriers to participation in travel surveys? These measures are discussed in terms of improving the coverage of samples, reducing non-response and improving survey instrument cognition.

8.4.1. Measures to Improve Coverage

Measures that can be taken to ensure that 'hard to reach' groups are included in sample frames are discussed in terms of general population surveys, and specific population surveys.

With respect to general population surveys, the problem of non-coverage in the main sampling frame can be addressed by multi-frame sampling (see Kalton & Anderson, 1986). For instance, taking care to eliminate or compensate for overlap, address frames might be combined with telephone number lists, and in developing countries with extensive informal settlements without street addresses and landline telephone connections, supplemented by a frame of dwelling units identified in recent aerial photography. Where applicable and available, further supplementary frames could take the form of registers or membership lists obtained from organisations representing the interests of particular 'hard to reach' groups (e.g. non-governmental organisations representing the disabled, refugees, etc.). A limitation in the use of multi-frame sampling, however, is that it tends to increase variance.

With respect to specific population surveys, in instances where an appropriate sample frame does not exist, a variety of rare population-sampling techniques can be applied (see Kalton & Anderson, 1986). One technique is to screen a sample in order to find certain types of respondents, but this can be an expensive exercise, particularly if the rare population sought represents a small proportion of the sampled population. Another technique is non-random multiplicity sampling (or 'snowballing'), in which a respondent who qualifies for the rare population sample is asked to provide the names of others from their network of acquaintances who also qualify (see Cowham, Webb, Dye, & Crowther, 2008 for an application of this method in a stated preference survey of persons with disabilities). Advertisements in a range of appropriate media can be targeted at specific groups to recruit the first set of qualifying respondents in this process. A further technique is to draw qualifying respondents from an extant access panel of recruited and managed individuals (see Stoop, 2005). A limitation of both multiplicity sampling and access panels is bias introduced by self-selection. It is difficult to check whether the respondent sample is representative of the whole rare population being studied.

8.4.2. Measures to Reduce Non-Response

Measures that can be taken to reduce non-response among 'hard to reach' groups are discussed in terms of data-collection protocols, recruitment techniques and non-response follow-up.

With regard to data-collection protocols, response rates can be increased by expanding the period over which data is collected in the case of cross-sectional surveys (continuous surveys have an obvious advantage here). Increasing the number of attempts to make contact increases the chance of contacting respondents who are frequently away from home (e.g. students, holiday makers, business travellers, seasonal workers, etc.). Notwithstanding the need for surveys to be conducted in the languages of all population groups of key interest whenever possible, a further measure to reduce the need for substitution is to permit proxy reporting as a way of overcoming communication barriers in the case of respondents within households who do not speak the language, or languages, in which the survey is conducted. This can also be helpful to include people with disabilities. Although, as Stopher, Wilmot, Stecher, and Alsnih (2006) argue, data collected by proxy should be coded as such in datasets so that any tendency towards greater underreporting can be identified in data analysis.

With regard to recruitment techniques, a variety of measures can be taken to provide advance warning of scheduled contacts and to incentivise response. Advance warning of scheduled contacts has been found in some surveys to result in increased response rates (see Contrino, McGuckin, Nakamoto, & Santos, 2008; Zmud, 2003). This can be achieved through media campaigns or letters that are customised to the 'hard to reach' groups in question. Endorsement of the survey by well-known local leaders or celebrities within the particular 'hard to reach' group could also encourage participation and allay fears around legitimacy.

Incentives are becoming more common in all kinds of surveys, and range from small pre-incentives to large lotteries. The incentive should be carefully designed, however, so that the impact helps response rates of the targeted 'hard to reach' groups. Careful testing of the impacts of incentives is required, as amongst some groups they have sometimes proven to have the opposite effect (see Singer, Groves, & Corning, 1999).

With regard to non-response follow-up, analysis of non-response should occur early in the survey process, so that patterns of, and reasons for, non-response can be identified with sufficient time to make any necessary changes to the survey protocol in order to increase 'hard to reach' group participation. This might involve follow-up contact to obtain further demographic information on the profile of non-responders, and their reasons for non-response. The common practice of analysing non-response at the end of the data-collection period does not facilitate amendment to survey protocols. In some surveys, a subset of specially trained interviewers, or non-response 'converters', are employed who follow-up with the 'hard to reach' non-responders encountered and attempt to induce a response (see Zmud, 2003). If the full survey cannot be completed, the follow-up contact may resort to capturing just essential demographic and travel-related information.

8.4.3. Measures to Improve Instrument Cognition

To address the variation among 'hard to reach' groups, multi-modal survey methods can be used. Different survey modes have different problems and strengths (e.g. telephone and Internet surveys have non-coverage problems, post surveys impose greater respondent burden and are easy to ignore, home interviews have contact refusal problems, etc.; see Morris & Adler, 2003).

Multi-modal surveying enables the most suitable survey instrument to be matched to different categories of 'hard to reach' respondents. It enables customised language and protocols to be targeted at particular groups to increase the likelihood of cognition and response. It also enables the allocation of greater resources per respondent to ensuring that 'hard to reach' groups are included. Caution and pre-testing are required, however, for even if the sequence and phrasing of questions is the same, the data collected using one survey mode may not be completely comparable with the data collected using another (see Bonnel, 2003). The mode effects should be analysed carefully so that final estimates can be made with confidence.

8.5. Conclusion

The foregoing discussion illustrates that the extent and nature of the problem of including 'hard to reach' groups in travel surveys varies considerably across contexts. Consequently, the identification of appropriate measures to ensure 'hard to reach' group participation will also be context specific, and will require an understanding of the nature of 'hard to reach' groups and how these groups are likely to respond to different forms of participation stimuli.

No 'one size fits all' approach can ensure that 'hard to reach' groups participate in a survey, and certainly the assessment of the significance of each special population varies between countries and from one context to another. Local practitioners will need to develop tailor-made solutions for overcoming the problem; these solutions will reflect the nature of 'hard to reach' groups in the particular target population and the coverage imperatives of the particular survey.

This raises the need for research that develops a better understanding of the extent and nature of 'hard to reach' groups in particular contexts, and of how this is likely to change over time. Such research could also usefully employ cognitive techniques to investigate the reasons for, and barriers to, non-response amongst different categories of 'hard to reach' groups. This improved understanding would enable better estimates of the extent of the problem in the survey design and budgeting phase, and better-informed decisions with respect to the need for the inclusion of mitigating measures in sampling strategies and instruments. The latter, in essence, involves an assessment of the trade-offs between, on the one hand, the ability to address coverage and response biases through the use of measures like multi-modal surveys and multi-frame and rare population-sampling techniques, and, on the other, the potential for the introduction of additional bias in survey data that these measures bring. Thus further research is also required to investigate potential design or mode effects, especially with respect to comparative unit non-response rates, and variations in respondent cognition. Opportunities exist here for comparative research into alternative survey methods, utilising experiment and control groups within different categories of 'hard to reach' groups as a means of isolating and measuring effects. In pursuing the above, a further general research need is to look beyond the travel survey field, as Riandey and Quaglia (2009) do, to see what lessons can be learned and what techniques might be adapted from survey practices in other domains.

Acknowledgements

The participants of this workshop were Kay Axhausen, Switzerland; Roger Behrens, South Africa; Heather Contrino, United States; Virginie Dejoux, France; Mark Freedman, United States; Mattias Gripsrud, Norway; Nancy McGuckin, United States of America; Sharon O'Connor, United States of America; Martine Quaglia, France; Benoît Riandey, France; Karl Sieber, United States; Anu Siren, Denmark; Juliane Stark, Austria; and Coralie Triadou, United Kingdom.

References

Bonnel, P. (2003). Postal, telephone, and face-to-face surveys: How comparable are they?. In: P. Stopher & P. Jones (Eds), *Transport survey quality and innovation*. Amsterdam: Pergamon.

Contrino, H., McGuckin, N., Nakamoto, H., & Santos, A. (2008). A re-examination of methods in the U.S. National Household Travel Survey. Presented at the 8th International Conference on Transport Survey Methods, Annecy, France.

Cowham, M., Webb, J., Dye, J., & Crowther, B. (2008). Prioritising street improvements for respondents with disabilities: Qualitative and quantitative research. Presented at the 8th International Conference on Transport Survey Methods, Annecy, France.

Kalton, G., & Anderson, D. (1986). Sampling rare populations. *Journal of the Royal Statistical Society A*, *149*(1), 65–82.

Morris, J., & Adler, T. (2003). Mixed mode surveys. In: P. Stopher & P. Jones (Eds), *Transport survey quality and innovation*. Amsterdam: Pergamon.

Riandey, B., & Quaglia, M. (2009). Surveying hard-to-reach groups. In: P. Bonnel, M. Lee-Gosselin, J. Zmud & J.-L. Madre (Eds), *Transport survey methods: Keeping up with a changing world*. Bingley, UK: Emerald.

Singer, E., Groves, R., & Corning, A. (1999). Differential incentives: Beliefs about practices, perceptions of equity, and effects on survey participation. *Public Opinion Quarterly*, *63*(2), 251–260.

Stoop, I. (2005). *The hunt for the last respondent: Nonresponse in sample surveys*. The Hague: Social and Cultural Planning Office of the Netherlands.

Stopher, P., Wilmot, C., Stecher, C., & Alsnih, R. (2006). Household travel surveys: Proposed standards and guidelines. In: P. Stopher & C. Stecher (Eds), *Travel survey methods: Quality and future directions*. Amsterdam: Elsevier.

Zmud, J. (2003). Designing instruments to improve response. In: P. Stopher & P. Jones (Eds), *Transport survey quality and innovation*. Amsterdam: Pergamon.

Chapter 9

Emerging Methods and Technologies for Tracking Physical Activity in the Built Environment

Sean T. Doherty

Abstract

Health scientists and urban planners have long been interested in the influence that the built environment has on the physical activities in which we engage, the environmental hazards we face, the kinds of amenities we enjoy, and the resulting impacts on our health. However, it is widely recognized that the extent of this influence, and the specific cause-and-effect relationships that exist, are still relatively unclear. Recent reviews highlight the need for more individual-level data on daily activities (especially physical activity) over long periods of time linked spatially to real-world characteristics of the built environment in diverse settings, along with a wide range of personal mediating variables. While capturing objective data on the built environment has benefited from wide-scale availability of detailed land use and transport network databases, the same cannot be said of human activity. A more diverse history of data collection methods exists for such activity and continues to evolve owing to a variety of quickly emerging wearable sensor technologies. At present, no "gold standard" method has emerged for assessing physical activity type and intensity under the real-world conditions of the built environment; in fact, most methods have barely been tested outside of the laboratory, and those that have tend to experience significant drops in accuracy and reliability. This paper provides a review of these diverse methods and emerging technologies, including biochemical, self-report, direct observation, passive motion detection, and integrated approaches. Based on this review and current needs, an integrated three-tiered methodology is proposed, including: (1) passive location tracking

Transport Survey Methods

ISBN: 978-1-84855-844-1

(e.g., using global positioning systems); (2) passive motion/biometric tracking (e.g., using accelerometers); and (3) limited self-reporting (e.g., using prompted recall diaries). Key development issues are highlighted, including the need for proper validation and automated activity-detection algorithms. The paper ends with a look at some of the key lessons learned and new opportunities that have emerged at the crossroads of urban studies and health sciences.

> We do have a vision for a world in which people can walk to shops, school, friends' homes, or transit stations; in which they can mingle with their neighbors and admire trees, plants, and waterways; in which the air and water are clean; and in which there are parks and play areas for children, gathering spots for teens and the elderly, and convenient work and recreation places for the rest of us. (Frumkin, Frank, & Jackson, 2004, p. xvii)

9.1. Introduction

The built environment, in the form of land use and transportation systems, has long been studied with respect to impacts on human activity and travel patterns, accessibility, equity, safety, environment, and congestion. More recently, we have become interested in how prevailing trends in the built environment influence the physical and dietary activities in which we engage, the environmental hazards we face, the kinds of amenities we enjoy, and the resulting impacts on our health. In particular, the sprawling, low-density suburbs that have dominated the landscape of North American cities since the 1950s, combined with new household technologies that require less manual effort, appear to discourage physical activity and encourage dependence on cars. This suburban sprawl is also not conducive to walking and cycling, limits accessibility to fresh food products for some, encourages fast-food consumption, limits social interaction, provides few natural amenities, and contributes to environmental problems; all of which have potentially adverse impacts on our health. This is in stark contrast to a century ago when cities were compact, mixed-use, highly walkable, and accessible by public transit. Physical activity was embedded in everyday life, from the chores that people did to the walking trips to local food markets. It is no surprise, then, that health scientists and urban planners have turned increasing attention to the influence that the built environment has on health in an attempt to establish designs that will return cities back to places that encourage healthy behavior.

Several recent books (Frank, Engelke, & Schmid, 2003; Frumkin et al., 2004), committee reports (Dannenberg et al., 2003; Humphrey, 2005; Transportation Research Board and Institute of Medicine of the National Academies, 2005), special issues of journals (e.g., Heath et al., 2006), and a wide range of reviews (McCormack et al., 2004; Boarnet, 2005; Booth, Pinkston, & Poston, 2005; Brownson, Boehmer, & Luke, 2005; Frank & Engelke, 2005; Handy, 2005; Davison & Lawson, 2006;

Papas et al., 2007) have firmly established the potential the built environment has to influence physical activity patterns. However, it is widely recognized that the extent of this influence, and the specific cause-and-effect relationships that exist, are relatively unclear and depend to a large extent on assessment methods and measurement issues. This state of affairs is no better summarized than by Susan Hanson reacting to the ground-breaking report "*Does the Built Environment Influence Physical Activity?: Examining the Evidence*" (produced by the Transportation Research Board and Institute of Medicine joint committee on "*Physical Activity, Transportation, and Land Use*"):

> The committee's findings can be summarized as follows: The built environment can facilitate or constrain physical activity. The relationship is complex, however, and operates through many mediating variables, such as socio-economic characteristics, personal and cultural variables, safety and security in the built environment, and the individual's decisions about time allocation. Empirical evidence shows a linkage between the built environment and physical activity, but causality has not been established, strengths of the relationships are not known, and the characteristics of the built environment that are most closely associated with physical activity are unknown. Also unknown are how the relationship between the built environment and physical activity varies by location (e.g., urban, suburban, rural) or by population subgroup (defined, for example, by age, sex, race/ethnicity) or how important different characteristics of the built environment are to total daily physical activity. The current literature reflects the lack of sound theoretical frameworks to guide empirical research, inadequate research designs (e.g., most studies are cross-sectional whereas longitudinal studies are needed to assess causality), and incomplete data (e.g., national surveys on physical activity lack data on location, and national data on travel, which do record location, neglect physical activity). (Hanson, 2006, p. S259)

This extensive multidisciplinary review clearly attributes our lack of understanding of the impacts of the built environment on physical activity to the lack of appropriate data collection, and focuses the solution on the need to collect *individual*-level data on *all daily physical activities* (not just utilitarian travel) over *long* periods of time linked spatially to *real-world characteristics* of the built environment in diverse settings (not just around the home), along with a wide range of personal mediating variables (especially residential location preference). This is obviously no easy challenge, especially as people are not able to self-report completely objective data of this detail and extent owing to reporting burden, self-reporting biases, limited spatial cognition, and lack of ability to measure physiological changes unassisted, implying a focus on alternative, objective and preferably passive measurement techniques.

Capturing and analyzing objective data on the built environment has benefited from wide-scale availability of detailed land use and transport network databases, combined with the data management and analytical power of geographic information systems (GIS) (for more extensive details, see Crane, 2000; Boarnet & Crane, 2001; Ewing & Cervero, 2001; Cromley & McLafferty, 2002; Frumkin et al., 2004). GIS software allows points (landmarks, intersections, etc.), lines (paths, walls, etc.), and areas (building footprints, natural areas, etc.) to be related to each other spatially. Each feature can have an associated set of attributes documenting such characteristics as types, sizes, purposes, flows, census data, etc. Aggregate features of neighborhoods or zones such as land-use density, network connectivity, and walkability can be calculated. To get a sense for what types of data are available and their accuracy, the layperson need only check out online maps such as Google Maps. Planning agencies, universities, governments, and other research institutes will have available additional datasets. However, a key issue will be balancing use of these existing data sources with the desire/need for more detailed or completely new features of the built environment at finer scales, including especially more qualitative information (e.g., how well lit or safe streets are). Regardless of data source, GIS represents the most promising technology for managing, analyzing, and comparing characteristics of the built environment.

The same cannot be said of human activity, for which a more diverse history of data collection methods exists and continues to evolve, owing to a variety of quickly emerging wearable technologies. At present, no "gold standard" method has emerged for assessing physical activity type and intensity under the real-world conditions of the built environment; in fact, most methods have barely been tested outside of the laboratory, and those that have tend to experience significant drops in accuracy and reliability. This paper provides a review of these diverse methods and emerging technologies in an attempt to narrow the focus and identify the most promising methods. Methods from both the urban studies (geography, planning, civil engineering, etc.) and the health sciences (kinesiology, physiology, public health, etc.) are intentionally covered and contrasted, particularly to highlight means to learn from each other and integrate the best of both worlds. Extensive illustrative examples, literature citations, and/or links to Web sites are provided as a resource for those who wish more detail with each method or technology. Based on this review, an integrated methodology drawing upon emerging techniques from both fields is proposed as most viable for future studies. Lessons learned and key methodological challenges for the future are then posed in the conclusion.

9.2. Physical Activity Measurement

Physical activity has been defined as "any bodily movement produced by skeletal muscles that results in caloric expenditure" (Welk, 2002a). It is most commonly characterized by:

1. Type
2. Duration
3. Intensity

Intensity can be a simple rating (low, moderate, high) or expressed more technically in terms of energy expenditure (EE). EE is most often measured in metabolic equivalents (METs), wherein one MET represents resting EE (about 3.5 ml/kg/min of oxygen consumption) and physical activity typically raises this by several magnitudes. EE can also be expressed in terms of calories burned. EE can be measured directly, but is often estimated based on the duration and type of physical activity using established classification schemes (most common, that established by Ainsworth et al., 2000). However, as noted by Ermes, Parkka, Mantyjarvi, and Korhonen (2008, p. 20) "Energy expenditure is only one important aspect of physical activity .... A more detailed analysis of physical effort can be obtained by activity recognition, i.e., by detecting the exact form of activity the subject is performing."

While the importance of physical activity is well established, more accurate and practical measures are still being sought by health professionals in order to better understand the specific amounts/types that are needed for health benefits (Welk, 2002a) and by a growing number of urban planners/researchers in order to better understand the impacts of the built environment. Particularly challenging have been the attempts to develop accurate, valid, and cost-effective techniques to quantify physical activity outside of the laboratory under free-living conditions. Overall, the more accurate the measurements of physical activity, the more power researchers have to detect changes or differences across various factors (which are many), and the smaller the sample needed to test the relationships.

Based on review of methods from published studies of physical activity, five main types of physical activity measurement methods are evident:

1. *Biochemical and physiological measurement* of EE (METs) using techniques such as doubly labeled water (DLW) or indirect calorimetry (IC), heart monitors, or skin responses.
2. *Self-reported* activities and travel, wherein reported activity or travel types, intensities, and durations are converted to episodes of physical activity and/or estimated METs using classification schemes.
3. *Direct observation*, analogous to self-reported surveys but completed by an independent observer.
4. *Passive motion detectors*, with varying capacities for providing objective measures of activity type, intensity, and duration; includes pedometers, accelerometers, and GPS.
5. *Integrated/hybrid approaches*, combining two or more of the above.

Each of the above methods is reviewed in detail in the following sections; some more than others, depending on their potential. In all cases, follow-up readings are provided as an additional resource for readers with specific interests.

9.2.1. Biochemical and Physiological Monitoring

9.2.1.1. Direct biochemical measurement Biochemical measurement involves measurement of chemical processes of the body. Physical activity involves EE that is directly related to the consumption of hydrogen and oxygen, carbon dioxide production, heat production, and other chemical affects. Several techniques have been developed for measuring these outcomes, and represent the most direct measurement of EE. The most common are through DLW and IC. The specifics of how these techniques work are covered in Starling (2002). Briefly, IC involves analysis of respiratory gases captured through a facemask, canopy, or whole room analysis. DLW involves analysis of the elimination of hydrogen and oxygen from urine samples. While these methods are the most accurate measure of physical activity energy consumption, they have tended to be too expensive and inconvenient for use in free-living conditions (Dale, Welk, & Matthews, 2002; Boarnet, 2005), and typically do not measure the qualitative nature of physical activity (such as type). For this reason, they are normally used only in a laboratory setting with small samples of subjects as a benchmark or "gold standard" from which to assess the validity the other measurement devices (Starling, 2002).

However, recent new devices are breaking cost, size, and usability barriers, such as the MedGem Indirect Calorimeter (see also McDoniel, 2007, or http://www.micro-lifeusa.com). The device weighs only 110 g and measures 5.5 × 5.5 × 11.5 cm. Users basically breathe into the portable device for up to 10 min (much shorter than the average 30 min of past devices). The device then measures oxygen consumption and estimates EE in the form of calories burned. Although recent studies have found it to be a reliable and valid instrument for the measurement of EE (Rubenbauer, Johannsen, Baier, Litchfield, & Flakoll, 2006; Fields, Kearney, & Copeland, 2006; Cereda et al., 2007; McDoniel, 2007), just as many studies appear to question the device's validity and usability (Hlynsky, Birmingham, Johnston, & Gritzner, 2005; Reeves, Capra, Bauer, Davies, & Battistutta, 2005; Van Loan, 2007; Fares, Miller, Masters, & Crotty, 2008) suggesting a lack of evidence to support its clinical usage. Additionally, there do not appear to be any published applications of this device for assessing the types and duration of physical activity, which presumably would require cumbersome repeated measurements. Thus, until such a device can provide continuous monitoring capabilities, its application is likely limited to serving as a field-based validation device, albeit more convenient and inexpensive than traditional IC methods.

9.2.1.2. Heart-rate monitors The development of lightweight, continuous heart-rate data recorders has improved the ability to detect the intensity, duration, and frequency of physical activity. Modern heart monitors use the electrocardiogram (ECG) signal to detect heart rate, can be attached to a small chest strap with integrated electrodes, and can transmit data for storage on devices as small as a watch (e.g., the Polar Heart Rate Monitor series shown at www.polarusa.com). Heart rate has been shown to increase in a predictable time-lagged linear fashion with moderate to vigorous aerobic physical activity, but can be poorly correlated with light and sedentary activity (such as TV watching or walking) or non-aerobic activity (for review, see Janz, 2002). Heart rate is also known to be affected by many other confounding factors, such as type of physical activity, physical fitness, hydration, caffeine usage, high temperatures, and emotional states. Because of these shortcomings, many researchers either dismiss them as impractical (e.g., King, Torres, Potter, Brooks, & Coleman, 2004) or combine them with accelerometers to improve prediction of EE (see also Strath et al., 2000; Brage et al., 2004).

9.2.1.3. Biometric monitoring devices One of the latest technologies for estimating EE from physical activity is through biometric/metabolic monitoring devices. One commercially available device assessed extensively in the literature is the SenseWear Armband, which weighs 85 g and measures 8.5 × 5.3 × 2 cm (see also www.sensewear.com). This small wireless monitor, worn on the arm for up to 5 days, collects minute by minute: (1) movements by a three-axis accelerometer; (2) heat flux by a thermocouple array; (3) skin temperature; and (4) galvanic skin response by two electrodes (an indicator of evaporative heat loss). Using proprietary algorithms, the manufacturer's software calculates and displays a subject's daily total EE (calories burned), number of steps, and duration of physical activity and sleep. Although only daily totals have been reported in the literature, it appears to have the capacity to provide highly detailed minute-by-minute reports.

Based on an earlier report by the manufacturer (Liden et al., 2002) and information on their current Web site (www.bodymedia.com), the prediction algorithm appears to first identify the subject's "context" (sleeping, resting, walking, driving a car), followed by estimation of calories burned (EE) in these contexts as a function of the accelerometer and heat sensing components (plus "other" unreported parameters). This is deemed particularly important for distinguishing four key scenarios: (1) low motion, high heat flux activity (e.g., weight lifting); (2) low motion, low heat flux (e.g., sitting a desk, watching TV); (3) high motion, high heat flux (e.g., running); and (4) high motion, low heat flux (e.g., driving a car). This addresses the shortcomings of using motion detectors alone (e.g., accelerometer), as highlighted by the manufacturer:

> Other devices currently being used for free living energy expenditure monitoring are not able to detect contextual differences, and, therefore, are not able to utilize such information in their calculations. As described previously, motion detectors, pedometers, and accelerometers share the disadvantages of being subject to the detection of false motion and the inability to accurately detect nonambulatory physical activity. Particularly critical to the accuracy of accelerometer based energy expenditure models are those times when subjects are traveling in motor vehicles. At these times the rapid and continued motion of the accelerometer may equate to high-energy expenditure. While the vehicle is expending high energy, the subject inside the vehicle is not. By incorporating a "motoring" context into an energy expenditure algorithm such false detection can be identified and corrected and, therefore, overall accuracy greatly improved. There are other significant contexts such as sleeping, resting, and walking, which,

> if known, can improve the accuracy of energy expenditure calculations. (Liden et al., 2002, p. 12)

Further testing by Fruin and Rankin (2004) and Jakicic et al. (2004) showed that the first version of the SWA provided reasonably reliable estimates of EE compared to IC. However, the EE of some activities, such as flat walking, stepping, and cycling exercise, were significantly underestimated by as much as 17%, whereas the EE of other activities, such as arm exercise, were overestimated (+ 29%). Others have found that the SWA can be significantly inaccurate for certain population groups such as the obese (Papazoglou et al., 2006; Malavolti et al., 2007) and children (Arvidsson, Slinde, Larsson, & Hulthen, 2007; Dorminy, Choi, Akohoue, Chen, & Buchowski, 2008). Jakicic et al. (2004) tested improved context-specific algorithms developed by the manufacturer and found that they produced EE estimates equivalent to calorimetry for all activities. Similarly, Dorminy et al. (2008) found that a simple adjustment for subject weight significantly improved the predicted EE.

Overall, like the MedGem Indirect Calorimeter, independent testing of this device has been largely limited to its capabilities in assessing total EE rather than its ability to detect activity types and duration (although it appears to have this ability). Additionally, one of the complaints about the SWA from researchers (e.g., see Papazoglou's response to the manufacturer's editorial in Andre, 2007) is that the details of the various algorithms and refinements are not revealed in detail. Although an early report by the manufacturer (see Liden et al., 2002) provides some conceptual level details on algorithm structure and development, data collection, context detection, and accuracy, moving forward may require more open testing of the device.

9.2.2. *Self-Reported Physical Activity*

Self-reported instruments rely on subjects' ability to recall the types, intensities, duration, and patterns of their own physical activity. They have been used extensively in the health sciences (for reviews, see Montoye, Kemper, Saris, & Washburn, 1996; Sallis & Saelens, 2000; Matthews, 2002) and urban studies (for reviews, see Richardson, Ampt, & Meyburg, 1995; Stopher & Jones, 2003). Although many variations exist, two basic types are evident: retrospective questionnaires and diaries.

9.2.2.1. Retrospective questionnaires Retrospective questionnaires pose a series of questions concerning a person's "typical" physical activity patterns in the recent past such as last week or month, by type, intensity, frequency, and duration. An example script from the 2005 California Health Interview Study (see also www.chis.ucla.edu):

> The next questions are about walking for transportation. Please only include walks that involved an errand or to get some place. I will ask you separately about walking for relaxation or exercise.

During the past seven days, did you walk for at least ten minutes at a time to get some place such as work, school, a store, or restaurant?
How much time did you [usually] spend walking on [one of those days/that day]?

These questions were then repeated for a wide range of physical activity types. Other examples include the Behavioral Risk Factor Surveillance System (BRFSS) that measures the frequency and duration of leisure time and occupational physical activity via phone interview; and the Youth Risk Behavior Surveillance System (YRBSS) that measures number of days per week with at least 20–30 min of vigorous/moderate exercise (administered by the Centre for Disease Control in the US; see www.cdc.gov/brfss/about.htm or www.cdc.gov/nccdphp/dash/yrbs/index.htm). A thorough "Collection of Physical Activity Questionnaires for Health-Related Research" is provided in volume 29(6) supplement of the *Medicine & Science in Sports & Exercise* journal. Other major national/regional surveys are reviewed in Boarnet (2005), whereas detailed assessments of validity can be found in Sallis and Saelens (2000) and Matthews (2002).

Overall, such studies tend to vary by target group (e.g., adult, children, elderly) and the physical activity categories of interest (occupational, household, sports, school-related, conditioning, transportation, recreational activities, etc.). The specific location of physical activities is not normally obtained, likely in part because of the awkwardness of asking location for "typical" behaviors. In lieu of this, aspects of the built environment surrounding a subject's home location are most often the focus of analysis.

9.2.2.2. Records/diaries with a physical activity focus Rather than "typical" behaviors, diaries are meant to capture "actual" episodes of physical activity as they occur during the day (or record them at regular intervals such morning, noon, and night), by types, intensity, start/end time, and/or duration. They tend to capture more detail about a diverse range of physical activities in daily life compared to retrospective questionnaires. They can also be easier for subjects to recall without the cognitive effort involved in generalizing about "typical" behaviors over long periods.

There are a variety of different diary formats and levels of detail, but they all tend to adopt a table or grid format with a temporal dimension. Stel et al.'s (2004) 7-day activity diary was completed every evening by the respondents, and covered the duration (but not time-of-day) of walking inside and outside, bicycling, gardening, light household activities, heavy household activities, and sport activities. Weston, Petosa, and Pate's (1997) Previous day physical activity diary involved recall of the previous day's activities and subjective intensities in 30-min time intervals across the whole day, including primary activity type (eating, sleeping/bathing, transportation, work/school, spare time, and physical activities) and subcategories. Bouchard et al.'s 3-day physical activity diary is similar, but broken into 15 min intervals and using a different activity categorization scheme (Bouchard, 1997; Bouchard et al., 1983). Wormald et al.'s (2003) physical activity diary asks participants to regularly (four times a day) record everything they do from getting up in the morning to going to bed at night, for 7 days. The actual start and finish times of each activity are recorded and a descriptive measure of physical activity intensity is indicated by circling a marker. Methods for operationalizing diaries of this type on hand-held computers or smartphones will most likely be the next generation of such designs, especially as it could simplify presentation of the often long categorical lists of physical activity types as fly-out or drop-down lists.

In health studies, the activity types/intensities and durations collected from the above-mentioned surveys are used to estimate EE using established activity-specific rates of EE multiplied by the duration. The most common and extensive coding scheme appears to be the compendium provided by Ainsworth et al. (2000) which includes a very highly detailed listing of 650 specific physical activity types in 21 major categories, followed by the METs estimate associated with it. Nonetheless, even these authors acknowledge that such tables do not take into account individual differences in body mass, body fat percentage, fitness level, or geographic environment that are known to alter EE, and that no such general correction is available.

9.2.2.3. Records/diaries with a travel-activity focus More recently, attempts have been made to elicit physical activity from daily activity and travel diaries deployed in urban and time-use studies. Extensive literature on such methods exists, exploring the variation in methods, administration, conduct, data quantity/quality, standardization, and use of emerging technologies (Richardson et al., 1995; Stopher & Jones, 2003). Generic to such techniques is the capturing of an ordered list of self-reported daily events (activities, trip) by time-of-day for one or more days, plus their attributes. Attributes normally include at least activity type (by main and subcategories, which vary widely), travel mode (auto, transit, walk, bike, etc.), start/end time, and location. Wide ranges of additional attributes tailored to the particular study purpose are also collected (involved persons, costs, frequency, when planned, flexibility, etc.). Example activity/travel diaries can be found in Barnard (1986), Mackett, Brown, Gong, Kitazawa, and Paskins (2007; see www.casa.ucl.ac.uk/capableproject/ActivityDiary.htm), Mackett, Brown, et al. (2007; see www.casa.ucl.ac.uk/capableproject/ActivityDiary.htm), or in the extensive "Travel Survey Manual" released in 1996 by Cambridge Systematics (currently available at www.travelsurveymethods.org; an updated online version is expected). Note that most recent large-scale surveys of this kind are often conducted via computer-assisted telephone interview (CATI).

The capturing of *location* characteristics is what distinguishes these diaries from those adopted in the health sciences, often including specific place names and geo-coordinates (longitude and latitude). For this reason, daily events can be precisely associated with characteristics of the built environment. However, in practice, complete details on the types, intensity, and duration of physical activity have not been captured in as much detail as in health studies, although theoretically they could in the form of more specific activity types and additional attributes. What can typically be identified is utilitarian walking and bicycling trips by duration (but not usually intensity), along with some capturing of recreationally active activities by duration (e.g., dog walking, jogging). Much depends on the detail in activity/trip-purpose types captured and, to a lesser extent, the location. As found by Forsyth, Oakes, Schmitz, and Hearst (2007) focus only on walking travel without full accounting of leisure/recreational walking and other exercise may unjustly lead to associating land-use attributes and exercise.

A recent example that that goes beyond use of just walk trips is by Copperman and Bhat (2007). They used the 2000 San Francisco Bay Area Travel Survey (BATS) to examine how physical activity of children is related to the built environment. While it was straightforward to identify utilitarian trips involving physical activity by mode type (i.e., walking, cycling vs. auto modes), they also identified two other types of physical activities through a combination of activity type and location:

- *Recreationally physically active activities* — recreational coded activities that occurred at locations *assumed* to be associated with physical activity. Of the 450 location categories, sixteen were identified as active locations (golf courses, in/outdoor sports facility, dance studio, miniature golf, bowling alley, ice rink, public park, health club, gymnasium, ski resort, campground, sports camp, skate park, martial arts centre, swimming pool).
- *Physically active recreational travel* — non-motorized travel episodes that begin and end at home without any stops in between (for example, walking or bicycling around the neighbourhood).

While 19% of children in their sample participated in utilitarian activity travel, a further 16 and 3% (respectively) participated in these latter categories, effectively doubling the amount of data available for analysis. The authors recognize that "a limitation of our classification procedure is that it is solely based on the location type of activity participation, and does not consider any measure of the physical intensity level or nature (structured vs. unstructured) of activity participation" (Copperman & Bhat, 2007, p. 69) and that such an approach ignores physical activity associated with other types of activities, such as during free play, because they cannot be identified in the survey data.

Recognizing these limitations, and the significant impact they have on validity, others are turning to time-use studies that provide much more specific detail on activity types. For example, Sener, Copperman, Pendyala, and Bhat (2008) used a time-use diary that detailed 28 types of "active leisure, sports, and exercise", 10 types of "attending sports events" (so as not to confuse them with activity sports), 12 types of game playing, 24 types of passive leisure, 12 types of household chores, and a range of travel modes. A particularly unique interactive activity type tree diagram is at http://psidonline.isr.umich.edu/CDS/time_diary/activity_tree_view.html, whereas full details of the survey design are available from the Institute for Social Research (2002; see also http://psidonline.isr.umich.edu/CDS/TDqnaires.html). Sener et al. aggregated these into eight distinct activity categories for analysis, by location (in/out-of-home), day of week (weekend/weekday), structure (organized or unstructured), and intensity of physical activity (active or passive). Note how physical activity is measured in only two categories, in parallel to the focus in health studies on estimating EE more explicitly.

Overall, existing activity-travel diaries used in the urban planning arena may be useful for assessing the physical activity involved with utilitarian walking and cycling trips and, with some assumptions, some physically activity recreational activities; however, it should be noted that the vast majority of American children and adults report very few (if any) walk/cycling trips to begin with, thus providing a very limited perspective on their overall physical activity levels. The type and intensity of physical activity assumed associated with other activity types (e.g., recreational activity) has obvious limitations with respect to actual intensity and duration; even further, the physical activity associated with the many other activities that people conduct on a daily basis is completely missed, such as in-home or at school. The sum of these concerns suggests that a priority in the near future would be to assess the validity of such an approach for assessing overall physical activity levels; until then, sophisticated modeling using such data and the drawing of strong conclusions about the association to the built environment from such studies should be viewed with some caution.

A third and less obvious methodological contribution from urban studies is the recent evolution of activity-*scheduling* surveys that target both observed activity-travel patterns and underlying scheduling decisions processes (e.g., Doherty & Miller, 2000; Doherty, Nemeth, Roorda, & Miller, 2004; Lee & McNally, 2006; Ruiz & Timmermans, 2006; Zhou & Golledge, 2007). Using computerized, Internet, and hand-held technologies, these surveys attempt to capture *when*, *how*, and *why* activities are planned, modified, and executed over time and space and across individuals, leading up to their observed outcome. On the surface, they appear much like a standard activity diary to subjects, but involve more in-depth querying and decision tracking typically over multiple survey days. Including a focus on physical activity in such surveys could provide wholly new insights into how it is planned and managed within a person's everyday life, barriers to its completion, and identification of ways to *fit physical activity into your schedule*.

9.2.3. *Direct Observation of Physical Activity*

Direct observation involves a second party shadowing/following subjects and making records of their behavior. Although regarded as labor intensive and tedious, direct observation is a low-tech method that exceeds other measures of physical activity in providing contextually rich quantitative and qualitative data, especially with regard to the what, when, where, and with whom it occurs (McKenzie, 2002). In the health sciences, this method has been almost exclusively applied to the study of children and adolescents in specific locales, including at school, during education classes, during leisure time, and at home. In a laboratory setting this method is often used to provide objective baseline measures of physical activity for validation or algorithm development purposes. For instance, Ermes et al. (2008) had research assistants use a simple hand-held computer program to record the activities of subjects doing a specified set of indoor and outdoor physical activity tasks (sitting, engaging in stationary exercise, walking, running, cycling, playing sports, etc.). Users simply checked a box beside the activity type category (walk, run, sit, etc.) and the location (indoors, outdoors, in-vehicle), which was appended to a log along with the time. Ermes et al. also experimented with having subjects use the same device to self-report their activities under free-living conditions. The result is not dissimilar to what would be produced from the diary methods reviewed above. The difference is in intended application — in this case, the physical activity events were plotted on a time line alongside accelerometer and GPS speed data for comparison and algorithm development purposes (see also Section 9.3) rather than for large-sample assessment of physical activity.

Although applied infrequently, direct observation is an important method to consider in context of this review. Firstly, it demonstrates the depth to which researchers will often go to obtain objective data on physical activity, especially for validation purposes — a crucial step in the application of new technologies and development of automated detection algorithms. Perhaps more importantly, an observer can make detailed notes not only on the location of a person's physical activity, but also on significant interactions with the built environment, such as with facilities, equipment, playgrounds, sidewalks, natural features, etc. Additionally, an observer can make detailed notes on the quantity and quality of the social context of physical activity — beyond just "who with" type questions — including observation of strong and weak ties, involved person characteristics, the nature of interaction, etc. And unlike all previous methods that focus on individuals, direction observation can be used to study groups of people in specified locales — such as in specific neighborhoods. For instance, McKenzie et al. developed the System for Observing Play and Leisure Activity in Youth (SOPLAY) to record youth activities in specified time intervals (McKenzie, Marshall, Sallis, & Conway, 2000; McKenzie, 2002; McKenzie, Cohen, Sehgal, Williamson, & Golinelli, 2006).

9.2.4. Passive Motion Detection Devices

A vast array of passive, wearable monitoring devices attempt to provide more objective measurement of physical activity including under free-living conditions. They are capable of measuring one or more of motion, acceleration, position, and speed. These measurements are then analyzed to estimate aspects of physical activity including type, intensity, duration, and EE. The most sophisticated devices can even be used to analyze posture and gait.

9.2.4.1. Pedometers Pedometers are small, inexpensive devices that have gained wide-scale acceptance among physical activity researchers for measuring steps from walking or running (Bassett & Strath, 2002). Normally worn on the waistband, pedometers contain a horizontal spring-suspended lever arm that moves up and down in response to vertical acceleration (for an inside look, see Bassett & Strath, 2002, p. 165). Estimates of distance and speed can be made if stride length is known. Extensive validity and reliability tests have been conducted on a wide range of devices (e.g., Bassett & Strath, 2002; Dale et al., 2002; Marsh, Vance, & Frederick, 2007). The best brands are fairly accurate at counting steps, but are less accurate for estimating distance and intensity/EE. Like other waist-mounted devices, they cannot detect other types of physical activities such as arm movements or external work, and may incorrectly count steps when users are in a moving vehicle. They also do not work well for certain groups, such as the frail who walk at slow speeds or obese subjects. For these reasons, their application is limited.

9.2.4.2. Accelerometers Accelerometers are a very popular device for assessing physical activity in free-living conditions owing to their small size (no larger than pedometers), noninvasive nature, low respondent burden, and ability to provide objective measures of physical activity intensity/EE, duration, and type over long periods of time. A good overall introduction to these devices and their validity is provided in Welk (2002b); the physics and measurement principles are covered well in Chen and Bassett (2005); and comparison of some of the most popular devices is covered in King et al. (2004). In general terms, an electronic component in the device measures acceleration of the body part to which it is attached in one or more specific directions (up/down, side-to-side, back–forth) at frequent sub-second intervals. Acceleration is defined as the change in speed with respect to time; if information on stride length is available, the data can be post-processed into speed and distance values. "Uniaxial" accelerometers measure acceleration in the vertical plane, whereas "biaxial" and "triaxial" accelerometers measure movements in two and three dimensions respectively. Figure 9.1 depicts a small sample of the latter type of data. Note that accelerometers do not measure overall body acceleration, only acceleration of specific body part it is attached to (typically hip, but also arms, legs, and feet) and that signals during steady-state movements (such as in a car) will typically be lower in magnitude. If worn on only one body part (e.g., hip), it will be incapable of detecting all body movements.

A wide range of literature reports on the ability of accelerometers to detect the type and intensity of physical activity (see reviews in Welk, 2002b; Aminian & Najafi, 2004; King et al., 2004; Mathie, Coster, Lovell, & Celler, 2004; Chen & Bassett, 2005). A significant number of laboratory studies have demonstrated accurate linear relationships between accelerometer "counts" and EE during physical activities such as walking and running, but frequent underestimation of EE (as high as 67%) during free-living conditions (Welk, 2002b; King et al., 2004). The development of mathematical models of EE from accelerometer data is an ongoing effort (Chen & Bassett, 2005). In terms of activity detection, Ermes et al. (2008) point out that accelerometer data have been used to automatically detect physical activity types in a laboratory setting, but that long-term out-of-laboratory monitoring potential is not as clear or well tested.

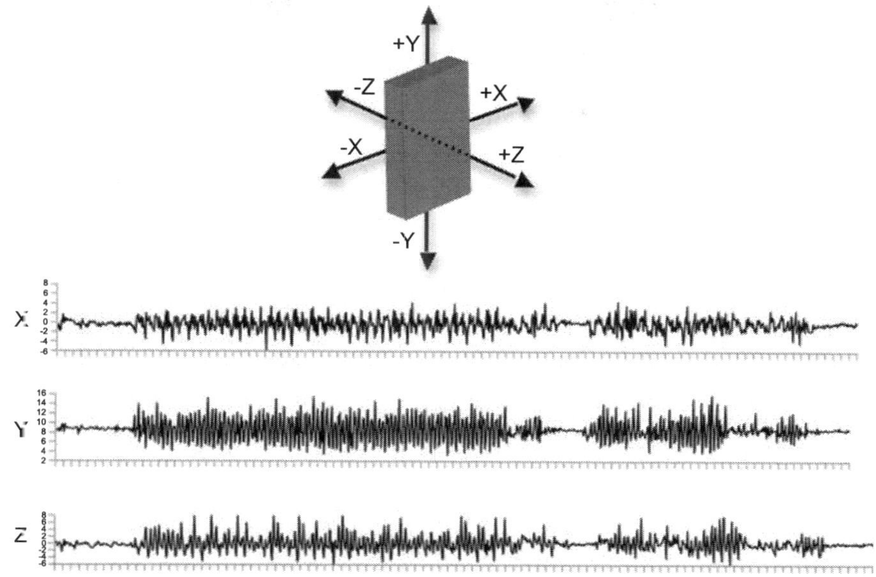

Figure 9.1: Example three-axis accelerometer data for 100-s period of walking. *Notes*: Illustration at top shows how each axis is oriented with respect to the device worn at the hip. Example is of person sitting down, then walking the dog, stooping to scoop (the pause about 2/3 way through), walking again, and then stopping. Measurements are in m/s^2 taken 10 times per second.

Recognizing various limitations of using single accelerometers, especially for free-living conditions, emerging approaches involve application of multiple sensors (e.g., on hip and other limbs), or combination with other physiological measurements (e.g., heart rate). In a short out-of-laboratory experiment with 24 subjects, Foerster, Smeja, and Fahrenberg (1999) placed accelerometers on subjects' sternum, wrist, thigh, and lower leg and had an observer take notes on nine physical activity patterns conducted by subjects in a laboratory and then under free-living conditions (sitting, standing, lying, sitting and talking, sitting and operating a PC, walking, taking stairs up, taking stairs down, and cycling). Based on these data, they developed an algorithm for predicting activity type obtaining an overall 96% accuracy prediction in the laboratory, but only of 67% under free-living conditions. Similarly, Parkka et al. (2006) developed a detection algorithm capable of detecting eight physical activities (lying, rowing, cycling, sitting, standing, running, Nordic walking, and walking) in a supervised free-living scenario, obtaining accuracy of 58–97%. Two key challenges to detecting physical activity type relate to the increased complexity/diversity of physical activity in daily life, and less control over being able to accurately "annotate" actual physical activity for verification and algorithm detection purposes.

An example off-the-shelf multi-accelerometer device is the intelligent device for energy expenditure and activity (IDEEA) portable monitor, which uses five integrated, thumbnail-sized sensors attached with tape to the chest, mid-thigh of both legs, and the soles of both feet (for images, see Maffiuletti et al., 2008). The sensors measure angles of body segments and accelerations in two orthogonal directions 32 times per second. Software that includes a neural network model outputs estimated EE, speed and distance, and an activity code that specifies the body position and activity being performed. This included 5 primary postures (sitting, standing, leaning, lying down, and limb movement), 27 secondary postures and limb movements, and 5 gait types (walking, running, climbing stairs, descending stairs, and jumping). Zhang et al. demonstrated breakthrough accuracy in a laboratory setting, finding that the IDEEA can detect the type, onset, duration, and intensity of most fundamental movements with 98% accuracy, near-perfect speed measurement, and EE estimates within 95–99% accuracy compared to biochemical measurement (calorimeter and a metabolic chamber) (Zhang, Werner, Sun, Pi-Sunyer, & Boozer, 2003; Zhang, Pi-Sunyer, & Boozer, 2004). Similar results have been found in subsequent studies under free-living conditions (Huddleston et al., 2006; Gardner et al., 2007; Maffiuletti et al., 2008). In comparison to the SWA armband monitor (see above) Welk, McClain, Eisenmann, and Wickel (2007) concluded that both are well suited for detecting free-living activity types and EE, but the SWA was less invasive and cumbersome for long-term monitoring compared to the multiple sensors and wires required of the IDEEA monitor.

9.2.4.3. Geo-location tracking In the last decade, several geo-location tracking technologies have become available that are small enough to be worn by a person, and provide accurate and frequent enough information on their location (typically longitude, latitude, and altitude) to allow analysis of motion and speed. The most popular are global positioning system (GPS) devices that receive information from a system of 24 satellites orbiting at high altitude allowing location to be calculated using triangulation (for a basic overview, see Bajaj, Ranaweera, & Agrawal, 2002). Others include: land-based tracking systems (e.g., time difference of arrival — see Shoval & Isaacson, 2006); cellular phone network-based location estimation (Asakura & Hato, 2004; Sayed, Tarighat, & Khajehnouri, 2005); WiFi fingerprinting (Retscher, 2007); and, to a limited extent, RFID tags (radiofrequency identification) that have shown some potential for indoor navigational aid (Szeto & Sharma, 2007). Various combinations of these technologies are also being experimented with to improve accuracy indoors and out, such as Assisted-GPS (GPS plus network-based location), GPS plus WiFi (Retscher, 2007), and GPS plus dead-reckoning via accelerometer. The most widely applied of these technologies to date is GPS, owing to its small size, accuracy, tracking frequency, low cost, and wide-scale availability. Many small off-the-shelf GPS receivers or GPS-enabled smartphones are now available that can track and store location to within 2–3 m accuracy on a second-by-second basis over multiday periods.

Early studies of the use of GPS in assessing physical activity focused on speed, distance, and gait. GPS receivers were found to provide highly accurate measure of the speed of walking, running, and cycling under open-sky test conditions to within about 0.08 km/h (Schutz & Chambaz, 1997; Schutz & Herren, 2000; Witte & Wilson, 2005; Townshend, Worringham, & Stewart, 2008), and distance to within about 3 m (Rodriguez, Brown, & Troped, 2005). Terrier et al. found that differentially corrected GPS with sub-centimeter accuracy is even capable of recording small body movements and thus is a promising tool for gait analysis (Terrier, Ladetto, Merminod, & Schutz, 2001; Terrier, Turner, & Schutz, 2005).

The further potential of geo-location technologies for assessing physical activity becomes apparent when examining traces of personal or vehicular movements over long periods of time (e.g., Murakami & Wagner, 1999; Wolf, Loechl, Thompson, & Arce, 2003; Stopher & Greaves, 2007). When viewed on a map (see example Figure 9.2), physical activity *types/intensities* and their *location* become much more readily apparent. Elgethun et al. overlaid children's GPS traces with aerial photos in a GIS and found they could manually determine subjects' *location* (inside, outside, in-building) and distinguish a variety of human activity *types* such as entering a retail store, walking on a sidewalk, traveling by car or bus, playing on a schoolyard, or

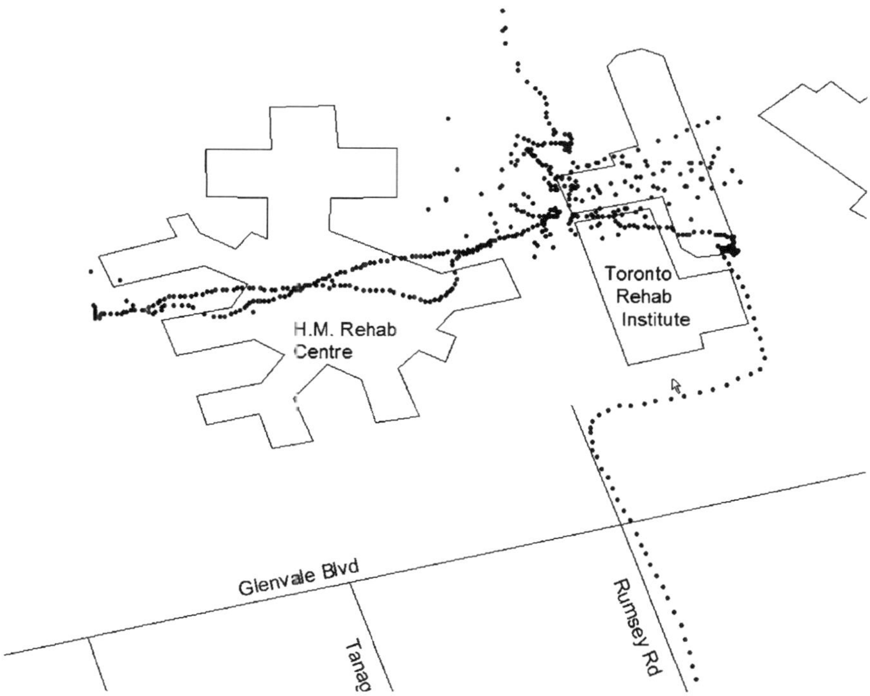

Figure 9.2: Example GPS data tracked every second, overlaid in a GIS. *Note*: Path on Rumsey Rd. is by bicycle, up to building edge where bike was parked. Remainder is walking or stationary activity. Note also how GPS track within the H.M. Rehab Centre more accurately portrays the walking that occurred, whereas GPS track within the Toronto Rehab Institute building exhibits more scatter indicative of a weaker GPS signal.

playing in and around a house (Elgethun, Fenske, Yost, & Palcisko, 2003; Elgethun, Yost, Fitzpatrick, Nyerges, & Fenske, 2007). Interestingly, when compared to parent-reported child activities via a manual diary, they found that parents misclassified 48% of their child's time. Rodriguez et al. (2005, p. s579) adopted a similar approach but used land use, road network, building footprints, and aerial photos. They were able to develop automated rules that determine the location of "bouts" of continuous physical activity (measured via accelerometer), classed as indoors, outdoors in the neighborhood, or outdoors out of the neighborhood. They found that 40.7% of bouts of physical activity had no GPS data, 14.4% had less than 30% of data, and 44.9% had more than 30% GPS data sufficient for accurately classifying the location. They suggest that increased accuracy may be gained by examining first and last successfully recorded GPS coordinates around a bout of physical activity to make inferences about its location.

One of the first attempts to explicitly validate GPS as a potential means to automatically detect walking versus resting episodes was reported by Le Faucheur et al. (2007). Subjects were given prescribed walking and rest routines in a park dictated to them orally via earphones, while being tracked with an off-the-shelf GPS receiver/logger with 1–3 m accuracy. They found initially that the GPS provided artificially high speeds on occasion (at start of walking, at other random times) and seldom provided true zero speeds during rest. However, with some automated data processing to correct these artifacts, 90% of walking and resting bouts could be automatically detected. A manual post-processing methodology using graphic analysis reached an accuracy of 97%.

Although not explicitly focused on physical activity, emerging from urban studies are additional attempts to use GPS to detect walking and cycling versus auto modes (Chung & Shalaby, 2005; Stopher, Jiang, & FitzGerald, 2005, 2008). Chung and Shalaby (2005) collected GPS traces for 60 simulated trips by bus, car, and walking, and kept a separate record of actual start/end times, mode, and route. After filtering out invalid or poor-quality data points, they developed sophisticated link-matching techniques in a GIS to identify routes, and automated rules for mode detection based on speed, speed range, and proximity of trip ends to transit stops. They report 79% classification accuracy on routes, and 92% of modes, using the same calibration data as test data. Recent work by Tsui and Shalaby (2007) has refined these techniques reporting 100% detection of stationary activities, and 91% of travel mode (based on a sample of nine people over 58 days of real-world GPS tracking, plus self-reported activities and trip annotation). However, details on the validation testing were not reported. Stopher et al. (2005) also used speed, acceleration, and route information along with a detail GIS of the road and transit network to predict travel mode. They first used extensive automated rules to process artifacts in the data, such as removing inaccurate GPS points, converting near-zero speed records to stationary points, and imputing values during cold starts and other signal losses such as in "urban canyons" and indoors. A series of rules were then applied to detect "trip ends" (no movement for 120 s or more) and mode of travel (car, walk, bicycle, bus, train). Although they report a success rate of about 95% in detecting modes, and 90% of trip ends, the data and validation methods are not reported.

Overall, while GPS appears to offer considerable potential for detecting physical activity under real-world conditions over long periods, few explicit studies of physical activity detection have been conducted, and the accuracy of detection algorithms has not been tested nearly as adequately as other physical activity technologies. In addition, significant data quality challenges remain, such as well-known signal outage issues indoors and in urban canyons. However, a unique advantage of GPS over other devices is the ability to obtain context-specific measures such as where and when the activity occurred, which obviously represents a key link to the built environment. This makes GPS particularly valuable as a supplemental device to other technologies (e.g., accelerometers, biometric monitors, heart monitors) that are more amendable to tracking a wider range of physical activities. Additionally, as GPS accuracy improves, especially indoors, finer patterns of movement may become detectable, such as moving about in the kitchen or garden or working out on stationary equipment. The example set by Terrier et al. (2001, 2005) using GPS with sub-centimeter accuracy to examine a person's gait suggests that potential exists on this front.

9.3. Moving Forward: A Three-Tiered Multi-Instrumented Approach

It seems clear from the review above that most individual methods of assessing physical activity have significant limitations with respect to their ability to detect the diverse types and intensity of physical activity, especially under free-living or real-world conditions. This would explain why many recent studies are experimenting with multiple instruments. Numerous studies in the literature have measured physical activity in laboratory settings using more than one instrument (Treuth, 2002; Chen & Bassett, 2005). Given the focus on the built environment, and the clear consensus reached by researchers from urban studies and health sciences over the need for a cost-effective method for comprehensively estimating physical activity in larger samples over longer periods of time under diverse, real-world built environment conditions (as identified in Section 9.1), a more limited combination/integration of methods appears viable.

The following three key generic components are thus proposed:

(1) *Passive location tracking*: Some form of person-based continuous location tracking technology appears necessary in order to establish meaningful, specific, and accurate links to environmental features encountered under free-living daily-life conditions over multiple days; GPS currently offers this ability, although continued improvements in indoor tracking sensitivity (via combination with other technologies such as WiFi or cell network towers) and artifact data processing is needed. Such technologies can also be used to assist with automatically detecting certain types of physical activity and intensities involving movement over space, such as walking and cycling (especially outdoors, but increasingly indoors).

(2) *Passive motion/biometric tracking*: A full accounting of physical activity will require the addition of some form of person-based continuous logging of EE or body motion, in particular to detect stationary and indoor physical activities. Multiple accelerometers, or single accelerometers combined with biometric tracking, appear most viable, although the latter device offers the advantage of being a single unit. Accelerometers can also be used to assist with location tracking by allowing dead-reckoning during short periods of GPS (or otherwise) signal loss.
(3) *Limited self-reporting*: An additional concern with the built environment is the desire to differentiate utilitarian and recreational types of physical activity, and capture other correlates of physical activity such as involved persons — attributes that most likely require some form of self-reporting. Keeping these to a minimum is key. *Integrating* self-reports and passively tracked data is a viable option, wherein passively detected patterns are displayed to users to prompt additional information, either *in situ* or after the fact.

In addition, early in application of these methods, additional objective direct observation methods may be needed to validate the method or develop algorithms for processing the data (e.g., for activity type detection).

As it turns out, several recent studies from both the health sciences and the urban studies realms reflect this three-tiered ideal, and perhaps serve as an indication of how best to integrate and move forward.

Ermes et al. (2008) combined three specific methods: (1) three-axis accelerometers located on wrist and hip; (2) GPS logging every 20 s; and (3) direct observation of physical activity type and start/end time using PDA software. They demonstrated how such data could be used to develop an algorithm for detecting physical activity type and duration during prescribed "supervised" itineraries (lying down, sitting/standing, walking, running, Nordic walking, rowing machine, stationary exercise bike, cycling with real bike, and playing football) and free-living conditions (only lying down, sitting/standing, walking, and cycling were observed), using a variety of explanatory variables derived from the combined sensors. These included peak frequency/range/spectral entropy of up-down acceleration and GPS speed. Using a combination of decision trees and artificial neural network classification models, they were successful in predicting 90% of prescribed physical activity types, and 72% of free-living physical activities — a significant improvement over past methods. Variation in classification accuracy was evident between activity types. They note in particular how GPS improved detection of cycling outdoors and how placement of accelerometer affects results (e.g., lack of leg sensor made differentiation of sitting and standing difficult). They suggest that the main challenge for the future is fine-tuning the activity-detection algorithms using more free-living data so that a wider spectrum array of real-life physical activities can be detected. Although not discussed, the existing use of GPS could be extended to allow aspects of the built environment to be examined and to assist with identifying additional explanatory variables for classification.

A similar but more extensive daily-life study of children with explicit focus on the built environment and the role of parents has recently been completed by Mackett et al. (Mackett, Brown, et al., 2007; Mackett, Gong, Kitazawa, & Paskins, 2007; Mackett, 2008). They combine: (1) a three-axis accelerometers recording at 1 min intervals; (2) a GPS recording about every 10 s; and (3) a self-reported activity diary. Methodologically, the authors highlight some of the complexities of combining such data, especially reconciling separately collected self-report diaries with GPS data. Two key problems occurred: unreported trips in the diary were detected by GPS, and GPS data were missing due to signal loss and battery failure. Conceptually, the authors demonstrate some new ways to map the data, including combining accelerometer and GPS data for multiple physical activity types and locations (for an interesting animated map, see www.casa.ucl.ac.uk/capableproject/maps/home.asp). The authors were able to subsequently explore EE in varying specific environments, including on land versus roadways (but not in private spaces such as indoors), and during play, in clubs, or during walking. They also note that much more work is to be done, including larger datasets, more specific land-use classification, and analysis of wider variety of environments.

One way to overcome the methodological complexities of combining GPS, accelerometer, and diary data identified by Mackett et al. is to *integrate* the passively detected activities and attributes (from GPS/accelerometer data) into the diary prior to subjects completing it, rather than collecting them in isolation. In the urban studies field, these hybrid surveys are known as "prompted recall diaries". There are several distinct methods/interfaces for conducting prompted recall surveys, including *in situ*, spatial, temporal/tabular, or some combination of these. They are reviewed in depth here as they represent a key way to more fully integrate the three 3-tiered multi-instrument approach identified above.

In-situ prompted recall would involve processing data in real time on the device, allowing opportunities for more "interactive" survey approaches. For instance, algorithms could be developed to automatically detect episodes of physical activity based on live streaming sensor data (be it GPS, accelerometer, pedometer, heart rate, biochemical, etc.). This could then "trigger" interactive queries for more qualitative data on the physical activity (intensity, type, pain, emotions, etc.) and contextual data (involved persons, natural/built environment) that are better captured *in situ*. Modern smartphones offer efficient means for capturing such data via textual prompts, voice recordings, or even via pictures and video of the surrounding environment. At this point, however, automated detection algorithms are likely not robust enough to support such an approach. A less ideal alternative would be to ask subjects to manually annotate their behavior immediately prior to or following a specified set of activities or events. One of the few large-sample examples of this approach was conducted by Murakami and Wagner (1999) who used GPS along with a hand-held computer to prompt for trip purpose and passenger information immediately prior to automobile trips. More recently, Zhou and Golledge (2007) pilot-tested a person-based system involving manual coding of activity and trip onset on a hand-held computer, followed by GPS tracking.

A temporal/tabular prompted recall diary would involve some form of time-ordered after-the-fact display of passively detected stationary activities and movements and their attributes (type, start and end times, location, route, etc.).

Subjects could then be asked to correct, confirm, and/or add supplemental information to such a display. An example of such a system is shown in Figure 9.3 (Doherty, Noël, Lee-Gosselin, Sirois, & Ueno, 2001; Doherty, Papinski, & Lee-Gosselin, 2006) pilot-tested recently by Clark and Doherty (2009). GPS-enabled BlackBerry smartphones were used to capture second-by-second data transferred to a central server for processing. An automated detection algorithm reads in GPS data, and outputs an event listing of stationary activities and movements/trips, including their type (travel mode in case of movement, indoor/outdoor in case of activity), and location (mapped, along with label derived from nearest land use and roadway). Interaction with the diary involves scrolling up/down in time and clicking on individual attributes to confirm, update, or add information. Event types are specified via fly-out list, locations via an interactive map, and time via + / − controls or stretching. Other attributes such as involved persons form additional columns. Current analysis is focusing on the accuracy and refinement of algorithms.

A GIS/spatial interface for prompted recall would involve the generation of a map showing a person's trip routes, activity stops/location, and an array of text boxes or other map attributes depicting such items as mode, speed, location name, start/end times, and trip and activity sequence (see Figure 9.4). All this would be overlaid on the road and land-use network for context. Again, the task of the user would be to correct, confirm, and/or add supplemental information, either directly on the map or via some alternative input means. This could be done using a paper-and-pencil map, or interactively on a computer screen. An early unpublished example is by Marca (2002) who used a simple map generator coupled with an HTML form for input of attributes such as destination name and involved persons for each trip segment displayed in the map. More recent applications are reported by Stopher et al. (2005), Stopher, FitzGerald, and Xu (2007), Li and Shalaby (2008), and Auld, Williams, Mohammadian, and Nelson (2009) who presented GPS-detected trips and stops/activities on a map and then prompted for additional information such as passengers, trip purposes, and travel costs. Stopher et al. used partially automated routines to initially detect trips, followed by manual preparation of tables and maps mailed to subjects. Tabular displays listed trips by end time, travel time, distance, location of stop, and a reference to a map depicting the trip. The GIS-produced maps display color-coded trips by route, direction of travel, and stop location, overlaid on the transport and land-use network. An eight-page questionnaire was then used to systematically prompt for stop purposes and travel modes for each trip detected, as well as queries for undetected stops and/or incorrectly identified stops. Alternatively, Li and Shalaby and Auld et al. (shown in Figure 9.4) used a Web-based interface to display maps and tabular information on automatically detected activities and trips.

Overall, prompted recall interfaces offer several key advantages for thoroughly exploring physical activity in the built environment:

- reduce respondent burden (passive tracking, limited self-reporting, memory jogging);
- are more accurate and detailed than self-reports (owing to passive location and body motion tracking technologies, assuming automated/manual detection is accurate and validated);

a) Initial diary with automatically detected activities/trips

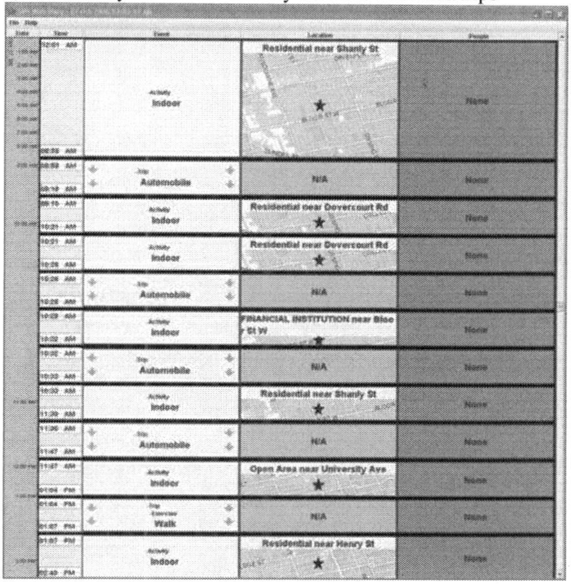

b) After prompted recall interaction with subject

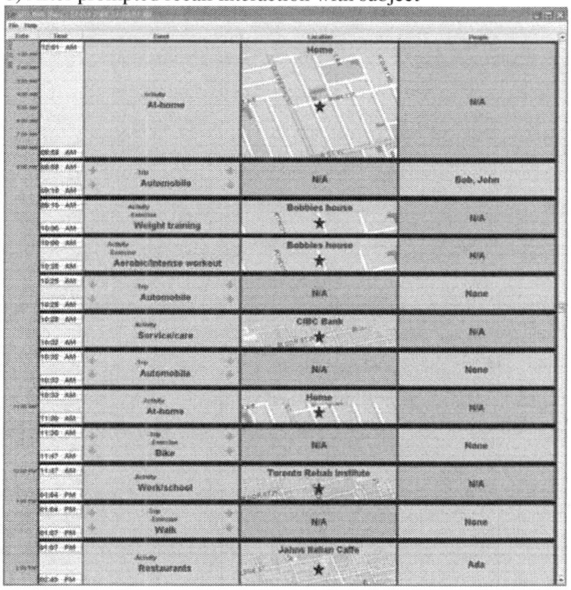

Figure 9.3: Example Web-based temporal GPS prompted recall diary screenshots. (a) Initial diary with automatically detected activities/trips. (b) After prompted recall interaction with subject. *Notes:* Original version is in color. Users can click on any attribute and modify/update interactively (times via +/−, event type via fly-out listing, locations via interactive map, and people via checkbox). Backgrounds convey the status of each attribute box (red: attribute requires manual entry; yellow: attribute can be updated with more detail; green: attribute completed). Raw GPS data are not shown to subjects; only locations of activities are depicted, centered on the spatial median of associated GPS points. Data shown are from a research assistant. Images provided by the author. See also Doherty et al. (2006) or Clark and Doherty (2009).

a) Initial screen for confirming accuracy of detected activities and trips

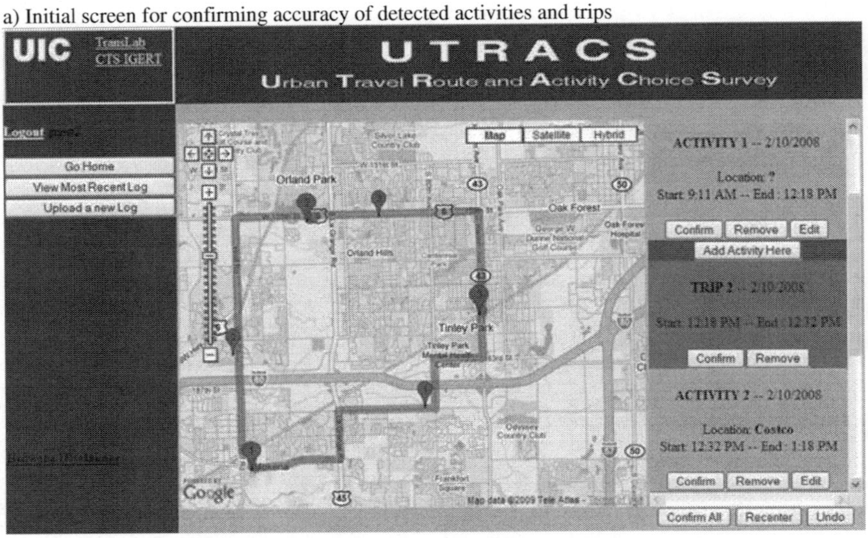

b) One of several screens to elicit further activity/trip attributes.

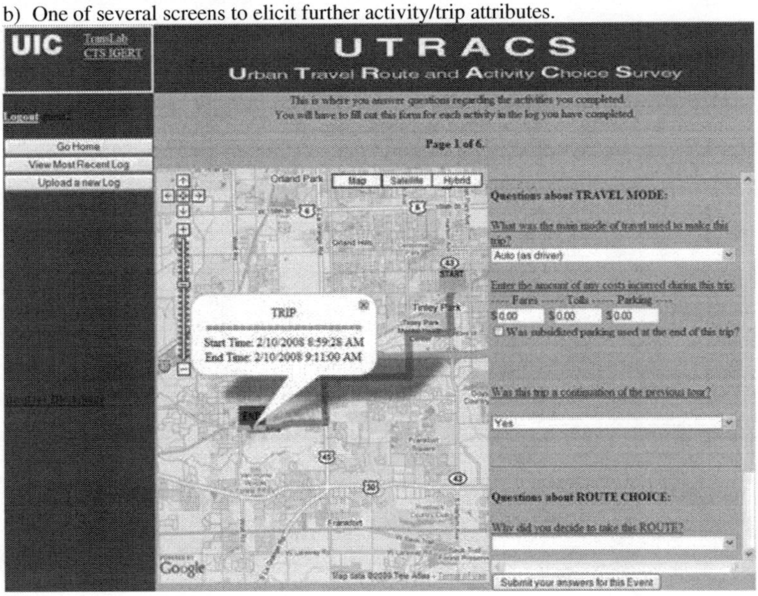

Figure 9.4: Example Web-based spatial GPS prompted recall diary screenshots. *Notes*: Original version is in color. The first screenshot (a) depicts smoothed GPS-tracked data with numbered callouts for each activity/trip — users first confirm that each activity and trip is correct. This procedure is followed by a sequence of screens that elicit further attributes of activities and trips, as shown in the second screenshot (b). Images provided for this paper by Josh Auld, University of Illinois at Chicago. See also Auld et al. (2009).

- allow additional attributes of interest to be optionally added manually by subjects;
- allow any cases of missing data or undetected activities to be correct/completed by the real "expert" (the subject), rather than imputed, allowing more thorough assessment.

It can be said, however, that such interfaces are still very much in development, as few large samples or thorough assessments of accuracy, validity, and reliability have been conducted. The key challenge will be the development of valid and reliable automated detection algorithms that incorporate the various streams of passively collected data for detecting the type, intensity, duration, and location of all physical activities conducted during daily life. Equally important is development of user-friendly interfaces for displaying temporal and spatial patterns of activity back to subjects for interaction purposes. Wireless transmission of the data to a central server may be needed to accomplish this (such as through cellular phone wireless network), both to allow easy access to interfaces via the Web, and to be able to effectively utilize large GIS databases or computationally intense data processing techniques beyond the capabilities of wearable devices.

9.4. Key Current Measurement Issues and Challenges

The following issues represent a prioritized list of challenges with respect to measuring physical activity in the built environment using the emerging three-tiered multi-instrumented methods outlined above.

9.4.1. Validation: An Immediate Need

Even integrated methods that combine the best available technologies are still in need of truly objective data to test their validity — the degree to which a measure truthfully reflects what it is intended to measure (in this case, the types, intensity, and location of physical activity). Simply put, if the instrument cannot be trusted as a result of poor reliability or validity, then it could very well give inaccurate or misleading information and cause one to draw invalid conclusions (Marrow, 2002). Despite their often high success rates under laboratory or simulated free-living conditions, all the devices reviewed here tend to perform much less accurately under real-world conditions, if tested at all. Additionally, most attempts at validation under real-world conditions are based on limited sample sizes, a limited set of physical activity types, pertain to a limited set of attributes, and/or do not include adequate details on validation methods. Many real-world tests also suffer from common data problems/complexities: equipment failures, user errors, signal complexities/outages, etc., most of which are not well documented or assessed.

Thus, a key immediate issue for emerging physical activity measurement methods, prior to drawing strong conclusions about their relationship to the built environment, is proper validation. Indeed, the lack of proper validation may be the single most important force hindering our ability to make solid conclusions thus far. For this, we should take our queue from the rigorous validation methods deployed in the health sciences. Extensive efforts have been made to test and compare the reliability and validity of any given physical activity measurement method in the health sciences (Mahar & Rowe, 2002; Marrow, 2002). Despite this, there is still a surprising amount of variability in the validity of all the methods. For instance, EE is often over/underestimated by 30–70% (Montoye et al., 1996; Dale et al., 2002; Morrow, 2002), leading Dale et al. (2002, p. 25) to conclude that "a major challenge in physical activity assessment research is the lack of a true gold standard of measurement". In comparison, in urban studies, we have been using self-reported travel diaries as the main source of data for decades, yet only recently has their validity been seriously examined using objective data derived from GPS, finding significant trip underreporting rates (Wolf et al., 2003).

If we take anything from this experience, it should be the importance of not putting blind faith in any given method. Applying such rigor to testing the validity of other aspects of physical activity, particularly including emerging methods attempting to detect activity type, duration, and location, is a critical development issue. This will mean that in the short term, we need to invest strategically in a reliable *fourth* component to the emerging three-tiered multi-instrumented approach (passive location tracking, passive motion/biometric tracking, limited self-reporting) that simultaneously provides truly objective data on physical activity types, intensities, and location. Techniques such as "direct observation" and "direct biochemical measurement", often ruled out as too tedious or expensive, should be deployed in the early stages at least for subsamples. In lieu of such direct measures, self-reported (or "annotated") physical activity *in situ* may be a viable alternative if implemented carefully. Hand-held devices and simple software would appear to offer potential on this front; getting consistent results under real-world conditions while minimizing respondent burden will be a key challenge.

9.4.2. *Automated Physical Activity-Detection Algorithms*

On a related note, validation efforts have a side benefit: they provide the "true" baseline measures of physical activity that are an essential ingredient to development of automated detection algorithms of the types, intensities, and location context of physical activity. Development of these algorithms is very much in its infancy, in both the urban studies and the health science disciplines, as shown above. In the health sciences, the focus is often more on detecting the intensity of physical activity in the form of EE; for making the connection to the built environment, location and activity type are more critical, as specific features of the environment are known to impact differentially on utilitarian and recreational types of physical activity. It turns out that there are also interesting synergies: detecting the type of physical activity first allows more accurate estimation of EE; measuring intensity of activity also assists in detecting type. Combining multiple types of sensors provides new opportunities for identifying explanatory factors that increase the accuracy of these algorithms, but we have barely scratched the surface of this potential. An exception is the work of Ermes et al., (2008) who used four explanatory variables derived from an accelerometer and one from GPS (speed) to detect activity type; those with more recent GPS experience (e.g., Chung & Shalaby, 2005; Stopher et al., 2005, 2008; Li and Shalaby, 2008) would suggest that many more GPS-derived variables are possible (e.g., post-processed proximity to roadway types). Bringing these efforts together would appear to be a key developmental opportunity.

9.4.3. Sorting Out the Causation Dilemma

> Existing research cannot disentangle, for example, if the observed association between certain neighborhood characteristics (e.g., high population density, mixed land use, good sidewalks) and higher-than-average levels of walking reflect the effects of the built environment on physical activity or the residential preferences among people who enjoy walking, which influence their decisions to live in such neighborhoods. The ability to answer this puzzle is important for policy: if the observed association between the built environment and physical activity is due to self-selection, then changes to the built environment, which can be very costly, are unlikely to yield the desired significant increase in physical activity (Hanson, 2006, p. S259).

As so clearly pointed out by Susan Handy, and widely recognized by others (e.g., Bhat & Guo, 2007), an *association* between a built environment attribute and a health-related characteristic does not imply *causation*, given the complicating residential self-selection bias. More simply put, "people who like to walk choose walkable neighborhoods" (Frank, Sallis, Saelens, Bachman, & Washbrook, 2005, p. 14). Two ways to deal with this are assessing residential self-selection bias through a questionnaire and subsequently controlling for it during analysis by using a variety of quasi-longitudinal designs (Frank, Saelens, Powell, & Chapman, 2007; Mokhtarian & Cao, 2008); or conducting true longitudinal studies before and after a person experiences a change in the built environment.

The implication for physical activity measurement is twofold: (1) more accurate and valid measurement would allow for smaller samples and (2) more passive and automated detection algorithms would support the longer observation periods required of longitudinal studies.

9.4.4. The Need for Cross-Cultural Comparison

No matter how valid and accurate the measurement, if significant variance in built environments and/or physical activities of the sample does not exist, there will be little opportunity for identifying healthy environments and people. Considering the large number of correlating factors, disentangling subtle differences in the built environments that exist across sprawling North American cities is surely a challenge; this is compounded by low rates of walking behavior in the first place. It would seem prudent to launch international cross-cultural studies, in particular in cities with contrasting built environments where physical activity continues to be woven into everyday life. Passive collection of physical activity data will assist in applying methods to diverse cultural groups and in different languages.

9.5. Conclusions: New Opportunities at the Crossroads of Urban Studies and Health Sciences

This paper draws upon methods from two loosely defined fields of inquiry: urban studies and health sciences. On close examination, some of the methods are not all that different (e.g., diary methods), whereas in other ways they contrast significantly, such as the focus on accelerometers in health studies and GPS in urban studies. In the end, it appears that researchers in both fields have arrived at the same conclusion: multiple methods are needed to assess physical activity in the built environment. This paper has identified three necessary components of this integrated method, building on the latest studies. It is fitting to end this paper with a look at some of the key lessons learned and new opportunities that have emerged at the crossroads of urban studies and health sciences.

Let's validate our methods before we draw conclusions: Reiterating the discussion above, the number one lesson from the health sciences is the priority they place on validation. No measurement device is ever accepted as a means to draw conclusions about health until it has been repeatedly validated with respect to what it is intending to measure, and is deemed reliable across varying people and situations. Despite the plethora of techniques reviewed here, no gold standard exists, even from the health sciences. Drawing strong conclusions about the impacts of the built environment on health should be viewed with considerable caution until we clearly establish the validity of the necessary emerging integrated methods. This will require more concerted effort in the urban studies, larger tests samples, and more consistent assessment means.

More accurate assessment means smaller sample sizes and less chance of erroneous conclusions: Recognizing the limitations of self-reports, both fields of inquiry have aggressively sought more objective, passive, automated technologies for assessing physical activity. These devices are surely only going to get smaller, less power hungry, and more sensitive, and will offer more integrated capabilities. Even biochemical measurement has emerged from the laboratory in the form of a wearable device, just as heart-rate monitors did decades before. We need to cautiously, but aggressively, keep up with these methods, initially with small sample experimentation, but quickly followed by validation studies and applications.

There is more to physical activity than just walking: Current urban studies are often limited to examining walking as an indicator of physical activity; few successful attempts have been made to examine recreational and stationary physical activities. Adopting methods from the health sciences that focus on assessing all physical activity types (moving and stationary) will surely expand this, and tie results more holistically to health.

There is more to daily life and health than physical activity: Obviously, you do not have to tell health scientists that there are other correlates of health, but the activity-based methods emerging from urban studies would surely broaden the scope of analysis beyond physical activities in isolation. This includes especially *social activities* that affect mental health, and *dietary activities* (eating in-home, preparing foods, eating out, grocery shopping) that affect weight and well-being, both of which have strong ties to the built environment and accessibility. Extending our observation to include these activities will provide for a more holistic assessment of health, especially in built environments with low walkability, poor access to fresh foods (or too many fast-food restaurants), and isolating social networks.

Similarly, there is more to the built environment than just the home: The field of urban studies has a long history of examining the location of activities, whereas in health sciences, the location "context" within which physical activity takes place has received only recent attention. Building on this, we need to refine the scale and extent of locations considered to impact physical activity using increasing accurate and longitudinal datasets, especially considering those around multiple anchor points of daily life beyond just the home. Additionally, a key challenge will be acquiring accurate and up-to-date data on the key features and attributes within these contexts that affect physical activity. While current GIS databases provide a reasonable accounting of the "quantity" of these attributes (e.g., densities, lengths, connectivity), the same cannot be said of the environmental "quality" attributes that can have a strong affect on behavior such as lighting and safety, which may require more direct observation.

The built environment is not the only constraint on physical activity: While health scientists often study physical activity at discre, prescribed moments in the laboratory, physical activity in the real world is part of a continuous decision-making process constrained not only by the external built environment, but also by personal constraints. Evolving from urban studies are new methods that provide more holistic insights into the spatial, temporal, and interpersonal scheduling constraints on daily activities. Applying these approaches to physical activity is likely to provide wholly new insights into how it is planned and managed within a person's everyday life and barriers to its completion.

Acknowledgements

The author would like to acknowledge the financial support received from the Canadian Institutes of Health Research (CIHR), the Social Sciences and Humanities

Research Council of Canada, and from the GEOIDE (Geomatics for Informed Decisions) Network of Centres of Excellence Program of the Canadian federal research councils.

References

Ainsworth, B. E., Haskell, W., Whitt, M., Irwin, M., Swartz, A., Strath, S. J., O'Brien, W., Bassett, D. R. J., Schmitz, K. H., Emplaincourt, P., Jacobs, D., & Leon, A. (2000). Compendium of physical activities: An update of activity codes and MET intensities. *Medicine & Science in Sports & Exercise*, *32*(9), S498–S516.

Aminian, K., & Najafi, B. (2004). Capturing human motion using body-fixed sensors: Outdoor measurement and clinical applications. *Computer Animation and Virtual Worlds*, *15*(2), 79–94.

Andre, D. (Ed.) (2007). Letters to the editor: Assessing resting metabolic rate using a multi-sensor armband. *Obesity*, *15*(5), 1337–1338.

Arvidsson, D., Slinde, F., Larsson, S., & Hulthen, L. (2007). Energy cost of physical activities in children: Validation of SenseWear Armband. *Medicine & Science in Sports & Exercise*, *39*(11), 2076–2084.

Asakura, Y., & Hato, E. (2004). Tracking survey for individual travel behaviour using mobile communication instruments. *Transportation Research Part C*, *12*(3–4), 273–291.

Auld, J. A., Williams, C., Mohammadian, A., & Nelson, P. (2009). An automated GPS-based prompted recall survey with learning algorithms. *Journal of Transportation Letters*, *1*(1), 59–79.

Bajaj, R., Ranaweera, S. L., & Agrawal, D. P. (2002). GPS: Location-tracking technology. *Computer*, *35*(4), 92–94.

Barnard, P. (1986). Use of an activity diary survey to examine travel and activity reporting in a home interview survey: An example using data from Adelaide, Australia. *Transportation*, *13*(4), 329–357.

Bassett, D. R., & Strath, S. J. (2002). Use of pedometers to assess physical activity. In: G. J. Welk (Ed.), *Physical activity assessments for health-related research* (pp. 163–177). Champaign, IL: Human Kinetics Publishers.

Bhat, C. R., & Guo, J. Y. (2007). A comprehensive analysis of built environment characteristics on household residential choice and auto ownership levels. *Transportation Research Part B: Methodological*, *41*(5), 506–526.

Boarnet, M., & Crane, R. (2001). *Travel by design: The influence of urban form on travel.* New York, NY: Oxford University Press.

Boarnet, M. G. (2005). *The built environment and physical activity: Empirical methods and data resources.* TRB Special Report 282. Washington, DC: Transportation Research Board and the Institute of Medicine, Committee on Physical Activity, Health, Transportation, and Land Use.

Booth, K. M., Pinkston, M. M., & Poston, W. S. C. (2005). Obesity and the built environment. *Journal of the American Diabetic Association*, *105*(Suppl. 5), 110–117.

Bouchard, C. (1997). Bouchard three-day physical activity record. *Medicine & Science in Sports & Exercise*, *29*(6), 19–24.

Bouchard, C., Tremblay, A., Leblanc, C., Lortie, G., Savard, R., & Theariault, G. (1983). A method to assess energy expenditure in children and adults. *American Journal of Clinical Nutrition*, *37*, 461–467.

Brage, S., Brage, N., Franks, P. W., Ekelund, U., Wong, M. Y., Andersen, L. B., Froberg, K., & Wareham, N. J. (2004). Branched equation modeling of simultaneous accelerometry and heart rate monitoring improves estimate of directly measured physical activity energy expenditure. *Journal of Applied Physiology*, *96*(1), 343–351.

Brownson, R. C., Boehmer, T. K., & Luke, D. A. (2005). Declining rates of physical activity in the United States: What are the contributors?. *Annual Review of Public Health*, *26*, 421–443.

Cereda, E., Turrini, M., Ciapanna, D., Marbello, L., Pietrobelli, A., & Corradi, E. (2007). Assessing energy expenditure in cancer patients: A pilot validation of a new wearable device. *Journal of Parenteral and Enteral Nutrition*, *31*(6), 502–507.

Chen, K. Y., & Bassett, D. R. J. (2005). The technology of accelerometry-based activity monitors: Current and future. *Medicine & Science in Sports & Exercise*, *37*(11S), 490–500.

Chung, E.-H., & Shalaby, A. (2005). A trip reconstruction tool for GPS-based personal travel surveys. *Transportation Planning and Technology*, *28*(5), 381–401.

Clark, A., & Doherty, S. T. (2009). Examining the nature and extent of the activity-travel preplanning decision process. *Journal of the Transportation Research Board*, *2054*, 83–92.

Copperman, R. B., & Bhat, C. R. (2007). An analysis of the determinants of children's weekend physical activity participation. *Transportation*, *34*, 67–87.

Crane, R. (2000). The influence of urban form on travel: An interpretive review. *Journal of Planning Literature*, *15*(1), 3–23.

Cromley, E. K., & McLafferty, S. L. (2002). *GIS and public health*. New York, NY: Guilford.

Dale, D., Welk, G. J., & Matthews, C. E. (2002). Methods for assessing physical activity and challenges for research. In: *Physical activity assessments for health-related research* (pp. 19–34). Champaign, IL: Human Kinetics Publishers.

Dannenberg, A. L., Jackson, R. J., Frumkin, H., Schieber, R. A., Pratt, M., Kochtitzky, C., & Tilson, H. H. (2003). The impact of community design and land-use choices on public health: A scientific research agenda. *American Journal of Public Health*, *93*(9), 1500–1508.

Davison, K. K., & Lawson, C. T. (2006). Do attributes in the physical environment influence children's physical activity? A review of the literature. *International Journal of Behavioral Nutrition and Physical Activity*, *3*(19), 1–17.

Doherty, S. T., & Miller, E. J. (2000). A computerized household activity scheduling survey. *Transportation*, *27*(1), 75–97.

Doherty, S. T., Nemeth, E., Roorda, M., & Miller, E. J. (2004). Design and assessment of the Toronto area computerized household activity scheduling survey. *Journal of the Transportation Research Board*, *1894*, 140–149.

Doherty, S. T., Noël, N., Lee-Gosselin, M. E. H., Sirois, C., & Ueno, M. (2001). Moving beyond observed outcomes: Integrating Global Positioning Systems and interactive computer-based travel behaviour surveys. *Transportation Research Circular*, *E-C026*, 449–466.

Doherty, S. T., Papinski, D., & Lee-Gosselin, M. (2006). An internet-based prompted recall diary with automated GPS activity-trip detection: System design. *Proceedings of the 85th Annual Meeting of the Transportation Research Board*, Washington, DC.

Dorminy, C. A., Choi, L., Akohoue, S. A., Chen, K. Y., & Buchowski, M. S. (2008). Validity of a multisensor armband in estimating 24-h energy expenditure in children. *Medicine & Science in Sports & Exercise*, *40*(4), 699–706.

Elgethun, K., Fenske, R. A., Yost, M. G., & Palcisko, G. J. (2003). Time-location analysis for exposure assessment studies of children using a novel global positioning system instrument. *Environmental Health Perspectives*, *111*(1), 115–122.

Elgethun, K., Yost, M. G., Fitzpatrick, C. T., Nyerges, T. L., & Fenske, R. A. (2007). Comparison of global positioning system (GPS) tracking and parent-report diaries to characterize children's time-location patterns. *Journal of Exposure Science & Environmental Epidemiology*, *17*(2), 196–206.

Ermes, M., Parkka, J., Mantyjarvi, J., & Korhonen, I. (2008). Detection of daily activities and sports with wearable sensors in controlled and uncontrolled conditions. *IEEE Transactions on Information Technology in Biomedicine*, *12*(1), 20–26.

Ewing, R., & Cervero, R. (2001). Travel and the built environment — A synthesis. *Transportation Research Record*, *1780*, 87–114.

Fares, S., Miller, M., Masters, S., & Crotty, M. (2008). Measuring energy expenditure in community-dwelling older adults: Are portable methods valid and acceptable? *Journal of the American Dietetic Association*, *108*(3), 544–548.

Fields, D. A., Kearney, J. T., & Copeland, K. C. (2006). MedGem hand-held indirect calorimeter is valid for resting energy expenditure measurement in healthy children. *Obesity*, *14*, 1755–1761.

Foerster, F., Smeja, M., & Fahrenberg, J. (1999). Detection of posture and motion by accelerometry: A validation study in ambulatory monitoring. *Computers in Human Behavior*, *15*(5), 571–583.

Forsyth, A., Oakes, J. M., Schmitz, K. H., & Hearst, M. (2007). Does residential density increase walking and other physical activity?. *Urban Studies*, *44*(4), 679–697.

Frank, L., Saelens, B. E., Powell, K. E., & Chapman, J. (2007). Stepping towards causation: Do built environments or neighborhood and travel preferences explain physical activity, driving, and obesity?. *Social Science & Medicine*, *65*(9), 1898–1914.

Frank, L., Sallis, J., Saelens, B., Bachman, W., & Washbrook, K. (2005). Travel behavior, environmental, & health impacts of community design & transportation investment: A study of land use, transportation, air quality and health in King County, WA. King County, Washington, DC.

Frank, L. D., & Engelke, P. (2005). Multiple impacts of the built environment on public health: Walkable places and the exposure to air pollution. *International Regional Science Review*, *28*(2), 193–216.

Frank, L. D., Engelke, P. O., & Schmid, T. L. (2003). *Health and community design*. Washington, DC: Island Press.

Fruin, M. L., & Rankin, J. W. (2004). Validity of a multi-sensor armband in estimating rest and exercise energy expenditure. *Medicine & Science in Sports & Exercise*, *36*(6), 1063–1069.

Frumkin, H., Frank, L., & Jackson, R. (2004). *Urban sprawl and public health: Designing, planning, and building for healthy communities*. Washington, DC: Island Press.

Gardner, M. J., Barker, J. U., Briggs, S. M., Backus, S. I., Helfet, D. L., Lane, J. M., & Lorich, D. G. (2007). An evaluation of accuracy and repeatability of a novel gait analysis device. *Archives of Orthopaedic and Trauma Surgery*, *127*(3), 223–227.

Handy, S. L. (2005). *Critical assessment of the literature on the relationships among transportation, land use, and physical activity*. TRB Special Report 282. Washington, DC: Transportation Research Board and the Institute of Medicine Committee on Physical Activity, Health, Transportation, and Land Use.

Hanson, S. (2006). Active living research in light of the TRB/IOM reports. *Journal of Physical Activity and Health*, *3*(Suppl. 1), S258–S266.

Heath, G. W., Brownson, R. C., Kruger, J., Miles, R., Powell, K. E., & Ramsey, L. T.Task Force on Community Preventive Services. (2006). The effectiveness of urban design and land use and transport policies and practices to increase physical activity: A systematic review. *Journal of Physical Activity and Health*, *3*(Suppl. 1), S55–S76.

Hlynsky, J., Birmingham, C. L., Johnston, M., & Gritzner, S. (2005). The agreement between the MedGem indirect calorimeter and a standard indirect calorimeter in anorexia nervosa. *Eating and Weight Disorders*, *10*(4), e83–e87.

Huddleston, J., Alaiti, A., Goldvasser, D., Scarborough, D., Freiberg, A., Rubash, H., Malchau, H., Harris, W., & Krebs, D. (2006). Ambulatory measurement of knee motion and physical activity: Preliminary evaluation of a smart activity monitor. *Journal of NeuroEngineering and Rehabilitation*, *3*(1), 21.

Humphrey, N. P. (2005). Does the built environment influence physical activity? Examining the evidence. *TR News*, *237*, 31.

Institute for Social Research. (2002). *2002 time diary questionnaires from the child development supplement (CDS) to the panel study of income dynamics*. Institute for Social Research, University of Michigan.

Jakicic, J. M., Marcus, M., Gallagher, K. I., Randall, C., Thomas, E., Goss, F. L., & Robertson, R. J. (2004). Evaluation of the SenseWear Pro Armband to assess energy expenditure during exercise. *Medicine & Science in Sports & Exercise*, *36*(5), 897–904.

Janz, K. F. (2002). Use of heart rate monitors to assess physical activity. In: Welk, G. J. (Ed.), *Physical activity assessments for health-related research* (pp. 143–161). Champaign, IL: Human Kinetics Publishers.

King, G., Torres, N., Potter, C., Brooks, T., & Coleman, K. (2004). Comparison of activity monitors to estimate energy cost of treadmill exercise. *Medicine & Science in Sports & Exercise*, *36*(7), 1244–1251.

Le Faucheur, A., Abraham, P., Jaquinandi, V., Bouye, P., Saumet, J. L., & Noury-Desvaux, B. (2007). Study of human outdoor walking with a low-cost GPS and simple spreadsheet analysis. *Medicine & Science in Sports & Exercise*, *39*(9), 1570–1578.

Lee, M., & McNally, M. G. (2006). An empirical investigation on the dynamic processes of activity scheduling and trip chaining. *Transportation*, *33*(6), 553–565.

Li, Z. J., & Shalaby, A. (2008). Web-based GIS system for prompted recall of GPS-assisted personal travel surveys: System development and experimental study. *Proceedings of the 87th Annual Meeting of the Transportation Research Board*, Washington, DC.

Liden, C. B., Wolowicz, M., Stivoric, J., Teller, A., Vishnubhatla, S., Pelletier, R., & Farringdon, J. (2002). *Accuracy and reliability of the SenseWear Armband as an energy expenditure assessment device*. Pittsburgh, PA: BodyMedia, Inc.

Mackett, R. (2008). Understanding the impacts of the travel behaviour and activity patterns of children on their physical activity and health. Paper presented at 8th international conference on survey methods on transport: Harmonisation and data comparability, May 25–31, Annecy, France.

Mackett, R. L., Brown, B., Gong, Y., Kitazawa, K., & Paskins, J. (2007a). Children's independent movement in the local environment. *Built Environment*, *33*(4), 454–468.

Mackett, R. L., Gong, Y., Kitazawa, K., & Paskins, J. (2007). Children's local travel behaviour — How the environment influences, controls and facilitates it. Paper presented at the 11th world conference on transport research, June 24–28, Berkeley, CA.

Maffiuletti, N. A., Gorelick, M., Kramers-de Quervain, I., Bizzini, M., Munzinger, J. P., Tomasetti, S., & Stacoff, A. (2008). Concurrent validity and intrasession reliability of the IDEEA accelerometry system for the quantification of spatiotemporal gait parameters. *Gait & Posture*, *27*(1), 160–163.

Mahar, M. T., & Rowe, D. A. (2002). Construct validity in physical activity research. In: G. J. Welk (Ed.), *Physical activity assessments for health-related research* (pp. 51–72). Champaign, IL: Human Kinetics Publishers.

Malavolti, M., Pietrobelli, A., Dugoni, M., Poli, M., Romagnoli, E., De Cristofaro, P., & Battistini, N. C. (2007). A new device for measuring resting energy expenditure (REE) in healthy subjects. *Nutrition, Metabolism, and Cardiovascular Diseases*, *17*(5), 338–343.

Marca, J. (2002). The design and implementation of an on-line travel and activity survey. The American Society of Civil Engineers. *7th International Conference on Applications of Advanced Technology in Transportation*, Cambridge, MA.

Marrow, J. R. (2002). Measurement issues for the assessment of physical activity. In: G. J. Welk (Ed.), *Physical activity assessments for health-related research* (pp. 37–50). Champaign, IL: Human Kinetics Publishers.

Marsh, A. P., Vance, R. M., & Frederick, T. L. (2007). Objective assessment of activity in older adults at risk for mobility disability. *Medicine & Science in Sports & Exercise*, *39*(6), 1020–1026.

Mathie, M. J., Coster, A. C. F., Lovell, N. H., & Celler, B. G. (2004). Accelerometry: Providing an integrated, practical method for long-term, ambulatory monitoring of human movement. *Physiological Measurement*, *25*(2), R1–R20.

Matthews, C. E. (2002). Use of self-report instuments to assess physical activity. In: G. J. Welk (Ed.), *Physical activity assessments for health-related research* (pp. 107–123). Champaign, IL: Human Kinetics Publishers.

McCormack, G., Giles-Corti, B., Lange, A., Smith, T., Martin, K., & Pikora, T. J. (2004). An update of recent evidence of the relationship between objective and self-report measures of the physical environment and physical activity behaviours. *Journal of Science and Medicine in Sport*, *7*(1), 81–92.

McDoniel, S. O. (2007). A systematic review on use of a handheld indirect calorimeter to assess energy needs in adults and children. *International Journal of Sport Nutrition and Exercise Metabolism*, *17*, 491–500.

McKenzie, T. L. (2002). Use of direct observation to assess physical activity. In: G. J. Welk (Ed.), *Physical activity assessments for health-related research* (pp. 179–195). Champaign, IL: Human Kinetics Publishers.

McKenzie, T. L., Cohen, D. A., Sehgal, A., Williamson, S., & Golinelli, D. (2006). System for observing play and recreation in communities (SOPARC): Reliability and feasibility measures. *Journal of Physical Activity and Health*, *3*(Suppl. 1), S208–S222.

McKenzie, T. L., Marshall, S. J., Sallis, J. F., & Conway, T. L. (2000). Leisure-time physical activity in school environments: An observational study using SOPLAY. *Preventive Medicine*, *30*(1), 70–77.

Mokhtarian, P. L., & Cao, X. (2008). Examining the impacts of residential self-selection on travel behavior: A focus on methodologies. *Transportation Research Part B: Methodological*, *42*(3), 204–228.

Montoye, J., Kemper, H., Saris, W., & Washburn, R. (1996). *Measuring physical activity and energy expenditure*. Champaign, IL: Human Kinetics Publishers.

Morrow, J. R. (2002). Measurement issues for the assessment of physical activity. In: G. J. Welk (Ed.), *Physical activity assessments for health-related research* (pp. 35–50). Champaign, IL: Human Kinetics Publishers.

Murakami, E., & Wagner, D. P. (1999). Can using global positioning system (GPS) improve trip reporting?. *Transportation Research C*, *7*(2/3), 149–165.

Papas, M. A., Alberg, A. J., Ewing, R., Helzlsouer, K. J., Gary, T. L., & Klassen, A. C. (2007). The built environment and obesity. *Epidemiologic Reviews*, *29*, 129–143.

Papazoglou, D., Augello, G., Tagliaferri, M., Savia, G., Marzullo, P., Maltezos, E., & Liuzzi, A. (2006). Evaluation of a multisensor armband in estimating energy expenditure in obese individuals. *Obesity (Silver Spring)*, *14*(12), 2217–2223.

Parkka, J., Ermes, M., Korpipaa, P., Mantyjarvi, J. A., Peltola, J. A., & Korhonen, I. A. (2006). Activity classification using realistic data from wearable sensors. *IEEE Transactions on Information Technology in Biomedicine*, *10*(1), 119–128.

Reeves, M. M., Capra, S., Bauer, J., Davies, P. S., & Battistutta, D. (2005). Clinical accuracy of the MedGem indirect calorimeter for measuring resting energy expenditure in cancer patients. *The European Journal of Clinical Nutrition and Metabolism*, *59*(4), 603–610.

Retscher, G. (2007). Test and integration of location sensors for a multi-sensor personal navigator. *Journal of Navigation*, *60*(1), 107–117.

Richardson, A., Ampt, E., & Meyburg, A. (1995). *Survey methods for transport planning*. Melbourne: Eucalyptus Press.

Rodriguez, D. A., Brown, A. L., & Troped, P. J. (2005). Portable global positioning units to complement accelerometry-based physical activity monitors. *Medicine & Science in Sports & Exercise*, *37*(11), 572–581.

Rubenbauer, J. R., Johannsen, D. L., Baier, S. M., Litchfield, R., & Flakoll, P. J. (2006). The use of a handheld calorimetry unit to estimate energy expenditure during different physiological conditions. *Journal of Parenteral and Enteral Nutrition*, *30*(3), 246–250.

Ruiz, T., & Timmermans, H. (2006). Changing the timing of activities in resolving scheduling conflicts. *Transportation*, *33*, 429–445.

Sallis, J., & Saelens, B. (2000). Assessment of physical activity by self-report: Limitations and future directions. *Research Quarterly for Exercise and Sport*, *71*, S1–S14.

Sayed, A. H., Tarighat, A., & Khajehnouri, N. (2005). Network-based wireless location finding. *Signal Processing Magazine, IEEE*, *22*(4), 24–40.

Schutz, Y., & Chambaz, A. (1997). Could a satellite-based navigation system (GPS) be used to assess the physical activity of individuals on earth?. *European Journal of Clinical Nutrition*, *51*(5), 338–339.

Schutz, Y., & Herren, R. (2000). Assessment of speed of human locomotion using a differential satellite global positioning system. *Medicine & Science in Sports & Exercise*, *32*(3), 642–646.

Sener, I. N., Copperman, R. B., Pendyala, R. M., & Bhat, C. R. (2008). An analysis of children's leisure activity engagement: examining the day of week, location, physical activity level, and fixity dimensions. *Transportation*, *35*(5), 673–696.

Shoval, N., & Isaacson, M. (2006). Application of tracking technologies to the study of pedestrian spatial behavior. *Professional Geographer*, *58*(2), 172–183.

Starling, R. D. (2002). Use of doubly labeled water and indirect calorimetry to assess physical activity. In: G. J. Welk (Ed.), *Physical activity assessments for health-related research* (pp. 197–209). Champaign, IL: Human Kinetics Publishers.

Stel, V. S., Smit, J. H., Pluijm, S. M. F., Visser, M., Deeg, D. J. H., & Lips, P. (2004). Comparison of the LASA Physical Activity Questionnaire with a 7-day diary and pedometer. *Journal of Clinical Epidemiology*, *57*(3), 252–258.

Stopher, P., FitzGerald, C., & Xu, M. (2007). Assessing the accuracy of the Sydney Household Travel Survey with GPS. *Transportation*, *34*(6), 723–741.

Stopher, P., & Greaves, S. P. (2007). Household travel surveys: Where are we going?. *Transportation Research Part A*, *41*(5), 367–381.

Stopher, P., Jiang, Q., & FitzGerald, C. (2005). Processing GPS data from travel surveys. *28th Australasian Transport Research Forum*, Sydney, Australia.

Stopher, P., Jiang, Q., & FitzGerald, C. (2008). *Deducing mode and purpose from GPS data*. Sydney, Australia: The Institute of Transport and Logistics Studies, University of Sydney.

Stopher, P., & Jones, P. (Eds). (2003). *Transport survey quality and innovation*. New York, NY: Pergamon.

Strath, S. J., Swartz, A. M., Bassett, W. L., O'Brien, D. R., King, G. A., & Ainsworth, B. E. (2000). Evaluation of heart rate as a method for assessing moderate intensity physical activity. *Medicine & Science in Sports & Exercise*, *32*(Suppl.), 465–470.

Szeto, A. Y., & Sharma, S. K. (2007). RFID based indoor navigational aid for persons with severe visual impairments. *Conference Proceedings IEEE Engineering in Medicine and Biology Society*, 6361–6364.

Terrier, P., Ladetto, Q., Merminod, B., & Schutz, Y. (2001). Measurement of the mechanical power of walking by satellite positioning system (GPS). *Medicine & Science in Sports & Exercise*, *33*(11), 1912–1918.

Terrier, P., Turner, V., & Schutz, Y. (2005). GPS analysis of human locomotion: Further evidence for long-range correlations in stride-to-stride fluctuations of gait parameters. *Human Movement Science*, *24*(1), 97–115.

Townshend, A. D., Worringham, C. J., & Stewart, I. B. (2008). Assessment of speed and position during human locomotion using nondifferential GPS. *Medicine & Science in Sports & Exercise*, *40*(1), 124–132.

Transportation Research Board and Institute of Medicine of the National Academies. (2005). *Does the built environment influence physical activity? Examining the evidence*. Special Report 282. Washington, DC: Transportation Research Board.

Treuth, M. S. (2002). Applying multiple methods to improve the accuracy of activity assessments. In: G. J. Welk (Ed.), *Physical activity assessments for health-related research* (pp. 213–225). Champaign, IL: Human Kinetics Publishers.

Tsui, S. Y. A., & Shalaby, A. (2007). An enhanced system for link and mode identification for GPS-based personal travel surveys. *Journal of the Transportation Research Board, 1972*, 38–45.

Van Loan, M. D. (2007). Do hand-held calorimeters provide reliable and accurate estimates of resting metabolic rate?. *Journal of American College of Nutrition*, *26*(6), 625–629.

Welk, G. J. (2002a). Introduction to physical activity research. In: G. J. Welk (Ed.), *Physical activity assessments for health-related research* (pp. 1–17). Champaign, IL: Human Kinetics Publishers.

Welk, G. J. (2002b). Use of accelerometry-based activity monitors to assess physical activity. In: G. J. Welk (Ed.), *Physical activity assessments for health-related research* (pp. 125–141). Champaign, IL: Human Kinetics Publishers.

Welk, G. J., McClain, J. J., Eisenmann, J. C., & Wickel, E. E. (2007). Field validation of the MTI actigraph and bodymedia armband monitor using the IDEEA monitor. *Obesity*, *15*(4), 918–928.

Weston, A. T., Petosa, R., & Pate, R. R. (1997). Validation of an instrument for measurement of physical activity in youth. *Medicine & Science in Sports & Exercise*, *29*(1), 138–143.

Witte, T. H., & Wilson, A. M. (2005). Accuracy of WAAS-enabled GPS for the determination of position and speed over ground. *Journal of Biomechanics*, *38*(8), 1717–1722.

Wolf, J., Loechl, M., Thompson, M., & Arce, C. (2003). Trip rate analysis in GPS-enhanced personal travel surveys. In: P. R. Stopher & P. M. Jones (Eds), *Transport survey quality and innovation* (pp. 483–498). New York, NY: Pergamon.

Wormald, H., Sleap, M., Brunton, J., Hayes, L., Warburton, P., Waring, M., & White, M. (2003). Methodological issues in piloting a physical activity diary with young people. *Health Education Journal, 62*(3), 220–233.

Zhang, K., Pi-Sunyer, F. X., & Boozer, C. N. (2004). Improving energy expenditure estimation for physical activity. *Medicine & Science in Sports & Exercise, 36*(5), 883–889.

Zhang, K., Werner, P., Sun, M., Pi-Sunyer, F. X., & Boozer, C. N. (2003). Measurement of human daily physical activity. *Obesity Research, 11*(1), 33–40.

Zhou, J., & Golledge, R. (2007). Real-time tracking of activity scheduling/schedule execution within a unified data collection framework. *Transportation Research Part A, 41*, 444–463.

Chapter 10

Physical Activity in the Built Environment: Synthesis of a Workshop

Kelly J. Clifton

10.1. Introduction

This chapter summarizes the discussions that resulted from a workshop focused on the links between transportation and health. The relationships between the physical environment and the levels of physical activity are of increasing interest to policy researchers as obesity and associated negative health outcomes are growing dramatically in many nations. Transportation choices, such as walking and cycling, can be a source of physical activity, and the approaches of travel and activity behaviour researchers have something to offer this emerging area of study. Yet, the questions posed here are interdisciplinary, intersecting the fields of public health, planning and design. Thus, transportation researchers can benefit from understanding the data requirements and approaches of the health professions, which are often quite different from our own. Of specific interest to the workshop participants are the salient questions being asked by these disciplines, the various kinds of information needed to advance the line of inquiry and the most promising approaches to gathering appropriate data.

10.2. The Resource and Contributed Papers

The workshop was anchored by the resource paper *Emerging Methods and Technologies for Tracking Physical Activity in the Built Environment* by Sean Doherty (in this volume). Doherty (2009) reviewed the existing methodologies and emerging technologies available to collect various types of physical activity data. This review emphasized the ways that technology can facilitate passive monitoring of location and activity and, when combined with traditional survey approaches, help

Transport Survey Methods

ISBN: 978-1-84855-844-1

to overcome some, but not all, of the unique challenges presented by this research. For example, accelerometry and global positioning system trackers can provide the intensity and duration of physical activity along with location (thus offering the ability to include spatial data in the analysis). Algorithms can be used to aid identification of the types of physical activity that participants are engaged in during the study period.

These methods of objective and passive data collection mitigate some of the difficulties associated with participant recall or diary methods and can reduce respondent burden. The paper also punctuated the need for validation of measurement methods, an area where transportation researchers could benefit from the attention given by the health sciences. As with the long line of research in land use and travel behaviour, the key issue of causation persists and requires longitudinal measurement of subjects, taking advantage of natural experiments, such as planned changes to the built environment, where possible. Doherty makes the argument that integration of technology can help here, as well. As measurement methods become more accurate, smaller sample sizes will be needed to capture variations in the population and reduced respondent burden will facilitate longer periods of study.

The presentation of the resource paper was followed by a discussant presentation by Chandra Bhat, who offered a critique of the state of the research in this area. He suggested that there is a lack of consensus of exactly what information we need (what measures, what resolution, what activities), in part, because there is a narrow and perhaps incomplete understanding of the fundamental relationships we are trying to capture. This includes the link between physical activity and health, which is generally thought to be better understood than the links between environment and physical activity. For example, there is no clear understanding about how much and how intense physical activity should be to support healthy outcomes. Physical activity behaviours should be placed in the context of daily activity and time-use decisions and with consideration of the interactions with other household members. The approaches used in travel behaviour research along these lines offer a promising framework to advance these questions.

Two contributed papers were presented at the workshop and aided discussion about how the study of specific population segments or activities, namely children and walking, places special requirements on data collection. Roger Mackett presented an extensive review of his work in *Understanding the impacts of the travel behaviour and activity patterns of children on their physical activity and health* (Mackett, 2008). Here he raised some general issues surrounding the study of children, such as the need for parental consent, adapting surveys to young respondents and successful recruitment by working through schools. In relation to physical activity, Mackett raised several important points. There are difficulties distinguishing between walking and play activity among children, although some studies have had success parsing the two. Intra-household interactions with respect to physical activity are important and the role of the parent as decision maker or at least as mediator of children's autonomy should not be overlooked. Finally he stressed the fact that there are enormous data needs if one is to study all aspects of children's physical activity, given the complexity of these household and social interactions and the nature of these activities in children.

Daniel Sauter and Martin Wedderburn presented: *Measuring walking: Towards internationally standardized monitoring methods of walking and public space* (Sauter & Wedderburn, 2008). Given the dearth of non-motorized data and the lack of consistency across the data that does exist, the aim of this paper is to establish a set of international guidelines for the collection, analysis and dissemination of quantitative and qualitative techniques for measuring walking. Here, the authors talk about the particular characteristics or dimensions of walking that make it a unique activity and raise several important points regarding measurement. In addition to the traditional trip diaries and pedestrian count data, they call for attention to sojourns without destinations, such as strolls in public places; the micro-scale built environment and perceptions of that environment; walking as a link between modes and other short episodes; route choices and motivations. Determining the influence of seasons and weather on walking behaviour adds to the list of justifications for more longitudinal studies.

These four presentations prompted a lively discussion and debate among participants over the three workshop sessions. Below is a summary of the major themes and debates that emerged and directions for future work.

10.3. Major Themes

One important theme is the integration of technology into data collection for passive monitoring of activity, location and other physiological information. Given that these physical activity studies add a health component missing in traditional travel research, the large amount and types of data collected can affect respondent burden. Yet, the cost and size of technical instrumentation is coming down and many instruments combine measurement technology (such as accelerometers and GPS data loggers), facilitating their use across larger samples and increasing the frequency of data collected. This passive monitoring, combined with automated algorithms to process data and identify activities, can reduce the amount of missing data, potentially increase data quality and transform the large amount of detailed information into a tractable data set.

Despite the opportunities that this technology provides, there are important limitations to its use in data collection. There remains a need to understand the socio-psychological factors, such as attitudes, perceptions and motivations, in the behaviour decision process. This type of information cannot be collected through technology alone. Mackett raises the point that the issues surrounding children's physical activity lend themselves to more qualitative approaches or direct observation. Likewise, the degree of mobility and physical activity has dominated the discussion but immobility and sedentary behaviours are equally important and may require different approaches. So while advances in technology can aid data collection, there is still a role for traditional approaches and use of multiple methods is recommended given the complexity and nature of the phenomena under investigation.

Validation, reliability and standardization of instruments and measures are important themes that emerged from the discussion. The health researchers tend to have more experience with validation as the measures of physical activity, health outcomes, biometrics and scales employed in health studies are routinely validated prior to use in analysis. Consistent with the call for more standardization in measurement, validation and reliability of transportation measures and instruments have often been neglected and are deserving of more attention. Because this area of research is multidisciplinary, there should also be emphasis on unifying the measures and reconciling the approaches across disciplines. An example of differing approaches comes from the literature on measuring and operationalizing the built environment. Planners have a variety of objective measures of the various dimensions of the built environment, although no standards or common definitions are in place. Health researchers, on the contrary, have relied heavily on subjective evaluations — how participants perceive their home environment. There is likely no agreement on the best measure or approach for any given problem; nonetheless, closer attention to the measures and instruments employed can improve the quality of data.

As with the links between travel behaviour and land use, questions about causality and self-selection bias haunt studies of physical activity and the built environment. Cross-sectional studies alone can only determine associations between behaviour and environment. Longitudinal studies are needed, preferably ones that adequately capture significant changes in the physical environment. Many researchers attempt to take advantage of natural experiments — capital improvement projects that add networks, facilities or other related infrastructure — and design pre-test/post-test intervention studies. Even if the project can be underway in a timely manner, the timing of waves of data collection is not straightforward. And often the changes to the physical environment are often not radical enough or wide-reaching enough to realize a measurable change in behaviour across a population. Furthermore, because walking, cycling and other physical activities are often rare or at least infrequent events, longer periods of study are required to capture them in significant numbers.

As the health field places emphasis on outcomes related to different demographic groups, physical activity studies will need to be sensitive to their particulars. The degree and kinds of activities engaged likely vary across different age, cultural, economic and ethnic groups. The methods of data collection will need to be sensitive to these population segments to ensure adequate representation of both participants and their preferred sources of physical activity.

The realm of physical activity is much larger than just active transport modes. Although travel surveys have made great strides to include more detail about walking trips, they are still unable to capture all types of physical activity tied to mobility, such as running, cycling, skating, rowing, etc., and are completely inadequate for capturing other sources of physical activity, such as organized or informal play, work- or housework-related physical exertion, swimming, etc. Survey methodologies will have to be expanded to consider the full range of sources of physical activity for these studies.

Similarly, the context of the built and natural environment where these various types of physical activities occur is very different. Measures of land use — density, diversity and design — have been theoretically and empirically tied to transportation behaviours but an expanded list of physical environmental data is needed to associate physical activity to spatial characteristics. Studies of travel behaviour also focus on outdoor environments, but indoor environments (home, gym, mall, work, etc.) can be locations for physical activity and their attributes merit a closer look.

One of the most provocative debates occurred around the issue of too much and/or too little information and not enough understanding about what is needed to answer the critical research questions about the links between environment, physical activity and health. Because the field is relatively new and 'borrowing' knowledge and methods from multiple fields, many gaps exist, leading to calls for more data. In addition, the complexity of these behaviours and need to capture them in their social, environmental and economic context makes heavy demands on the amount and kinds of data required. But some question the need for more information absent a clear theory guiding the research. Questions also exist about the appropriate level of resolution and accuracy required for these data.

10.4. Conclusions

The complexities of this new and exciting field of study provide no shortage of research topics. The merger of disciplines interjects new approaches and raises questions about traditional approaches in travel behaviour research. Likewise, health professions can gain from considering the theories and methods of travel behaviour. Specifically, time use and activity analyses hold promise for studies of physical activity because they place behaviour in the context of daily decisions. Yet these studies are ambitious for their scope and will undoubtedly continue to face many challenges as the field matures. The exposure to other fields has been a healthy one and transportation will gain from its participation. However, more interaction with health researchers is needed because the differences between the disciplines are still greater than the similarities.

Acknowledgements

The workshop participants were Chandra Bhat, United States; Mark Bradley, United States; Andrew Clark, Canada; Kelly Clifton, United States; Sean Doherty, Canada; Randi Hjorthol, Norway; Michelle Lee, United States; Roger Mackett, United Kingdom; Daniel Sauter, Switzerland; Martin, Trépanier, Canada; and Martin Wedderburn, United Kingdom.

References

Doherty, S. (2009). Emerging methods and technologies for tracking physical activity in the built environment. In: P. Bonnel, M. Lee-Gosselin, J. Zmud & J.-L. Madre (Eds), *Transport survey methods: Keeping up with a changing world.* Bingley, UK: Emerald.

Mackett, R. (2008). Understanding the impacts of the travel behaviour and activity patterns of children on their physical activity and health. Presented at the 8th International Conference on Transport Survey Methods, Annecy, France.

Sauter, D., & Wedderburn, M. (2008). Measuring walking: Toward internationally standardized monitoring methods of walking and public space. Presented at the 8th International Conference on Transport Survey Methods, Annecy, France.

Chapter 11

Acquiring Data on Travel Behaviour During Emergencies and Exceptional Events

Earl J. Baker

Abstract

Surveys of the public have been conducted to document and explain evacuation behaviour in a wide range of threatening events during the past half-century. Many of the behaviours are directly applicable to transportation modelling and management: whether people evacuate, when they depart, where they go, the routes they employ and the number of vehicles they use. Data have usually been collected by telephone interview or mailed questionnaires. Traditional survey methods should be supplemented by Internet surveys, traffic counts and GPS tracking. More real-time data collection should be employed to document a wider range of behaviours during a threat more accurately and to better understand the dynamics of evacuation decisions.

11.1. Introduction

For more than half a century, survey methods have been applied to studies of how people respond to warnings of potential disasters. For some types of threats, responses include evacuation behaviours: whether people evacuate, how promptly they depart, where they go and how they get there. Many of the studies have been undertaken specifically to provide inputs to transportation modelling aimed at calculating the time required to complete a successful evacuation. Other studies have been motivated by the need to understand how to manage evacuations by increasing or decreasing the number of people leaving, for example. Although the latter group of studies has not been aimed specifically at transportation applications, many of

Transport Survey Methods

ISBN: 978-1-84855-844-1

them have implications for transportation. A variety of data-collection methods have been employed and applied to a wide range of hazards. This review traces the beginnings and evolution of survey research applications to disaster evacuations, summarizes the behaviours addressed in the studies, lists methods that have been employed, discusses challenges posed to data-collection efforts and recommends a number of topics for research. Most examples cited in the review deal with hurricanes in the United States which is the author's own field of expertise, and more studies have been conducted about evacuation behaviour in hurricanes than in other hazards. Many hurricane evacuation behavioural studies have been conducted specifically to support transportation modelling and management.

11.2. History and Evolution of Evacuation Surveys

During the 1950s, federal civil defence officials in the United States were concerned about how American citizens and emergency organisations would respond in the event of a nuclear attack. The closest analogy was response to disasters, so a research programme was undertaken to document how the public and emergency groups behaved when faced with warnings of a potential disaster and how they functioned during and after the event (Barton, 1969). Of particular concern was whether people would take warnings seriously and take appropriate self-protective actions. A related issue dealt with how people would respond to a second or third warning for the same kind of hazard, after earlier warnings had not been followed by occurrence of the event. This was called the 'cry-wolf' syndrome and continues to be a policy concern today. Officials worry that people will become complacent and fail to take warnings seriously after a number of 'false alarms'. Fewer people evacuating would result in fewer trips being generated, from a transportation standpoint. Surveys were conducted with the public following both false alarms and actual disasters to document how people responded. Other focuses of the research centered on the warning process: credibility of the warning source, how the warning was worded, how the warning was disseminated and characteristics of recipients of the warning. Early studies followed floods, hurricanes, tornadoes and chemical accidents. Most of the early surveys concentrated on the warning process and whether people took protective action. One exception was a survey following hurricane Carla, in which respondents were also asked whether they left their community when they evacuated (Moore et al., 1963).

Many of the issues deemed relevant for nuclear attack planning in the 1950s are relevant to preparing for other sorts of disasters as well. The next phase of evacuation survey studies was motivated by learning about how people responded to threats of floods, tornadoes and hurricanes for the sake for preparing better for future floods, tornadoes and hurricanes. The US National Weather Service was a principal user of the information and funded at least a couple of studies themselves. Much of the emphasis of studies conducted in the 1960s and 1970s was how to maximise response rate — that is, the number of people responding to flood warnings, for example (Mileti & Beck, 1975; Gruntfest, 1977). In the case of hurricanes and floods, it meant getting everyone told to evacuate to evacuate. But it was also during this period that behaviours were included in surveys that dealt with a greater range of evacuation variables. Specifically, surveys began asking about when evacuees departed, where they went and the transportation they used (Wilkinson & Ross, 1970).

It was also during the 1960s and 1970s that academic social science researchers began studying warning response and evacuation as research specialties, beyond a response to the need for policy applications. Sociologists saw disaster research as a special case of how individuals and organisations behave and interact under stress. In geography, hazards research became a stand-alone field of study, either as a subset of how societies interact with the natural and built environments or as a type of spatial behaviour. Much of the research focused on aspects of warnings and the warning process and on the experience, demographics and personality attributes of the warned populace (Drabek, 1986; Mileti & Sorensen, 1987).

In 1979, an evacuation occurred at the Three Mile Island (TMI) nuclear power plant in Pennsylvania. There were no plans in place for evacuating people living around nuclear power plants in the United States at the time. Decision making by public officials was uncoordinated and uncertain, there were no inventories of numbers of people living within various distances of TMI, and when evacuation recommendations were issued, many people not told to evacuate did so on their own initiative. The spontaneous evacuation of people living beyond the areas told to evacuate was labelled shadow evacuation, and it persists as an evacuation concern today, both for nuclear power plants and other hazards. Several surveys were conducted with people living around TMI to document their warning response and to explain the reasons for their actions (Lindell et al., 1985).

Predictably, all the nuclear power plants in the United States were required to develop evacuation plans soon, and the plans involved transportation modelling to calculate the times required for populations at risk to reach safety. The modelling required assumptions about the number of people evacuating, how quickly they would leave, where they would go and how they would get there. For a few nuclear plants, surveys were conducted with residents to gauge their evacuation intentions (Lindell et al., 1985).

At nearly the same time two US agencies, the Federal Emergency Management Agency (FEMA) and the US Army Corps of Engineers (USACE) embarked on a series of collaborative studies to provide technical data that state and local emergency management officials could use to develop better evacuation plans for hurricanes. The effort is usually referred to as the Hurricane Evacuation Study (HES) programme. The HES is a comprehensive endeavour that provides computer simulation of hurricanes to identify the areas in coastal communities that would need to evacuate (hazard analysis), provides inventories of the number of households and other facilities that would need to evacuate (population analysis), provides estimates of the demand for public shelters and identifies facilities that could be used to house evacuees before and during the storm (shelter analysis), and calculates the time required to clear the road network of evacuating vehicles (transportation analysis).

Both the shelter and transportation analyses require assumptions about how the threatened population will respond, and in most HES projects, the local population is surveyed to gather information that will help predict those behaviours (behavioural analysis). The HES programme has also conducted post-storm assessments, in which local officials are interviewed and the public are surveyed, in part to assess the accuracy and utility of HES products previously provided and in part to gather information that will help refine the HES products. A great deal of the survey work conducted on evacuation behaviour has been a result of the HES programme, which continues (Baker, 2000).

Today survey research about evacuation behaviour is a well-developed field of study with both academic and applied motives. Studies exist not only for hurricanes, floods and nuclear power plants, but volcanoes (Perry & Greene, 1983), tsunamis (Anderson, 1969), wildfires (Benight, Gruntfest, & Sparks, 2004), hazardous material accidents (Rogers & Sorensen, 1989; Mitchell, Cutter, & Edmonds, 2007) and dam failures (Rosenkoetter, Covan, Cobb, Bunting, & Weinrich, 2007), among others. Tierney, Lindell, and Perry (2001) provide a recent synthesis of some of the more salient research findings concerning warning response and evacuation behaviour, although it is not focused on behaviours most relevant to transportation issues.

Sample sizes have gotten larger and the range of behaviours addressed has expanded. In 1999, hurricane Floyd prompted the largest evacuation in US history and resulted in lengthy, problematic 'commute times' for many evacuees from south Florida through North Carolina. One response to Floyd was an increased interest among transportation researchers outside the social sciences in modelling evacuation events, creating an increased demand for evacuation behavioural data. The broadened community of researchers has brought a healthy new perspective to analyses of behavioural surveys dealing with evacuations.

11.3. Behaviours Addressed in Evacuation Surveys

It is not feasible to list all the topics and behaviours addressed in evacuation surveys, but there are certain behaviours that are typically included and certain issues that recur. The main behaviours about which people are asked in surveys are evacuation participation rate, evacuation timing, type of refuge, location of refuge and vehicle use. The inclusion of these behaviours is driven partly by the demand for this information in transportation and shelter analyses.

Participation rate refers to the percentage of a population that evacuates. Evacuation means leaving one's residence or lodging to go to a safer place. Sometimes this behaviour is referred to as trip generation, but at least in the HES context it is a multiplier used as an input into trip generation. Participation rates vary as a function of actions taken by public officials, location vulnerability within a community, severity of the threatening agent, housing, and certain demographic, socio-economic and psychological factors. A persistent finding is that too few people evacuate from the most vulnerable locations and too many evacuate from relatively safe locations. In planning for evacuations and calculating the time required for an evacuation, analysts often ignore the most probable participation rates and assume that 100% of the population being told to evacuate will do so. They reason that the plan should provide sufficient time and shelter capacity in case everyone at risk does evacuate. Even in that case shadow evacuation (people leaving from areas not told to evacuate) must be accounted for. Consequently, 'predicted' evacuation clearance times and shelter use usually exceed observed values.

Evacuation timing refers to when evacuees depart their residences or other origins. From the emergency management perspective, it is not sufficient to convince people to evacuate, because if too many wait too long to depart, there will be insufficient time for at least a portion of the population to reach safety. This is usually displayed as a cumulative response curve, showing the cumulative percentage of eventual evacuees who have left by a particular time. In many evacuations, the cumulative response curves are 'S' shaped logistic curves, although their slopes can vary significantly among evacuations (Rogers & Sorensen, 1989; Fu, Wilmot, Zhang, & Baker, 2007).

The curve is compared to the timing of other events such as the issuance of evacuation notices and the onset of dangerous conditions. In most evacuations, few people evacuate before evacuation notices are issued, and then they leave as quickly as they believe they need to leave. If an evacuation notice is issued days before anticipated onset of dangerous conditions, and there is no urgency of immediate departure communicated by officials, evacuees will take the entire two-day period to evacuate. Some of the later evacuees wait to see if evacuation will actually prove to be necessary, because the threat might diminish. When prompt evacuation is urgent and officials communicate that necessity successfully, evacuees leave much more quickly, resulting in a much steeper response curve (Baker, 2000; Sorensen, Vogt, & Mileti, 1987). Evacuation plans usually focus on the minimum amount of time necessary to complete a successful evacuation, given the size of the evacuating population, the roadway network and other factors. Therefore, clearance time modelling does not usually incorporate the longer-duration response curves if they reflect earlier-than-necessary, precautionary timing of evacuation notices.

In most evacuations the majority of evacuees go to the homes of friends and relatives, followed by hotels and motels and then public shelters operated by government and disaster organisations (Drabek, 1986). This variable is usually called type of refuge. In one sense it is of greater interest to public safety officials than to transportation analysts. Government has a responsibility for public safety and attempts to ensure that enough public shelter space is provided to accommodate the demand. Once demand is projected, organisations strive to identify and manage safe, suitable facilities. However, the locations of types of refuge will affect trip assignments. In some communities one or more categories of refuge can be provided locally, but in other communities evacuees will need to leave the local area to reach the refuge.

Location of refuge indicates the geographical location where evacuees will seek refuge. In most evacuations a mixture of local and out-of-town destinations are used. Trip assignments in transportation analyses need to identify the distribution of trips to specific local places or areas as well as ascertaining the out-of-town locations to which evacuees will travel. It is common, at least in hurricanes, for many evacuees to travel distances that are much greater than necessary to reach safety, but in some locations people must travel long distances to reach places that offer both safety and refuge. A related behaviour is the evacuation routes used by evacuees. It is asked about in some surveys, but many transportation analysts infer routes from origins and destinations. Route choice can also be managed by public officials by closing some routes, using contraflow or giving right-of-way to designated streets and roads (Dow & Cutter, 2002).

Transportation analyses need to know the number of vehicles on roads, not the number of people, so surveys typically measure vehicle usage. This can be expressed as an average number of vehicles per household (e.g. 1.5) or as a percentage of the vehicles available to the household (e.g. 70%). Sometimes surveys also identify how many evacuating households pull trailers or take motorhomes to account for vehicles that might have an unusual impact on traffic flow or lane volume or vulnerability (e.g. instability in strong winds). For hurricanes, in the great majority of evacuations, between 65% and 75% of the available vehicles are used by evacuees. Some people do not have transportation of their own, and some of those individuals will need assistance from public agencies or organisations. Surveys usually identify the extent of those needs (Baker, 2000).

The bulk of survey instruments are made up of questions about variables that will refine, explain or predict variations in the behaviours enumerated in the previous discussion. The specific variables that are included depend on the application of the data, the user of the information and the perspective of the researcher. For example, local public safety officials might be interested in how they can educate the public to change unwanted behaviours, how they can word and disseminate evacuation notices to achieve desired results, how they need to manage traffic to overcome behavioural tendencies and what sorts of special accommodations they might need to make for evacuees needing medical care or for those who will bring pets. A few examples of specific issues addressed in surveys include effect of probability information on evacuation, the relationship between home safety and evacuation, effect of pet ownership on evacuation, implications of evacuation fatigue following multiple evacuations, evidence of the cry-wolf syndrome, evacuation expenditures, preference for and effect of refuges of last resort.

11.4. Approaches to Data Collection

Surveys about evacuation behaviour are conducted both before and after evacuations using a variety of techniques. The following discussion describes several approaches and lists some of their advantages and disadvantages.

The earliest evacuation surveys were conducted almost exclusively with door-to-door interviews. Questionnaires (i.e. interview schedules) were structured with both open- and closed-ended questions. It is not clear how individuals were selected for interviewing, although addresses could have been chosen at random. It is more likely that a spatially systematic scheme was employed (e.g. every third house). Response rates (i.e. people agreeing to participate in the survey) were generally good. The face-to-face nature of the interviewing sometimes fostered interaction between interviewer and respondent that facilitated follow-up questions and in-depth explanations and elaborations. Door-to-door interviewing is still practiced occasionally, but less and less both because of concerns about cost, time requirements and safety of interviewers. If no one is home the first time an interview is attempted at a residence, interviewers should go to the same address on a different day of the week and different time of day, up to perhaps four times, before replacing it with another address. That can be prohibitively expensive. American society is different today than it was in the 1950s, 60s and even 70s. Residents are more cautious about strangers knocking on their doors, and interviewers face greater risks in certain neighbourhoods. Still, some populations might be accessible only by going to their residences, and certain types of questionnaires (e.g. involving graphics) require the respondent to view objects before responding.

Occasionally people are interviewed in high-traffic-volume venues of convenience, hence the name convenience samples. Interviewers have approached patrons in parking lots of 'big box' department stores, for example. The motivation is usually cost, and the technique is usually justified by the argument that the clientele of the establishment represents a 'cross-section' of the community. Socio-economic information can be gathered about respondents so the sample can be compared to the general population, and the sample might be weighted to make it more representative. Other examples of convenience samples are interviewing tourists on boardwalks, interviewing attendees of an event such as a 'hurricane expo' where vendors exhibit products of interest to residents of hurricane-prone communities and interviewing evacuees at rest areas along evacuation routes.

One of the first surveys conducted specifically to provide data for hurricane planning was a newspaper survey (Southwest Florida Regional Planning Council, 1983). Questions were printed about whether people would evacuate, where they would go and when they would leave. Readers were asked to cut out the survey, complete it and mail it to an address. The technical data report for the HES acknowledged the non-random nature of the newspaper survey, but employed the results for planning anyway. One seldom sees newspaper surveys in printed media today.

Mail surveys are probably the second most commonly employed data-gathering technique for evacuation surveys today. Researchers employing the mail surveys are confident that they provide reliable results and yield larger samples than telephone surveys, at the same cost. A certain level of education is required, which might exclude the less educated from the survey, although demographics of respondents can be compared to the general population. The length of the questionnaire must be limited, and branching questions (e.g. 'if…go to') are difficult to include. One of the greatest concerns about mail surveys is that they require a high level of motivation to respond. There are follow-up methods that can eventually get a completion rate that might be comparable to telephone surveys, but it is still possible that the survey participants are people whose views and behaviours are atypical with respect to the survey topic. People who evacuated, for example, might be more likely than others to participate in the survey, possibly because they are interested in sharing their opinions and experiences during the evacuation or simply because the same characteristics that led them to evacuate are more likely to motivate them to talk to an interviewer about evacuation. Demographics are not generally good predictors of most evacuation behaviours, so there is no way to use demographics to correct the behavioural bias. Mailed questionnaires do allow respondents to consider their answers more carefully than if respondents were being interviewed by telephone, and therefore mailed questionnaires might sometimes yield more accurate measures of behaviour, particularly prospective behaviours.

The great majority of surveys conducted about evacuations today are done by telephone. There are many firms and centres in the business of conducting telephone surveys on a daily basis about political and consumer issues, and their resources can be applied to evacuation subjects. In many cases random-digit dialling can be employed, but in other instances sampling is allocated spatially, to reflect vulnerability to the hazard in question. Although vulnerability levels can be assigned after data collection is finished if address information is gathered, there might be too few responses in certain locations. Many interviewing organisations employ Computer Aided Telephone Interview (CATI) systems that reduce interviewer error on branching questions and create the database as the interview is finished. Although it is possible to conduct interactive, open-ended interviews by phone, in which responses are categorized for analysis after the fact, most telephone surveys are extremely structured and provide little opportunity for probing and free-form explanations. Non-response rate is an increasing problem for telephone surveys. Even though interest and co-operation is typically greater for evacuation surveys than for consumer and political surveys, non-response overall is still a concern. In some locations, 12 phone numbers are needed for every completed interview. Some numbers are simply out-of-date or incorrect, but many people do not answer at all, letting their answering machines screen their calls. After someone answers the call, the completion rate is roughly 34%. As with mail surveys, one worries that the people agreeing to answer questions about evacuation are atypical of the general population with respect to how they evacuate.

Internet surveys are uncommon in the evacuation field, but will probably gain in popularity in the future. They contain sampling biases, which are generally acknowledged. They are confined to people who at least have access to the Internet, and most likely include people who use it routinely. Solicitations to participate can be sent to email lists but can also be distributed by Web site postings, mailings, newspaper advertisements and even telephone calls. Internet surveys afford an opportunity to employ sophisticated graphics, conjoint methods and complex question branching. In addition to the sample bias, there is still the non-response problem that also plagues other data-collection methods.

Sometimes respondents are contacted a second time or even more frequently to compare their responses over time and from one evacuation event to another (Dow & Cutter, 1998). The respondents are often called panels, but sometimes the response sets are simply called longitudinal data. It is certainly valuable to know what the same person did in more than one evacuation. However, that can frequently be accomplished in a single interview if more than one evacuation preceded the interview. The panel approach offers certain advantages, but it raises certain concerns as well. One survey might ask respondents what they would do in an evacuation, for example, and if an evacuation ensued, contacting the same person again would permit comparison of the intended and actual responses. Panels also provide data about responses in multiple evacuations that had not occurred at the time of the initial data collection. However, one concern is that respondents become atypical after being interviewed even once, and almost certainly twice. If they are told that they will be contacted again in the future, they might become even more atypical. Part of the apprehension is similar to the 'Hawthorne effect' in which experimental subjects behave differently when aware that they are participating in an experiment. Another worry is that the very act of talking with a respondent for 10–20 min about evacuation causes the person to become sensitive to issues they were unaware of before the interview. Their subsequent evacuation behaviour could change as a result. One hurricane researcher reported that her panel was becoming more 'hurricane savvy' with repeated evacuations. Other observers worried that the panellists might be becoming more 'survey savvy' as a result of repeated interviews.

Many evacuation surveys ask people what they would do rather than what they actually did. This is necessary in locations that have not experienced an evacuation and might be relevant in places that have. Most of the social science literature calls this survey information 'intended-response' data but in the transportation field it is sometimes called 'stated preference' data. It can be collected using any of the methods described above but a distinction is made here simply to differentiate between surveys that ask about intended as opposed to actual behaviour. From the earliest days of intended-response surveys on all subjects, there have been reservations about the correspondence between what people say and what they do. Although there have been reports of close agreement in some pre-event and post-event evacuation surveys (Kang, Lindell, & Prater, 2007), most of the evidence suggests significant disagreement. In hurricane surveys, for example, people overstate the likelihood that they will evacuate in low-threat scenarios, without evacuation notices being issued. They also overstate the likelihood that they will use public shelters. On the other hand, intended vehicle use matches well with actual vehicle use (Baker, 2000). A persistent policy issue is whether the sort of shadow evacuation that occurred around TMI will occur in other nuclear power plant evacuations. Stated preference surveys suggest that it will, but there is the possibility that it might be suppressed with effective public information (Lindell et al., 1985). A nuclear power plant in New York was constructed but never operated, in part because of concerns about shadow evacuation. Some of the earliest HES efforts took intended-response survey data at face value and used it without modification as inputs into the transportation and shelter analyses. More recent attempts to derive planning assumptions for those analyses employ a mix of actual response data, modified intended-response data and statistical models that capture general relationships between behaviours and certain predictor variables.

Another valuable source of data about evacuations comes from mechanical observation, not surveying. Traffic counters are used to record traffic volume over time at numerous roadway locations during evacuations. A couple of decades ago, traffic count data were relatively sparse and unreliable. Counters sometimes relied on electricity, which occasionally was disrupted before the evacuation was complete or the data was not stored due to electrical failure. In some instances, software automatically discarded traffic count data during an evacuation because it was out of the range of values considered reliable. Today, traffic counts are monitored during evacuations, revealing traffic volumes, average speeds and trends in departures. In Florida real-time traffic count data is compared to results of HES products to monitor whether evacuations are proceeding as anticipated.

11.5. Challenges to Conducting Effective Evacuation Surveys

Many challenges facing data collection for evacuation surveys are the same as those facing other sorts of surveys. Other challenges are more specific to the topics and issues that need to be addressed in evacuation surveys.

Non-response bias continues to cast doubt about the representativeness of evacuation survey data. Because demographic variables are not well correlated with evacuation behaviours, adjustments to results cannot be made confidently on that basis. All forms of data collection suffer from non-response bias, and non-response is getting worse as a result of the profusion of telemarketing and political polling. People who participate in evacuation surveys are likely to be people who are most interested in the subject, the most informed and people who have given the topic the most thought.

There are subgroups of populations that are more difficult to reach than others. Many poor individuals do not have telephones and do not have the educational skills required to respond to mail questionnaires. Language is often a barrier, especially in certain neighbourhoods of major cities (Lindell & Perry, 2004). Any single hard-to-reach subgroup of the population might not constitute a large enough percentage of the total population to significantly affect overall clearance times and shelter demand, but collectively, they might. Moreover, public officials have responsibilities for the safety of each segment of the population.

Sample sizes have gotten larger and larger in evacuation surveys, as users have insisted on more and more disaggregation of the data. Whereas this is not a challenge in and of itself, it results in higher costs. A recent telephone survey in Florida dealing with evacuation cost $29 per completed interview, not including survey design or report preparation (Downs, 2008, personal communication). Each coastal county had 400 completed interviews, and each non-coastal county had 150. The survey cost for data collection alone was $545,000 (U.S). The client required 'statistically valid' data in each county.

Cell phones have created a challenge for certain types of telephone surveys as well as for telephone-based emergency notification of evacuations. Evacuations for many events are based on spatial variations in vulnerability to the events. Hurricane evacuations and riverine flood evacuations are ordered primarily for areas that would flood dangerously, and some flood-prone areas would flood more dangerously than others. Nuclear power plant evacuations would be called for within certain distances of the facility, referred to as emergency planning zones (EPZ). Many evacuation surveys are designed to meet quotas of responses from predetermined evacuation zones. To do that by phone, telephone numbers within the target zones are identified prior to calling. This has traditionally been accomplished by using reverse telephone directories, in which an address can be entered and the directory will display a phone number for the address. More recently geographical information system (GIS) overlays of evacuation zone boundaries have been used in conjunction with address and phone databases to yield phone numbers in certain zones. However, cell phone numbers do not usually match a physical address as land lines do. As more and more people rely on cell phones exclusively, they will be excluded from spatially targeted evacuation surveys by phone.

Tourists and other transient populations pose particular problems for data collection. They can be interviewed prior to an event, but it is extremely difficult to contact them after an evacuation. They do not have local residences where the evacuation occurred, and tourists have returned home by the time an evacuation survey can be mounted. At least one study identified tourists who were present during an evacuation by inspecting registrations at public attractions and then attempted to contact them by mail (Drabek, 1996). Attempts to obtain contact information from accommodations have been largely unsuccessful due to concerns about privacy.

A secondary challenge associated with evacuation surveys is the use to which the data will be put. Social science research on evacuation was initially concerned with identifying factors associated with evacuation behaviour and warning response, with the aim of discovering ways to enhance the level of appropriate responses. The application to clearance time calculations and projections of shelter demand have led to the need for prediction of very precise numerical values. In some jurisdictions, such as North Carolina and Florida in the United States, regulations governing whether new residential developments will be allowed to occur are tied to hurricane evacuation clearance times and shelter demand. Thus, the quality of data upon which predictions are made must be legally defensible.

11.6. Research Needs

As with most topics, there is no shortage of candidate ideas for 'further research'. The list compiled by one person might vary greatly from that of another, based on the experiences, perspectives and disciplinary blinders of each. Moreover, even from a single perspective, no list will be exhaustive. The following list is offered with those caveats.

The collection of real-time survey data, during an evacuation, is almost totally absent. Traffic counters do provide real-time data, but for a limited range of variables and for only certain points on particular routes. Surveys following evacuations often ask respondents to explain certain behaviours: why they did or did not evacuate, for example. The surveys are often conducted months or even years following the event, and interviewee responses are subject not only to memory failure but to 'reconstructions' of their own experiences and decisions, based on discussions with others and media accounts of the event. Real-time data collection would permit accounts of behaviours and decision processes as they occur.

Real-time data collection would also yield more complete and accurate measures of the information that people are receiving during the threatening event. This includes media broadcasts, but also official pronouncements by public officials, which can be used to help explain variations in certain evacuation behaviours.

Real-time data collection might also be the best way to ascertain more detailed information about evacuation travel behaviour than that which is typically gathered. Little is known about local trips that contribute to background traffic during an evacuation. Although it is known that not all vehicles available to households are used in evacuations, there is no data on whether the unused vehicles are moved locally to safer locations prior to evacuating. Dynamics of how evacuees make evacuation decisions during a threat or alter their trips during an evacuation could be best measured with real-time monitoring.

Finally, real-time survey data could provide information useful to public officials managing the evacuation. This might provide information about certain subgroups of the population that are not responding as officials believe they need to respond and why.

Survey techniques employed for most evacuation surveys today are designed to achieve large sample sizes at minimal cost. They rely greatly on closed-ended questions that lend themselves to ready data entry suitable for statistical analysis. The methods seldom allow for in-depth, interactive follow-up questions and probing for insights not anticipated prior to the survey instrument being designed. 'Qualitative' research is probably an abused term, sometimes used to legitimise anecdotal data, but good qualitative methods are underutilized in evacuation research. When coupled with large-sample, structured survey data, the addition of focus group data and unstructured interview data might provide insights to patterns in survey data that otherwise go undetected. At least one study used ethnographic survey methods to construct a decision tree for evacuation decisions (Gladwin, Gladwin, & Peacock, 2001).

Both intended response and actual response surveys about evacuation behaviour are common, and there have been a few comparisons between results obtained by the two methods. Some of the better-established generalisations were cited previously. One type of behaviour that merits more comparative work as well as explanatory modelling work is destination choice. Many surveys ask people where they would go or ask people where they went with respect to geographical destinations, particularly the distribution of local trips versus out-of-town trips, and the percentage of out-of-town trips to various cities or counties. A better, more systematic comparison of the two data sets is needed to indicate the extent to which the stated preference data can be utilised to make trip assignments. If the intended-response data does not predict well or if it is absent, then more statistical modelling of actual destination distributions is needed to assess the ability to make trip assignments with predictor variables.

The main reason people do or do not evacuate is that they do or do not believe they would be unsafe staying in place. Misconceptions about vulnerability contribute to shadow evacuation and also contribute to under-evacuation from high-risk locations. Surprisingly little is known about why people believe they would be safe or not, beyond having confidence in the relevant forecast information. Belief that one's home would or would not be safe in a hurricane, for example, is a strong predictor of evacuation, but many people have misconceptions about the safety of their homes (Baker, 2002), and researchers cannot explain why. A better understanding of this belief could help reduce shadow evacuation and enhance evacuation from high-risk areas.

A related issue is the failure to hear evacuation notices from public officials. Authorities vary with respect to being able to compel evacuation (i.e. mandatory evacuation orders), but people who say they heard from officials that they should or must evacuate are much more likely than others to leave. Many residents in areas told to evacuate, however, say they did not hear that they were supposed to go. In some instances the failure to hear evacuation notices or to comprehend to whom they applied is understandable. But in other cases it seems incomprehensible that residents did not hear, either directly or indirectly, that they were being told to leave. This too is a strong predictor of evacuation participation rate (Baker, 2002), and research has not explained why so many people fail to hear evacuation notices.

Experimental methods, even 'pencil-and-paper' experiments, are uncommon in evacuation research. There have been experimental survey designs to assess the likelihood of vertical evacuation during hurricane threats (Ruch et al., 1991), evaluate the effect of hurricane probability forecasts on evacuation response (Baker, 1995), gauge the effect of colourised satellite imagery of hurricanes on response (Sherman-Morris, 2005) and to model bushfire evacuation decisions (Stopher, Rose, & Alsnih, 2004). In general, though, few studies have attempted to present sets of experimentally controlled attributes when eliciting intended-response survey data. These methods have the potential to replicate the sorts of decisions and tradeoffs that people have to make in evacuation situations. In such cases, they should provide insights into the effect of those attribute variables on evacuation decisions and facilitate better predictions from intended-response data.

More creative methods or combinations of data-collection methods need to be employed. Wilmot (2004) has proposed a variety of approaches used to study transportation behaviours other than evacuation and described how they could be used for evacuation research. All of the proposals have a great deal of merit. If combined with more traditional methods, they could broaden both the scope and depth of understandings about evacuation behaviour, while being benchmarked against measures whose reliability is already known.

Finally, is a strong appeal for collaborative research. Social scientists and transportation engineers and other transportation specialists often have different but complementary perspectives and tool sets for conducting evacuation research. The state of knowledge can only improve as a result of interdisciplinary collaboration and fresh looks at old problems.

References

Anderson, W. (1969). Disaster warning and communication processes in two communities. *Journal of Communication*, *19*, 92–104.

Baker, E. J. (1995). Public response to hurricane probability forecasts. *The Professional Geographer*, *47*, 137–147.

Baker, E. J. (2000). Hurricane evacuation in the United States. In: R. Pielke, & R. Pielke, (Eds), *Storms* (Vol. 1, pp. 306–319). London: Routledge.

Baker, E. J. (2002). Social impacts of tropical cyclone forecasts and warnings. *World Meteorological Bulletin*, *51*, 229–235.

Barton, A. H. (1969). *Communities in disaster: A sociological analysis of collective stress situations*. Garden City, NY: Doubleday.

Benight, C., Gruntfest, E., & Sparks, K. (2004). *Colorado wildfires 2002*. Quick Response Report #167, University of Colorado Natural Hazards Center, Boulder, Colorado.

Dow, K., & Cutter, S. L. (1998). Crying wolf: Repeat responses to hurricane evacuation orders. *Coastal Management*, *26*, 237–252.

Dow, K., & Cutter, S. L. (2002). Emerging hurricane evacuation issues: Hurricane Floyd and South Carolina. *Natural Hazards Review*, *3*, 12–18.

Drabek, T. E. (1986). *Human systems responses to disaster: An inventory of sociological findings*. New York: Springer-Verlag.

Drabek, T. E. (1996). *Disaster evacuation behavior: Tourists and other transients*. Boulder, CO: Natural Hazards Research and Information Center, University of Colorado.

Fu, H., Wilmot, C. G., Zhang, H., & Baker, E. J. (2007). Modeling the hurricane evacuation response curve. *Transportation Research Record: Journal of the Transportation Research Board*, *2022*, 94–102.

Gladwin, C. H., Gladwin, H., & Peacock, W. G. (2001). Modelling hurricane evacuation decisions with ethnographic methods. *International Journal of Mass Emergencies and Disasters*, *19*, 117–143.

Gruntfest, E. (1977). *What people did during the big Thompson flood*. Working Paper 32. University of Colorado Institute of Behavioral Science, Boulder, Colorado.

Kang, J. E., Lindell, M. K., & Prater, C. S. (2007). Hurricane evacuation expectations and actual behavior in hurricane Lili. *Journal of Applied Social Psychology*, *37*, 887–903.

Lindell, K., et al. (1985). *Planning concepts and decision criteria for sheltering and evacuation in a nuclear power plant emergency. AIF/NESP-031*. Bethesda, MD: Atomic Industrial Forum.

Lindell, M. K., & Perry, R. W. (2004). *Communicating environmental risk in multiethnic communities*. Thousand Oaks, CA: Sage.

Mileti, D. S., & Beck, E. M. (1975). Communication in crisis: Explaining evacuation symbolically. *Communication Research*, *2*, 24–49.

Mileti, D. S., & Sorensen, J. H. (1987). Natural hazards and precautionary behavior. In: N. D. Weinstein (Ed.), *Taking care: Understanding and encouraging self-protective behavior* (pp. 189–207). Cambridge: Cambridge University Press.

Mitchell, J. T., Cutter, S. L., & Edmonds, A. S. (2007). Improving shadow evacuation management: Case study of the Graniteville, South Carolina, chlorine spill. *Journal of Emergency Management*, *5*, 28–34.

Moore, H. E., et al. (1963). *Before the wind: A study of response to hurricane Carla*. NAS/NRC Disaster Study #19. National Academy of Sciences, Washington, DC.

Perry, R. W., & Greene, M. R. (1983). *Citizen response to volcanic eruptions: The case of Mount St. Helens*. New York: Irvington Publishers.

Rogers, G. O., & Sorensen, J. H. (1989). Warning and response to two hazardous materials transportation accidents in the US. *Journal of Hazardous Materials*, *22*, 57–74.

Rosenkoetter, M. M., Covan, E. K., Cobb, B. K., Bunting, S., & Weinrich, M. (2007). Perceptions of older adults regarding evacuation in the event of a natural disaster. *Public Health Nursing*, *24*, 160–168.

Ruch, C., et al. (1991). *The feasibility of vertical evacuation*. Program on Environment and Behavior Monograph #52. University of Colorado Institute of Behavioral Science, Boulder, Colorado.

Sherman-Morris, K. (2005). Enhancing threat: Using cartographic principles to explain differences in hurricane threat perception. *The Florida Geographer*, *36*, 61–83.

Sorensen, J. H., Vogt, B. M., & Mileti, D. S. (1987). *Evacuation: An assessment of planning and research. ORNL-6376*. Oak Ridge, TN: Oak Ridge National Laboratory.

Southwest Florida Regional Planning Council. (1983). *Hurricane evacuation study*. North Ft. Myers, FL: Southwest Florida Regional Planning Council.

Stopher, P., Rose, J., & Alsnih, R. (2004). *Dynamic travel demand for emergency evacuation: The case of bushfires*. Working Paper ITS-WP-04-16. Institute of Transport Studies, Sydney.

Tierney, K. J., Lindell, M. K., & Perry, R. W. (2001). *Facing the unexpected: Disaster preparedenss and response in the United States*. Washington, DC: Joseph Henry Press.

Wilkinson, K. P., & Ross, P. J. (1970). *Citizens' responses to warnings of hurricane Camille*, Social Science Research Center Report 35. Mississippi State University, State College, Mississippi.

Wilmot, C. G. (2004). Data collection related to emergency events. Paper presented at the 7th International Conference on Travel Survey Methods, Los Suenos Mariott Hotel, Costa Rica, August 2004.

Chapter 12

Capturing Travel Behavior during Exceptional Events: Synthesis of a Workshop

Chester Wilmot and Matthieu de Lapparent

12.1. Introduction

Most travel data collection in the past has been targeted at identifying regular or routine travel. In urban travel surveys, cross-sectional data collection has dominated under the implicit assumption that travel behavior observed over a limited period of time is representative of travel behavior in general. However, exceptional events generate irregular travel that far exceeds regular travel in volume and importance. Without data of behavior during exceptional events, we will be unable to model and develop contingency plans to deal with them. Thus, the purpose of this workshop was to make the case for data collection of travel generated by exceptional events and to promote its implementation. We concentrated on data collection of person movement and omitted freight movement.

Exceptional events are all occurrences that generate irregular travel. The events may be good or bad, isolated or recurring, planned or unplanned. They generally generate traffic over relatively short periods of time and affect limited areas. Depending on the level of forewarning and the importance of the event, the traffic generated by an exceptional event could totally superimpose itself on regular traffic, or replace it. Because of the great diversity of exceptional events, their importance and impact, it is useful to establish a typology that will help establish data needs, data collection procedures, and research needs associated with this topic.

12.2. Typology of Exceptional Events

Exceptional events can be described in terms of their characteristics or features. The features describe the different dimensions on which exceptional events typically

Transport Survey Methods

ISBN: 978-1-84855-844-1

differ, and the range over which this occurs. The main features of exceptional events identified by workshop members, and the extent to which values on them vary, are listed in Table 12.1.

In Table 12.1, the first listed feature, the nature of the event, draws attention to the fact that exceptional events range from desirable, attractive events to totally undesirable, sometimes life-threatening experiences. Thus, people are attracted to some exceptional events and repelled by others. Special sport and entertainment events are examples of attractive exceptional events; natural or man-made disasters are examples of the latter.

The next feature noted in Table 12.1 is the frequency with which exceptional events occur. On one hand, they can be single or very rare events, and on the other they can occur on a regular, recurring basis. Two aspects of this feature have a bearing on data collection. First, as events become rarer, all else being equal, they become less important because the probability of them occurring is small. Second, with rare events, revealed behavior surveys can rarely be performed, and little historical information is, subsequently, available. In contrast, recurring exceptional events such as the annual Mardi Gras in New Orleans provide rich historical data and good opportunity for surveys.

Speed of onset is the degree of advance warning received of an event. It varies from no warning, such as in the case of a terrorist attack, to several years when planning for a special event such as the Olympics. The big difference in these two extremes is that advance warning permits pre-event action to mitigate negative impacts of the event, while the other only allows post-event response.

Degree of impact refers to the extent and severity of an event and its impact on travel. For example, some events, such as chemical spills, are often localized in extent, while others, like hurricanes impact large areas. The severity of a threat can vary from mild to severe, resulting in discretionary response at first but progressing toward mandatory response as severity increases. This impacts data collection because response decisions must be modeled with milder cases but are redundant in severe threats.

Duration refers to how long an event lasts. Among exceptional events, it varies from a few minutes in the case of an earthquake, to days in the case of a hurricane. Duration has an impact on data collection because it influences whether it is

Table 12.1: Typology of exceptional events.

Feature	**Limits**	
Nature	Desirable	Undesirable
Frequency	Single	Recurring
Speed of onset	Fast	Slow
Degree of impact	High	Low
Duration	Short	Long
Direction	Centrifugal	Centripetal

important, or possible, to collect data during the event, and whether response during the event can lead to improved operation or mitigation of damage.

The last listed feature in Table 12.1, the direction of travel in response to an exceptional event draws attention to the fact that while there may be a dominant direction of travel, there is often travel in the opposite direction as well. For example, in a hazardous situation where inhabitants are evacuating an area, there is a need for emergency personnel to enter the area at the same time. In addition, when evacuation occurs, there is always the re-entry phase as inhabitants return after the threat is dispelled.

12.3. Data Needs

Data on exceptional events are needed for planning and operation purposes. Data for planning requires information that will allow the development of models to estimate behavioral response to events. Thus, they require establishment of relationships that can be used in estimating response to alternative scenarios. Operations data are related to observation of what is occurring in real time, and, possibly, making short-term predictions based on past and recently observed conditions.

Currently, we place little emphasis on the dynamics of the factors that influence travel, but in exceptional events the dynamics of travel response, and the conditions prompting travel, are very important. For example, in evacuation from a hurricane, the threat level steadily rises as the hurricane approaches, local conditions change as businesses close, congestion develops, contraflow operations are introduced, flooding occurs, and some roads are perhaps closed. In addition, local emergency managers issue evacuation notices, local leaders make statements and assess the danger, and television broadcasters describe the storm and report estimated surge levels and wind speeds. All these developments occur over time, produce individual perceptions of the threat it poses to them personally, and result in changing evacuation response and network flows. Little attention has been given in the past to this dynamic aspect of exceptional events, and its implication for data collection.

Frequency of occurrence has a significant impact on data collection both in terms of data needed and the opportunity for data collection. Using extreme cases of regular versus rare events illustrates the point. For example, snowstorms are regular events in some areas, and authorities rely almost entirely on historical information and experience to plan for them and manage their clean-up once they are in progress. On the other hand, a rare event such as hosting the Olympics in a particular city has no local historical data on which to depend, and experience from other cities is only as transferable as the cities are similar. However, if there is sufficient forewarning of an event, the opportunity for a unique type of pre-event data collection exists. For example, if an event such as the Olympics is going to be staged in a particular city and authorities are investigating ways to reduce local traffic on the transportation network during the event, one experiment that could be conducted would be to observe what impact a holiday has on reducing local traffic. Another possibility would be to host other popular events prior to the main event, to test the transportation, parking, crowd control, security, and reliability of the services planned for the Olympics.

The amount of forewarning of an event is very influential in what data can be collected. As noted in the previous paragraph, with long periods of forewarning, experiments, simulations, and trial runs can be conducted prior to the event. These are new types of data collection that have not been conducted extensively before. However, as forewarning is reduced the opportunity to act in advance of the event disappears. With no forewarning, the best that can be done is to obtain real-time information to inform those making operational and management decisions on the fly. Real-time data is routinely collected as part of Advanced Traveler Information Systems (ATIS) and Advanced Traffic Management Systems (ATMS) in major urban areas, and could be used to provide real-time information during exceptional events as well.

An issue that distinguishes travel during an exceptional event and regular day-to-day travel is that while much travel may be discretionary in regular circumstances, some exceptional events demand travel in a short window of time, and demand it of all inhabitants. For example, the sick, infirm, handicapped, institutionalized, and incarcerated generally travel very little, but in an emergency they may be required to evacuate. In addition, because these people are tended by others, their attendant's travel behavior is likewise affected. This also applies to people tending animals (personal pets requiring attention or not accepted at shelters, animals in zoos, nature reserves and laboratories, fish or birds kept in captivity, etc.). This restricted-mobility group needs to be surveyed to identify their extent, needs, constraints, possibilities, and preferences regarding transportation during an evacuation.

Another group of the population that warrants data collection is emergency personnel and those used to implement emergency plans. These include emergency management, service, security, and transportation personnel that monitor, manage, and execute evacuation, rescue, recovery, and return activities. Attention must also be given to their shelter and accommodation needs, and the safekeeping of their dependents.

Among emergency-type exceptional events, transients and tourists can pose a particular problem because they will generally behave differently to local residents, and the means of gathering data from this group is likely to be quite different from that of local households. Depending on the season and the city, transients and tourists can make up a significant portion of those requiring evacuation.

12.3.1. Data Collection

Data collection of behavior during exceptional events poses unique challenges due to the dynamic nature of exceptional events, their relative infrequency, limited advance warning, short duration, conflicting movements, diversity of population, and the serious consequences that can result from incorrect decisions. Another challenge is that since data are required for both planning and operational purposes; there is a need to evaluate alternative scenarios in advance of events, as well as provide real-time information while an event is in progress.

The conditions prevailing during an exceptional event and the diverse purposes for which the data is required suggest that data should probably be collected by means of a variety of procedures rather than by one procedure. For example, travel behavior would probably be best collected using a time-use type of home interview survey, but the stimulus for travel and the travel conditions prevailing over time that would inhibit travel, need to be collected by other means. It is expected that data collection on exceptional events would include data on weather, traffic conditions, traffic control, transit service, announcements made, road closures, and contraflow operations.

Collection tools that workshop members identified as likely playing an important role in the future were remote sensing devices, and opinion surveys. Remote sensing in terms of GPS, cell phone tracking (GSM), ITS vehicle-detection systems, Bluetooth-equipped instruments, and even the use of manned and unmanned aircraft, was seen as having great potential in future data collection during exceptional events. Spatial location technology, such as GPS and GSM, automatically incorporate time logging so that spatio-temporal data such as exact time of departure, speed, route, delay, and time of arrival are all obtained accurately with little respondent burden. Current ITS systems are capable of estimating travel times, and detecting and reporting incidents in near real-time, so their use during the operational phase of exceptional events is potentially of tremendous value. Bluetooth signals between equipment placed in vehicles and strategic places around a city, such as transit stations and key intersections, could provide valuable supplemental information on the movement of vehicles in the system.

The increased use of opinion surveys is expected to occur in terms of a greater use of stated choice surveys and focus groups. This is because they provide the means to understand the evacuation decision-making process better, and they can be conducted at any time and are not tied to the occurrence of an event. Visualization can play a vital role in presenting realistic scenarios in stated choice surveys and promoting communication and understanding in focus groups.

12.3.2. Research Needs

Exceptional events typically generate exceptional travel which has been poorly planned for and managed in the past. Data are needed to build models that can be used to estimate travel conditions under alternate events, policies, and infrastructure changes. They could be applied pre-event to produce contingency plans for different events. At the same time, data are needed to inform decision makers during an event for which there is no contingency plan, or in which conditions vary from those in a contingency plan. These data should provide information on current conditions and allow short-term predictions of conditions in the transportation network with and without policy and operational changes.

Dynamic data are essential for planning and operational modeling of exceptional events, because peaking is more pronounced and variable during exceptional events than in regular travel. Thus, dynamic data for planning and operations purposes are needed. The first research need in this regard is identification of official sources of data that can serve these purposes. Recognizing that travel behavior in exceptional events depend on the traveler, the stimulus for travel, and the conditions in which travel will occur, a wide range of sources could be considered ranging from meteorological records, through traffic counts, to policy or management decisions.

Another research need, related to the first, is identification of low mobility groups in the population and establishing travel opportunities for them. Included among these groups are those that are very reluctant to, or cannot, move such as the sick, infirm, handicapped, institutionalized, incarcerated, and those that tend to them. Associated with this group, are those dependent on transit or limited in mobility due to poverty, language barriers, or concern over illegal immigrant status. A third group is tourists and transients whose mobility, destination options, and familiarity with the local environment are distinctly different from those of local people.

12.4. Concluding Remarks

Workshop members felt that the perception of risk and other factors playing a role in the evacuation decision are currently poorly understood. Included among these factors are the risks and costs of travel, as well as the impact of work responsibilities and responsibility for others (e.g., relatives, neighbors, and friends) in making a decision to travel during an exceptional event.

Research is needed into the full use of remote sensing in gathering data on travel during exceptional events. Spatio-temporal passive data collection devices such as GPS and GSM have the potential to provide information that is very important in characterizing travel during exceptional events such as time of departure, route, speed, delay, time taken to reach safety, and so on. Other devices, such as real-time traffic counters, wireless communication, and aerial reconnaissance options should also be investigated.

Because exceptional events are often rare occurrences and provide relatively little forewarning, most surveys related to exceptional events have been post-event surveys in the past. This restricts the surveys to locations where the event occurs, and to the time when it occurred. Greater flexibility of location and timing of a survey would be obtained if the revealed behavior approach were replaced by stated choice. Research is needed to develop a dynamic stated choice procedure for travel during exceptional events. The stated choice procedure could include visualization to enhance the presentation of alternate scenarios.

Chapter 13

Tourist Flows and Inflows: On Measuring Instruments and the Geomathematics of Flows ☆

Christophe Terrier

Abstract

The importance of tourism for France's economy and society means that proper knowledge of tourism flows is essential. But designing a measuring system and periodic gathering of statistical data raise several difficulties. First, tourism is, by definition, based on movement, and all phenomena involving movement are difficult to measure. Second, there are many different forms of tourism, including holidays and business trips, short and long stays and so forth. Third, the notion of tourism flows has different meanings for those in charge of road, rail or air traffic management, and for those in charge of tourist visits.

This paper first discusses the ambiguities of the notions used in tourism studies. It emphasises the distinction between tourist flows along transportation routes and those in specific places. It then reviews the proper calculation rules for each of the geographical objects used for measuring tourism phenomena, which are primarily lines and areas. It also addresses some of the problems raised by the failure to comply with these rules in published information.

Third, this paper presents the various systems used to measure tourist flows and inflows, and discusses their usefulness and limitations, before discussing some new developments in the field.

☆ An earlier version of this paper appeared in French in *FLUX, International Scientific Quarterly on Networks and Territories*, no. 65, July–September 2006, 47–62, Paris.

Transport Survey Methods

ISBN: 978-1-84855-844-1

Finally, it examines the potential value of modern communication technologies for mobility studies. More specifically, it raises the issue of striking the right balance between statistical accuracy and individual freedom.

13.1. Introduction

In December 2004, when a tsunami wreaked havoc on Asian coastlines that attract tourists from all over the world, the French government wanted to know how many French tourists were in the area, in order to mobilise appropriate resources immediately. When a mosquito-borne disease spread on the island of Réunion, the government wanted to know what the usual tourist flows were for the period, how many tourists had cancelled their trips and what the shortfall was for local businesses and communities, in order to provide the necessary assistance for their survival. The same question comes up with each oil spill. The international body that examines applications for compensation demands estimates based on indisputable statistics. Therefore, the statistics required for these operations need to be provided. These statistics deal with tourist inflows measured over several years and the revenues generated by these inflows.

Notwithstanding these exceptional cases, the importance of tourism for France's economy and French society means that proper knowledge of tourist flows and the tourist industry is essential. Tourism plays a key role in France's economy and in French society. It concerns more than three quarters of the members of the French population who take a trip at least once a year. In economic terms, tourism accounts for nearly 7% of GDP (gross domestic product), and more importantly; it provides the main positive contribution to France's foreign trade balance (Direction du Tourisme, 2007). This warrants an attempt to determine the pattern of tourist flows as accurately as possible. But the many different forms of tourism make it difficult to come up with a system for measuring tourist flows and to gather systematic data.

Who Is a Tourist?

According to the international definitions used by the UN Statistics Commission, a tourist is anyone who takes a trip that involves spending at least one night away from home. This definition does not imply any specific activity. It covers business travel, holidays and journeys of discovery, even though tourists are most often thought of only as the sunburnt, camera-toting variety. The word tourist does not have a very positive connotation in France, and many "travellers" reject the label of tourist that statisticians want to put on them when they are travelling "on business." This distinction may lead to major discrepancies between "tourism" statistics and statistics dealing with "holidays" or "leisure."

The definition requires that a tourist spend at least one night away from home. This minimum means that day trips are not considered to be tourism.

This distinction may lead to major discrepancies between "tourism" statistics and "transport" statistics. Even within the tourism category, distinctions are drawn based on the length of the trip and the purpose of travel. More specifically, "holidays" cover only pleasure trips lasting at least four nights. Once again, tourist flows are often confused with flows of holidaymakers.

Tourism statistics always measure "nights" and not "days." It is presumed that travellers stop moving to spend the night in a given place. Therefore, travellers are not counted as being in a place unless they stop there to sleep. It is also easier to survey travellers in the places where they stop for the night.

Therefore, tourists are normal individuals who, at a given moment, are in a specific spatial and temporal context that involves taking a trip away from home and returning home after a minimum length of visit. A minimum stay is required to qualify travel as "tourism," but there is no minimum distance to be travelled. This means that people travelling to the other side of the world or just a few miles down the road are all qualified as tourists.

An "international tourist" is any person who is not resident in a country in which he spends at least one night during his trip. This final definition is an important one because it is used to measure international tourist flows.

If we consider that a tourist is someone who can come from anywhere and go anywhere else, it is easy to see why it is so hard to translate all these movements into statistics. It is always easier to count things that stay in one place than things on the move. The figures produced by counting these movements have to be processed using appropriate calculation rules, which vary according to the geographical object involved; hence, the term *geomathematics of flows*. Major institutions do not always comply with these rules. Moreover, tourist flows and inflows are often estimated on the basis of the same surveys and measurements, which may lead to inaccuracies and even assessment errors that could be troublesome for the various players concerned with the development of tourism.

There are many such tourism players, and their information needs are different. They can be split into two broad sectors, each of which covers several institutional or private players, and some players may be active in both sectors:

The "transport" sector (transport operators, transport infrastructure producers and managers, along with various government agencies managing traffic) relates primarily to tourist travel and thus needs "flow" statistics measuring traffic of vehicles and people. A trip has a starting point and a destination and involves one or more modes of transport. Yet flows are not everything. If we want to know about traffic along a route, a measurement of flows is not enough; we need to know the date and time of travel: the same quantity of vehicles may move smoothly if the flow is regular, but may become congested during peak times. Tourist flows are not the only flows, even over long distances: they come on top of other flows of shorter duration, flows over shorter or longer distances and flows of other types, such as flows of goods.

The "inbound tourism" sector (hotels, various forms of accommodation and catering, various levels of local government, tourism-related activities, such as sports and cultural activities, along with various types of trade, etc.) is concerned with the tourist's "visit." We use the term "inflow" (or visits) to measure tourists' presence in France (Terrier, Sylvander, & Khiati, 2005a; Terrier, Sylvander, Khiati, & Moncéré, 2005b; Terrier, 2006). A visit implies the tourist's presence in a given place. Once again, it is important to know the date and length of the visit. Then we look at the tourist's activities and, naturally, their spending; but this last point is not of direct concern to us in this paper.

Therefore, traffic advisory centres, transport network managers, mayors of tourist towns, hotel managers, tourist site managers and leisure centre managers are all going to want to know about tourist flows, but this term will not have the same meaning for each of them and the same statistical system will not provide the answers to all of their questions.

The borderline between the private sector and the public sector is not always very clear-cut in the tourism sector. Local and central governments make great efforts to promote the attractions of their localities for domestic and international tourists. This means that the demands on the public statistics system exceed what would seem to be its legitimate mandate and, in any case, its capacities, given the resources dedicated to it. This paper deals with the public statistics system and does not attempt to define the ideal statistics system for tourism. Instead, it reviews the information sources available for measuring tourist flows, along with their contributions and limitations, and then considers some avenues for future development.

13.2. The Geographical Objects Used: Points, Lines and Areas

When we speak of measurement instruments, we need to start by specifying which geographical objects are used as a basis for measurements. The geographical objects used may be lines (traffic routes) or areas (territories). Points may also be used as limits for lines (origin or destination) or as proxies for areas, as when one locality is used to represent an entire local area.

Traffic follows given routes (roads, railways and air corridors), which means that the natural geographical object for measuring flows should be a line, whereas an area should be used to measure inflows, or the presence of tourists in a given territory. In practice, we use areas most of the time, for the simple reason that all social and demographic statistics refer to areas. More specifically, these statistics do not refer explicitly to a geographical object, but to a political or administrative entity that has power over a clearly delineated physical territory. Most of the information derived from conventional statistics is therefore based on areas. This situation, as we shall see, calls for a number of precautions when measuring flows and making the related calculations.

13.2.1. Area-Based Measurements

Several population surveys provide data on origins and destinations. These data are often specified at the level of municipalities (communes). The municipalities of departure and destination are known in the information drawn from the population census, such as data on commuting, and it is also true of most surveys dealing with tourists.

Population surveys provide results by administrative entity, as we have said. Sometimes the main locality of an area is used as a proxy for the whole area. This is common practice when calculating distances travelled, with the distance from one main locality to another being used. These distances may be calculated "as the crow flies," or using commercially produced driving-distance tables.

13.2.1.1. Area-based calculations The method of operation is always the same when the reference geographical object is an area. Any movement within the limits of the area counts for nothing; only movements that cross the limits of the area are counted. Such movements may be "inbound" or "outbound," depending on whether the individual is entering or leaving the area. Flows in "transit" are not usually counted, unless there is a stop of a given length of time in the area, such as stay of one night or more in the case of tourism. Surveys generally provide information about the area of departure and the destination area, but no information about the itinerary, the areas crossed or even the route travelled.

The distinction between internal flows (stable), which are not counted, and external flows ("inbound" and "outbound"), which are the only flows counted, is the determining factor for measuring flows between areas. The data are biased because movements over short distances are counted if they cross the limits of an area, but movements over similar distances are not counted if they take place within the limits of the area. The smaller the areas used, the weaker the bias. The impact of the bias is insignificant only if bulk of the movements measured involve distances that are greater than the diameter of the area. This means that the size of the reference areas used has a direct impact on the volume of flows measured. We would obtain different figures for the same volume of flows, depending on the size of the areas used. Furthermore, if the sizes of the areas vary significantly, the bias will not be uniform. Under these circumstances, it is not right to compare flows measured using areas of different sizes or levels. Yet this is what often happens in international comparisons, where only the area of a whole country is taken into account. This results in disproportionate comparisons between geographical entities, for example placing the entire United States on the same level as individual European countries. This is a widespread problem in territorial analysis (Grasland, Mathian, & Vincent, 2000; Terrier, 2000), but it is especially significant when analysing tourist flows, particularly international tourist flows, as we shall show below.

13.2.1.2. Changes of scale The main problems in calculating flows based on areas come up when we want to change scale in order to transform data obtained at one level of areas into data about another level of areas. This operation is feasible, subject to certain conditions, when changing from smaller areas to larger areas (by aggregating smaller areas into a larger area), but it is impossible in the opposite direction.

13.2.1.3. Changing from a smaller area to a larger area This change needs to be made, for example, when flow data between municipalities (known in France as "NUTS 5" zones) are available and we want to calculate flows between Départements ("NUTS 3" zone), or when data are available on flows between countries and we want to calculate flows between continents.

The new area must be made up of contiguous smaller areas in order to be valid. The calculation involves eliminating flows within the new larger area and recalculating inbound and outbound flows. In order to do so, we must consider flows between the smaller initial areas that make up the larger area to be internal flows that need to be eliminated. In our example of a change from the municipal level to the Département level, all of the flows measured between municipalities within the new larger area making up the Département become internal flows and are no longer counted. Then we add up the outbound flows from each of the initial areas to destinations outside of the new larger area. Finally, we add up the inbound flows to each of the initial areas from outside the new larger area.

We must have detailed information about flows between each of the initial smaller areas to carry out this operation. At the very least, we must be able to distinguish between the initially measured flows that become internal flows in the new larger area and those that remain external flows. If the only information we have for each smaller area is the sum of the inbound flows and the sum of the outbound flows, we will not be able to derive information about a larger area made up of these smaller areas.

Combining areas to make a larger area generally leads to a decrease in the flows measured since some of them become internal flows. Flows between Départements will always be less than the sum of flows between municipalities, and flows between continents will always be less than the sum of flows between countries.

13.2.1.4. Changing from a larger area to a smaller area A change to a smaller area is impossible to calculate. If we do not have any detailed information about the smaller area, we could attempt to model the flows by making assumptions based on information that may be available from other sources about the average distance travelled and the likely location of outbound flows, such as the population of places of residence, or inbound flows, such as the accommodation capacity of tourist areas. The accuracy of the figures obtained would depend on the quality of the supplementary information available and the assumptions used to build the model.

These problems relating to size and changes of size in areas give rise to a great deal of uncertainty in existing estimates of foreign tourists in a given country and, consequently, in estimates of international tourist flows around the world.

How Can We Measure International Tourist Flows Around the World?

The number of foreign tourist arrivals in a country is measured each year. "Foreign" designates any person who does not reside in the country in question. Such a person is considered to be a tourist if they spend one or more nights in the country. If the person makes 10 trips in the year and spends one or more nights each time, these trips are counted as 10 foreign tourist arrivals. Each country provides this information to the UN World Tourism Organisation once a year. The UNWTO collects these statistics in order to analyse tourism around the world and publishes its findings annually (UNWTO, 2006).

First observation: countries are not at all uniform in terms of size. A resident of Brussels who makes a one-hour-and-twenty-minute trip on a high-speed train to spend a weekend in Paris will be counted as a foreign tourist, whereas a New Yorker who travels to San Francisco will not. This means that an area made up of little countries will generate a higher count of foreign tourist arrivals than a country as big as a continent.

Second observation: in addition to its size, the geographical position of a country relative to other countries also plays a role. Centrally located countries are much more likely to be crossed by tourists than border areas or peninsulas. This observation is especially germane for a country like France, which is located at the crossroads between Northern and Southern Europe and which is large enough to warrant an overnight stop when crossed by car.

Final observation: in view of the requirements set out above for changing from information about smaller areas (countries) to information about larger areas (continents), it is impossible, using the data that each country provides to the UNWTO, to calculate the inbound flows of foreign tourists to a continent. Such a calculation would count only those people who are not residents of the continent and who spend one or more nights there. We may have information about the aggregate number of foreign tourist arrivals for each country, but we do not have detailed information about the origins of foreign tourists visiting a country, such as their country of residence or the countries visited previously in cases where travel involves stays in several different countries. And yet world tourism analysts add up the figures provided by each country. This leads them to state, for example, that 52% of the tourist flows in the world have Europe as their destination. These figures count all tourists from other European countries as "foreign" tourists, whereas such tourists should not be counted when talking about foreign tourists in Europe. However, the UNWTO is unable to make such a calculation as things stand at present.

However, the statistics that member states provide to Eurostat should make such a calculation possible, since each country is asked to distinguish between tourist flows arriving from Europe and tourist flows from the rest of the world.

But there is still another major problem regarding the treatment of tourists from another country who make a tour of Europe. At present, an American, a Japanese or a Chinese tourist who visits 10 European capitals and spends one night in each country is counted as a foreign tourist arrival in each country visited.

This means that when the figures for each of the European countries are added together to talk about tourism in Europe, these tourists will each be counted 10 times. There is no practical way of obtaining figures on tourism in Europe from the statistical systems of each country, since these systems record only the tourist's country of origin and not the country previously visited. A survey or a recording and monitoring system implemented at the European level would be the only way of obtaining information about such tourist flows without double counting. This is only one of the many difficulties encountered when making international comparisons. We have already mentioned the first problem, which stems from the uneven size of the reference areas, which means that measurement data relating to international flows are not comparable.

The second problem is partially linked to the first and stems from the fact that we never count domestic tourism. This eliminates a good portion of the flows in France, where two-thirds of the tourists are French. The effect is even more pronounced in the United States because the country spans the whole continent.

13.2.2. Line-Based Measurements

The lines we are looking at here are transport routes. Measuring flows means counting the elements, or vehicles, that move along these routes. Yet these elements have varying degrees of autonomy, depending on the type of route and the mode of transport. This leads us to distinguish between "closed" lines and "open" lines.

We say that a line is closed if an element moving along it has no autonomy with regard to its line of travel. A railway line is the purest form of such a line. It is materialised by two strips of steel that the vehicles must stay on. Travel along the railway line is very predictable from a departure point to an arrival point. But the line is only really closed from a single departure point to a single arrival point. A change of trains during a trip marks a break in the line, and a train stopping at local stations, even though it stays on the railway line, should not be seen as travelling along a closed line, but along a succession of closed lines placed end to end.

Airline routes are defined by their departure and arrival points in practice. Knowing the real flight path of an aircraft and the air corridors used is of no real use when measuring flows. This means we can consider airline routes to be closed lines.

Roads are an example of very open lines. The road network is very complex, making it possible to travel from one point to another along several different routes. Furthermore, the vehicles that use roads have a high degree of autonomy and are able to turn off, stop or turn around at any time. This means that there is no clear relationship between lines and trips. On the other hand, motorways can channel traffic over certain stretches between tollbooths or between on-ramps, and thus can be seen more as closed lines.

Only closed lines really lend themselves to measuring flows. For one thing, travel along closed lines is usually managed and recorded, primarily through ticket sales, which means it is pointless to undertake other types of counts. In addition, the information is the same whether measured at the starting point, the destination or any point along the way. If we want to do more than simply measure flows, we can conduct a passenger survey. The best place for conducting surveys varies depending on the mode of transport. Air passenger surveys are generally conducted in the boarding lounge, whereas train passengers are usually surveyed once they are aboard, since there is usually no significant waiting time before the train leaves.

There are fewer opportunities for measuring flows along open lines. Such lines are characterised by the freedom of movement of the vehicles that use them. Vehicles can enter, exit, stop, turn and backtrack. We can measure flows past a given point, using automatic counters, tolls or counts made by survey takers, but the flow measured at one point along the way does not enable us to make a direct estimate of the flow at another point. In practice, roads are broken up into segments where traffic is more or less uniform. The flows measured at one point along the segment are attributed to the segment as a whole. This system cannot be used for urban areas, where the lines are too open.

There are not really any calculations that can be used for lines other than simply adding up the flows measured successively along a single segment. However, we could attempt to extrapolate the flows measured at one point to a line or an area. As is the case with any extrapolation, this would require us to make assumptions about patterns, based on further information from other sources, whenever possible. As noted above with regard to such an approach, the accuracy of the results will depend on the accuracy of the information from other sources used to develop plausible assumptions about patterns.

13.3. Tourist Survey Methods

One good way of measuring tourist flows and inflows is to conduct surveys of tourists. Such surveys may be conducted in the respondents' homes, at the places they are visiting or at points along the way when they are travelling. We shall not describe all of the tourism statistics systems here; readers can refer to more comprehensive papers on the subject (Bernadet, 2003). We shall merely review the main principles of existing statistical systems and try to point out the contributions and limitations of each survey method when it comes to increasing our knowledge of tourism and measuring tourist flows and inflows.

13.3.1. Household Surveys

Two national tourism surveys in France are based on the principle of surveying respondents at home: the Tourism Demand Survey (SDT) and the Holiday Survey. The former is conducted once a month by TNS-Sofres for the Directorate of

Tourism. It is based on a panel of 20,000 respondents who are surveyed monthly by mail. The latter is conducted by France's National Statistics Institute (INSEE) every five years and involves face-to-face interviews in respondents' homes.

The principle behind these surveys is simple. They take place after tourists return home, when they are no longer tourists, and they involve questions about travel during the reference period. The information sought can be complete, since the travel has been completed. On the other hand, some of the people in the sample are bound not to be tourists, since there is no way of knowing whether they have made a trip during the reference period before interviewing them. That being said, knowledge about respondents who stay home is considered important for social analysis (Rouquette, 2001; Chevalier, 2004).

The survey method used is the key to the level of detail in the information gathered, along with the quality of the sampling. Survey methods include telephone interviews, mailings or home visits. This type of survey encounters two main types of problems: memory problems and capturing the complexity of the trip. Memory problems are easy to understand; the longer the survey comes after the trip, the more uncertain memories are. This is not much of a problem for major trips that leave big impressions, but little trips are often quickly forgotten. A face-to-face interview or a mail survey is less likely to encounter this problem since respondents can take their time to answer and even look at their diaries or other documents, in contrast to telephone surveys, which require off-the-cuff answers.

The complexity of a trip may vary depending on the succession of movements and visits. The simplest case is a trip that involves a single return journey to a single location. This is the case for 90% of the trips made by the French. The matter becomes more complicated when the trip involves several movements and locations.

The questionnaire for the SDT, which covers overnight travel completed in the previous month, leaves room to describe three trips and two consecutive visits for each of the trips described, as well as the travel between locations. After that, there is a loss of detail about travel and visits. This means that information about the most nomadic tourists is incomplete; a long meandering trip with stays in different places each night cannot be fully described.

In the Holiday Survey, which covers the previous year, details are required only for holidays, meaning visits of four or more nights for pleasure. Respondents are also asked about visits of two or more nights, but in less detail. This information is gathered by survey takers, who make appointments to call on respondents in their home. This arrangement makes it possible to ask more complex questions.

The household surveys make it possible to cross-reference places of residence and holiday destinations. This provides us with an estimate of flows between French regions and between France and other countries. The SDT also makes it possible to study absences and presences in an area on a given day simultaneously. Comparing the number of persons present in a given area to its resident population makes a huge contribution and gave new impetus to inflow analysis.

Tourist Inflows: Estimating the Population Present
Tourism Leads to Major Variations in the Population Present in a Place

As mobility increases, particularly tourist mobility, people are not always where we expect them to be, and the population "present" in a place at a given time may be significantly different from the "resident" population revealed by the general population census (Terrier, 2005a, 2005b). For example, tourism surveys show that on 15 August, nearly 14 million French residents are away from home; 2 million of these are outside of France. If we count the 4 million foreign tourists in France on the same date, we get nearly 16 million people in France who are away from home.

Proper public management and planning calls for an estimate of the population present. Some amenities need to be calibrated according to the maximum population that may be present in a place at a given time. This means we need to be able to estimate the peak inflows. Other services need to be calibrated for the average population present, which also needs to be estimated.

The Tourism Directorate has calculated a day-by-day estimate of the population present in each Département in response to the public authorities' concerns. The estimate starts with a day-by-day evaluation of absences, meaning the number of residents of the Département who are travelling outside of the Département. Then the presence, in the Département, of foreign tourists and French tourists who do not live in the Département, is estimated on the same day.

Another View of Spatial Planning

This research has led to another view of spatial management and planning issues. It has also provided a basis for a new approach to local economics, called presence economics, based on the principle that consumption and, consequently, economic activity stem from the presence of people in an area at a given time (Davezies & Lejoux, 2003).

The results show how important it is to distinguish between the population "present" and the "resident" population. Some Départements see their population double at certain times of the year, while in others, the population present is nearly always smaller than the resident population. If we wanted to have sufficient vaccine stocks in each Département to immunise the entire population present in an emergency, it is estimated that 71 million doses would be necessary (i.e. 6 million more than the population of France), in view of the fluctuations in population present.

In 2003, which was the reference year for the survey, the Hautes Alpes Département set a record with a population present at the end of July that was 2.7 times greater than the resident population. The ratio in early November was only

0.96. Over the year, the population present in this Département, calculated in permanent resident equivalents, is one and a half times greater than the resident population. The Départements in the Ile-de-France region posted the highest absence rates. The population present in Paris ranges from 109% of the resident population at the beginning of December to 73% on 16 August. The record was set by the Hauts-de-Seine Département, where the population present on 15 August was only 56% of the resident population.

Variations in Space and Time

The difference between the resident population and the population present varies over time and from place to place. This difference stems from the dual movements of outbound residents and inbound tourists. Tourist inflows nearly everywhere in France are at their highest in August. But the simultaneous resident departures and tourist arrivals mean that the present highest population figures are reached on different dates in different Départements: around 15 August in the Savoie and Morbihan Départements, and early December for Paris and the Bas-Rhin Départements. Cities, and the largest cities in particular, post the highest number of departures, especially in the summer and during the school holidays. Cities have large populations and these populations travel more. On the other hand, tourist destinations each have their own seasons. For example, seaside resorts draw large crowds in the summer, and the mountains attract skiers in the winter and hikers and climbers in the summer.

Some highly urbanised areas are characterised by large flows of inbound and outbound tourists. This is obviously true of Paris, where the very large numbers of inbound French and foreign tourists are never greater than the numbers of absent Parisians travelling elsewhere at any time of the year. This phenomenon occurs in other Départements, but to a lesser degree, especially at the seaside or in the mountains, where there are cities that produce large flows of outbound tourists, but are also resorts that attract large inflows of tourists.

Inflows Are Relative

People who worry about the 75 million foreign tourists overrunning our small country and its 65 million residents should note that these tourists do not all descend on France together. The peak day is reached in the summer, with slightly fewer than 4 million foreign tourists in metropolitan France. At the same time, some 1 million French residents are travelling abroad, which means that the excess population attributable to tourism is never more than 3 million.

Source: From Terrier (2005) in "Le Tourisme en France" (INSEE-Références, December 2005).

The current prevalence of position determination technology opens up opportunities for new technologies that could supplement the contributions made by the household surveys. One example of using this technology is provided by BVA, which conducts telephone surveys for the outdoor advertising audience measuring firm Affimétrie using a high-performance geographical information system (GIS). Respondents are questioned about their movements on the previous day. The respondents tell the interviewers the addresses of their departure points (e.g. their homes) and the addresses of their destinations (e.g. their places of work). The interviewers then call up detailed maps on their computers that enable them to have the respondents specify the exact routes travelled, street by street. In this case, we are measuring more than just the flow between the departure point and the destination; we have the exact route taken. When this information is supplemented by means of questions about the departure and the arrival times, we obtain precise space and time information about the movements. If this system were added to existing household surveys on tourism, it would greatly enhance the information about tourists' movements.

Surveys conducted in the respondents' homes are obviously good for studying French residents, but they do not provide any information about foreign tourists. This limitation is particularly unfortunate for a country like France, in view of the numbers of foreign tourists visiting the country. Extending the scope of the survey would call for a representative sample of the entire population of the world, which is obviously inconceivable.

Some entities, such as IPK International, try to gather the findings of similar surveys of outbound tourists in as many countries as possible. These comparisons encounter problems because the surveys used are not all designed the same way. More specifically, these surveys do not all use the same definition of tourism. Many cover only stays of four or more nights, whereas the general trend is towards a greater number of shorter trips. Some of the surveys deal only with leisure travel, as in the case of the Holiday Survey. In some cases, the surveys deal only with the destination country and often ignore stays of one or two nights in countries crossed *en route*. These restrictions of survey fields, which are warranted because of the major cost savings involved, make it a delicate matter to compare the figures from surveys that differ in content. This has even given rise to some pointless controversies (Terrier, 2003). Nevertheless, such an approach may produce fairly good results for overall analysis and for measuring changes in major trends.

13.3.2. Other Surveys of Tourists

Other types of surveys have been developed to enhance our knowledge of foreign tourists. Such surveys are often not as detailed at surveys conducted in respondents' homes, and they raise many methodological issues. Nevertheless, subject to taking a number of precautions, they can be used to obtain data about foreign tourist flows and inflows. Proper knowledge of these flows and inflows is critical for France and its regions.

13.3.2.1. Accommodation surveys Based on the, sometimes theoretical, principle that a tourist stops for the night to sleep, we consider that the location of the overnight stay is the place of accommodation. In France, we rely mainly on the hotel and campground occupancy surveys conducted by the National Statistics Institute (INSEE) and the Directorate of Tourism. These surveys monitor occupancy from day to day, which means we can use them to estimate inflows, but not flows. They provide information about the geographical origin of tourists, thus enabling us to distinguish French tourists from international tourists.

The problem is that these surveys cover only two types of accommodation, which account for less than 20% of the stays in France by French tourists. Some other types of commercial accommodation (self-catering houses and flats) are surveyed in a few regions, but there is no survey covering all types of tourist accommodation in the whole of metropolitan France, instead of just hotels and campgrounds. The other types of commercial accommodation, including self-catering houses and flats, tourist residences and holiday villages and non-commercial accommodation (family, friends and holiday homes), cover such a wide variety of individual and collective establishments that it is very hard to design exhaustive surveys.

However, it is possible to use the findings of other surveys, such as the survey of foreign tourists conducted at border crossings in 1996, to establish correlations and to develop a model for estimating tourist inflows from the occupancy of hotels and campgrounds. But this approach would rely on extrapolations that would be valid only if the overall patterns have remained the same. In view of the age of the main 1996 survey to be used as a basis for the extrapolations and the fact that it cannot be reproduced (see below), the validity of the estimations would diminish over time.

13.3.2.2. Border surveys (cordon-line surveys) Cordon-line surveys, which are called border surveys when the limits of the survey area are national borders, are the surest way to determine the presence or inflow of tourists in an area on a given day. They are based on the principle of establishing a cordon line around the survey area. By questioning the people crossing the line into or out of the area, or a representative sample of them, we can determine the population present in the area at any time. If we question tourists leaving the area, at airports, at seaports, on trains or at border posts on the roads, we can gather comprehensive information about their visit to the area. But the elimination of road borders in Europe under the Schengen Agreements and the practical impossibility of using law enforcement officers to stop vehicles in order to survey the occupants mean that this type of survey is now unfeasible for tourists travelling by car. This is particularly unfortunate in France, where most tourist traffic enters and leaves by road (Christine & Vassille, 2004).

As noted above, the most recent border survey providing detailed data on foreign tourists dates back to 1996. Several test surveys have been conducted since then, but none has produced useable findings. The methodology of such border surveys had to be adapted for tourists travelling by road. In the new Foreign Visitor Survey (EVE), foreign drivers are interviewed in motorway rest areas. Naturally, this raises major methodological issues about the validity of the sample.

Cordon-line surveys, also called "flow" surveys, have also been used in France's top tourist regions (PACA, Bretagne, Midi-Pyrénées, Rhône-Alpes, Aquitaine and Languedoc-Roussillon) for more than 20 years now. The quantitative findings are produced for each region or for each Département. These surveys provide helpful information on domestic tourism, and they can be used to estimate the population present in different sub-national geographical units, and these estimates are more reliable than those produced by other surveys (Carreno & Marchand, 1999).

13.3.2.3. On-site surveys A new type of survey has emerged recently, called a "shared-weight" survey. It follows major research into methodology suited for measuring tourism and holds out great hope (Deville, Lavallee, & Maumy, 2005; Deville & Maumy, 2005; Lavallée, 2002, 2007). INSEE had already used this methodology for a survey of the homeless. This methodology surveys tourists in the different places they go, such as museums and beaches, as well as in bakeries, newsagents and hotels. This type of survey can be used to count tourists (including foreign tourists), as well as to gain more detailed knowledge about their behaviour. One of the problems involved in this type of survey, which is conducted at different sites, is to make a fair estimate of the actual number of tourists, since some tourists visit only one site, whereas other, more mobile tourists can be encountered in more than one survey site. The respondent should therefore be asked whether they have visited, or will visit, each of the other sites where they are susceptible to be surveyed. This would enable a variable to be created to measure the number of times that a person could be reached by the survey. This would then be taken into account when recalculating the weighting assigned to that respondent. The use of these "shared-weight" methods makes it possible to construct the probabilities of the breakdown of tourist numbers by site, without knowledge of the parent population.

This principle is being tested in Brittany for a visitor survey (Deville & Maumy, 2005; Morgoat, 2005). It is still too early to obtain any concrete results, but this survey method does seem to hold out more promise for understanding tourism in a given area. If such surveys live up to expectations, public and private sector players in the survey area can use them to obtain high-quality information about the tourists who visit their region. The national systems are no longer able to provide such information with the requisite level of detail.

This rapid review enables us to state that the data provided by the various existing surveys at the national level can be used to produce fairly reliable estimates of tourist inflows and seasonal variations in such inflows for different geographical units. This means that the surveys provide measurements of the most relevant items for the various "inbound tourism" players. However, there is still substantial uncertainty about estimates of foreign tourist numbers. The major comprehensive survey that serves as a reference is the 1996 Border Survey. It is the foundation for the extrapolations and assumptions used to develop various models. The 1996 Border Survey is an old one, and it can no longer be reproduced.

Given the economic importance of foreign tourism in France, particularly in terms of the trade balance, very substantial resources are currently invested in the EVE. There is no denying that this type of survey raises some major methodological problems that we are striving to overcome. The preliminary results should be published soon, and they have proven to be consistent with the results of the last border survey. However, it is critical for the future to be able to develop the new survey techniques, currently being tested, that seem to be capable of producing substantial improvements in the quality of tourism measurements.

As we can see, none of these surveys provides "real-time" knowledge of outbound tourism. Detailed surveys are conducted in retrospect, once the tourist has returned home. Furthermore, the sample may be large, with 20,000 people surveyed each month for the SDT, but not all of the respondents have taken trips in France or abroad during the survey period. This means the SDT is ill-suited to obtaining detailed knowledge of flows by destination. Consequently, it is impossible, with any of the surveys, to know the exact number of French tourists present in a given place in the world at a given time.

However, we shall see that the development of techniques for locating and monitoring individuals and vehicles on the move opens up new possibilities.

13.4. Measuring Flows Without Direct Surveys: Monitoring and Measuring

There have long been ways of locating and counting tourist flows that do not rely on conducting tourist surveys. The development of these systems is very closely linked to the various types of monitoring, such as the monitoring of the use of transport modes and infrastructures. The oldest system is the one based on ticket sales.

We are seeing the rapid development of possibilities for locating and tracking vehicles and people on the move, stemming from new passive and active position determination technologies, which can be used to record movements and measure the resulting flows in great detail. The development of the systems for making such measurements is clearly part of the trend towards closer tracking of individuals and vehicles (or certain vehicles) by governments. Using the data produced for statistical purposes in order to improve knowledge of tourist flows should at the very least result in the drafting of ethical rules to govern such use and measures to ensure compliance with these rules.

13.4.1. Ticket Sales and Measuring Flows: Air and Rail Travel

If there is a charge for the use of a transport mode or an infrastructure, the operators concerned have to develop various practices to ensure that the user has actually paid. For a long time, the sole purpose of a ticket was to make it possible to ensure that a passenger had paid the fare. But the systems have changed over time, and many of them can now be used to gather information about passenger flows and passenger characteristics. As the market grows more competitive between companies and modes of transport, operators have had to refine their management of passenger flows to avoid empty seats and to provide services that meet their customers' expectations and suit their habits.

Air travel occupies a special place in this respect because passengers are closely monitored. Individual passengers must provide proof of their identity, which means that airlines are able to know the exact numbers and types of passengers on a given flight. Comprehensive information about the flow of passengers on a flight is available, if the data gathering and processing system is properly organised. But for the purposes of understanding tourism, the passenger information that the airlines actually collect is incomplete. Unless a special survey is conducted, we cannot know the characteristics of the passengers (age, gender or nationality), or the purpose of their trip (business, holiday, etc.) Nor can we know how the various trips by air or other modes of transport are combined in the tourists' overall itinerary. In the case of rail travel, there are several systems for recording and collecting data based on ticket sales. Some systems merely cancel the ticket, while others read the information contained in the magnetic stripe or the chip. Stand-alone readers check that the ticket is appropriate for the trip. Networked readers can feed an information system used to monitor the flow of passengers in real time or after the fact. Furthermore, mandatory reservation systems and tickets that are valid only for a specific train provide a good proxy for measuring flows.

Reservations are required for high-speed and intercity trains, but they are optional for other mainline trains, and no reservations are taken for travel on regional rail networks, which means that there is no way of automating the measurement of passenger flows. Some travel cards issued to individuals provide greater knowledge of passenger behaviour. The "Grand Voyageur" card issued by the French railways (SNCF) is a prime example. Customers holding the card obtain a few benefits, and in exchange, the railway obtains comprehensive information about the customers' rail travel.

Identification of individual transit pass holders is possible in some cases, such as that of the Navigo pass issued by the Paris transit authority (RATP). Such identification could be used, for example, to distinguish whether a flow of 100 trips was made by 100 different passengers or by 50 passengers making two trips each. It would also make it possible to quantify the chaining of different trips requiring connections between the Métro, trains and buses and could be used to determine travel patterns. However, this type of system seems to be better suited for transport used for commuting rather than for tourism.

Collecting information is not the same as disseminating it. The major transport operators have implemented more and more sophisticated systems for closer monitoring of passenger flows and obtaining more detailed knowledge of their customers, but they guard these data jealously, deeming them to be strategic in an increasingly competitive market.

13.4.2. Road Travel: Tollbooths and Radar

The road network presents the greatest difficulties when it comes to determining flows. There is no way of monitoring traffic on surface roads or toll-free motorways.

Toll systems provide partial and patchy information only. Some motorway routes are broken up into toll segments, and motorists pay a few euros at each successive tollbooth, which entitles them to drive to the next tollbooth. This toll system does not provide any more information than a counting station. Other motorway routes use a ticket system, where motorists take a ticket on entering the motorway and hand it in when they exit the motorway. This makes it possible to calculate the distance travelled and the corresponding toll. This system seems to provide a good measurement of flows, since the points of entry and exit are known, as well as the dates and times, but, in practice, the stretches of motorway using the system are too dispersed to provide a basis for a complete flow measurement system. For example, the system would not make it possible to estimate the number of vehicles that drive straight through France from north to south when travelling from Germany to Spain.

Brief mention should also be made of traffic data measurements carried out by the technical staff of the Ministry of Public Works. The system is based on counting stations called SIREDO installed on publicly operated roads. There are currently 1021 such counting stations installed along the 28,000 km of the publicly operated road network. The counting stations can be accessed remotely, and batch processing of the data collected provides the basis for measuring road traffic. Real-time processing of the data provided by two-thirds of the counting stations is also carried out to regulate traffic. The data are gathered and imported by the MELODIE system, and the main statistical processing to produce usable data for various agencies is carried out using the ARPEGES software. All of this musical-sounding system will change as part of the SICOT computerised traffic knowledge system project being carried out by the Road and Motorway Technical Research Unit (SETRA) to rebuild the entire system around a centralised architecture with decentralised management.

One method now being tested on foreign tourists counts the vehicles driving through motorway tollbooths and determines the nationality of the passengers from the credit cards that are used to pay the tolls (Provansal & Houée, 2003). Credit card use varies from one country to the next and with the type of toll. Therefore, the information is supplemented by spot manual counts of foreign licence plates. This technique can be used to estimate flows along a route and to break them down by geographical origin. When the geographical circumstances permit, we can also attempt to derive estimates of tourist inflows in a region, but such extrapolations call for a great deal of caution. The counts are anonymous, which means that there is no way of distinguishing whether a car leaving the motorway at a given point is going to return to the motorway within the hour or if it will stay in the region for a while or if it will continue on its way to another region using surface roads.

Optical recognition protocols have now achieved performances that open up the possibility of using cameras to read the licence plate numbers of moving cars. This technology was developed to back up the radars that check speeds, making it possible to automatically ticket the owner of a speeding car. As is often the case, the monitoring systems came before the measurement systems. However, it is conceivable that these expensive systems installed along roads and motorways could be used for non-law-enforcement purposes to monitor traffic flows. Such a system could be used, for example, for real-time measurement of the number of cars travelling from Germany or Belgium directly to Spain and their travel times. A network of cameras installed along major roads could also be used to distinguish through traffic from local traffic. This system would make it possible to conduct an advanced form of cordon-line surveys to determine how much time elapses between the arrival of a vehicle in an area and its exit. This would give us the length of stay, and the licence plate would give us the geographical origin.

13.4.3. Tracking Individuals' Movements

The new technologies for monitoring, recording and measuring flows of travellers mentioned above primarily involve what is known as passive positioning. This means the system does not transmit a signal to locate the vehicle or the passenger. The presence of an individual is recorded when they pass in front of a fixed sensor. Chips are increasingly replacing magnetic stripes in these systems, which are becoming more widespread, particularly in places of work, where each individual has to have their own access card that allows them to enter only the areas for which they have clearance. Credit cards are another passive positioning system, which, as we have seen, can be used at motorway tollbooths. Credit cards can also be used to keep a record of the holder's presence and, more importantly, the holder's spending, as long as the holder uses them to make payments.

Today, the development of active geographical position determination technology that provides accurate tracking of individuals' movements and their presence in a given place opens up new possibilities for measuring flows, as well as monitoring, since these two aspects are linked to each other. Active position determination systems can be used to locate the target individual or vehicle precisely, wherever they are. This requires both the system and the target to transmit signals.

The most common type of equipment found in this category is the mobile telephone. When it is switched on, it exchanges signals with nearby cells. Telephony operators have recently started marketing real-time telephone position tracking service. This service, which naturally requires the prior consent of the person being tracked, is primarily aimed at businesses that wish to track their employees' movements on the road so as to optimise their ability to respond quickly to customer calls, for urgent on-site repairs, for example. This GSM positioning method is not very accurate (to the closest 700 m or 20 km, depending on the density of the cell network), but it requires little investment in equipment.

Another means of tracking tourists using their mobile telephones is offered by the geographical range of networks. In practice, users change networks when they change countries. The operators detect the change, and the new operator takes over providing service in its own territory, under the terms of bilateral agreements between operators. If all of the operators agreed to provide this information, it could be used to estimate foreign tourist inflows. These measurements would have their limitations. They would cover only mobile telephone users with international calling plans. They would exclude tourists from countries where the telephone networks are not compatible with ours, such as Japan. They would also exclude tourists who change telephones when changing countries in order to save money by calling at local rates.

After the tsunami, there were suggestions that procedures for locating mobile telephones be used systematically in emergencies to locate individual travellers. The technical feasibility and legal issues involved in such a measure still needs to be established.

Using Mobile Telephones to Count the Population Present

One example of the use of mobile telephones to measure tourist inflows is provided by an ongoing project by Bet F. Marchand, which consists of quantifying the total population present in an area at a given time using mobile telephone data. The method is called Info-mobility, and it can be used to estimate the various components of the population present (residents, visitors and transients), using information supplied by the different telephony operators concerned. The data are processed totally anonymously. The method counts the people present, but does not identify them. This method can be used as long as the area to be analysed is covered by a mobile telephone network. The results obtained for each area under study could then be aggregated to meet data needs at the level of the local community, the Département or the region.

Global Positioning Systems (GPS) have been available since the 1990s. They can give a position plus or minus a few metres, which is much more precise than the GSM positioning methods based on mobile telephones. GPS units are not transmitters. If they are to be used for tracking, they need to be combined with a recording system for batch processing of positioning data, or with transmitters for real-time tracking. The latter solution would require substantial investment in equipment.

The various active position determination systems are increasingly used to track goods using transmitter tags. Road vehicles are sometimes equipped with onboard transmitters. These systems are also used to regulate bus systems and provide information for passengers. They are being made mandatory for heavy goods vehicles in Germany, which have to pay licence fees based on distances travelled. They are found in Argos beacons for use in rescues at sea.

The technology is currently used for tracking people under specific circumstances (radio bracelets for prisoners, tracking elderly patients, tracking children in amusement parks, etc.) Widespread use of this technology would enhance the accuracy of statistics but is bound to stir controversy about its social impact.

This technology is evolving rapidly, and the surge in the use of a new technology known by the generic name of RFID tags (radio frequency identification) can no longer be ignored (Culnaert, 2006). These tags were developed for the US Army, and they have bridged the gap between active and passive systems. The tags are passive components, but they can still be read at distances of about 100 m, which opens up some impressive possibilities. The starting price for such tags for civilian uses was about one euro in 2005. Tags were incorporated into the tickets sold to spectators at a major football match in Germany in order to monitor fans more closely. Tests were carried out in a California school, using the tags to track children and keep them safe. The United States requires that, in the future, people entering the country carry passports containing such tags. The tags are already widely used to track goods. In a field related to ours, RFID tags are being used in London to track the movements of a panel of drivers and, more specifically, to measure how often they drive by advertising hoardings.

13.5. Conclusion: Striking the Right Balance Between Statistical Accuracy and Freedom

Readers can see that movements are hard to translate into statistics. Accurate and comprehensive mobility measurement would require powerful and effective tracking systems. It would not be an exaggeration to say that good flow statistics cannot be obtained without a good policing system. The present trends in this matter are contradictory.

On the one hand, every effort is being made to facilitate free movement, without any constraints or checks. Under the terms of the Schengen Agreements, border controls have been eliminated inside Europe. In France, hotel guests are no longer required to fill out police forms. The elimination of border controls and police surveillance has obviously weakened our ability to establish good statistics on the movement of people. The French National Council for Statistical Information (CNIS) has recently upheld the principle that the use of law enforcement agencies to stop vehicles for the purpose of statistical surveys is inconsistent with individual freedom. The elimination of the traditional methods used for traffic surveys and border surveys has made life more difficult for statisticians dealing with these sectors.

On the other hand, and especially since the attacks of 11 September 2001, some countries, including the United States, have instituted more restrictive monitoring systems. The United States now requires a level of information about travellers entering the country that violates individual rights previously recognised under French and European law.

For the time being, only animals are required to have an electronic tag inserted under their skin containing all of the information necessary to verify their identity and state of health. But our rapid review shows that all of the technology is available for real-time tracking of vehicles and persons. This technology is currently used primarily to provide services that users want (Belleil, 2004). However, we should be careful to strike the right balance between the precision sought with regard to flow data and the protection of individual freedom, even if there is a price in terms of statistical accuracy.

Furthermore, improving the accuracy of measurements is not the answer to every problem, because the difficulty of establishing the numbers is compounded by the equally important problem of explaining them. One of the lessons of this all-too-brief review of instruments for measuring tourist flows is that we must always be prudent when interpreting the data. The measurement of tourist flows may make statistical methodologists and modellers happy by giving them several interesting challenges, but it can also cause a great deal of confusion for those who need to use the data to carry out tourism development projects, or simply to enhance the management of their communities. Tourism industry professionals are often laymen when it comes to statistics. When they are faced with apparently contradictory figures, they are rarely able to distinguish between the discrepancies that are attributable to differences in the definitions of the geographical units covered by the survey and those resulting from differences in the measurement instruments employed. The effects of this complexity are compounded when the data are interpreted owing to the ambiguity of the terms used. For those of us in the business of measuring tourism, it is striking how nobody wants this word to stand for the same thing. Transport analysts see "passengers" and, consequently, study trips, the activities they serve, how they are structured and how they are made, whereas experts analysing areas focus on the temporary residents of a place, the length of their visit, the reason for the visit and the circumstances. Then there is the use that is made of the figures in institutional communications about tourist flows. International bodies, such as the UNWTO, Eurostat and the OECD, have announced ambitious plans to improve and set global standards for the methods for measuring and analysing tourist flows as part of a new methodology for the tourism satellite accounts. But efforts still need to be made to prevent the communication of these figures from creating confusion.

References

Belleil, A. (2004). La traçabilité de l'individu mobile: Vers la surveillance désirée. In: D. Kaplan & H. Lafont (Eds), *Mobilités.net* (pp. 372–376). Paris: LGDJ, coédition FING-RATP, collection Questions numériques.

Bernadet, M. (2003). Les dispositifs statistiques d'observation de l'activité touristique au niveau national: Présentation et évaluation. Paris: Observatoire National du Tourisme.

Carreno, M., & Marchand, F. (1999). L'évaluation par les flux: Outil d'analyse touristique territoriale. *Analyse et prospective du tourisme*, No. 56, September, Paris: Observatoire National du Tourisme.

Chevalier, B. (2004). Voyageurs et non partants en 2002. *Les essentiels du tourisme*, No. 20, May, Paris: Observatoire National du Tourisme.

Christine, M., & Vassille, L. (2004). Les enquêtes au bord des routes dans les domaines des transports et du tourisme: Quelles alternatives? *Rapport CNIS (Conseil National de l'Information Statistique)*, No. 82, February, pp. 1–64, Paris.

Culnaert, E. (2006). Identification par radio fréquence — Promesses et menaces des puces et étiquettes RFID. *AEC (Aquitaine Europe Communication) — Dossier de veille*, No. 20, February, Bordeaux.

Davezies, L., & Lejoux, P. (2003). Un train peut en cacher un autre-derrière l'économie productive, attention à l'économie présentielle. *Colloque de l'Association de Science Régionale de Langue Française*, September, Lyons.

Deville, J. C., Lavallee, P., & Maumy, M. (2005). Composition, factorisation et conditions d'optimalité (faible, forte) dans la méthode de partage des poids. Application à l'enquête sur le tourisme en Bretagne. *Journées de Méthodologie Statistique de l'INSEE*, March, Paris.

Deville, J. C., & Maumy, M. (2005). Extension de la méthode d'échantillonnage indirect et son application aux enquêtes dans le tourisme. *Journées de Méthodologie Statistique de l'INSEE*, March, Paris.

Direction du tourisme. (2007). *Mémento du tourisme* (2007 édition), Paris.

Grasland, C., Mathian, H., & Vincent, J. M. (2000). Multiscalar analysis and map generalisation of discrete social phenomena: Statistical problems and political consequences, *Statistical Journal of the United Nations ECE*, 17, IOS Press, 1–32.

Lavallée, P. (2002). *Le Sondage Indirect ou la Méthode Généralisée du Partage des Poids*. Éditions de l'Université de Bruxelles (Belgique) et Éditions Ellipses (France).

Lavallée, P. (2007). *Indirect sampling*. Springer editor (ISBN 978-0-387-70778-5).

Morgoat. (2005). Une nouvelle méthode d'enquête et de sondage dans le tourisme. *Vent d'ouest*, No. 90, pp. 26, 27, Comité Régional du Tourisme de Bretagne, Rennes.

Provansal, C., & Houée, M. (2003). Le trafic autoroutier comme vecteur de connaissance de la fréquentation étrangère: Application au réseau de la SAPRR. *Analyse et prospective du tourisme*, No. 82, March, Paris: Observatoire National du Tourisme.

Rouquette, C. (2001). Départs en vacances: La persistance des inégalités. *Economie et statistique*, No. 34, November, Paris.

Terrier, C. (2000). Les Zonages et l'Europe. *Annales des ponts et chaussées*, No. 93, pp. 68–72, Paris.

Terrier, C. (2003). Territoires, tourisme et statistique: Le défi de la méthode, le délire de la parole. *Colloque de l'Association de Science Régionale de Langue Française*, September, Lyons.

Terrier, C. (2005). Les tourismes en France. In: M. A. Le Garrec (Ed.), *Le tourisme en France* (pp. 11–24). INSEE-Références, November, Paris.

Terrier, C. (Ed.) (2006). *Mobilité touristique et population présente, les bases de l'économie présentielle des départements*. Paris: Direction du Tourisme.

Terrier, C., Sylvander, M., & Khiati, A. (2005a). En haute saison touristique, la population présente double dans certains départements. *INSEE-Première*, No. 1050, November, Paris.

Terrier, C., Sylvander, M., Khiati, A., & Moncéré, V. (2005b). Population présente: Méthodes de détermination à partir des enquêtes sur les touristes. *Journées de Méthodologie Statistique de l'INSEE* — 15 March, Paris.

UNWTO (World Tourism Organisation). (2006). *UNWTO world tourism barometer*, Vol. 4, No 1, January.

Chapter 14

Surveys of Tourists and Transients: Synthesis of a Workshop

Carlos H. Arce and Alan Pisarski

14.1. Introduction

This chapter summarises discussions from a workshop meant to examine how to measure the impact of tourists and transients (temporary residents such as business people, out-of-town shoppers or other visitors) on travel in urban areas, including the nature, timing and location of travel within an urban area. The travel behaviour of these people is likely to be quite different from that of regular residents. Therefore, the whole approach to travel survey methods needs to be reconfigured for this purpose. With this challenge in mind, the workshop participants identified different groups of tourists/transients in terms of their expected travel behaviour, and how they could be sampled, recruited and surveyed. An important associated issue was consideration of how the data from tourists/transients could be combined with regular travel survey data (i.e. passenger and freight) to obtain a more complete picture of travel in an urban area. In addressing this latter issue, the workshop participants recognised that the conference theme of improved harmonisation of data-collection protocols and focused communications among professionals in urban travel behaviour and tourism research were central concerns.

14.2. Key Concepts and Definitions

To provide a common framework for the discussion, Alan Pisarski, workshop chair, provided a brief summary of the work of the United Nations World Tourism Organisation (UNWTO) regarding fundamental concepts and definitions that have

Transport Survey Methods

ISBN: 978-1-84855-844-1

Table 14.1: United Nations forms.

	Domestic	Inbound	Outbound
Internal tourism	✔	✔	
National tourism	✔		✔
International tourism		✔	✔

been adopted and promulgated by the United Nations. 'Tourism' is travel for recreational, leisure or business purposes. The UNWTO defines 'tourists' as people who travel to and stay in places outside their usual environment for more than 24 h and not more than one consecutive year. The UNWTO distinguishes between types of tourists and categories of tourism. Key types of tourists are:

- Domestic tourists — residents of a given country travelling only within that particular country,
- Inbound tourists — non-residents of a given country travelling to and within that country and
- Outbound tourists — residents of a given country travelling to and within another country.

Using these forms of tourists, the United Nations classifies three basic categories of tourism:

- Internal tourism, which comprises domestic tourism and inbound tourism;
- National tourism, which comprises domestic tourism and outbound tourism and
- International tourism, which consists of inbound tourism and outbound tourism.

These tenets of the definitions are shown in the table above, and it was the sub-area of internal tourism that was deemed most connected to the interests of travel behaviour researchers, as it is most associated with the nature, timing and location of travel within an urban area (Table 14.1).

14.3. The Resource and Contributed Papers

Because the nexus between travel behaviour research and tourism research is a relatively new area of study, the workshop was grounded by three paper presentations. The first was the resource paper *Tourist Flows and Inflows: On Measuring Instruments and the Geomathematics of Flows* authored by Christophe Terrier (2009) and presented at the workshop by Thomas LeJennic. This paper provided an extensive treatment of the subject of appropriate definitions for measuring tourist flows and gathering systematic data. While the paper was related to the French context, the author pointed out that the dismal state of definitions and procedures for data collection were applicable to most countries in the world. Among the critical issues identified in the paper that impact our ability to harmonise travel survey and tourism data were:

- Definition variability and national idiosyncrasies;
- Barriers to data access — intercepts/cordon, proprietary nature of valuable information and privacy concerns;
- Unrealistic search for a 'one-stop' solution — the bureaucratic desire for a highly centralised solution to a highly decentralised phenomenon.

The presentation of the resource paper was followed by two contributed papers, which in their narrow foci illustrated the critical need for methodological guidance in terms of definitions and data-collection methods. The first, *Alternative Ways of Measuring Activities and Movement Patterns of Transients in Urban Areas: International Experiences* by Borgers et al. (2008), among others, described a procedure for obtaining information from visitors to shopping venues. It assessed varying techniques for developing information on visit frequency, activities, duration of visit and spending. It very carefully identified the statistical limitations of the procedure with regard to sampling design and weighting or expansion procedures. While it demonstrated a promising direction for research on a single micro-sector of the tourism phenomenon, it was agreed that it was a helpful contribution for at least this relatively limited area.

The second paper, *Estimating Foreign Visitors Flows from Motorways Toll Management System* by Mr. Michel Houee and Claudine Barbier (2008), described a promising technique for using toll road records to calculate the origins and destinations of visitors. This same technique was shown to be applicable to freight flows in another workshop session. While the paper contained both elegant and precise techniques for adjusting and expanding data, there were information gaps, such as the use of rental cars, which limited the validity of observations. Overall it was seen as a potentially powerful technique particularly when used in conjunction with other data sets, and thus, emphasised the need for data harmonisation and data fusion techniques, as did other workshops at the Conference.

Antonio Massieu, Chief of Statistics and Economic Measurement of the Tourism Statistics Section of the UNWTO, provided a detailed overview of the work of his agency which added depth and breadth to the introductory comments and papers. The UNWTO has been attempting to harmonise the definitions and protocols for exchange of data among nations. The traditional focus of tourism statistics has been on National Accounts, which provide a complete and consistent conceptual framework for measuring the economic activity of a nation. National accounts broadly present the production, income and expenditure activities of the economic actors (i.e. corporations, government, households) in an economy, including their relations with other countries' economies, and their wealth. While sharing many common principles with business accounting, national accounts are firmly based on economic concepts. As such, tourism estimates have figured significantly as important contributors to imports and exports. In this area, the UNWTO was responsible for the ground-breaking work in developing satellite accounts that better identify travel and tourism contributions to important national economic benefits.

Another use has been in the marketing area as countries seek to attract visitors from abroad. This can be seen to be in contrast to the typical interests within the transportation field, regarding investment needs, facility planning as well as environmental and safety concerns. These discussions identified such important gaps in the professional linkages in the respective fields that needed to be addressed systematically. As a possible bridge to such gaps, the focus of national tourism statistical agencies has turned more and more to regional domestic and inbound tourism activities defined as internal tourism above, which exactly coincides with the interests of transportation and travel behaviour professionals.

These five presentations, three papers plus remarks by Pisarski and by Massieu, prompted lively discussions and debates among participants over the nexus of these two interrelated disciplines. Below is a summary of the major deliberations and findings.

14.4. Deliberations and Findings

The group addressed the question of whether the central concern was to develop travel and tourism statistics (i.e. estimates) or to examine and understand the travel behaviour of tourists. The conclusion was that both were needed. The group further considered the distinction between inbound/outbound statistical measures versus internal travel at the metropolitan level (i.e. getting there vs. getting around). Again the group recognised that both were crucial for different purposes and agreed that future measurement systems needed to focus on addressing the latter.

It was deemed a great challenge that, despite the formally defined statistical terminology of the UNWTO and its adoption by the United Nations, national idiosyncrasies and definitional variability persisted from country to country either due to specific local needs or the lack of funding and training in the areas of interest. The gradual disappearance of border formalities (i.e. in the European Union), while clearly positive in most regards, has severely affected statistical capabilities, which had been further eroded in some cases by new stringent security procedures.

The ability to distinguish visitors from local residents and their different system and economic impacts even when they are doing exactly the same thing remains a great concern. The more precise distinctions between same-day visitors and tourists; and also between those typically counted and those excluded are critical tasks ahead. Some of the key issues to be addressed in future research efforts are identified below:

- Distinguishing between international versus domestic tourists and their travel behaviour;
- Distinguishing between visitors versus others who are not residents (immigrants, border workers, refugees, transit passengers, etc.) and their travel behaviour and impact on urban areas;

- Distinguishing between overnight visitors (tourists per UNWTO) versus same-day visitors and their travel patterns and impacts;
- Distinguishing between business versus leisure travel and their travel patterns and impacts;
- Defining and investigating travel involving visiting former home (family visits) in both domestic and international settings;
- Inbound versus outbound direction of trip making and the dynamics of such travel.

Rather than being able to depend on 'one-stop shopping' to meet data and information needs it was deemed more likely that the future will have to include data mining, integration and fusion. It has to be expected that this will be an area of iterative adjusting, balancing and expansion. Meta-analysis will often become the standard.

It was recognised that our travel survey methods tools are changing: both getting more difficult, as in phone survey opportunities; and also moving toward new opportunities using the new technologies such as use of remote sensors, GPS, cell phone detection and monitoring of credit card transactions. Many of these new tools need to be vetted, further tested and subjected to statistical scrutiny.

14.5. Conclusions

Because tourism is important to many countries and regions of countries as a generator of revenues and a promoter of better human understanding, we can view the future very positively. Countries, sub-areas and private industry will all have great interest in better understanding and quantification of tourism activities and their effects. In this environment, research must be framed as the tool that can improve greater understanding of tourism for both the public and private sectors and for improving the overall tourism experience for the traveller. Greater collaboration from all sectors of the tourism experience, together with travel behaviour researchers, will be crucial. Finally, the promise of research producing benefits to stakeholders was seen as an attractive solution to funding the needed research on this topic.

Acknowledgements

The workshop participants were Carlos Arce, United States; Linnea Abramowski, Sweden; Earl Baker, United States; Marco Diana, Italy; Jean-Paul Hubert, France; Tobias Kuhnimhof, Germany: Thomas Le Jeannic, France; Jon Inge Lian, Norway; Antonio Massieu, Spain; Alan Pisarski, United States; Francoise Potier, France; Valentina Rappazzo, Italy; Arne Rideng, Norway; Silvaramakrishnan Srinivasan, United States; Claude Weis, Switzerland; Oliver Horeni, Netherlands; Lissy LaPaix, Spain; Marie-Odile Gascon, France and Michel Houee, France.

References

Borgers, A., Joh, C., Kemperman, A., Kurose, S., Saarloos, D., Zhang, J., Zhu, W., & Timmermans, H. (2008). Alternative ways of measuring activities and movement patterns of transients in urban areas: International experiences. Presented at the 8th International Conference on Transport Survey Methods, Annecy, France, May 2008.

Houee, M., & Barbier, C. (2008). Estimating foreign visitors flows from motorways toll management system. Presented at the 8th International Conference on Transport Survey Methods, Annecy, France, May 2008.

Terrier, C. (2009). Tourist flows and inflows: On measuring instruments and the geomathematics of flows. In: P. Bonnel, M. Lee-Gosselin, J. Zmud & J.-L. Madre (Eds), *Transport survey methods: Keeping up with a changing world.* Bingley UK: Emerald.

PART IV

FREIGHT AND TRANSIT PLANNING

Chapter 15

How to Improve the Capture of Urban Goods Movement Data?

Daniele Patier and Jean-Louis Routhier

Abstract

This paper provides an extensive review of surveys and data-collection programmes focused on urban goods movement (UGM). Surveys investigating passenger urban travel have a decades-long tradition. The same is not true for UGM. The first specific UGM surveys appeared about 10 years ago in response to the rapid growth of car traffic, congestion, pollution and lack of space. Most of the time, these surveys have been carried out to resolve specific, local problems concerning traffic. Only a few of them have taken a global approach to urban logistics by including all logistics operators (own-account and carriers), all delivery vehicles (heavy and light vehicles), all deliveries and pick-ups (from express to full payload) and an entire metropolitan area and surroundings. Due to various European programmes, an inventory has been created to analyse urban goods data collection according to spatial level and methodology of capture. With this inventory, European urban freight indicators can be described, along with the units in which they are measured and their purposes. The relevance of urban goods transport surveys lies in their capacity to give decision-makers an account of urban freight transport functioning, ratios and data, so as to help in formulating planning, regulation and forecasting. It appears that focusing on the movement (delivery/pick-up), as the unit of analysis in establishment-driver surveys is the most efficient approach to describe the generation of vehicular flow in the city. This fact is revealed in the French UGM surveys, which take into account the complexity of urban logistics.

Transport Survey Methods

ISBN: 978-1-84855-844-1

15.1. Why Urban Goods Movement Surveys Today?

Surveys specifically focused on gaining detailed data on urban freight transport appeared about 10 years ago. While there are many reasons, the main one is that the issues and requirements encountered by urban authorities have changed. For a long time, their main aim had been to develop transport infrastructure to meet the goal of economic growth, in areas without much traffic congestion. Thus, efforts were essentially focused on long-haul transport. The problematic aspects of goods transport at a local level were partially taken into account by city planners and researchers. Until recent years, the integration of goods transport in models of total urban flows was estimated by applying a multiplying factor to car traffic. Delivering goods was not considered to be a concern.

As car traffic grew rapidly in cities, policy goals changed too. Top priorities now included the fight against traffic congestion, the management of lack of space (including shipment consolidation and storage), the desire to reduce local environmental impacts and global externalities (energy saving, reduction of greenhouse gas emissions) and handling the economic valuation of city centres (under the pressure of slowing economic growth).

All these changes took place in a context in which room for manoeuvre was limited by factors such as congestion, concerns about quality of urban life and budget restrictions. This resulted in a growing anxiety regarding the freight transport industry on the part of city authorities, the latter having little or no data, methods or references in order to elaborate a satisfactory policy framework.

Urban goods transport has specific characteristics linked to:

- products: numerous contributors, influence of upstream firm logistics, rapidly changing aspects of logistics management (i.e. decreases in parcel size, changes to downstream orders);
- urban logistics: the dominance of transport via road, the great diversity of delivery-vehicle types (ranging from two-wheel vehicles to 44-tonne trucks), the different travel patterns (round trips and direct trips), the large segment of the industry comprised of own-account (i.e. owner-operator) transport, the lack of foresight of the actors (complicated by subcontracting and moonlighting), complex last mile movement, congestion, strong constraints on road sharing and the intervention of local authorities for regulation and planning.

These specific characteristics require that researchers recognise urban goods transport as a specific field of research: urban logistics. Such a field has not yet been well examined, so there is no consensus on the definition of urban goods transport. Therefore it is necessary to precisely state what is taken into account in urban goods movement (UGM).

Figure 15.1 represents the global field of urban goods transport.

All vehicles carrying goods are involved:

- delivery vehicles used for inter-establishment trade (commercial, industrial, services), home deliveries, road works deliveries,

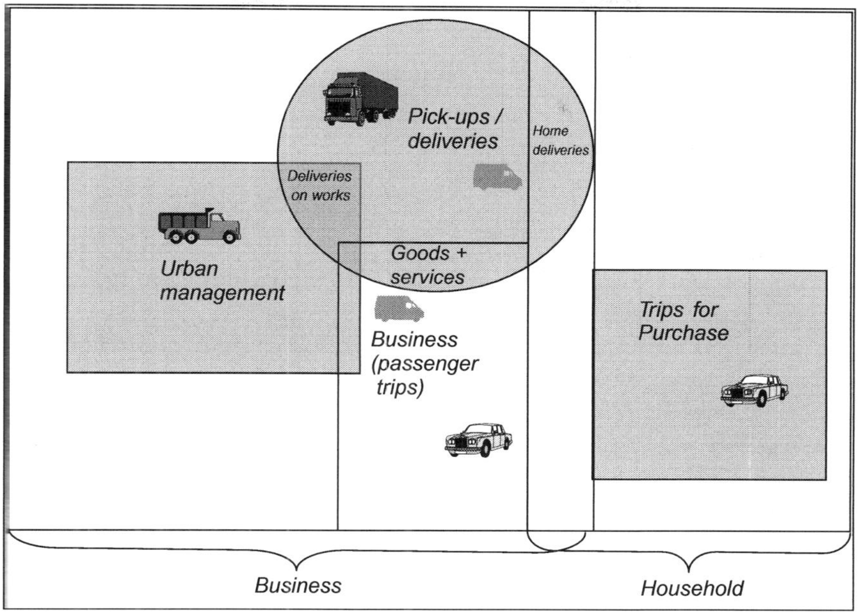

Figure 15.1: The components of urban goods and business traffic.

- vehicles for business trips (i.e. craftsmen carrying goods from depot to site),
- own-account and professional transport,
- from two-wheel vehicles to articulated lorries,
- private cars carrying goods (e.g. household purchasing),
- vehicles involved in urban management (i.e. waste collection, postal service, hospital, public works, building works).

The grey pattern indicates urban goods traffic (UGT). The whole of urban logistics is comprised of these three parts in order to take into account the whole urban logistics.

The current meaning of urban goods transport is related to *pick-ups and deliveries* carried out by own-account or third party. It includes goods and materials for industry, home deliveries and goods movement associated with business services.

Purchasing trips have to be integrated in surveys concerning urban goods transport insofar as households carry goods in their car. Their contribution is more than a half of vehicle × km car unit of road occupancy compared to all components of urban goods transport (Patier & Routhier, 1998). Purchasing trips cannot be investigated in the same survey with inter-establishment trade. Specific household transport surveys take place about every 10 years. It is nevertheless difficult to have an unbiased estimation of the origin–destination trips for purchase with such surveys (Routhier, Segalou, & Durand, 2001).

Specific surveys are necessary for *urban management*. Building and public works take place in various areas and intermittently. Postal distribution concerns both businesses and households, so it is impossible to take into account the global traffic it generates in such surveys. The post office is able to give roughly how many vehicles of each type are operating in the city and how many kilometres are travelled. This data may be included in the global results. The same thing goes for waste collection.

Business traffic consists of goods transport performed by heavy trucks (more than 3.5 tonnes) and business trips realised by light vehicles (less than 3.5 tonnes). Trips by light vehicles include more personal trips (covered by a household survey) than goods movement; this includes, for example, trips performed for commercial function without any goods movement. About one-third to half of business passenger traffic is counted in the framework of the household surveys. In fact, delineations are vague between goods, professional and individual trips. All inbound and outbound goods movement, as well as goods movement within the city, are included.

This paper will focus on the core of UGM data collection (see movements generated by the activities inside the circle of Figure 15.1); that is, the freight transported for deliveries and pick-ups, including home deliveries, works delivering (e.g. construction), and simultaneous business and goods supplying (e.g. a photocopy repair company delivering and installing spare parts). The following state of the art in such surveys presents a non-exhaustive but illustrative review, showing the diversity of the survey methods and case studies recently carried out. It is based, notably, on the work of the BESTUFS II European project (Bestufs.net) and on our own experience of French surveys.

15.2. State of the Art

Reviews on freight data collection have taken place in a few of the countries surveyed, as well as in other countries outside the EU. Such reviews are typically used to establish what data is being collected, why and how it is collected and the extent to which the data being collected meets the data requirements in terms of factors such as supporting freight policy decision-making and freight modelling. However, where such reviews have taken place, they tend to be concerned with freight data at a national level, rather than specifically at an urban level.

15.2.1. The Reviews

In Germany, an inventory of all data collected concerning commercial traffic was carried out between 1997 and 2000 by a special research team on behalf of the German Ministry of Transport. The objective was to review the available data on commercial traffic, to identify possible extensions to existing data collections and also to produce recommendations to overcome possible deficits in the existing data collected (Binnenbruck, 2006).

In the United Kingdom, the Department for Transport commissioned ‘The Review of Freight Modelling Project’, which took place between 2001 and 2003. This project considered data requirements and data sources currently available in the United Kingdom for freight modelling purposes (WSP & Katalysis, 2002). Much of the review was at the national and regional scales, but urban scale was considered. Work by the University of Westminster for Transport for London (TfL) has reviewed freight data sources for London (Browne, Allen, & Christodoulou, 2004).

In France, a review carried out by LET on behalf of the French Ministry of Transport in 1994 concluded that there was a major lack of urban freight data collection. More recently, a comparison of the objectives, methods and results of UGM in nine industrialised countries of Europe, North America and Asia was presented in Ambrosini and Routhier (2004). This review shows a similarity of objectives, but different approaches in data-collection methods and models. This highlights the lack of comprehensive surveys on UGM and the need to take account of urban logistics in the broadest sense.

Articles by Garrido (2001) and Wermuth, Neef, and Steinmeyer (2004) pointed out that there was a lack of in-depth studies for freight and commercial transport survey methodology. They make some recommendations for future international conferences of the ISCTSC in their critical analysis of current surveys. Wermuth et al. (2004) made a thorough review of business traffic surveys in Germany.

The BESTUFS I EU project has also previously examined urban freight data in 17 selected European countries, in terms of the availability of such data (BESTUFS I, 2000, 2003). In the BESTUFS II (2004–2008) EU Co-ordination Action, Browne and Allen (2006) edited an extensive review of urban goods data collection in 11 European countries. In the United States, a review of national freight data needs was carried out by the US Transportation Research Board (TRB) in 2003.

The 2003 report by the OECD on urban freight transport (OECD, 2003) is fully devoted to the topic of urban goods delivery in urban areas. It notes the lack of data and analytical tools for evaluating effectiveness of urban distribution policy measures. It suggests there is a need for standardisation of data in order to understand and monitor urban freight transport to improve comparability and consistency. Studies by Eurostat (2006) about road freight data are typically published at a national level but contain some data about urban freight transport activity within it.

15.2.2. Urban Freight Oriented Projects with Data-Collection Issues

Numerous European network initiatives devote themselves to harmonising the different approaches of UGM data collection.

The EU COST 355 WATCH (changing behaviour towards a more sustainable transport system, 2004–2008) includes the 'Freight transport and energy consumption' work group 1. This research focused on the calculation of the fuel consumption and CO_2 emissions of goods transport all along the chain, from the source to the consumer (including collection of raw material, processing, consolidation, wholesale and retail sale). A rich assessment was carried out at a very detailed level, focusing on textiles and fresh dairy products and including calculations of various last-kilometre impacts. This research indicated that it is possible to know the energy consumption levels in the different links of the supply chain. The main problematic issues remaining have been identified as: analysis of the last kilometre, the main indicators of change and their data elements and above all, does data already exist?

EU project City Freight (2006–2009) undertook a comparative analysis of the effects of the main efforts for different cities and situations in Europe. The socio-economic and environmental impacts of changes in freight transport and door-to-door delivery in a variety of European metropolitan areas were analysed in a systematic and innovative way. The research analysed selected supply chains schemes in Europe and evaluated their impacts in an urban context, making use of a common assessment methodology.

The EU project START (Short-Term Actions to Reorganise Transport of Goods, 2006–2009) was developed in the framework of the Intelligent Energy Europe Programme. Its aim is to shape consolidation of deliveries, access restrictions and incentives towards sustainable development in five European cities. It looks specifically at the data issues involved in doing so.

Several EU Interreg III B projects, under the leadership of Regione Emilia-Romagna in Northern Italy, have a link with surveying goods transport in urban areas:

- *City Ports* is an Interreg III-CADSES project. The main features of the project's methodology concern the systematic analysis of the supply chains behind the haulage of goods, a careful identification of the most critical conditions within the cities, the identification of specific solutions according to an integrated approach, comprehensive *ex-ante* evaluation and attention paid to all stakeholders with an interest in the policy measures. This approach requires a large collection of data through surveys in several middle-sized Italian cities.
- *Mataari* is an Interreg III-MEDDOC project for the improvement of the accessibility of transport facilities in urban centres (consolidation centres) and inter-modal platforms. The aim is to reduce the conflicts between the various traffic components and to promote inter-modal freight transport.
- *Merope* is also an Interreg III-MEDDOC project that focuses on the telematics oriented to logistic management in the urban and metropolitan areas. It was developed in response to the need for a review of the different applications of the new ICT within the framework of urban logistics.
- *Sugar* (Sustainable Urban Goods logistics Achieved by Regional and local policies) is an Interreg IV project, which aims to test the transferability of the more successful best practices developed in European cities to other cities. Here, data collection is directed to the evaluation of projects.
- *Citylogistics* biennial conferences, driven by Pr. E. Taniguchi of Kyoto University, favour the development of a worldwide academic network on urban freight transport analysis. Many papers have been presented analysing case studies and modelling aspects of city logistics, referring often to data-collection gaps. However, we found no significant references to thorough, innovative UGM surveys.

15.2.3. BESTUFS II: A Recent Survey on Urban Goods Data Collection

In BESTUFS II, the Laboratoire d'Economie des Transports (Lyon) have conducted a research programme, entitled 'Urban freight data collection harmonisation and modelling'. Through a questionnaire delivered to 70 European experts (35 respondents) in 11 European countries, it was possible to compare urban freight data collected on a national basis. The work consisted of linking the urban data collected in each country with the objectives of the collection, the methodology used, the frequency, the unit observed, sample size and many others items, as described in Table 15.1.

The freight experts surveyed were asked about the availability of the following categories of urban freight data in their country:

- Name of data-collection survey
- Observation unit (vehicle, commodity, establishment, depot, delivery/pick-up etc.)
- Name of organisation collecting data
- Reason for data collection
- Whether the data is used for modelling
- Frequency of data collection
- Last time data was collected
- Type of data collected
- Method of data collection
- Sample size
- Units of measurement used
- Geographical area over which data collected
- Difficulty involved in extracting urban data (if data set is greater than urban)

The diversity of the responses presented in Table 15.1 highlights the need to harmonise definitions and units of observation if we expect to be able to compare the outcomes in relation to the survey's objectives. In some surveyed countries, such as Hungary and Portugal, there have been few efforts to collect urban freight data in the past 10 years. However, this situation is expected to improve in Portugal over the next five years as a result of increasing congestion and concern about environmental problems (Browne et al., 2007).

15.2.3.1. Freight data is collected for a wide range of reasons

- Investigating specific projects and initiatives
- To produce national and local estimates
- Government monitoring and performance measurement
- For freight transport modelling and forecasting (with goals such as improving transport efficiency and reducing negative impacts)
- To meet requirements of EC Directives
- Commercial monitoring (i.e. company vehicle operating and marketing data)
- Helping implement regulations
- Legal requirement for licensing and safety controls
- Crime investigation (e.g. speeding and loading offences)

Table 15.1: Inventory of urban goods data collection in 11 European countries.

Type of data collection	Countries	Level	Concerns
Commodity flows (O/D)	Belgium (Be)	NS	Exchanges between regional areas
	Sweden (Swe)	NS	
	Switzerland (Swi)	NS	
Site/land use/establishment surveys	Be	NS	Movement generation
	Germany (Ge)	NS	
	France (Fr), UK	SUS	
	Netherlands (Nl)	RS	
	Spain (Sp)	OUS	
Goods vehicle activity surveys (including driver diary surveys)	All countries (11), except	NS	Vehicles use and traffic
	Hungary (Hu), Nl	SUS	
Shipper surveys	Fr, Swi	NS	All sendings
	Be	OUS	
	Ge, Sp	SUS	
	Italy (It)	CD	
Receiver surveys	Be, Fr, Ge, It	SUS	All deliveries
	Nl, Sp, UK	SUS	
	Swi	NS	
Good vehicle fleet licensing data	All, except	NS	All vehicles
	Hu	SUS	
	Sp	RS	
Traffic counts	Ge, Portugal (Po)	NS	All vehicles
	Swe	NS	
	Be, Fr, UK	AUS	
	Hu, It, Nl	SUS	
	Sp, Sw	SUS	
Distribution industry surveys	Ge, It, Nl, UK	CD	Logistics chain
Vehicle operating cost surveys	Be, Fr, It	NS	Cost
	Ge, Nl, Swi	CD	
	Sp	RS	
Loading/unloading/parking infrastructure data for goods vehicles	Be, Hun	OUS	Method of delivery
	Fr, Po, Sp	SUS	
	Nl	AUS	
Data on road accidents involving goods vehicles	All, except	NS	Security
	Hu, Sp	OUS	
Data on lorry/lorry load thefts	Be, Fr, Nl, UK	NS	Security
	Ge, Swi	CD	

Table 15.1: (*Continued*)

Type of data collection	Countries	Level	Concerns
Employment surveys in freight transport and logistics industry	All, except Hu, Swe, Swi	NS ?	Employment
Land use databases for town/city needed for freight modeling	Fr, Ge, UK It Po	NS OUS SUS	Location, road occupancy
Port freight traffic data in urban areas	Nl, UK Be Fr, Ge	NS OUS CD	Contribution of port to UGM
Rail freight traffic data in urban areas	UK Nl Ge	NS OUS CD	Modal share of UGM
Inland waterway freight traffic data in urban areas	UK, Nl Fr, Ge	NS CD	Modal share of UGM
Airport freight traffic data in urban areas	Be, Fr, Ge Nl, UK	CD NS	Contribution of airport to UGM
Freight 'NTIC' data (from cameras, sensors and other automatic data capture devices)	Nl, UK	CD	Movements of vehicles, traffic

Notes: ?, uncertainty exists about whether freight data is collected; NS, national survey/data collection; SUS, survey in some urban areas; RS, regional survey/data collection; OUS, survey in one urban area; AUS, survey in all urban areas; CD, data collected by companies, trade associations or other commercial organisation.

It is fair to say that the variety of motives for collection also make for diverse surveys, which can be very difficult to compare across countries. Urban freight data may be collected according to different methodologies. The information provided by freight data experts has indicated the breadth of different techniques that are currently being used to collect urban freight data. These techniques include:

- Interviews with freight transport company managers
- Interviews with establishment managers (receivers/shippers)
- Interviews with drivers (on consolidation platforms or roadside)
- Questionnaires sent to freight transport company managers
- Questionnaires sent to drivers
- Questionnaires sent to receivers
- Questionnaires sent to shippers
- Accompanied trips with goods vehicle drivers

- Group discussions (including discussions with drivers, representatives from a single supply chain, representatives from different supply chains)
- Parking and loading activity surveys (i.e. observation surveys)
- Parking and loading infrastructure/inventory surveys
- Traffic counts (manual and automatic)
- Data collection using new technology, including satellite tracking data (GPS), roadside camera data (including automated number plate recognition (ANPR) data) and weigh-in-motion (WIM) technology to measure axle weight of a moving vehicle

Obviously the type of data that is being collected and the use to which it is being put influence the technique used to collect data. For instance, data used to provide a quick snapshot of an existing situation is likely to be collected using a different methodology and sampling approach than data used as an input to a freight model.

15.2.3.2. Addressing gaps in urban freight data collection A wide range of urban freight data gaps have been identified through research conducted by the freight experts participating in this study (Browne & Allen, 2006). The most commonly mentioned data gaps include:

- Data about light goods vehicle activity (generally vehicles below 3.5 tonnes gross vehicle weight)
- Data about the supply chain as a whole
- Data about freight and logistics infrastructure to and from which urban freight activity takes place
- Data about loading and unloading operations and infrastructure for goods vehicles
- Geographical data about goods vehicle trips in urban areas
- Data about trips carried out by consumers for the purposes of shopping
- Speed and route data for goods vehicles
- Data for non-road modes

Specific, large urban freight surveys have been rare in Europe, having mostly been carried out in Germany, Italy and France. In most cases, there surveys have been on a national scale, due to the lack of appreciation of the need for urban freight data by national, regional and urban governments.

15.2.4. Common Surveys

This review shows that the regular national approaches are often similar and are not specifically urban oriented. We present, nevertheless, their positive and negative aspects relating to urban freight objectives.

15.2.4.1. Automatic counting (magnetic loops) These permit tracking of the number of heavy vehicles (more than two axles) and cars using a road in a given time period. This is useful for monitoring and provision of traffic, for validating the results of specific surveys, and for calibrating models for traffic generation and O/D matrices. In all the European countries, traffic data are usually collected by means of traffic counts. The device is either a double cable or a magnetic loop buried under the pavement. It is used by the authority in charge of the road; gathering data is easy and inexpensive.

15.2.4.2. Roadside surveys (or 'cordon' surveys) Such surveys permit the calculation of the number of different types of vehicles using the road. When they are made around an area ('cordon' survey), they permit the calculation of total through traffic and internal/external traffic for that area. These surveys involve stopping vehicles on the roadside and interviewing the drivers about their trip, including origin, destination, different stops and purpose. Only a few questions are asked (to avoid unduly inconveniencing drivers). It is not possible to link industry classification with the purpose of the trip or the freight carried. Internal traffic within the cordoned area is not measured. Such surveys permit the validation and calibration of the results of specific surveys and models. The management of those surveys is not easy: they often cause traffic jams and incur concerns about road safety. As a result, a number of truckers were found to avoid survey sites, for example by taking alternate routes. In France, for these reasons, such surveys have fallen out of use in the last 15 years.

15.2.4.3. Vehicle-based surveys The most recent surveys entail selecting a sample from the vehicle registers. In France, as in many other developed countries, these surveys are carried out by national authorities. They are sometimes known as national Light Goods Vehicles and Heavy Goods Vehicles surveys.

The freight road transport business vehicles surveys. These surveys permit the evaluation of the traffic performance of large goods vehicles, including the tonnage of different types of goods (according to 12 Transport Statistical Nomenclature NST) and the vehicles travelling by road from one region to another annually. Such surveys are the main basis for calibration of inter-urban freight traffic models (commodity flow models). In France, these surveys are carried out annually by the Service of Statistics (SESP) of the French Ministry dealing with transport. They conform to the European standard CE 1172/98. A questionnaire is sent throughout the year to the owners (or for-hire users) of vehicles over 3.5 tonnes. The respondent fills a log about the transport performed with the vehicle in one day during the previous week. The results are updated on an annual basis, thanks to a quarterly sampling of the national heavy vehicles register. In France, 15,000 rigid trucks and 70,000 articulated lorries are surveyed each year. The survey is of general interest and is compulsory. Short trips (< 50 km) travelled within densely populated areas and those including several stops are generally not surveyed.

Periodic surveys on light commercial vehicles. These surveys permit the evaluation of the traffic performance of light goods vehicles by periodically evaluating the tonnage and types of goods (according to 12 NST) transported by such trucks from one region to another. Few countries carry out periodic surveys on light commercial vehicles (LGV). In France, SESP carries out a national survey on the use of the LGV (<3.5 tonnes) every five years. A representative sample of 20,000 vehicles is picked from the national LGV register. Each owner receives a postal questionnaire and gives information on the average yearly usage of the vehicle (including what fraction of usage is individual or occupational, what fraction is urban or inter-urban and what fraction of usage involves transport of goods, tools, material and individuals). The survey is compulsory for the occupational users. Since 2000, a diary log, covering two days, has been added; this enables a more accurate description of the spatial and occupational use of the vehicles. However, with this addition, the postal survey method reaches a threshold: the response rate has been decreasing with the increasing complexity of the questionnaire.

Vehicle-based surveys provide a picture of national goods flows. The KID 2002 survey in Germany is an example of comprehensive vehicle-based survey at national and regional levels (Wermuth et al., 2004). At the national (federal) level, a representative sample of 100,729 vehicles was picked from among the vehicle population in 2001, giving a net (after data collection) sample of 51,778 vehicles. Another 25,019 net cases from regional surveys were added to the total federal survey. In total, 4249 commercial owners of cars and commercial and private owners of trucks up to 3.5 tonnes, as well as 1131 owners of semi-trailers and trucks under 3.5 tonnes have been included. Stratification was vehicle-based (by type and age of vehicle), owner-based (sorted by private and commercial use, and by 17 types of activity) and spatial (4 urban types, 3 urbanised areas, 2 rural areas) for representative sampling. A minimum of 240 vehicles per stratum was provided. The survey documents include a questionnaire on the vehicle and owner and a travel diary on reference day (gross weight of load, type of goods, address and kind of destination, end of trip and distance travelled). The KID 2002 survey greatly improved the information base available for German traffic statistics. It was possible to have an estimation of the amount of private, business and commercial traffic all across Germany, including the annual performance of each vehicle group (number of trips and number of kilometres). Indicators are calculable: percentage of mobile vehicles, traffic involvement per mobile vehicle, goods transport performance per mobile vehicle (tkm/veh/day) and trip-related characteristics such as mean volume of goods transported per load trip (kg). To our knowledge, local and urban data have not been calculated. Furthermore, stops are not described precisely, and it is not possible to have a thorough picture of the route, weight and type of goods transported. Surveys of this type do not describe the variety of traffic flows generated in the urban areas. That is the reason why, despite the additional questions, extensive surveys are requisite.

15.2.4.4. Shipper surveys The aim of this category of surveys is to identify decision-making factors in order to analyse the behaviour of the parties involved in the transport chain. Information is collected directly from the shippers, carriers, customers and also the forwarders or logistic services providers concerned.

In France, two main shipper surveys have been carried out by INRETS (The French National Institute for Transport and Safety Research). A Shipper survey took place in 1988 and the Echo survey (*Envois, CHargeurs-Opérateurs de transport*) was conducted in 2006 (Rizet, Guilbault, van Meijeren, Bijster, & Houee, 2004). The main feature of these surveys is to track the shipments all along the transport chain, from the shipper to the consumer. Supply and demand of goods are observed through tracking shipments carried out by the shippers. A representative sample of 2935 establishments was picked from among 78,000 (industry, wholesale, mail-order companies with more than 10 employees). Then 9742 shipments among 10,462 have been tracked from the shippers to the final destination. The resulting sample was stratified according to activity and type of business. For each shipper examined, the last 20 shipments were surveyed. The less-used modes (sea, air, train, road-rail containers, river) and international shipments were over-represented, in order to ensure a good description of them. Five consecutive questionnaires were submitted: a postal establishment questionnaire for transport activity; an interview with a relevant person of the same establishment (focused on economical characteristics and relation to partners including carriers, customers, forwarders or logistical services providers); a dispatch questionnaire (focused on the relationship between shipper and customer, as well as the organisational and physical characteristics of the shipment); phone interview (to follow up the dispatch, examining the role of price and the purposes of the modal choice of each operator in the chain); and trip questionnaires, completed by the carriers (providing a breakdown of the shipment if several vehicles are needed, detailing different stops, processing of the goods and description of the route).

Shipper surveys enable many opportunities for research on freight demand analysis and modelling: traffic generation, spatial distribution, transport organisation, subcontracting etc. A logit modal choice model enabled researchers to estimate the modal split for each segment and to predict the new split in the case of demand changes. It is already possible to analyse the evolution of the transport chains between 1988 and 2005 and to compare the shipments in different sizes of urban areas (Dufour, 1995; Beaudoux, 2006). However, it is not possible to consider the last miles travelled in rounds; in addition, dispatches from the retailers are not considered. Weightings are made for the expansion of the sample to the whole country. It is not easy to use the results of those surveys at the urban level, because the sample size is too small.

15.2.4.5. Conclusion Among the more or less common approaches described above, none were designed to answer the global specificity of the urban freight transport. Apart from the KID survey, which takes every type of vehicle into account, other surveys consider only one type of vehicle or describe the start point of the transport chain of the sending (e.g. shipper surveys). For those reasons, the observation units used in those surveys (vehicles, shippers) do not allow an exhaustive description of the variety of the urban goods transport flows. Those current approaches have to be compared with some more recent approaches: establishment-based surveys and tour-based surveys.

15.2.5. An Evolution of Methods

We present the approaches that appear, in our opinion, to promise the most efficient analysis of the specific issues of urban goods transport.

15.2.5.1. Business and vehicle travel regional surveys in Hamburg and Dresden These surveys attempt to give a good description of the entirety of business traffic generated in relation with their different types of trip purpose. Separate business surveys with vehicle travel diary surveys were carried out Hamburg and Dresden and surrounding areas in 2000 (Steinmeyer, 2003). In Hamburg, it included a mail-out component and face-to-face interviews; in Dresden it included only face-to-face interviews. An establishment survey collected the characteristics of each business, including owned or hired vehicles and employment. Then, for each commercial vehicle, a travel diary was filled in order to survey the travel of the employees in commercial vehicles. In Hamburg, among the businesses contacted by post and telephone, the response rate of the mail-out was 36%; it was 40% for the face-to-face interviews. The response rate was 42% in Dresden. The response rate of the vehicle diary survey was about 30%. The results permit the counting of the own-account and for-hire vehicle movements generated by different activities. Because only the owned commercial vehicles were surveyed, it is not possible to calculate the ratios of deliveries and pick-ups in a given establishment. Furthermore, it appears that the organisation of the various types of operations (i.e. round or direct trip, sizes of rounds, business activity at the delivery or pick-up point) is difficult to assess. Such surveys are not specially oriented to logistical behaviour analysis or to the assessment of the impact of the business activity on urban freight traffic.

15.2.5.2. One-shot surveys about light goods vehicles in Paris These surveys were essentially used to analyse various aspects of delivering goods with LGV in the Paris area. In the last several years, significant surveys addressed the analysis of deliveries with light vehicles. In the Ile de France (Region of Paris), a survey was carried out by IAURIF (Institut d'Aménagement et d'Urbanisme de la Région Ile de France) concerning the use of the LGV (IAURIF, 2004). The goal was to investigate the extent of the use of LGV in order to complete the results of the TRM survey (see above). Between 2000 and 2002, one survey was carried out at the place of delivery and surveyed drivers; a consignee establishment survey was also conducted; and another survey was carried out at the loading locations. In total, 2950 drivers and 3240 establishments were surveyed. The consignees had to describe their receipt in the form of frequency of delivery with LGV and HGV per day, week and month, and also provide data on the share of products for sale, material and components. Duration of stocking time and conditions of deliveries were surveyed. The driver survey described the current loading or unloading operation, with a description of the three subsequent delivery locations. Data from the on-board devices and on traffic conditions were also captured. The sample of establishments was designed to focus on specific streets. The characteristics of the establishment (size, specific activity) were not noted. It was therefore impossible to reliably project from the sample to the whole urban area.

15.2.5.3. A survey in the Netherlands In order to uncover the relationship between firm characteristics and freight traffic, Iding, Meester, and Tavasszy (2002) built a model based on firm-level survey data from a heterogeneous population. A supporting large-scale survey was carried out in the Netherlands. Postal interviews were sent to an initial sample of 10,000 firms (with at least 5 employees) selected on the type of activity and size. The questionnaire was quite short, with questions on core business, site and floor area, number of employees, average number of trucks per day bringing and taking out freight per type of vehicle, transport mode, loading units and organisation of transport. The target response rate was about 15%. A regression analysis of incoming and outgoing trucks was made for each sector of industry, according to two variables: site area of the firm and the number of employees. The survey yields good information on the traffic per site. The effect of the number of employees on incoming freight vehicles is often different from its effect on the outgoing vehicles in several sectors. Regional differences for the freight traffic indicators are not significant. We notice that the survey was not exhaustive: small establishments (less than five employees) are missing. (Note that the French surveys showed that the establishments with less than five employees comprised about 50% of the total deliveries and pick-ups in French urban areas.)

15.2.5.4. CityPorts: the initiative of Emilia-Romagna CityPorts (EU INTERREG III/B 2003–05), co-ordinated by the Emilia-Romagna Region (in Northern Italy), proposed a general methodological framework for the design and evaluation of City Logistics Actions (Gentile & Vigo, 2007). The collection method focused on the different supply chains and their impact on the different zones of the urban area. For each zone, the main supply chains (SC) are defined on the basis of a stratification of the various industries operating in the zone. A survey has been carried out in several cities of Emilia-Romagna in order to assess the generation of urban freight transport (operations) and to obtain the distribution of different attributes on demand generation, attraction and flows per operations per SC. The sampling of establishments has been based on the ATECO (only large companies) economic classification and the number of employees. The questionnaires included questions on operations generated, parking type, time of service, timetables, type of vehicle etc.) (Table 15.2).

In each city, the size of the samples was between 250 and 500 establishments. According to the authors, the methodological approach was inspired by the French survey methodology. It is thus easy to compare the results of the Italian and French surveys: the breakdowns of parking duration are quite similar (mode is about 5 min), and peak hour is between 10 and 11 h. Parking types are different and service days

Table 15.2: Description of the surveys in the CityPorts project.

Surveys of	Description
Flow generators (establishments)	Private information and supply chain characteristics Comments on city logistics situation and perspectives Information about loading and unloading
Operators : local transport companies	Private data and 'static' information about supply chains, fleet and infrastructure 'Dynamic' information on transport weekly activities
Hauliers	Vehicle and goods information and supply chains Travel information about journey and route
Investigation of urban structure	Identification of homogeneous zones in terms of type of activity (distribution, production) Information about urban freight traffic, urban and extra-urban road network etc.

are more regular in France. From this data collection it was possible to develop a logistic demand generation model (Gentile & Vigo, 2007).

15.2.5.5. Canadian business surveys: Edmonton and Calgary About 3000 business establishments in the Calgary Region in the year 2000 and 4500 business establishments in Edmonton Region in 2001–2002 were interviewed concerning the commercial movements they generated on an assigned travel day. Information was collected on the full range of commodities being transported, including any goods (physical item) or service (action performed) produced by a business establishment, together with a description of vehicle movements arising with this transportation activity. An establishment-based process was used, analogous to the household-based approach used in personal travel surveys. It was possible to distinguish goods and services in all vehicle movements (Hunt, Stefan, & Broownlee, 2005; Canadian Council, 2001; Krüger, 2004).

The survey objectives were the investigation of generation of goods movements, distribution patterns, influence of employment size and type on commodities shipments, investigation of vehicle movement arising with commodity movements (pattern of delivery for a given commodity flow, use of depots, all the trips needed to accommodate a commodity flow, the types of vehicles used). Three basic categories of data are collected, as described in Table 15.3.

Generally, the establishment forms were completed by the initial contact person (or his or her designee); the shipment forms were completed by the shipper at the establishment; and the vehicle forms were completed by drivers. Such a survey consists of providing information about only the shipments of production leaving the establishment. Information about the shipments arriving at the establishment was not requested. This was adopted to reduce respondent burden.

Table 15.3: Description of the Canadian business surveys.

Questionnaires	Data collected
Establishment form: location, category, employment and total movements of goods and services produced at the establishment and leaving it, by category. *Establishment depot form*: goods and services entering the depot.	*Establishment data*: employment of 25 categories based on industrial sectors, relating to the nature of production process performed at establishment and to the space occupied.
Goods shipment form: quantities, values, routings and final destination for goods leaving the establishment. These were distributed to vehicle drivers taking goods from the establishment on the collection day. *Goods depot form*: quantities, origins, destinations, vehicles used for goods entering and leaving the depot.	*Shipper behaviour data*: included locations of destinations for goods and services on survey day, broken down by size, type and value of shipments to each destination.
Goods vehicle form: description of the pattern of all stops made by the vehicle departing the establishment carrying goods produced at the establishment. *Service vehicle form*: description of the pattern of stops made by vehicles carrying goods and services provided by the establishment.	*Vehicle movement data*: type of stops, location of stops, timing of stops, size, type and value of shipment exchange at stops; vehicle configuration.

Survey process recruitment and collection. The establishment sample was developed from data supplied by Alberta Treasury, and included business name, address and industry classification. In Calgary, the lack of telephone number made it necessary to spend extra time. As with the French register, a large number of duplicate and out-of-business establishments were included in the sample, complicating the process. The recruitment of establishments had three steps: first contact by mail; a follow-up phone call soliciting number of employees of the establishment and nature of the establishment (shipping goods/services or depot) and verifying acceptance of the survey. Then, a formal letter was sent by mail or fax, and the contact person was called against in order to send the appropriate forms.

The collection of data on the survey day was carried out according to three levels of assistance:

Minimal: The person participating was called before and after the survey day and the establishment was later visited in order to review the completed form.

Partial: A first visit was made at the start in order to provide initial guidance as the survey got underway. A call was made after the survey day, followed by a visit, as above.
Full: Someone working for the survey firm remained on site for the entire survey day.

The endorsement of the major shipper and trucking associations were significant in facilitating survey success. However, among the 16,000 establishments initially contacted, only 3000 completed the survey.

Sample expansion and weighting. The individual scaling factor was determined according to three variables: establishment size (number of employees), standard industry category and geographical location. In the Edmonton Region, among 27,478 establishments contacted, 13,792 were found to be shipping establishments and 4324 completed the full survey.

Validation of the survey and results. In order to check the results of the expansion process, and potential respondent bias, the expanded set of vehicle movements were loaded to the node-and-link modelled network in the regional travel model. Assigned truck flows were found to match very closely the observed values from vehicle classification counts. Some results are not too far from the results obtained with the French urban goods surveys (e.g. share of movements, share of light vehicles). Light vehicles were found to be used in about two-thirds of all commercial stops in a typical weekday. This is a higher proportion than is appreciated in 'conventional wisdom'. About a third of all commercial stops in a typical weekday are for service deliveries. Again, these service deliveries are often ignored in studies of freight movement or commodity flow.

The data collected has been used to develop a novel form of tour-based micro-simulation of commercial vehicle movements, which has proved very successful in practice. It has been integrated with a conventional aggregate, equilibrium model of household movements, and has the potential to work with the emerging state-of-the-art tour-based/activity-based models of household travel. As such, it leads in a new direction for urban commercial movement modelling. Its development required the data obtained in this survey, and surveys of the sort described here would be required as part of developing this sort of model elsewhere (Hunt et al., 2005). This entails a necessary but costly survey expense: in Calgary, the total cost was CAD $600,000 (from survey planning and design to verification and expansion). In Edmonton it was CAD $800,000.

15.2.6. The Establishment-Driver Surveys in France

A pervasive feeling of unease in France led to the national experimental and research programme on urban goods transport, launched in 1993 by the Ministry of Transport and ADEME, the French public agency for the environment and energy. The programme pointed out an important lack of any research in this field for about 20 years. As far as the problems of urban goods transport were concerned, neither suitable statistics were available nor systems to gather information from the various economic players involved, nor analytical method, nor methodological tools, nor monitoring of experiments. The first task was therefore to carry out comprehensive surveys in a number of different fields. There was no research body available to city authorities, and from the outset, it was quite clear that the human and financial resources required exceeded the capacities of any single conurbation. Moreover there was a common need for a corpus of research that could be used by anyone, not just for a specific city. The Laboratoire d'Economie des Transports conducted and analysed three large, original surveys between 1995 and 1997, in order to provide the quantitative elements useful for demand forecasting and vehicle flow generation in French towns. These were carried out in the French cities of Marseilles (population 1.1 million), Bordeaux (population 750,000) and Dijon (population 240,000) (Patier et al., 1997; LET, 2000). They became known as UGM surveys (Patier, Cholez, Routhier, & Ambrosini, 2004).

The results have notably been used as a database for the Freturb model calibration, and to aid in making decisions for the transport master plans of several large and medium-sized French cities. In each city the same approach was carried out, based on an establishment-based survey and a tour-based (drivers) survey.

15.2.6.1. Methodology We note the methodology issues that need to be addressed in an effective UGM survey. The optimum number of vehicles sampled depends on the type of goods carried, their packaging, the frequency of deliveries and the type of delivery vehicles used. Required data do not exist in commonly gathered statistics (at local, regional or national levels). Generally known data are: population, employment, number of establishments with the type of their activity. In order to know the relationship between the activity and the flows of vehicles they cause, it is necessary to survey trip generators, sending or receiving goods. Only these trip generators know precisely how many vehicles stop every day in their establishment, the size of the latter, the management of the delivery trips and the constraints of management regarding the type of goods.

Researchers need data that is relevant for a good representation of reality. In urban areas, the knowledge of freight flows is not reduced to knowing the routes of the delivery vehicles, because about 75% of pick-ups and deliveries are carried out in more or less complex rounds. This is the reason why it is advisable to re-examine the traditional four-step approach usually used in long-haul modelling (generation of the freight flows, modal shift, building of a freight and vehicle O/D matrix, assignment on the network). In order to focus the survey objective, the most important question about goods movement should be considered: road occupancy by the goods vehicles that are in competition with individual cars. This objective would address congestion and accessibility alike.

In the standard analysis of urban passenger mobility, the aim is to reach a good understanding of daily individual trips. The authors think that the same cannot be said for urban goods transport. An urban goods transport origin–destination matrix has no meaning in terms of transport: 1 tonne from zone i to zone j can be carried in a single payload in direct trip between i and j as well as in hundred small parcel deliveries, some of them being delivered straightforward and some other delivered in complex rounds with light goods vehicles. Because the current modelling approaches are not efficient at the urban scale, the underlying hypotheses do not fit.

What is the most quantifiable observation unit? In order to observe the different travel patterns of a goods-transport vehicle, several statistical units may be considered: monitoring of *a street segment* during a defined period may provide the parking places, time and movement of the goods vehicles working on this segment. Surveying the *routes* of goods vehicles provides a thorough description of the delivery patterns. Through surveying *the shippers*, all the pick-ups and deliveries could be registered. Each of these aforementioned observation units has drawbacks: the rules of sampling of street segment are difficult, the routes may not be linked to the land-use characteristics and shipper surveys do not easily provide route characteristics.

The 'movement of vehicle', in the sense of deliveries or pick-ups conducted at establishment, is recommended as the most efficient statistical unit for the following reasons:

- A movement of the vehicle can be captured (as the trip and the on-road parking time), which assures the measurability of the objective.
- The movements may be described through an establishment survey, in order to identify with appropriate precision the useful characteristics of each movement. In particular, the space occupancy of the stationary vehicles may be precisely measured.
- In order to explain and quantify the road occupancy when a vehicle moves from the depot to the delivery zone or between two delivery points, it is necessary to complete the establishment survey with a deliveryman survey. This is a self-administered questionnaire that is filled out by the drivers who have visited the surveyed establishments. In this way, it is possible to recreate the route containing the movement initially surveyed.

If the size of the sample is sufficient, such a survey may give an accurate picture of urban goods vehicle traffic. The traffic is considered in the general sense of road occupancy for both trips and parking on the road. This choice also allows us to circumvent the difficulties inherent in identifying the origin/destination flows, which are one of the priority aims of the approaches usually encountered (Bonnafous, 2000).

The first major survey of this type was launched in Bordeaux (population 750,000) with the active participation of city authorities. Tracking was carried out by means of three interconnected surveys:

An establishment survey was carried out from establishments shipping or receiving goods (industrial, commercial, tertiary sectors).

A driver survey targeting the driver delivering goods or picking up goods in an establishment, for own-account or third party.

A transport companies survey of the most frequently cited transport companies described: the company's activity (express, consignment, national, international, fleet and number of employees), the organisation of the transport chain, the frequency of deliveries (hourly, weekly, monthly and annually), the fleet of vehicles allocated to deliveries in the city, truck movements, the number of pick-ups and deliveries, the number of rounds per day and per time period, the type of vehicle etc. The organisation of the activity was defined as location of most frequently used terminals, main logistics chains, number of rounds, number and type of vehicles involved etc.

15.2.6.2. Choice of cities Carrying out a survey of this type depends on the commitment of a body responsible for managing an urban transport system. In the first survey case, Bordeaux city officials were involved. The geographic delineation for the survey (i.e. perimeter of the urban area) took into account the density of economic activity and the existence and availability of local databases. The survey perimeter was defined in accordance with the individual household surveys areas, in order to have available data about car and public transport road occupancy. It corresponded approximately to the French National Institute of Statistics (INSEE) definition of the conurbation. The survey area was divided into zones similar to household trips survey zones, in order to ensure correspondence between population indicators and movements of goods.

15.2.6.3. Collection method In order to control the reproducibility of the results of the Bordeaux survey, two similar large surveys were carried out in Marseilles (population 1,100,000) and Dijon (population 240,000) in 1996–1997. The approach and objectives were similar, but the collection method was changed. Different modes of data collection were used by level (establishment, goods, driver, carrier) and by cities (Bordeaux, Dijon, and Marseilles).

15.2.6.4. Bordeaux

Establishment survey. A questionnaire provided data through interviews of the establishment manager, examining the establishment's activity, fleet of vehicles, storage capacity, parking facilities and the surroundings of the establishment. A weekly log (or diary), kept by the person responsible for logistics, provided data on all incoming and outgoing movements of goods over a week. The log consisted of a set of data sheets. Each data sheet contained information relating to one movement in the week, including pick-up or delivery data (place, time, type of vehicle, time of delivery), the carrier's name, the frequency of pick-ups or deliveries and data on the product (type of product, packaging, weight, origin and destination). A removable driver questionnaire was attached to each movement data sheet.

Driver survey. This was in the form of self-administered questionnaires; they were given to the drivers in the establishment where they were making a delivery/pick-up and were returned by post. The questionnaire solicited a description of the round, that is the number of stops in the city, the type and weight class of the vehicles used, the type of handling equipment used, the distance covered and the type of each establishment to which a delivery was made. The route was drawn on a map of the city and information was gathered on the number, location, schedule and parking time of stops.

Carriers survey. This was conducted by a transport scientist via a face-to-face interview.

15.2.6.5. Marseilles and Dijon

Establishment survey. The survey of establishments was intended to be conducted by Computer Assisted Telephone Interviews (CATI method), covering 2000 establishments in Marseilles and 1000 in Dijon. As the length of the interview was too long, this method was quickly replaced by a phone call and postal questionnaire.

Driver survey. The driver survey of Bordeaux was satisfactory, but some questions were missing. In order to gather the description of the activity of the different places delivered, as well as the description of the type, weight and packaging of the different loadings and unloadings, the driver survey of Marseilles and Dijon used a different approach. Instead of giving the questionnaire to the driver when he loaded or unloaded in establishments, the surveys took place at the company in charge of deliveries (carrier, consignee or consignor identified in the place of each establishment). For direct trips and simple rounds, the surveys were made by interviews of the deliverymen on the carriers' or forwarders' platforms or consolidation centres. In order to capture the data on the complex rounds (more than 30 delivery points), on-board surveys were conducted by a pollster accompanying the driver. The number of valid questionnaires was 800 in Marseilles and 500 in Dijon.

The total database of the three cities contains 4500 establishment questionnaires and 2200 driver questionnaires.

15.2.6.6. Sampling The French exhaustive establishment register 'SIRENE' of INSEE is easily accessible. It contains a high degree of invaluable information on each establishment, including activity, number of employees, nature of the premises etc. In the town of Bordeaux the sample was picked from 38,507 of the 40,466 establishments of the urban area (95%). Public services (schools, local authorities, post offices and hospitals) were covered by separate studies and were not included in the survey. In the conurbation, these activities, which generate relatively few goods movements, account for 25% of employment, meaning that the sample covered only 75% of employment. The survey had to cover a large number of establishments in order to obtain acceptable results on the generation of movements in each stratum: 1500 establishments were surveyed.

Each establishment provided five data sheets on average, concerning the same number of different movements undertaken by the same carrier in an establishment (6600 data sheets in total). Dummy data sheets, one per establishment on average, were established for regular movements that were not carried out during the survey week. A precise description of movements during a week (nature and path of the goods of each shipment, packaging, weight) was described in 8300 product lines. It was shown that a majority of movements are involved in a single shipment.

The sample of establishments was picked from this database. A stratified sampling method was used so as to ensure adequate representation of different categories of establishments according to their type and size with regard to the generation of movements. An *a priori* partition into 37 classes was employed in Bordeaux, according to industry class, number of employees and a comprehensive geographical coverage of the conurbation.

The following table shows the 37 classes (Table 15.4).

Of the 6000 driver questionnaires that were given out by the establishments taking part in the survey, 903 (a rate of 17%) were returned by the drivers, validated and linked to the surveyed establishments. This included 69 of the most frequently used carriers.

A preliminary pilot study of 150 establishments was carried out in Lyons. Strategic variables were systematically monitored during the survey. Despite these precautions, numerous follow-up phone calls were needed in order to finally clear the various surveys.

15.2.6.7. Expansion and weighting Expansion of the sample results allowed researchers to extrapolate the observed data to the whole urban area. This involved weighting each entity measured.

Concerning the generation of movements, the number of movements is mainly sensitive to the type of industry and the size of the establishments. The hypothesis is that the fluctuations linked to other characteristics cancel each other out when the movements are aggregated for large groups of establishment.

Weighting of establishments. This weight w_s is specific to each strata s on which the sample is based. It is determined by the ratio r_s (number of establishments of the strata in the conurbation)/(number of establishments of the same strata in the sample).

$$w_s = \frac{1}{r_s}$$

Weighting of movements. In the diary log, some movements were forgotten or incorrect. It was necessary to amend each weighting at the establishment level in order to make possible the extension of the results from each data sheet of movements. It was observed that the number of movements obtained by the sum of the frequencies of movements in the data sheets were 20% lower than the actual total described by the establishment interviewed. To do the correction, the scaling factor c_e of the characteristics of each movement of e is such that:

$$c_e = \frac{m_{se}}{\sum_{i \in e} f_i}$$

Table 15.4: Activity stratification in the establishment survey in Bordeaux.

Nature of the activity	Size of the establishment
Agriculture	
Building construction	
Heavy industry	>2 employees
Chemical industry	
Industry of producers and intermediary	
Industry of consumer goods	<3 employees
Industry of consumer goods	>2 employees
Transport	
Wholesale: intermediary producer goods	
Wholesale: non-food consumer goods	<3 employees
Wholesale: non-food consumer goods	>2 employees
Wholesale: food consumer goods	
Hyper-market, department store	
Supermarket	
Mini-market	
Retailers: clothing, shoes and leather	
Retailers: butchers	
Retailers: grocery, foodstuffs	Without employees
Retailers: grocery, foodstuffs	With employees
Bakeries and pastry shops	
Café, hotel, restaurant	
Chemist	
Retailer: hardware	
Retailer: furniture	
Stationers and booksellers	Without employees
Stationers and booksellers	With employees
Other retail trade	<3 employees
Other retail trade	>2 employees
Open-air market	
Various repair shops	<3 employees
Various repair shops	>2 employees
Tertiary offices	Without employees
Tertiary offices	1–2 employees
Tertiary offices	>2 employees
Other tertiary services (rental, sale)	
Other services	

where m_{se} is the number of movements of an establishment e in the stratum s during a week and Σf_i the number of movements obtained by the sum of the frequencies f_i of data sheet movements of the establishment e.

The total number of movements M_s of vehicles generated by the establishments of the stratum s is:

$$M_s = \sum_e m_{se} \times w_s$$

For each stratum s, each movement i of e in the sample has a weight of mv_{sei} movements in the whole study area, such as:

$$mv_{sei} = w_s \times c_e \times f_i$$

The hypothesis is that in each stratum, the distribution of establishments of the sample according to their number of generated movement is similar to the distribution of the establishments in the whole conurbation. In some cases, the elimination of only one establishment or its change of stratum could modify noticeably the number of movements generated by one stratum in the conurbation. This is true in the case of the big generators, which had to be surveyed separately.

Weighting driver trips. Two sources of bias were found in the driver sample:

- For each data-collection mode, there is a disagreement between the number of movements generated by individual driver delivery reports versus the number of movements generated by the entire route inventory coming from the establishment sample.
- Drivers who belonged to the surveyed firm answered more frequently than did third-party drivers: the rate of questionnaires sent back was different according to the relationship between delivery driver and the establishment receiving the shipment.

In order to amend those biases, the following weighting of a tour of the type (m,a) was used:

$$r - \text{tour}_{m,a} = \frac{Nbmvt_{m,a}}{nbmvround_{m,a}}$$

is the weight (in weighted rounds number). Where a is the activity branch (industry, craftsman, wholesale, retail, large store, tertiary services, warehouses; agriculture) and m the management mode, in five classes: own-account realised by the establishment as consignee, own-account not realised by the establishment as consignee, own-account realised by the establishment as forwarder, own-account not realised by the establishment as forwarder, third party (carriers).

$Nbmvt_{m,a}$ is the weighted number of movements calculated in the urban area for the (*m,a*) establishments, $nbmvround_{m,a}$ the number total of stops of the (*m,a*) drivers trips in the driver sample.

If *nbstop* is the number of stops of a round for a driver of type (*m,a*), the weight (in number of movements) of this driver is:

$$nbmvdriv_{m,a} = nbstop \times r - tour_{m,a}$$

15.2.6.8. Period and duration of the survey diary The survey was conducted across one week (monday–sunday) (an ordinary week, without holidays). We observed that the frequency of supplying an establishment is not the day, but the week. In the city there exist so many different actors with so many different logistic organisations that it is necessary to observe all days of the week. Moreover, it is impossible to calibrate data from only one day without knowing the weight of this day in the global week. It is important to have data on seasonal changes. This is the reason why the questionnaire included questions about the weight of the surveyed week regarding the others weeks of the current month, and the proportion of this month in the year.

15.2.6.9. Budget Each survey needed a funding of about 500,000 € (in current €). It was shared between the Transport Ministry, the ADEME and the city authorities.

15.2.6.10. Primary results

The Bordeaux example

Number of deliveries and pick-ups per week	270,000
Average ratio number of movements per employee	1.1
Ratio delivery/pick-ups	61–39%
Portion of own-account deliveries/pick-ups	56%
Delivery/pick-ups with vehicles <3.5 tonnes	52%
Ratio of movements realised in rounds	75%
Ratio of rounds among total involved vehicles	25%
Share of retail trade (number of deliveries/pick-ups)	33%
Proportion of small establishments<employees in vehicles movements	50%

Goods vehicle generation. For each category of activity, the number of deliveries and pick-ups carried out by vans, rigid vehicles and articulated vehicles is calculable. Average size of the rounds is 13 delivery points (19 in third party, 11 in consignor-own account, 5 in consignee-own account).

Road occupancy In the centre of Bordeaux, the time of road occupancy by delivery vehicles that are double-parked is twice that of running delivery vehicles. In order to compare the road occupancy of all actors in the city (residents, firms, services, city government etc.), we propose as the statistical unit the vehicle kilometre in car unit. A light delivery commercial vehicle (van) is equivalent to 1.5 cars, a truck (rigid) equals 2 cars, an articulated lorry equals 2.5 cars. In this way, the components of urban goods transport are as follows:

- deliveries and pick-ups (total establishments): 40%
- purchasing trips (by car): 50%
- city government (building sites, networks maintenance, waste collection): 10%

According to the town, the share of UGT in total traffic is: from 9% to 15% of trips, 13% to 20% of vehicle × km, 15% to 25% of vehicle × km (in car unit).

One of the main contributions of the surveys was in the knowledge of rules of urban demand management (the links between activities, operation mode, management mode, type of vehicles used, distance covered, number of served establishments, running time, parking place and parking time) that are directly linked with the type of activity, the number of deliveries and pick-ups, the operation structure, the management mode and the organisational mode.

The main relationships are between:

- management mode and organisation mode (own-account carries out mostly direct trips, third-party transport uses rounds)
- type of vehicle and management mode (own-account uses more LGV, third-party transport uses trucks)
- distances covered and management mode and organisation modes (the third-party transport's round are longer than that of own-account)
- distance covered between two stops and size of the tour (the longer the tour, the more likely the distance between two stops is short)
- stop timing and size of the round (the greater the number of stops, the more likely the delivery time at each stop is short).

A very important result is that the same relationships were observed in the three surveyed cities: the comparison of the results of the three French freight transport surveys has shown a notable similarity in terms of indicators; the number of deliveries caused by each type of activity; and, within each class of activity, the number of deliveries realised by type of vehicles, the share of own-account shipping and the share of the different types of rounds. This proves that, for each class of activity, the overall economic and logistic structure overrides the size and geography of particular cities or towns. This result has permitted the authors to transfer the knowledge issued from those surveys to others towns and to build a general model applicable to French towns and even towns throughout Europe. In conclusion, the specificity and the contribution of the French establishment-driver surveys consists of:

- taking into account the diversity of urban goods transport,
- coupling methodology of survey and objectives (diagnosis, knowledge of demand, link between generation and development of flows, simulation of policy-oriented scenarios),

- weighting that permits researchers to calibrate the method for planning and modelling (generation of deliveries and pick-ups, distance covered according to each activity sector, each management and organisational mode).

15.2.7. Methods and Technologies for Data Capture

This last section consists of a critical examination of the various collection methods employed, in order to try to ascertain which ones fit the relevance and the efficiency of the survey. In any case, the objectives of describing urban logistics in a global way need heavy surveys. Wermuth et al. (2004) presents the different types of surveying methods and their advantages and disadvantages for business surveys, comparing written (postal) interviews, personal (face-to-face) interviews, telephone interviews, Internet interviews and traffic counts using mobile phones. During the course of the surveys carried out in France, various interview procedures were tested

- in establishment surveys (face-to-face, with or without diary, CATI, phone-mail, Internet etc.)
- in driver surveys, description of the routes and itineraries (by the drivers themselves, GPS, integration of a Web-based survey with in-vehicle tracking and embarked surveys).

These techniques of capture may be analysed according to the cost of the survey, the objectives, the coherence of the sources and the possibility of weighting of the data.

Regarding the quality of results, it appears that the establishment-driver survey method used in Bordeaux worked better than the method used in Marseilles and Dijon. According to the establishment surveys, face-to-face surveys were more efficient, because of the complexity of the questionnaire and the need for a diary covering one week. Face-to-face surveys are, however, very expensive. The CATI method did not allow a quick control of coherence (due to the impossibility of linking one answer with another answer and coming back to modifying the answer). The questionnaire was too extensive and the CATI method was given up.

The embarked pollster (observer) method seemed to work well, but it showed some drawbacks: its high cost and the impossibility of keeping the link between the establishment visited by the driver and the route surveyed by the embarked pollster. This underlines the importance of enforcing pollster training and ensuring follow-up in real time.

15.2.7.1. Aerial photography Aerial photographs have been sometimes used to give an instantaneous picture of street occupancy by vehicles in urban areas. A campaign of photographs was carried out above the city of Bordeaux (Routhier & Araud, 1997). By this means it was possible to measure the number of cars, vans and rigid and articulated heavy vehicles that were parked or running on the road network and in outdoor parking places. About 80% of the occupied pavement was occupied by private cars. This work was costly, because the recognition and counting of vehicles was done manually. Today, satellite photographs are sufficiently precise to distinguish the different outlines of vehicles. This could provide a solution for the measurement of road occupancy of different types of vehicles.

15.2.7.2. The use of IT for survey assistance A lot of experiments and IT applications are available to make more attractive or more accessible the collection of specific data. We give some information about two of them: Internet surveys and GPS route capture.

A survey of consumers using Internet to make purchases (e-commerce) can be approached via filling an 'e-questionnaire'. Thanks to an open online survey management platform named LOGISKOP, several surveys have been carried out in various studies by the Laboratoire d'Economie des Transports. For example, a survey entitled 'practices of food supplying of the e-consumer' was completed by 85 individuals. Another one, entitled 'e-commerce and mobility', was made easily at a very low cost, thanks to the LOGISKOP platform. Among 1625 connections to the site, 1037 forms were totally completed.

The main advantages are the possibility to build dynamic questionnaires, which may be filled by any e-consumer and the answers immediately processed. The surveys may be open to the public or restricted by password. In the first case, it is possible to obtain answers effortlessly: e-consumers from all around the world may reach the site and take part in the survey. However, several weaknesses limit the use of this process: in the case of open share, it is impossible to control the honesty of the respondent; if there is a password control, it is not easy to find the e-mail address of the individual interested in answering the survey. This platform was developed in 2003 using PHP and Java Script languages. Several similar freeware tools are now available (Alligier, 2007).

GPS is also a very attractive way of capturing the routes of vehicles. It is easy and not very expensive to fit a vehicle with a GPS transmitter. If the pulse frequency is around 1 per second, recording the co-ordinates of the vehicle's position is easy. Nevertheless, this method may present several drawbacks: the geo-location of the vehicle must be coupled with a interactive system by which the driver or the on-board pollster must add information, such as the nature of stops (e.g. GPS does not differentiate between a stop at a traffic light and a stop for delivery), the description of the delivery, the characteristics of the establishment etc. Interactive GPS device are expensive and require training, which limits their usefulness for GPS embarked surveys. Moreover, some artefacts in the capture (echo or loss of the signal) may cause erroneous route recording (Marchal, Hackney, & Axhausen, 2005.

15.2.8. Conclusion of the Review

This comprehensive review presents the great variety of methodological approaches to surveying urban goods transport. Current surveys are essentially vehicle surveys, focussed on the estimation of vehicle performance in wider areas (region, state) than cities. They make it possible to assess commodity flows from one region to another. Generally, they are not designed to analyse urban goods flows. Another type gathers establishment information, the main objective of which is to calculate the amount of total business traffic (including passengers, goods, services etc.). The corresponding questionnaires do not include thorough questions about goods movement at a local level.

In European countries in which urban freight data availability has been improved, we found both new national freight surveys and one-off projects and data-collection efforts at an urban scale.

The surveys specifically oriented towards urban goods vehicle flows are few. The establishment-driver surveys developed in France seem to be the most focused on capturing the entire scope of UGM and to the assessment of the relationships between the logistic behaviour of the freight movement generators and the transport system.

15.3. From Diagnosis to Simulation to Decision-Making

Decision-makers need to know the distribution of flows generated by each activity in order to understand and control the functioning of urban freight transport and its impact on the environment. For this reason they need to know the share of each activity (*shippers*: industry, handcraft, trade, wholesale; *operators*: hauliers, integrators, warehouses, own-account and finally households for their shopping activities) in the generation of the flows. As showed above in Section 15.2.6, establishment-driver surveys allow the estimation of such indicators. We attempt to establish below the ability of such surveys to integrate the various policy measures to answer the main concerns of the stakeholders.

15.3.1. Indicators

Common long-distance transport indicators as tonnes, tonne/km, energy intensity or empty running seem not very relevant at the urban level. For example, energy intensity (fuel consumed per tonne-kilometre) is relevant in the measurement of the impact of the freight transport between two regions. In order to carry goods inside cities, the types of vehicles and frequencies are very much varied and the unit of measurement depends on the type of vehicles used; it is very difficult to measure and to calculate the flows without a totally disaggregated approach. The same goes for the empty running, calculation of which requires the knowledge of the load charts of each round carried out in the city. Nevertheless, the following indicators have been recorded from the results of the establishment-driver surveys (Table 15.5).

This table shows the wealth of indicators that can be calculated thanks to establishment-driver urban goods surveys. As shown in the third column, each indicator is useful to make a comprehensive diagnosis of urban goods transport and

Table 15.5: Urban freight transport indicators used in research projects in France.

Title and description of the urban freight indicator	Units in which the indicator is measured	Purpose of the indicator
Ratio: number of loading/unloading	Number of deliveries and pick-ups per week per employee in each activity	Make possible a fast appraisal of the generation of deliveries and pick-ups in a city without any survey. The contribution of each industry sector is calculable.
Loading/unloading density	Number of deliveries and pick-ups per km^2 in a zone	Measures the importance of the goods flows in a zone.
Loading/unloading intensity per activity	Number of deliveries and pick-ups per activity in a zone	Measures the contribution of each industry sector to the goods flows.
Loading/unloading time	Number of hours of on-street double parking for delivery or pick-up in a zone, per vehicle, per activity	Measures the contribution of each industry sector to road congestion by on-street double parking deliveries.
Distance covered for Loading/unloading	Number of kilometres covered for one delivery or pick-up in a zone, per vehicle, per activity	Measures the contribution of running vehicles to road congestion as they make deliveries in support of each industry sector.
Average length of the first leg from platform to delivery area	Kilometre	Measures the impact of the location of the platform delivering goods relating to its market radius.
Average distance travelled per collection/delivery	Kilometres per collection or delivery	Measure of the contribution of one delivery/pick-up to urban traffic (per type of involved vehicle).
Total distance travelled on roads in urban area by HGV, rigid lorries and LGV (<3.5T) used.	Total vehicle km per week in urban areas	Measure of the contribution of total industry activity on the traffic.
Average time taken per delivery (per activity type, per vehicle, own account, for hire etc.)	Minutes per delivery	Measure of the time taken for delivering in a tour, on a street, for an industry activity.
Average speed per round (including and excluding stops to make deliveries) km/h	km/h	Measures the performance of rounds for each type of organisation, type of vehicle.

Table 15.5: (*Continued*)

Title and description of the urban freight indicator	Units in which the indicator is measured	Purpose of the indicator
Average weight per kilometre	Tonne (or kg) km per tour, per activity, per type of vehicle	Measures the performance of rounds for each type of organisation, type of vehicle.
Greenhouse gas and pollution according to zone, vehicle, activity, management	Gram pollutant per km Gram CO_2 per km Litre of fuel per km	Measure of the impact of urban goods movement on energy consumption, local and global nuisance and greenhouse gas.

helps to estimate the involvement of the main components of UGM: industry activity types, methods of management (own-account, third party), organisation (size of vehicles, size of rounds). We show below that this approach is a very promising tool to improve knowledge of the links between activity and goods transport at the urban scale, including the main components of sustainability (Segalou, Routhier, Albergel, & De Rham, 2006).

15.3.2. Modelling

The comparison of the results of the three French freight transport surveys shows a great similarity in terms of indicators. For example, the number of deliveries generated per employee in various activities is quite similar in the three cities. Other indicators are also quite similar: the share of the different sizes of vehicle (<3.5 tonnes, rigid heavy vehicles, articulated vehicles), the share of own-account and third-party deliveries and the different size of rounds. This tends to prove that, in French urban areas, economic concerns supersede location concerns. Consequently, it is possible to link together into a model the main relationships between economy and transport system, as revealed by the surveys.

The policy-oriented model Freturb has been developed (Routhier et al., 2001; Routhier & Toilier, 2007) on the basis of the establishment-driver surveys, which had been conceived and carried out in accordance with the structure of the model. On the basis of an establishment data register, and some geographical data, the model simulates the number of vehicle movements generated in each zone, the impact of the economic activity on the road occupancy (traffic and also parking time) and several indicators for diagnostic use, such as time average for delivering goods for each size of vehicle. The recent development of this model permits researchers to calculate the distribution of flows in urban areas. It takes into account the round organisation of vehicle routes. An environmental impact module will soon be available.

The main advantage of this model is that no localised, specific survey is necessary for its calibration. It requires only administrative data available in all French cities: establishment registers and geographical and network data (zoning, road network, speed average on links). With these data, it is possible to simulate the effects of several indicators of urban logistics:

- outcome of the activity location (firms and households), its density, new regulations such as time windows for delivering zones,
- new ways of supplying (home deliveries, e-commerce),
- urban logistics facilities and organisation (urban distribution centres, private–public-partnership, spatial co-operative system etc.).

15.4. General Conclusion

The relevance of urban goods transport surveys lies in the capacity to give decision-makers an account of global urban freight transport functioning, ratios and data, in order to facilitate planning, regulation, forecasting etc. So, these surveys have to take into account all components, including own-account and third-party transport, all goods exchanged between establishments, encompassing small firms as well big firms and the gamut from LGV to maxi-code trucks. Based on these requirements, we suggest several conditions of efficiency of urban goods surveys.

A stratified sample and post-stratification statistical analysis will permit researchers to improve the precision of the coefficient of delivery and pick-up generation. The expansion of the results is possible only if the size of the sample is large. According to the preceding review, the association of business (establishments) surveys and vehicle travel diary survey (drivers) have been carried out in several cases. It seems to be a very effective approach, in order to answer the following issues:

- the need to find a good compromise between bias and accuracy;
- the need to explain transport conditions by the functioning of the logistics establishments and to feed a policy-oriented model;
- the need to separate road occupancy into parked and running vehicles (with impacts on congestion, to provide a tool of management for decision-makers).

In that case:

1. In our opinion, the most relevant observation unit is the *movement* (delivery and pick-up).
2. It is important to pick a sample from a *reliable* and comprehensive *register of establishments*, containing, for each establishment, at least its address, detailed goods activity code and number of employees.
3. The *choice of the city* is linked to the involvement of local authorities (e.g. urban community, Chamber of Commerce and businesses). Support from the government department responsible for road maintenance and issuing building permits

can be useful (as in the case of Bordeaux). Such assistance will bring information on the particulars of the city as well as help in managing the survey.

4. The *choice of consultant* for the implementation of the survey is very important. Transparency and total collaboration are necessary. The quality of the results depends on the capability to react quickly if problems arise. Moreover, the consultant must be well versed in the research approach rather than the marketing approach.
5. *Collection method*: Face-to-face is the more efficient technique, but it is expensive. In order to reduce the cost, new methods such as Internet surveys can be investigated. These consist of filling questionnaires on the Internet, with a direct control by the researcher.
6. The *cost of such a survey is high*. The financing has to be well estimated (expenditure). Unlike the household travel surveys, urban goods surveys are one-off surveys. It appears difficult to carry out such surveys in a periodic way, due to their high cost and the lack of involvement of the stakeholders and funding authorities.
7. *Recruiting pollsters* is fundamental. These surveys are intensive and complex. The pollsters have to be adaptable, reactive and have to understand the stakes. Good training and tests are required.
8. The representativeness depends on a good stratification of the sample.

Several constraints must be taken into account: the scale of management of the survey does not permit the inclusion of comprehensive joint business trips. The cost is high and must be assumed by local and national partnerships. Several collection methods may be carried out for the same sample, according to the size and the activity of the establishments, taking into account the transport operator or the size of the rounds for the driver interviews.

As an extension, it would be useful to combine trade establishment surveys with shop customer surveys. This would permit the measurement of the relationship between stores supplying and household purchasing behaviour. A tool for modelling interaction between business traffic and individual traffic should be very useful.

We noted that methodological papers about those surveys have been rare, with even fewer receiving international study. References are generally limited to national papers, in their national language (e.g. in French or German). Few of them cite international research. This situation limits the distribution of knowledge within the scientific community and among persons in charge of carrying out surveys.

Fortunately, the situation is changing quickly, as the TRB special session and initiatives in European programmes (BESTUFS, INTERREG) show. These initiatives made possible links between different researchers of different countries and with community planners. We hope this paper will contribute to transferability of experiences from all over the world in order to direct the survey technical choices regarding the targeted objectives. Methodological discussions in the previous conferences and in this paper are laying the foundation for a more mature methodological debate. The remarks of Meyburg and Mbwana (2002) are relevant: ‘It is generally observed that data collection, storage, and distribution are expensive activities, and [we urge] data users to make full use of available data. Any effort to collect new freight data should be preceded by an understanding as to why the new data are needed'.

These types of activities make it possible to harmonise survey methods, as has been done for passenger household surveys, and at last understand the comparability of the results obtained in different countries.

References

Alligier, L. (2007). *Mesurer l'impact du commerce électronique sur la logistique urbaine*. Ph.D. thesis on Transport Economics, LET, Université de Lyon, 234pp.

Ambrosini, C., & Routhier, J. (2004). Objectives, methods and results of surveys carried out in the field of urban freight transport: An international comparison. *Transport Reviews, 24*(21), 57–77.

Beaucoux, E. (2006). *Exploitation de l'Enquête ECHO sur le transport de marchandises en ville*. Master thesis, Paris-Dauphine University, Paris.

Binnenbruck, H. H. (2006). *Report on urban freight data collection in Germany*. Report produced for workpackage 3.1 BESTUFS project.

Bonnafous, A. (2000). Les marchandises en ville: le problème méthodologique de l'appréhension statistique. In: L'intégration des marchandises en ville dans le système des déplacements urbains, *LET Entretiens Jacques Cartier Proceedings*, Montreal (pp. 85–92).

Browne, M., & Allen, J. (2006). D 3.1 Best practice in data collection, modelling approaches and application fields for urban transport models I. Theme: Urban freight data collection — Synthesis Report, Best urban Freight Solutions II — EU-FP6 Co-ordination Action, 119pp.

Browne, M., Allen, J., & Christodoulou, G. (2004). *Freight transport in London: A summary of current data and sources*. Final report for Transport for London, University of Westminster.

Browne, M., Allen, J., Woodburn, A., Patier, D., Routhier, J. -L., & Ambrosini, C. (2007). Comparison of urban freight data collection in European countries, 11th WCTR, Berkeley, CA.

Canadian Council. (2001). 1999 National Roadside Study Project Report.

Dufour, J. G. (1995). Exploitation de l'enquête "Chargeurs" de l'INRETS, en termes de marchandises en ville, CETUR.

Eurostat. (2006). *Road freight transport methodology Volume 1: Reference manual for the implementation of Council Regulation No. 1172/98 on statistics on the carriage of goods by road*, Eurostat.

Garrido, R. A. (2001). Insights on freight and commercial vehicle data needs. *ITSC Costa Rica Conference Proceedings*.

Gentile, G., & Vigo, D. (2007). *Movement generation and trip distribution for freight demand modelling applied to city logistics*. Working Paper.

Hunt, J. D., Stefan, K. J., & Broownlee, A. T. (2005). An establishment-based survey of urban commercial vehicle movements in Alberta: Survey design, implementation and results, TRB annual meeting, CD-Rom.

IAURIF. (2004). Le transport de marchandises par VUL en Ile de France, 190pp.

Iding, M., Meester, W. J., & Tavasszy, L. A. (2002). Freight trip generation by firms, 42nd European Congress of the Regional Science Association, Dortmund.

Krüger, D. (2004). Planning process for urban goods movement, (iTrans consulting), 2004 Annual Conference of the Transport Association of Canada, Quebec City, Quebec.

LET. (2000). Transport de marchandises en ville: Enquêtes quantitatives réalisées à Dijon et Marseille, 2 final reports MELT-DRAST, LET, 2 × 120pp.

Marchal, F., Hackney, J., & Axhausen, K. W. (2005). Efficient map-matching of large GPS data sets: Tests on a speed monitoring experiment in Zurich. *Transportation Research Record*, *1935*, 93–100.

Meyburg, A. H., & Mbwana, J. R. (Eds). (2002). *Data needs in the changing world of logistics and freight transportation-conference synthesis NYSDOT*. Washington, DC: Transportation Research Board.

OECD. (2003). Delivering the goods: 21st century challenges to urban goods transport, OECD. Available at: http://www.oecdbookshop.org/oecd/display.asp?TAG = X8OWU8X X4X7X9945977HWN&CID = &LANG = EN&SF1 = DI&ST1 = 5LMQCR2K2N5H

Patier, D., Cholez, C., Routhier, J.-L., & Ambrosini, C. (2004). Transport de marchandises en ville quelles spécificités, quelles méthodes — volet 3: Méthodologie des enquêtes tournées dans les agglomérations urbaines, Rapport pour la DRAST-MELT, LET, 36pp.

Patier, D., & Routhier, J. L. (1998). Transport de marchandises en ville: une méthodologie d'enquêtes quantitatives adaptée. *Transport, Environment, Circulation*, *1997*(12), 8–14.

Patier, D., Routhier, J. -L., Ambrosini, C., Durand, S., Bossin, P., Gelas, P., & Le Nir, M. (1997). Transport de marchandises en ville : Enquête quantitative réalisée à Bordeaux, final report MELT-DRAST, LET, 230pp. + annexes.

Rizet, C., Guilbault, M., van Meijeren, J. C., Bijster, E., & Houee, M. (2004). Tracking along the transport chain with the shipper survey. *ISCTSC Costa Rica Conference Proceedings*, Elsevier, Oxford.

Routhier, J.-L., & Araud, C. (1997). Tentative de modélisation du transport de marchandises en ville-Exploitation des photos aériennes à Bordeaux, rapport pour la DRAST-MELTT.

Routhier, J.-L., Segalou, E., & Durand, S. (2001). Mesurer l'impact du transport de marchandises en ville — Le modèle de simulation FRETURB, Guide technique MELT, ADEME, LET, 104pp.

Routhier, J.-L., & Toilier, F. (2007). FRETURB V3, a policy oriented software tool for modelling urban goods movement, 11th WCTR, Berkeley, CA.

Segalou, E., Routhier, J.-L., Albergel, A., & De Rham, C. (2006). Méthodologie pour un bilan environnemental physique du transport de marchandises en ville, ADEME *ed. Connaître pour agir*, 88pp.

Steinmeyer, I. (2003). Kenndaten der Verkehrsentstehung im Personenwirtschaftsverkehr — Analyse der voranschreitenden Ausdifferenzierung der Mobilitätsmuster in der Dienstleistungsgesellschaft. Dissertation an der Technischen Unversität Hamburg-Harburg.

TRB. (2003). A concept for a national freight data program committee on freight transportation data: A framework for development. *Transportation Research Board Special Report 276*, Washington, DC.

Wermuth, M., Neef, C., & Steinmeyer, I. (2004). Goods and business traffic in Germany. In: P. Stopher & C. Stecher (Eds), *Travel survey methods: Quality and future directions*. Oxford: Elsevier.

WSP & Katalysis. (2002). Report B4 — Review of data sources, review of freight modelling, report produced as part of review of freight modelling project with University of Westminster, University of Leeds, Rand Europe, MDS-Transmodal, Oxford Systematics, Parsons Brinckerhoff and Imperial College. Available at: http://www.dft.gov.uk/stellent/groups/dft_econappr/documents/page/dft_econappr_507986.pdf

INTERNET SITES

http://www.bestufs.net
http://www.citylogistics.org
http://www.COST355.Inrets.fr
http://www.docapolis.com

Chapter 16

Surveys on Urban Freight Transport: Synthesis of a Workshop

Matthew Roorda, Arnim Meyburg and Michael Browne

16.1. Introduction and Purpose

Workshop participants were given the task of defining the state of the art in urban freight surveys, identifying key issues and challenges and providing advice for continued research. This topic is of growing interest for urban and regional policy-makers, as evidenced by the increasing number of surveys being undertaken worldwide. This trend is likely to continue, given pressures for improved air quality in urban areas, pressures to regulate and reduce the emissions of greenhouse gases and pressures imposed on logistics chains by increasing fuel prices.

The 19 participants of this workshop represented a diverse range of expertise from 12 countries including Brazil, Canada, Finland, France, Germany, Ireland, Luxembourg, Nigeria, S. Africa, Spain, the United Kingdom and the United States. The geographical and topical diversity of the participants presented a clear opportunity for workshop participants to share both similar and different experiences, to identify areas where significant progress has been made in urban freight survey research and to identify critical gaps in knowledge. The diversity also underscored the need to develop improved harmonisation of both data collection and communication about data, the theme of the ISCTSC 2008 conference.

16.2. Summary of Resource and Contributed Papers

Three papers were presented in this workshop, including the resource paper and two contributed papers.

The resource paper, prepared by Patier and Routhier and entitled 'How to Improve the Capture of Urban Goods Movement Data' (2009), set the stage for the

Transport Survey Methods

ISBN: 978-1-84855-844-1

workshop. This paper outlines the rationale for goods movement data collection, common strategies for data acquisition/collection, survey instruments and key gaps in the current state of practice. The paper focuses on examples of urban goods movement studies in the urban areas of Bordeaux, Dijon and Marseilles to describe the evolution of methodology in France towards combined surveys of both the establishment and the drivers that move the goods. The paper also reviews the relationship between urban goods movement data collection and decision-making through the development of indicators and models, and comments on some of the challenges associated with maintaining that relationship.

The contributed paper by McCabe, Roorda, and Kwan (2008), 'Comparing GPS and Non-GPS Survey Methods for Collecting Urban Goods and Service Movements', complements and extends this discussion through a discussion of the Region of Peel Commercial Travel Survey, undertaken by the University of Toronto. This pilot study uses a similar approach to the French studies in that it combines observations of both establishments and drivers. However, the study also evaluates the use of electronic onboard recorders as a supplement to the driver survey for the purposes of obtaining more detailed global positioning systems (GPS) location data and engine data. The survey covers both the movement of goods and the delivery of commercial services that require transportation (such as tradespersons visits). Underreporting of stops and practical implementation issues are explored in detail.

The contributed paper by Arndt (2008), 'Combination of Quantitative and Qualitative Methods for the Research of Freight Transport', discusses the use of two methods, online surveys and expert interviews, for analysing potential for goods movement planning instruments such as tolls, business co-operation for more efficient urban logistics and e-business. A German case study is used to describe the level and quality of response, the conditions of success and potential for further use of these two types of survey instruments.

16.3. Observations of the Current State of Urban Freight Data Collection

Observations that were made in the workshop are organised below in the following categories: purpose and scope, definitions and harmonisation, technology, data from existing sources and the relationship between data and modelling.

16.3.1. The Importance of Purpose and Scope

Urban goods and services movements are very complex. A commercial movement may involve multiple decision-makers in multiple firms. There are many decisions that are influenced to a greater degree by global economics, supply chains and other external factors than by the decision-makers in local firms or as a result of local infrastructure or policy. Additionally, commercial vehicle flows in urban areas do not necessarily follow the same paths as good and services flows because of the prevalence of tours that may include multiple stop points for delivery and/or collection.

This complexity and heterogeneity means there is an extremely large amount of information required to fully describe every aspect of urban goods movement. For this reason it is impossible to collect all of it in a single survey. Given limitations on data collection imposed by respondent burden, by respondent knowledge and by survey budgets, it is very important to articulate the purpose (i.e. formulate the policy questions) of any goods movement survey at the outset. With a carefully defined purpose, it is possible to appropriately decide upon the 'universe' of commercial travel that needs to be understood to fulfil the purpose, with careful attention paid to the boundaries of the universe. For example, policy questions related to the impacts of e-business on total vehicle flows may require that shopping trips in personal vehicles be included as one component of commercial travel — a component that could be replaced by truck delivery trips.

The scope of a particular survey must be coherently related to the universe of commercial travel and to other contributing data sources that help describe that universe. The scope should be defined to minimise data gaps, double counting and the potential for other inconsistencies.

16.3.2. Definitions and Harmonisation

One of the key barriers to innovation in urban freight surveys is the multiplicity of definitions and classification systems between nations and between data sources within nations (see keynote address by Simo Pasi). While this is problematic for all types of travel survey data, it is particularly so for goods movement, because freight information is derived from many sources (e.g. national economic accounts (input/output tables), roadside surveys etc.). Furthermore, commercial travel involves movements of goods and people that are non-homogeneous and are not always easily categorised (e.g. shall we consider a plumber's visit that requires a van load of tools to be a service trip, but a plumber's vehicle that includes delivery of a washer to be a goods delivery trip?). Although there may be appropriate differences in definitions to suit the variety of individual study purposes, it is critical that researchers and practitioners can communicate effectively, that study results can be compared and that data collection methods can be shared. Data fusion techniques (see Bayart, Bonnel, & Morency, 2009; Polak & Cornelis, 2009, both in this volume) that are popular for urban freight modelling applications would also be highly beneficial if greater harmonisation existed.

The workshop participants recommend that a glossary of definitions be developed to enhance communication, comparison, data sharing and data integration. Such a glossary should be intended as a resource, rather than a prescriptive standard to be applied to all studies.

16.3.3. Technology

The use of technology for goods movement data collection is likely to expand. Technologies such as global positioning systems, radio frequency identification (RFID) and onboard sensors can provide highly precise, rich and wide-ranging information. Use of technology also has the potential to reduce respondent burden significantly. While this is also clearly true for passenger travel surveys (see Wolf, 2009; Stopher, 2009), one clear difference is that most large fleets already make use of technology extensively for fleet and product management.

Whether the data from existing fleet/product management technologies are obtained, or technological devices are provided to subjects (as for the contributed paper by McCabe, Roorda, & Kwan, 2008), it is clear that privacy of such rich and proprietary data is of critical concern to industries of all types. The data collection community needs to find ways to address the privacy concerns of industry.

16.3.4. Data from Existing Sources

It is clear that primary data collection is expensive and difficult to justify when large commercial datasets already exist. Researchers, therefore, need to be more entrepreneurial in their efforts to obtain data from carriers, third-party logistics firms, large retailers, ports, technology providers, toll authorities, governments etc. Examples were noted in the workshop of successful partnerships with private organisations. However, these relationships require effort on the part of the researcher to demonstrate that data sharing tangibly benefits the private partner, resulting in a 'win-win' scenario.

16.3.5. Data and Modelling

Finally, the workshop participants recognised the integral relationship that exists between urban goods movement data collection and modelling. The provision of data to support the estimation of a particular modelling framework is a common and legitimate purpose for data collection. The French case study discussed in the resource paper by Patier and Routhier is an excellent example of this. However, workshop participants were clear that a mathematical model is not a prerequisite for urban freight data collection. Descriptive statistics can be as valid a basis for data collection as modelling.

16.4. Some Research Challenges

Based on the workshop participants' collective observations of the current state of urban freight data collection, the following research challenges were identified. The list is not to be considered to be comprehensive; many more research opportunities clearly exist. However, they are considered to be areas of potentially fruitful research.

16.4.1. Capturing Links between Global Supply Chains and Urban Goods Movement

While the current state of practice in urban goods movement does attempt to relate goods flows (establishment based) to vehicle flows (driver based), much more work is needed to reflect the impact of global trends and supply chains at the local level. This includes mode shifts, which are usually ignored in an urban context, but may have important urban area influences near intermodal terminals. Advances in this area will require harmonisation of data definitions, as well as new efforts to establish appropriate boundaries and develop improved interfaces between models and data collection efforts at very different geographic and organisational scales.

16.4.2. Understanding Substitution between Consumer and Business Goods Movement (e-Business)

New approaches to data collection at the interface of goods movement with consumers are necessary. To evaluate the impact of new distribution channels (such as direct-to-home delivery) on their traditional counterparts (passenger pick-up) there must be better information obtained on the commodities moved in the shopping trip than are collected in conventional household travel surveys. Additional data will need to be collected: for example, weight of purchases and better understanding of multi-purpose trips. Definitions and measurement of vehicle utilisation will also need to be enhanced.

16.4.3. Complexity of Sampling with Diverse Economic Structures

Sampling for freight is inherently more complicated than for passenger travel surveys, simply because establishment attributes, trip rates, relationships between firms etc., are more diverse than for households. For example, firms range in size from one to many thousands of employees, and even within a single industry code, two businesses may produce products that have completely different handling characteristics. Business establishments may also be interconnected, either through long-term contracts/alliances or because they are franchises within the same firm. Finally, a large proportion of urban commercial travel is made by firms that are not physically located within the urban area. Research is needed into appropriate sampling structures.

16.4.4. Data Collection Directed Towards Understanding the Implications of Public Policy Decisions

The workshop participants identified that some public policy decisions could be evaluated directly through specific surveys (in contrast with urban goods movement surveys intended to support regional transportation models, or surveys that attempt to simply 'improve understanding of freight movement'). More research needs to be done to develop policy-directed survey methods. Such methods could include, for example, stated preference surveys and before/after studies.

16.4.5. New/Alternative Techniques for Data Collection

As shown in the contributed paper by Arndt, there are many alternative techniques for collection of urban freight data that offer potential to provide more in-depth understanding, improved response rates or reduced burden and cost. Further research into the effectiveness of methods such as online surveys and qualitative surveys is desirable. This research could draw upon experiences in passenger travel surveys. However, it is also likely that entirely new techniques would be appropriate given the differences in the nature of business establishments and households.

The benefits of comparing approaches between different countries were evident from the workshop. More emphasis on international comparisons and the potential lessons and good practices should be encouraged. National research is always likely to dominate in terms of urban freight data collection but funding for international comparisons can be particularly helpful in such a fast-changing field where the technology developments may profoundly influence the data quality that is available in the next few years.

Acknowledgements

The workshop participants were Wulf-Holger Arndt, Germany; Michael Browne, United Kingdom; Alban Igwe, Nigeria; Nina Karasmaa, Finland; Kara Kockelman, United States; Stephan Krygsman, South Africa; Jacques Leonardi, France; Richard McMahon, Ireland; Julius Menge, Germany; Arnim Meyburg, United States; Simo Pasi, Luxembourg; Danièle Patier, France; Pedro José Pérez Martinez, Spain; Alan Pisarski, United States; Christophe Rizet, France; Matthew Roorda, Canada; Jean Louis Routhier, France; Imke Steinmeyer, Germany; and Paulo Ueta, Brazil.

References

Arndt, W.-H. (2008). Combination of quantitative and qualitative methods for the research of freight transport. Presented at the 8th International Conference on Transport Survey Methods, Annecy, France.

Bayart, C., Bonnel, P., & Morency, C. (2009). Survey mode integration and data fusion. In: P. Bonnel, M. Lee-Gosselin, J. Zmud & J.-L. Madre (Eds), *Transport survey methods: Keeping up with a changing world*. Bingley, UK: Emerald.

McCabe, S., Roorda, M. J., & Kwan, H. (2008). Comparing GPS and Non-GPS survey methods for collecting urban goods and service movements. Presented at the 8th International Conference on Transport Survey Methods, Annecy, France.

Patier, D., & Routhier, J.-L. (2009). How to improve the capture of urban goods movement data? In: P. Bonnel, M. Lee-Gosselin, J. Zmud & J.-L. Madre (Eds), *Transport survey methods: Keeping up with a changing world*. Bingley, UK: Emerald.

Polak, J. W., & Cornelis, E. (2009). Best practices in data fusion: Synthesis of a workshop. In: P. Bonnel, M. Lee-Gosselin, J. Zmud & J.-L. Madre (Eds), *Transport survey methods: Keeping up with a changing world*. Bingley, UK: Emerald.

Stopher, P. R. (2009). Collecting and processing data from mobile technologies. In: P. Bonnel, M. Lee-Gosselin, J. Zmud & J.-L. Madre (Eds), *Transport survey methods: Keeping up with a changing world*. Bingley, UK: Emerald.

Wolf, J. (2009). Mobile Technologies: Synthesis of a Workshop. In: P. Bonnel, M. Lee-Gosselin, J. Zmud & J.-L. Madre (Eds), *Transport survey methods: Keeping up with a changing world*. Bingley, UK: Emerald.

Chapter 17

The Collection of Long-Distance Road Freight Data in Europe

Alan McKinnon and Jacques Leonardi

Abstract

This chapter examines the systems used to collect data on the long-distance movement of freight by road for modelling, market research and other purposes. It begins by listing the features of an ideal long-distance freight data set that would fully meet the needs of policy makers, business analysts, academic researchers and freight operators. It then reviews the existing data-collection systems in Europe, highlighting their shortcomings and how they might be overcome. The third section analyses the strengths and limitations of several alternative approaches to collecting and analysing road freight data, some of which are at an early stage of development.

17.1. Introduction

Detailed and accurate statistics on the movement of road freight are required for public policy making, market assessments and academic research. This chapter examines the methods currently used to compile these statistics and suggests ways in which the collection of data might be improved to enhance our understanding of road freight transport. It focuses on the movement of freight over longer distances.

One of the main issues to be addressed is how new sources of long-distance freight can be most effectively incorporated within the existing framework of data collection and analysis. It would be too ambitious at this stage to envisage widespread adoption of some of the novel approaches to road freight data collection that have been tested in some countries. Gaining national and international approval for statistical

Transport Survey Methods

ISBN: 978-1-84855-844-1

innovations and harmonising new data-collection standards can be a time-consuming process. By assessing the strengths and weaknesses of the various data sources and survey methods, this chapter shows how the collection of long-haul road freight data might be upgraded over the next decade or so.

The term 'long-distance' freight transport is not precisely defined. The minimum length for 'long' distance movements is sometimes considered to be around 150 kilometres (km), though this partly depends on the size of the country and spatial distribution of its population and economic activity. In many smaller European countries, such as Belgium or the Netherlands, 50 km might be regarded as the lower limit. So it does not make sense to impose a strict distance threshold in the definition of long-distance freight. No such threshold is currently used in the collection and presentation of road freight data in Europe, nor is specific reference made to a separate category of 'long-distance' road freight movement in the publication of official transport statistics at national or international levels.

Arguably the easiest way of defining long-distance road freight transport is on an exclusionary basis, in other words saying what it is not. By the term 'long distance', the transport statisticians usually mean interurban goods movement, as opposed to freight deliveries within urban areas. The 'long-distance' label could therefore be assigned to all freight movements other than those confined to urban areas. Even this definition is unsatisfactory, however, as a truck starting a journey just outside a town or city may run almost the entire delivery distance within the urban boundary, while in large conurbations (or city regions) vehicles can run 70–100 km within what is essentially an urban area.

More fundamentally, if the transport of goods is understood as a supply chain starting at the point of production and ending at the home of the consumer, the 'long' distance road trip could be defined as that part of the trip connecting the regions of production and consumption. There are also practical difficulties with this definition, however. For instance, container hinterland transport to and from ports would not conform to this classification: the freight movement might be deemed to be urban if the port is close to a large city, like Rotterdam/Randstad agglomeration, Hamburg or Antwerp; and long-distance if the harbour is more remote from its main hinterland, like Felixstowe for Greater London/English Midlands or Le Havre for Ile de France.

Another pragmatic approach to characterising 'long' distance is in terms of the size of the vehicle. Eurostat collects from the 27 member countries annual survey data for goods vehicles that are above 3.5 tonnes carrying capacity or 6 tonnes gross vehicle weight (Eurostat, 2008a). Note that smaller, lighter vehicles are mainly involved in local distribution and thus less relevant for the common European transport policy, much of which relates to long-haul, cross-border traffic. A definition based on vehicle weight is also problematic, however, as many European countries adopt a different minimum gross weight for trucks in official road freight surveys (Eurostat, 2008b). Furthermore, as a result of the rapid growth of van traffic (vehicles with gross weights of less than 3.5 tonnes) these smaller and lighter vehicles are attracting increasing attention from policy makers.

As none of these other definitions are completely satisfactory, we propose that the term 'long-distance' be defined with respect to the main objective of data collection: to provide guidance for decision makers with statistics that are comparable, reliable, harmonised, consistent through time and sufficiently detailed. Most European countries are collecting long-distance freight data according to national legislation and European directives (Eurostat, 2008a). Since the current European rules were established in 1998, there has been no ongoing research investigating what type of road freight data should be collected and by what means. However, due to the current debate about the decarbonisation of freight transport, and other recent policy initiatives concerning congestion charging and the internalisation of environmental costs, freight data requirements are changing. Studies have had to collect original data to compensate for deficiencies in official freight statistics. One can suggest, therefore, that the definition of 'long-distance' freight and related statistical requirements will be dependent on the particular context and objectives of the study.

This chapter begins by listing the set of data requirements, which in an ideal world would be satisfied by long-distance road freight statistics. It then reviews the existing data-collection systems in Europe, highlighting their various deficiencies and considering how these might be overcome. The final section discusses the strengths and weaknesses of a range of possible approaches to collection and analysis of road freight data.

17.2. Data Requirements

In an ideal world, freight operators, transport researchers and policy makers would have access to detailed, accurate and up-to-date data on 12 aspects of long-distance road freight:

1. *Level of activity:* This would be measured in tonnes-lifted, tonne-km, vehicle-km (loaded and empty) as well as volumetric- and value-based estimates of the quantity of freight movement.
2. *Commodity mix:* This would include a disaggregation of tonnes-lifted, tonne-km and vehicle-km by commodity type.
3. *Vehicle types:* This would include a disaggregation of tonnes-lifted, tonne-km and vehicle-km by type of vehicle (rigid, articulated, drawbar units, long combination vehicles and different body types).
4. *Utilisation of transport capacity:* This would be measured by the ratio of actual loading (by weight and volume) to maximum possible loading.
5. *Consumption of resources:* These resources should be expressed both in physical terms (e.g., litres of fuel, number of drivers) and monetary values (i.e., vehicle operating costs and freight rates). These data would permit productivity calculations and international comparisons of road transport costs.
6. *Externalities:* These would include emissions of atmospheric pollutants and greenhouse gases (GHG), noise levels, accidents and other impact indicators

surveyed in accordance with the new European external costs calculation methodology, and be related to the Euro emission standards of the vehicles.

7. *Energy use and emission reductions measures:* As in Japan, it could be made compulsory for large companies (those with more 200 employees) to reduce transport energy use and to report annually on the effects of energy-efficiency measures taken at company level.
8. *Scheduling:* This would cover average transit times, delivery reliability and time utilisation of vehicle equipment.
9. *Supply-chain structure:* This would go further down or up in the chain and indicate how freight journeys were inter-connected into a supply (or transport) chain from the producer to the final consumer and differentiating them with respect to land uses at origins and destinations.
10. *Intermodal links*: Where road freight operations interface with other transport modes, these modes would be specified and some indication given of the complete door-to-door movement, permitting analysis of the 'transport chain'.
11. *Market structure*: This would distinguish own-account from hire-and-reward operations, identify cabotage and cross-trade activity and, possibly, differentiate carriers into several categories.
12. *Infringements:* As the long-distance road freight sector has a poor reputation for observing regulations, data on the nature, severity and frequency of infringements would contribute to a wider understanding of the industry.

Some of these variables could be used to calculate composite values such as fuel efficiency (km per litre), energy intensity (tonne-km per litre) and vehicle productivity (tonne-km per annum). By cross-tabulating and analysing all of the original data and composite indices, one would gain a very detailed insight into the workings of the long-distance freight market.

In practice, however, most of these data are currently lacking, both at EU and country levels. Traditionally, the main focus of official data collection in this field has been the level of activity expressed in tonnes-lifted and tonne-km, usually split by commodity and vehicle type, with some reference to type of operator. The inadequacies of this system are examined in the next section.

17.3. National Surveys of Road Freight Movement

The main purpose of national surveys of road freight movements has been to measure the level of freight activity in tonnes-lifted and tonne-km, indicating areas of origin and destination, type of commodity and vehicle specification, empty running and load factors. Within Europe, this must be done in accordance with statistical directives to ensure that the figures are collected on a consistent and comparable basis. The rules establish the minimum data requirements, but individual governments can supplement them with additional questions (e.g., on fuel efficiency) (Eurostat, 2008a). Large samples of trucks are randomly sampled from vehicle registration databases and their operators instructed to complete a postal questionnaire for a period of one week detailing the journeys they make and the loads carried. A distinction is normally made between journeys comprising more or less than five collections/deliveries. Most long-distance freight movements will make fewer than five drops and in most cases only one. Separate surveys are used for domestic and cross-border road freight movements. In both cases, the surveys are administered at the national level and relate to vehicles (in the case of domestic haulage) and operators (in the case of international haulage) registered within the country. Eurostat acts as the 'clearing house', assembling survey data from all the EU member states to compile aggregate measures of domestic road haulage activity and estimates of the amounts of cross-border freight movement and cabotage.

This long-established system of data collection is deficient in many respects:

1. *Lack of a supply-chain perspective*: As the unit of analysis is the journey, it is impossible to track the movement of individual consignments across a supply chain comprising several linked journeys. Dividing the tonnes-lifted figure by a measure of the weight of goods produced or consumed within an economy gives a crude indication of the number of links in the chain (the so-called 'handling factor'). In the United Kingdom, this averages around 3.4 for all road freight. If it were possible to analyse the interconnection between these multiple links, one could gain a better understanding of the supply networks within which road freight moves. As much of the growth in road freight traffic is the result of the restructuring of supply chains (McKinnon & Woodburn, 1996), this would help to place freight forecasts on a firmer causal basis. The commodity flow studies undertaken regularly in the United States provide a supply-chain perspective and could serve as a model for European countries to emulate (Southworth, 2006).
2. *Lack of data on the cubic volume of freight moved*: The quantity of freight moved is expressed solely in terms of weight, preventing analysis of the cubic volumes of freight being transported. This is significant as it is generally acknowledged that the density of freight is declining through time partly through the substitution of lighter materials for heavier ones, and also because of the growth of packaging and handling equipment. The utilisation of freight capacity is invariably expressed in weight terms, as the ratio of actual tonne-km moved to the maximum that could have been moved had the vehicle been fully laden (by weight). A large and increasing proportion of loads 'cube-out', however, before they reach the maximum weight carrying capacity. Since the late 1990s, national surveys of operators have asked whether vehicle loads are subject to weight and/or volume constraints. This has shed some light on volume utilisation, but only at the extreme end of the load distribution. No information is available on the average cube-utilisation of the truck. Nor can we take a two-dimensional view of capacity utilisation and measure the average proportion of vehicle floor space occupied by a load. As interest mounts in 'transport optimisation' and opportunities for decarbonising freight transport operations, it is unfortunate that the available data permits only a partial analysis of vehicle load factors. The lack of two-dimensional and three-dimensional data on space utilisation also seriously inhibits

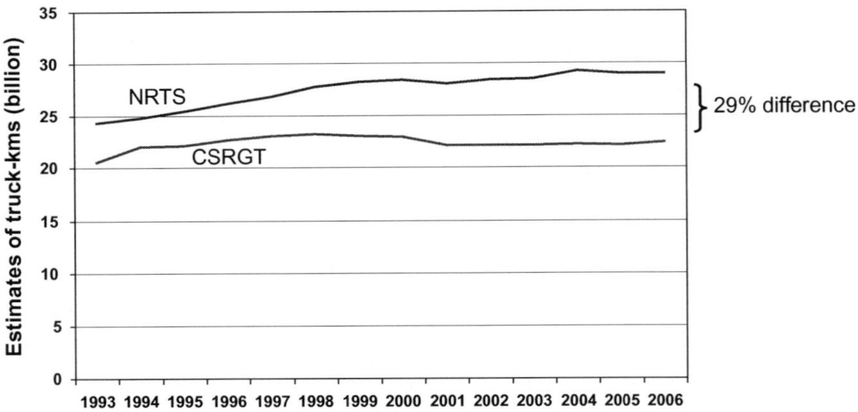

Figure 17.1: Divergent estimates of total truck-km in the UK: 1993–2006 (*Source*: Department for Transport, 2007a, 2007b).

analyses of the economic and environmental impacts of increasing the maximum size and weight limits on lorries, as recently exemplified by studies of longer and heavier vehicles in the United Kingdom (Knight et al., 2008) and at an EU level (Transport & Mobility Leuven, 2008).

3. *Inconsistencies in the estimation of truck-km*: Estimates of the distances that trucks travel are derived in two ways: from questionnaire surveys of operators and roadside traffic counts (which can be either manual or automated). There is often a significant difference between these two sets of truck-km estimates. This is well illustrated in the case of the United Kingdom. Figure 17.1 shows the trends in truck-km based on questionnaire survey data (from the Continuing Survey of Road Goods Transport [CSRGT]) and the traffic count data (from the National Road Traffic Survey [NRTS]) for the period 1990–2006. In 2006, the NRTS estimate was 29% greater than the CSRGT estimate.

This gap has roughly doubled since 1996 and appears to be the result of three factors:

- Exclusion of foreign-registered lorries from CSRGT: It has been estimated that this might account for between 15% and 40% of the discrepancy (McKinnon, 2007).
- Misclassification in the NRTS of vans and rigid vehicles around the 3.5-tonne weight threshold. As van traffic has grown sharply in recent years and much of this growth has been in the largest/heaviest category of vans, the opportunities for misclassification have increased.
- Under-reporting by CSRGT respondents of the number and/or length of trips that sample trucks undertake during the survey week. A separate analysis of the tachographs of a sample of vehicles surveyed by the CSRGT did show significant under-estimation. This occurs despite the fact that there is a separate question in

the questionnaire enquiring about the annual distance travelled and validation of the trip-based distance values using vehicle routing software (the Autoroute package).

Similar discrepancies exist in other European countries. This seriously undermines confidence in the validity not only of the vehicle-km figures but also of other statistics derived from these figures, such as tonne-km, percentage of empty running and fuel efficiency. In the United Kingdom, an internal enquiry within the Department for Transport has investigated this issue and recommended the adoption of an algorithm to reconcile the two sets of truck-km values. It is understood, too, that at a European level, Eurostat will soon be announcing significant improvements to the way in which vehicle-km data are collected and presented.

4. *Inconsistency in the estimation of cabotage penetration*: Cabotage is domestic road haulage work undertaken by foreign-registered carriers. It was fully legalised in the EU in 1998, following the gradual liberalisation of the practice using a quota system. Over the past decade, the level of cabotage has been steadily increasing though, in most countries, from a very low base (Sciullo & Smihily, 2006). The level of cabotage penetration is generally expressed as the % of total domestic road tonne-km carried by foreign operators. Eurostat compiles this information from the results of surveys of international hauliers conducted in each EU member state. To work out the total amount of cabotage undertaken in a particular country, Eurostat sums the estimates of cabotage reported by international hauliers registered in all other EU states. It is acknowledged that cabotage data for the period 1990–1998 were relatively poor, though 'from 1999 onwards data on cabotage transport are considered to be reliable' (Ecorys/Ernst & Young, 2006, p. 16). Some doubt has been cast on the accuracy of this method of collating cabotage data by a separate survey of the activities of foreign vehicles in the United Kingdom in 2003 (DfT, 2003). This suggested that the level of cabotage was slightly less than half the Eurostat estimate. As the UK survey has only been conducted once and there have been no reports of similar surveys in other EU countries, it is not known if this inconsistency is more widespread.
5. *No explicit monitoring of intermodal freight movements*: All the annual surveys are both mode- and trip-specific. Where a trip ends at a freight terminal, no account is taken of any subsequent transfer of the load onto another mode. It is not possible, therefore, to distinguish intermodal traffic, which uses more than one transport mode in the course of a door-to-door journey from distribution entirely by road. As intermodal transport features prominently in the 'sustainable distribution' strategies of the European Commission and national governments, this lack of data on intermodal traffic is particularly serious.
6. *Incompatibility of commodity, industry and trade classifications*: Most of the official freight surveys employ the NST [R] commodity classification. NST [R] system does not correspond directly to either Standard Industrial Classification (SIC) system of industrial classification or the Standard International Trade Classification (SITC) of trade flows. Attempts have been made to 'map' the NST [R] categories across to SIC sectors and SITC classes (such as in the EU REDEFINE

project (Netherlands Economic Institute, Heriot-Watt University, & TFK and Service Economiques et Statistique, 1997), though the results are fairly crude. This makes it difficult to relate freight transport trends to changes in production systems and patterns of trade flow.

7. *Lack of data on the tare/unladen weight of the vehicles*: Without empty weight data, it is difficult to determine the energy used in empty running and to analyse the relationship between payload and gross vehicle weight. For example, according to official French road freight statistics, heavy trucks are loaded with an average of 16 tonnes, but no indication is given of their tare weight. Depending on the truck and trailer configuration, this tare weight can vary from 12 tonnes (7 tonnes tractor and 5 tonnes trailer) to 20 tonnes for standard articulated trucks. Special trucks like construction vehicles can have higher empty weights. In the absence of information about tare weight it is not possible to express the load factor as the percentage of average load weight to the available load capacity in vehicles in differing gross weight classes. If such data were available, it would be possible to conduct more detailed analysis of vehicle utilisation trends, such as the decline in weight-based load factors in European road parcel networks.
8. *Need for more geographically-disaggregated data:* The size of the origin and destination zones used in the questionnaire determines the geographical scale at which the pattern of road freight flow can be analysed. Sample sizes also tend to be too small to permit the required degree of spatial disaggregation within acceptable statistical confidence limits.
9. *Failure to exploit telematics data*: The electronic tracking of road freight vehicles has created major new data-collection opportunities. It offers the prospect of quicker, cheaper and easier collection of more accurate long-haul freight data. There has been debate in Europe recently about the use of GPS and other telematics data to assess the infrastructural needs of international and national freight road traffic at a much higher degree of detail than currently available. It is still not clear, however, how telematics and GPS will be used to supplement, or possibly replace, current questionnaire surveys of surveys of road freight movements. This will require much wider diffusion of this technology across the truck fleet and a willingness by companies to share the data generated by their vehicle tracking systems.
10. *Delay in the compilation of international statistics*: There is generally a time lag of two to three years in the publication of international sets of road freight data. Analysis of trends is therefore several years out of date. Delays in the submission of data by some countries hold up the compilation of statistics aggregated at an EU level.

It is not possible at present to correct all these deficiencies with a single revision to the methodology. This will require a combination of improvements to the standard method of data collection in Europe and the more widespread application of other approaches and techniques that have so far only been implemented in a few countries.

17.4. Alternative Methods of Collecting and Analysing Long-Distance Road Freight Data

17.4.1. Complementarities and Differences in Other Survey Approaches

Rizet (2008) has differentiated three approaches to the collection and analysis of road freight data:

1. *Vehicle approach*: This applies when the survey unit is the individual vehicle and data are collected on individual trips. The official road freight surveys (Eurostat 2008a; DfT, 2008a, 2008b) and Transport Key Performance Indicator (KPI) surveys presented below fall into this category.
2. *Shipper and transport chain approach*: This approach takes a broader overview of shippers' transport operations and can examine the inter-relationship between transport configuration, modal changes and logistics structures. This approach is well-developed in the United States, where the Commodity Flow Survey focuses on shipments and could potentially serve as a model for future freight data collection in Europe. There are, nevertheless, numerous problems in implementing the US CFS and analysing its results (Aultmann-Hall & Drumm, 2006). The French Shipper Survey ECHO remains the only one of its type in Europe (Rizet, Guilbault, van Merijeren, Bjister, & Houée, 2001; Zmud, 2006).
3. *Supply-chain approach*: This approach traces the movement of particular products through a supply chain from the producer to the final consumer, sometimes including the final leg from shop to home, and sometimes monitors the amount of transport generated, energy consumed and emissions released.

Examples will be given of each of these alternative methods of collecting road freight data, which can be used to supplement the traditional vehicle-based surveys of road freight activity. These methods can provide greater insight into the workings of the road freight system and address some of the shortcomings of the official surveys listed above.

17.4.2. Transport KPI Benchmarking Surveys

Over the past decade, the government of the United Kingdom has sponsored a series of synchronised surveys of road freight fleets in which efficiency is measured against a set of KPIs. These have become known as the 'Transport KPI' surveys (McKinnon & Ge, 2004), and are now organised as part of the government's 'Freight Best Practice' programme (www.freightbestpractice.org.uk). The surveys have had several objectives:

- To promote standard methods of performance measurement in road freight transport.
- To allow companies to benchmark the efficiency of their road transport operations.

- To provide the government with a range of transport variables excluded from its main road freight survey (the CSRGT), including volumetric and time-based measures of vehicle utilisation, delivery schedules and delays.
- To calculate the potential for improving transport and energy efficiency across industry sectors.

It was recognised that the large trip databases generated by these KPI surveys could be used for retrospective analysis of patterns of road freight flow (McKinnon & Ge, 2006). Since 1997, ten surveys have been undertaken in the food, automotive, non-food retail, pallet-load (less-than-truckload or LTL), air cargo and building sectors. Five sets of KPIs have been used in all the surveys:

1. *Vehicle fill by weight and volume*: measured by payload weight, pallet numbers and average pallet height. In sectors producing and distributing low-density products, vehicle loading is constrained much more by the available deck-area and/or space than by weight. As the vast majority of loads are unitized on wooden pallets, roll cages, dollies or stillages, 'space-efficiency' is expressed as the ratio of the actual number of units carried to the maximum number that could have been carried. Where products are transported in non-unitized form, conversion factors have been used to translate the load data into a pallet-equivalent measure. This yields a two-dimensional measure of the utilisation of vehicle floor space. The measurement of load fill is extended into the vertical dimension by asking companies to estimate the proportion of trips on which the average height of pallet loads falls into one of four intervals (<0.8 m, 0.8–1.5 m, 1.5–1.7 m and over 1.7 m). This has permitted the calculation of cube utilisation. Data have been collected on the maximum carrying capacity of trailers and rigid vehicles (by weight, pallet numbers and height) and the loading expressed as a proportion of these maxima.
2. *Empty running*: the distance the vehicle travelled when empty. This excludes the return movement of empty handling equipment where this prevents the collection of a backload. These movements are separately recorded as a form of loaded trip.
3. *Fuel consumption*: for motive power units and any onboard refrigeration and mechanical handling equipment.
4. *Vehicle time utilisation*: This has been measured at hourly intervals over the 48-h period for all the vehicles surveyed. In the food sector, the survey units have been either the trailer of an articulated vehicle or a rigid vehicle. In other sectors, the activities of tractors have also been separately monitored. A record is made of the dominant activity of the vehicle over the previous hour. Time is classified into seven activities depending on whether the vehicle is: running on the road (including legal breaks), on the road but stationary during the daily driver rest period, being loaded or unloaded (including time spent on manoeuvring/paperwork), preloaded and awaiting departure, delayed or otherwise inactive, undergoing maintenance or repair or empty and stationary.
5. *Deviations from schedule*: Companies have been asked to log all significant delays and attribute them to six possible causes: problem at collection point (responsibility of the consigning company), problem at delivery point (receiving

company's responsibility), own company actions, traffic congestion, equipment breakdown or lack of a driver.

17.4.2.1. Method of data collection Unlike the official surveys of road freight operations, which are mandatory, these KPI surveys are voluntary. They must be quite intensively marketed. As they require a significant level of support and staff time, companies must be convinced of the benefits of participation. The process of recruiting companies for the survey can be greatly assisted where it is endorsed by a trade association. The commitment of a few large players in an industry to take part can also increase the credibility of the survey and promote wider participation.

Companies joining the survey are asked to assign appropriate staff to the collection and collation of the transport data. They are then invited to briefing sessions at which the data-collection procedure is outlined in detail, definitions clarified and advice given on ways of inputting the data. All the data must be entered into special spreadsheets which are tailored to the needs of particular sectors. This can either be done manually or transferred from existing company data files. As the KPI project has developed over the past decade, an increasing proportion of participating companies has been able to download the necessary data from their IT systems. Responsibility for organising the survey resides within the company. It must also decide how many fleets and vehicles it wishes to survey over the specified period. Some large companies with fleets based in different locations have used the KPI data to undertake internal benchmarking of operational efficiency.

Once the data have been collected, the spreadsheets can be used to calculate average values for all five sets of KPI values. Before these values are calculated, the spreadsheet automatically checks the data set for consistency and highlights any anomalies. To take part in the benchmarking exercise companies must return the raw data to the survey organisers for further checking, data pooling and analysis. Once this process is completed, companies receive reports showing how their fleet(s) compared with those of similar businesses in the same sector.

17.4.2.2. Advantages of the KPI benchmark surveys This survey method has several advantages over the official questionnaire-based surveys of road freight operators:

(a) It collects data on a larger set of variables, offering the level of detail needed in order to deliver better solutions to the problems of freight transport efficiency and fuel use.
(b) By monitoring the operation of many vehicles across a fleet, rather than individual sample vehicles, it permits the benchmarking of vehicle productivity and service quality at company, business unit or depot levels.
(c) The sectoral focus permits more detailed analysis of freight transport operations.

17.4.2.3. Problems and limitations The research and consultancy teams that have undertaken the transport KPI surveys over the past decade have encountered a number of problems.

1. *Difficulty of securing adequate company involvement*: The KPI surveys completed to date have varied widely in the level of industrial support. Success in recruiting companies appears to be related to the following factors:

   - Support of a trade association
   - Backing of industrial 'champions' who are respected within the industry
   - Continuing engagement with the companies during the preparatory phases of the survey
   - Participation of a few major companies against whom other business are keen to benchmark their transport operations.
2. *Failure to survey logistical activities other than transport*: Companies that appear to have poor vehicle-fill and energy-inefficient operations may be behaving rationally, sacrificing transport efficiency for greater gains in the management of production, inventory or warehousing. It would be beneficial to explore the logistical trade-offs that companies make.
3. *Non-random sampling*: Companies participating in the transport KPI surveys have not been chosen randomly. The aim of the marketing has been to recruit as many companies as possible in the targeted sectors. The aggregate results may therefore not be representative of particular sectors and sub-sectors.
4. *Difficulty of measuring particular KPIs*: The benchmarking of vehicle utilisation and energy efficiency in the automotive sector, for example, was inherently much harder than in other sectors because of the diversity of handling equipment and large proportion of non-unitized freight. In the automotive sector, most loads comprise relatively low-density products, making volumetric measurement much more important than weight-based measures.
5. *Limited analysis of the reasons for observed differences in performance*: Beyond the KPI survey itself, benchmarking is more likely to have a beneficial impact on behaviour when the causes of under-performance are diagnosed and guidance offered on improvement measures.
6. *Difficulty of finding well-matched comparators*: Even within industry sub-sectors there can be significant differences between companies' distribution operations. Some of the variation in KPI values can be ascribed to differences in the nature of the customer base, the product range and production scheduling — all factors outside the logistics manager's control. Particular circumstances can cause a company's transport efficiency to deviate from the average of its benchmark group.

17.4.3. Questionnaire Survey of Vehicle Activity

Leonardi and Baumgartner (2004) used a driver-based questionnaire survey in Germany to obtain data on the amount of vehicle activity undertaken between refuelling stops. Information was collected about distances travelled, load weight between each stop, percent volume-fill (within numerical intervals), vehicle empty weight and empty running. Duration of the record diary between two tank fillings varied between one trip and four days. This was supplemented by general company data relating to vehicle type, vehicle numbers, company type and the nature of efficiency measures adopted. Unlike the UK Transport KPI surveys, the questionnaires were randomly distributed to trucking businesses. The results were subsequently validated using 'digital records... of three companies with vehicles equipped with onboard units'. The data generated by this survey were used to calculate various composite indices, particularly the CO_2 efficiency of road freight operations (expressed at tonne-km per kg of CO_2 emitted). It was shown that this CO_2 measure varied widely between transport operations and that there was high potential for improving average CO_2 efficiency. The survey method was designed to answer specific research questions, illustrating how one-off road freight surveys can complement official national surveys used for the compilation of more general data sets.

17.4.4. Surveys of Shippers to Monitor the Through-Movement of Individual Consignments

INRETS conducted two large shipper surveys in 1988 and 2004, and two smaller ones in 1999, which focused on the movement of consignments (or 'shipments') along transport chains to customers. The ECHO (Envois Chargeurs Opérateurs) surveys comprised 5000 shipments in 1988 and 10,000 in 2004 (Rizet, 2008). It included data on commodities, origins and destinations, vehicle type and, overall, more than 1000 secondary indicators needed for modelling purposes. These transport chains generally contained several links and involved the routing of products through transhipment points. Unlike official road freight surveys which collect data for individual trips, these surveys traced the through-movement of consignments along chains made up of several trips. Operating and energy efficiency could thus be measured at a supply-chain level and the links between freight transport and logistics structures more fully explored. The US Commodity Flow surveys, undertaken roughly every 4–5 years, collect similar consignment data on a supply-chain basis. Around 50,000 companies participated in this survey in 2002, making it by far the largest data-collection exercise of this type (Zmud, 2006).

17.5. Harmonisation of the Collection of Long-Distance Road Freight Data in Europe

The national statistics agencies of European countries continue to collect data on an annual basis from operators of road freight vehicles using similar methods. Pasi (2008) notes that the main purpose of gathering national and European statistics is to monitor annual changes in freight transport demand and market structure. Little consideration is given to how transport researchers are actually using the data. The availability of road freight data from official sources strongly influences the nature and scale of research efforts in this field. If the range of variables monitored and sample sizes could be expanded, new research opportunities would emerge. This, however, would impose a heavier burden on road freight operators, possibly increasing the proportion of non-response and eroding some of the goodwill upon which the data-collection process is founded.

Harmonisation is an enduring objective of European organisations and Eurostat. It is still a long way from being achieved in the collection of long-distance freight data. For example, data on many key indicators such as the volume of intermodal shipments, truck capacity, load factors and fuel use are still not available at the European level in 2008. These data are available only in a few countries for all years, in other countries for some years and in most countries not at all. Even in the case of the most commonly used indicators, such as tonne-km, harmonisation principles are not sufficiently applied. Given that the definition of tonne-km is fairly standard and the methods of data collection are very similar, the total tonne-km figures for domestic road transport published by the UK, France and Germany should match the national statistics published by Eurostat. Recent inspection of these published data sets, however, reveals that this is still not the case. Significant disparities exist even among key variables in the data sets. The consistency and reliability of European long-distance road freight data could be improved if there was greater interaction between officials compiling these statistics at national and EU levels, particularly on the issue of data harmonisation.

Harmonisation is a particularly pressing issue for the collection of data relevant to public policy interventions such as congestion charging and carbon abatement in the freight sector. For researchers performing other types of surveys (e.g., adopting a supply-chain perspective), the focus is more on collecting and processing data in a form that is compatible and complementary with the results of official national surveys and KPI surveys. Researchers are likely to continue to design one-off surveys in order to answer specific questions. Such diversity is inevitable and desirable, though it will frustrate attempts to fully standardise the collection of road freight data.

17.6. Conclusion

Much of our understanding of the system of long-distance freight movement in Europe has been based on tonnes-lifted and tonne-km data collected by questionnaire surveys of road freight operators undertaken by national governments and relating to the activities of samples of vehicles. As a consequence, this understanding has been rather superficial. To obtain a deep insight into the workings of this system, we would require data on a much broader range of variables disaggregated to a greater degree and of comparable quality for domestic and international road haulage operations. It is no longer enough simply to know how much freight of different commodities is being moved in different types of vehicle. There is now much greater interest in the efficiency, reliability and environmental impact of long-distance freight movement by road. Given these new priorities, currently available official data on road freight transport are deficient in many respects: the absence of a supply-chain/logistics perspective; the failure to measure cubic volume of freight transported; limited cross-national data on average fuel efficiency by vehicle type; a lack of data on the routeing and scheduling of freight flows; no explicit monitoring of intermodal freight movements; incompatibility of commodity, industry and trade classifications; limited data on the tare/unladen weight of the vehicles; lack of geographical granularity; failure to exploit telematics data and delay in the compilation of international statistics.

Efforts have been made to overcome some of these deficiencies by collecting freight data in different ways. Even if the vehicle approach is retained, it is possible to survey whole fleets of vehicles operated by particular companies and get the companies to provide information on cube- and time-utilisation of vehicles. The offer of a benchmarking service can incentivise companies to participate in such surveys. An alternative approach is to survey shippers and enquire about the movements of samples of consignments along transport chains composed of several freight journeys. Such surveys tend to be expensive and are only carried out at long intervals, preventing short-term time-series analysis.

Arguably the most serious shortcoming at present in the collection of long-distance road freight data in Europe is its reliance on nationally based surveys. As the proportion of cross-border traffic has grown and as cabotage has increased its share of domestic road haulage markets, the need has arisen for a truly pan-European system of data collection.

References

Aultmann-Hall, L., & Drumm, S (2006). *Improving the use and accessibility of the 2002 CFS.* Transportation Research Circular: Commodity Flow Survey Conference, July 2005, Boston. TRB (ed.), pp. 184–186, available at: http://onlinepubs.trb.org/Onlinepubs/circulars/ec088.pdf

DfT-Department for Transport. (2003). Survey of foreign vehicle activity in GB — 2003 London.

DfT-Department for Transport. (2007a). *Transport statistics Great Britain 2006.* London: TSO.

DfT-Department for Transport (2007b). *Road freight statistics 2006.* London.

Ecorys/Ernst & Young. (2006). Study on road cabotage in the freight transport market. European Commission, DG TREN, Brussels.

Eurostat. (2008a). *Reference manual for the implementation of Council Regulation No 1172/98 on statistics on the carriage of goods by road.* 2008 ed. Road freight transport methodology, Luxemburg, available at: http://circa.europa.eu/Public/irc/dsis/transport/library

Eurostat. (2008b). *Methodologies used in surveys of road freight transport in member states and candidate countries*, Luxemburg, available at: http://circa.europa.eu/Public/irc/dsis/transport/library

Knight, I., Newton, W., McKinnon, A. C., Palmer, A., Barlow, T., et al. (2008). *Longer and/or longer and heavier vehicles (LHVs): A study of the likely effects if permitted in the UK — Final report*. TRL Ltd., Crowthorne.

Leonardi, J., & Baumgartner, M. (2004). CO_2 efficiency in road freight transportation: Status quo, measures and potential. *Transportation Research Part D*, *9*, 451–464.

McKinnon, A. C. (2007). *CO_2 emissions from freight transport in the UK*. Paper prepared for the Commission for Integrated Transport, London, available at: http://www.cfit.gov.uk/docs/2007/climatechange/pdf/2007climatechange-freight.pdf

McKinnon, A. C., & Ge, Y. (2004). Use of a synchronised vehicle audit to determine opportunities for improving transport efficiency in a supply chain. *International Journal of Logistics: Research and Applications*, *7*(3), 219–238.

McKinnon, A. C., & Ge, Y. (2006). The potential for reducing empty running by trucks: A retrospective analysis. *International Journal of Physical Distribution and Logistics Management*, *36*(5), 391–410.

McKinnon, A. C., & Woodburn, A. (1996). Logistical restructuring and freight traffic growth: An empirical assessment. *Transportation*, *23*(2), 141–161.

Netherlands Economic Institute, Heriot-Watt University, TFK and Service Economiques et Statistique. (1997). *Analysis of collected data and selection of goods flows*. Deliverables 1 and 2, REDEFINE project, European Commission, Brussels.

Pasi, S. (2008). *Trends in road freight transport 1999 – 2006*. Statistics in Focus, Transport 14/2008, Eurostat, Brussels.

Rizet, C. (2008). *3 Approaches of freight transport energy analysis*. Presentation to the IEA Workshop on 'New Energy Indicators for Transport: the Way Forward', Paris, January 2008.

Rizet, C., Guilbault, M., van Merijeren, J. C., Bjister, E., & Houée, M. (2001). Tracking along the transport Chain via the shippers survey. International Conference on Transport Survey Quality and Innovation, 2001, available at: http://www.its.usyd.edu.au/

Sciullo, C., & Smihily, M. (2006). *Road freight transport 1999–2004: Cabotage and transport with non-EU countries*. Statistics in Focus 8/2006, Eurostat.

Southworth, F. (2006). Filling gaps in the US commodity flow picture. Using the CFS with other data sources. Workshop Resource Papers, Transportation Research Circular: Commodity Flow Survey Conference, July 2005 Boston. TRB (Ed), 29–46. http://onlinepubs.trb.org/Onlinepubs/circulars/ec088.pdf

Transport & Mobility Leuven. (2008). *Effects of adapting the rules on weights and dimensions of heavy commercial vehicles as established within directive 96/53/EC*. European Commission, Brussels.

Zmud, J. (2006). Commodity flow survey. Improving methods to enhance data quality and usefulness. Workshop Resource Papers, Transportation Research Circular: Commodity Flow Survey Conference, July 2005, Boston. TRB (ed.), pp. 98–112, available at: http://onlinepubs.trb.org/Onlinepubs/circulars/ec088.pdf

Chapter 18

The Acquisition of Long-Distance Freight Data: Synthesis of a Workshop

Kara Kockelman, Michael Browne and Jacques Leonardi

18.1. Introduction

The workshop was largely devoted to methods for improving the practice of long-distance freight-transport data acquisition, in order to address near- and long-term planning needs and policy objectives. Twelve participants engaged in an active dialogue, quickly filling the workshop's 8 h. Starting from a review of existing data sets and methods, various survey designs were assessed and suggestions for overcoming current data gaps were provided. Discussions focused on what needed to be changed in relation to survey design and implementation in order to achieve the diverse goals of policy-makers, industry and the research community. Within this discussion both organisational and practical questions were addressed. The key issues were the importance of clarity about the main purpose for data collection, the differences and similarities in survey methods, the sampling and, last but not the least, the motivation and incentives for companies to participate in freight surveys. In the following workshop report, the structure of this discussion is broadly maintained. It starts with an overview of key topics and a summary of the papers presented including a short discussion of the current status of long-distance freight surveys. The chapter continues with the identification of major gaps and challenges and ends with a review of opportunities and recommendations including some suggestions for innovative solutions to the problems discussed.

18.2. Key Topics

Freight transport data collection was clearly a priority of the ISCTSC conference, as this workshop complemented an earlier workshop, which focused on intra-urban

Transport Survey Methods

ISBN: 978-1-84855-844-1

truck movements. Key topics of discussion in this latter, long-distance freight workshop included survey needs and uses, optimal methods of data acquisition, status of current freight surveys around the world, and sampling issues. Major public- and private-sector *motivations for freight transport data collection* include an appreciation (and forecasting) of the following:

- Traffic loads for operations management and infrastructure investment (including pavement maintenance, capacity expansion, etc.),
- Trade linkages and economic interactions across firms, industries, regions and nations (to anticipate opportunities for enhancing such linkages and facilitating efficient operations),
- Vehicle utilisation (including driver and fuel use, as well as capacity utilisation by weight and volume),
- Logistical operations (including scheduling, trans-shipment activities, intermodality and other supply-chain linkages),
- Safety in freight movements and
- Other areas for policymaking and planning, including the introduction of new technologies (e.g. new vehicle designs), regulation (e.g. border control, fuel economy standards and driver rest policies) and pricing (via gas taxes, road tolls and registration fees, both fixed and variable).

A greater appreciation of these many variables across shippers, carriers and commodities enables an evaluation of firm- and system-based performance, competitiveness, costs, benefits and environmental impacts. This includes opportunities for new, more sophisticated and hopefully more robust modelling efforts, thereby facilitating forecasting traffic, revenues and other impacts, for more comprehensive evaluation of transport and trade policies, including service provision and energy standards. Policymakers, transportation planners and regulators stand to benefit greatly from access to detailed data sets on freight transport, assuming sampled movements fairly represent the travellers of interest.

In recent years, multiple *mechanisms for data collection* have been added to the set of established freight transport surveys, and this variety was well presented and discussed during the workshop. Unfortunately, none of the existing survey methods is presently without limitations. Past and present approaches for data acquisition include the following:

- Vehicle-based surveys, such as EUROSTAT's extensive, continuous survey for EU countries and its intra-national complements, including:
  - the United Kingdom's Continuing Survey of Road Goods Transport (CSRGT) (DfT, 2009),
  - Spain's Encuesta Permanente Transporte Mercancías Carretera (EPTMC) (Pérez-Martinez, 2008),
  - France's TRM and SiTRAM surveys (SESP, 2008),
  - the United Kingdom's Key Performance Indicators (KPI) survey (McKinnon 2007),

  - and the United States' recently terminated Vehicle Inventory and Use Survey (VIUS) (US Census Bureau, 2003),
- Shipper-based surveys, such as the United States' Commodity Flow Survey (CFS) (US Census Bureau, 2009),
- Carrier-based surveys, such as the DLR freight company survey presented by Julius Menge at the COST355 final conference in Annecy (Menge & Lenz, 2008),
- Supply-chain interviews, data collection and case studies, as presented at COST355 final conference by Michael Browne (Browne, Rizet, Leonardi, & Allen, 2008),
- Roadside-interview surveys (rather common throughout the world, including a four-nation collaboration for long-distance freight crossing the Alps and Pyrenees [Houée & Spiegel, 2008]),
- Shipment-based surveys (where an item is tracked using electronic tags or other information from point to point, start to finish, like the French ECHO survey)[1] and
- Other, often existing, complementary opportunities, such as transponder tag reads on an instrumented (typically tolled) highway system, tax data required on transactions and customs data on imports and exports.

Discussions noted the rise in third-party logistic firms (3PLs), to manage shippers' transport needs. Such firms can be key in obtaining proper data, and may store a great deal of the needed details. Of course, *sampling and other issues* emerge, with the rise in 3PLs, prompting concerns about privacy protections, sample coverage, burden and non-response. For example, to avoid any opportunity for unique identification of a firm's trading, the CFS shipments are bundled by state or super-region, resulting in dramatic geographic-information losses. Commodity flow data at the county or small-region level (with industry-level aggregation, to preserve anonymity of responses) would allow the analyses that modellers envision while providing the information base decision-makers truly need for infrastructure planning and the application of new policy. In addition, survey scope definition is critical in determining sample frame and questionnaire design. For example, are service trips in company vehicles considered as freight or as passenger transport data? Are public sector vehicles included? How about pipeline transport?

Another key issue is the *motivation of companies and decision-makers for participation* in data collection. For example, there is a need to maintain interest in the United Kingdom's KPI surveys (McKinnon, 2007), by providing indicator values directly back to responding firms, thereby supporting the within-firm competitiveness evaluation process. In time-consuming, repeat government-required surveys, it was noted that some firms are choosing to pay the fine for non-response. In this volume, McKinnon and Leonardi (2009) note that support of a major trade association, the

1. ECHO stands for Enquête CHargeurs Opérateurs. Undertaken in 2003/2004, this French Shipper Survey sampled 3000 companies and monitored roughly 10,000 shipments from sender to final receiver, across vehicle types, illuminating entire supply chains. Non-road mode shipments were over-sampled, to ensure higher confidence on related estimates.

backing of respected industry champions, and a continuing relationship with sampled firms are key ingredients of a relatively high response rate. One key conclusion of the workshop is that the data providers need to understand the value of aggregate data sets, ideally in a personalised way (e.g. firm-based KPIs tabulated relative to competitors). Reporting is burdensome, and, unless firms are willing to allow electronic identification of their vehicles and shipments or share their entire delivery database with researchers or a third party (who can "scrub" identifying information and tabulate needed values), it is likely to remain burdensome.

18.3. Presented Papers

The workshop was informed by a resource paper, complemented by a discussion paper, two contributed papers and stimulating dialogue. Highlights of these papers are summarised below.

Titled "The Collection of Long-Distance Road Freight Data in Europe", the resource paper (McKinnon & Leonardi, 2009) emphasised data requirements for long-distance road-based freight transport. The authors described various data collection systems used for statistics, models, market and scientific studies on long-distance road-based freight transport. They presented first a list of what would be required in an ideal data set, to enable high-quality research while providing reliable guidance for freight operators and policymakers. They then introduced data collection systems emerging in Europe, and discussed some essential shortcomings that need to be addressed. A key criticism relates to the limited number of freight indicators collected in national surveys across Europe. Such feedback is not consistent with the freight market's complexities and does not address various important policy questions. Finally, the third part of their resource paper evaluates the strengths and limitations of alternative approaches to collecting and analysing road freight data.

Gaps include a lack of detail on vehicle fuel consumption and cargo-space utilisation, and on intermodal movements. The paper ended by proposing that there is considerable value to be gained from adopting a KPI-style data collection approach, enhanced by lessons learned in the KPI process. These lessons include involving major professional organisations (to help promote participation in the data collection process), showing sampled firms past survey outcomes and derived analyses, and demonstrating how such information can complement company decision-making. One particularly valuable option is the benchmarking of company performance values, for comparisons within sectors along with potential explanations for differences, focused on improving the freight transport system and fuel efficiency.

The contributed paper titled "The progressive elaboration of a multinational harmonised database for freight transit through Alps and Pyrenees" (Houée & Spiegel, 2008) highlighted the benefits of adopting a common data approach across France, Spain, Switzerland and Italy. The paper identified limitations in such surveys (including a lack of roadspace for stopping fully 1% of all trucks at certain times of the day and year, and the need to abandon some questions such as initial origin and final destination of goods[2]) and scope for improvements. These include the combination of such results with data on maritime transport (which serves as a substitute mode to the truck and rail data obtained) and inclusion of better data on passenger transport in the region (an emerging issue, as congestion in these alpine corridors grows). The current survey takes place every five years — in future, by combining data from other sources, more frequent updates may be possible.

The contributed paper titled "The vehicle approach for freight road transport energy and environmental analysis in Spain" (Pérez-Martinez, 2008) used 1997–2003 EUROSTAT data[3] for analysis of freight transport's carbon footprint, by vehicle type, across Spain. The work reviewed several key indicators of vehicle efficiency and performance. The EUROSTAT data are based on a stratified random sampling, and over the seven-year period studied, represent an audit of 41,600 vehicles per year (each over the course of a week). Although the study data exhibit wide variation in absolute values between 1997 and 2003, energy requirements per tonne-kilometre fell just 0.2% over the seven-year period. The analyses suggest new approaches for assessing vehicle utilisation and fuel efficiency, along with the result that larger commercial trucks are delivering far more tonne-kilometres per gram of carbon dioxide than smaller trucks.[4]

18.4. Status of Long-Distance Freight Surveys

Dialogue regarding the status of long-distance freight surveys took as its starting point Christophe Rizet's discussion notes. These focused primarily on the comparison of the EU's Heavy Goods Vehicle (HGV) vehicle-based survey and the United States' CFS (a shipper-based approach).

The CFS started in 1954 and targets shippers. Europe's various national HGV surveys began with France in 1952, and were harmonised over several decades. The CFS is undertaken every five years, in league with the US Economic Census of businesses, while the EU HGV is continuous in nature. Both rely on mandatory reporting requirements of goods shipments, but only the EU HGV requests information on actual price paid (in for-hire transport). Over the years, the CFS sample size has varied between 50,000 and 200,000 establishments (each assigned

2. True origins and final destinations are often not known by drivers, and will be reported with error. Related to this, detailed route information also is difficult to obtain, due to roadside-interview time constraints and driver memory limitations.

3. Spain satisfies EUROSTAT data requirements via its Encuesta Permanente Transporte Mercancías Carretera (EPTMC) survey.

4. Estimates of tonne-kilometre per kilogram of CO_2 are 14 for trucks over 26.1 tonnes versus 3.0 for those in the lightest weight category (3.5–7.1 tonnes). Of course, energy efficiency is influenced by several factors, including engine technology and transport optimisation strategies.

four one-week reporting periods), while the HGV obtains a sample of 85,000 vehicles (for an entire week of use).

Each survey exhibits various strengths and limitations. The HGV survey offers better information on vehicle activity and utilisation (including flows and load factors), but nothing on transport chains or shipper activity levels. There is no way to identify the true origin and destination of HGV cargos, and there is nothing on non-truck modes (such as rail, pipeline, air and maritime transport). Advantages of the CFS include mode choice and some view of transport chains (though these are often inferred by third-party analysts, at Oak Ridge National Laboratory). However, there is no route information or supply-chain information across shippers. Completing the CFS survey may be viewed as a burden by respondents (possibly influencing firms' selection of reported shipments) and there are some trips omitted (e.g. links to farms as a source of raw materials/foods).

Possibilities exist to complement both types of surveys. In the case of the CFS it would be feasible to perform carrier surveys with the goal of enhancing inference of route, intermodality, trip chaining and other trade-system attributes. In addition, smaller update surveys could be completed to enhance data timeliness. In the case of the HGV survey (and specifically in France) it is possible to include data on other modes (by means of the SitraM database[5]).

In addition, shipper surveys were carried out to improve the knowledge of transport chains and the links with the economy.

Of course, both types of survey provide complementary information, and there is no winner here. Workshop participants support a policy of developing and delivering both types.

18.5. Key Challenges

As suggested, no single survey can do it all. Workshop participants commented several times on the importance of identifying the key questions of interest before selecting a sample frame and sample size. Major gaps and challenges were identified, as follows:

1. *Tying survey purpose to instrument and frame:* It was agreed that there was a need for greater clarity and precision in defining the purpose of the surveys discussed. The various surveys discussed clearly serve different purposes, so approaches and frames will/should differ. Some surveys (such as the EU's HGV survey and the United States' CFS) may be able to serve several purposes, and researchers can use them for a range of analysis. Nevertheless, the group felt there is a need for

5. The SitraM database (Système d Information sur les TRAnsports de Marchandises) was created by the French Ministry of Transport in 1975. It contains data on transport flows between every pair of districts in France, in tonnes, by road, rail and inland waterway, and at NST 3 level. Coastal shipping, pipelines and air transport are not included (SESP, 2008).

caution when using some of the results for forecasting and other initiatives. For example, there is little evidence that improvements of load factors in one country or in one sector will be transferable to another. The details of business operations and product designs often hinder the simple transfer of specific solutions.

2. *Variety:* There is great variety in the types of questions one may ask and uses to which a data set may be put. There is great variety in the types of commodities transported, and the drivers, vehicles and modes that move them, along with involved industries, locations and routes. Freight surveys are profoundly affected, in terms of the range of companies that need to be surveyed and the complexity of the interactions of decisions (e.g. mode choices, trans-shipment points, route choices and trip chaining). The different and evolving responsibilities within a supply chain all have implications for the survey design, robustness of the results and subsequent uses of the data sets obtained.
3. *Missing information:* Participants agreed that current gaps in freight transport information are substantial, despite the scale of some of the existing surveys. Among the most important missing elements are values and volumes moved; shipper costs and prices by mode; international and intermodal ties; use of public vehicles; vehicle fuel consumption; use of small vans for long-distance trips; the impact of new technologies, investments, policies and organisational structures on performance; and the costs and benefits of different policy options.
4. *Speed of change:* Technology and supply chains, both local and global, are evolving rapidly. Within a region or nation there may be stability in terms of the total freight flows (volumes and distances); however, as a result of changes in company sourcing and/or company logistics and supply-chain strategies, there may be dramatic differences in the freight flow patterns over the network. The speed of such change has increased with rising globalisation, and there are many implications for data acquisition, assembly and distribution, including frequency of surveys, and timeliness in data delivery. There also is an evolution in the technologies available for surveys (such as global positioning systems [GPS] and global system for mobile communications [GSM]), which should change the design of surveys.
5. *Survey continuation:* The desirability of maintaining data collection over the long run was discussed by workshop participants, with significant benefits to be gained, such as robust and consistent trend comparisons. Nevertheless, many nations are experiencing an increased reluctance by government departments to meet the costs of surveys and to burden industry with mandatory (statutory) surveys to complete. For example, the United States' Vehicle Inventory and Use Survey (VIUS) was recently scrapped. One solution may be great diffusion and use of automatic data processing devices, on-board and in companies, for continuous, low-marginal-cost data collection.
6. *Data linkages:* The need for and importance of linking survey results effectively across nations, levels of spatial and industry detail, carriers and shippers, modes and acquisition methods seems clear. Workshop participants agreed that survey methods should seek to facilitate such linkages, in order to exploit the

complementary nature of freight surveys at different levels, using different approaches. In particular, many data sets can be enriched by acquisition of data through new technologies, such as roadside toll-tag readers, GPS records and other types of on-board and off-board devices. However, privacy concerns, cultures and customs tend to slow such adoption.

18.6. Opportunities and Recommendations

Many avenues exist for enhancing long-distance freight data set acquisition, and participants chose to highlight the following:

Complementary data sets: Most EU countries appear to be lacking establishment-based data sets and records of non-truck shipments, while the United States is now lacking carrier- or vehicle-based surveys. Whatever new data are obtained, there generally is scope to use complementary approaches, thereby enriching existing data sets. However, in order to do this, pilot initiatives are needed, to show what can be done and what is cost-effective (and robust, in research terms), along with a multinational research exercise comparing the establishment- and vehicle-based survey designs and results.

A more global view, up and down the supply chains: By not recognising complete supply-chain linkages, from source (of raw materials) to consumers, we run the risk of collecting much data but having too limited an understanding of critical variables (such as shipment scheduling, trans-shipment decisions and ultimate destinations). Increasingly, transport decisions are taken in a logistics and/or supply-chain context (down to the retail store or end user via home deliveries), where the decision is influenced by organisations upstream and downstream of the intermediate decision-makers. There needs to be a way to acquire more information on these upstream and downstream influences and cross-actor linkages. Moreover, final consumers ideally should be made aware of their purchasing decisions' implications (e.g. carbon impacts) by providing supply chain and freight transport information through to their home site.

Sharing instruments and harmonising data sets: Workshop participants agreed that major opportunities exist in harmonising future survey instruments and sharing existing instruments. This includes formal agreement on and consistency in terminology (such as a glossary of terms, for researchers and respondents) and greater collaboration across agencies and countries.

Mechanisms for protecting confidentiality while releasing useful results and products: Given the wealth of data that may become available from new technologies being widely implemented in the transport and logistics arenas (e.g. electronic toll collection, high-resolution satellite images and GPS on-board units) there is an urgent need to find appropriate and robust ways in which commercial confidentiality can be assured. At present, data that could be valuable in a policy context is not being released by companies because of privacy concerns. Use of a third-party intermediary to scrub identifying information and prepare data in accordance with analyst and agency needs may be highly desirable and merits thoughtful examination.

Demonstrate value and use of data collected, educating stakeholders and students: Funding, coverage, data quality and response rate issues could all be addressed, to some extent, by demonstrated value in data collected. Researchers and other users need to become better at marketing the benefits of their work with long-distance freight survey data. At present, many companies view data provision as a burden and do not appreciate the benefits of improved transport policymaking and infrastructure management decisions. The information and other benefits that result from the analysis of freight surveys need to be communicated to the many stakeholders and to students in a more proactive and timely way. Researchers need to ensure that students on transportation programmes (who represent the next generation of collectors and users of freight data) are also familiar with the scope for improvement and the value of high quality and insightful data collection activities. By doing this the research community will strengthen the willingness of companies to participate in the surveys exercises.

18.7. Summary and Conclusion

To summarise, the workshop participants and transportation community at large look forward to future research activity and data acquisition innovations in the freight transport arena, both long- and short-distance, across modes and nations, to facilitate economic analysis, environmental policy, transportation system management and the like. There was great benefit in comparing and sharing information and ideas within the workshop's international group, and the ideas and suggestions emerging from our session will inform future freight-related surveys. Heavy-duty truck transport accounts for upwards of 15% of a nation's vehicle-miles travelled and most pavement damage. Both trucks and trains congest tunnels and track, and many seaports and airports regularly reach capacity. Freight transport consumes roughly 10% of a developed nation's petroleum, resulting in a significant carbon footprint along with other regulated emissions. Good data are needed now, for modelling, planning and policymaking.

Acknowledgements

Workshop participants were Mike Browne, United Kingdom; Bastian Chlond, Germany; Michel Houée, France; Kara Kockelman, United States; Jacques Léonardi, United Kingdom; Mariame Mbacké, France; Julius Menge, Germany; Arnim Meyburg, United Kingdom; Peter Ottmann, Germany; Danièle Patier, France; Pedro José Pérez-Martínez, Spain; and Christophe Rizet, France.

References

Browne, M., Rizet, C., Leonardi, J., & Allen, J. (2008). Analysing energy use in supply chains: The case of fruits and vegetables and furniture. In: A. C. Lyons (ed.), *Supply chain innovations: People, practice and performance, Proceedings of the Logistics Research Network Annual Conference 2008 University of Liverpool, UK,* 10th–12th September 2008 (pp. 395–401).

DfT — Department for Transport. (2009). *Road freight statistics 2007*. London. Available at: http://www.dft.gov.uk/pgr/statistics/datatablespublications/freight/goodsbyroad/roadfreight statistics2007

Houée, M., & Spiegel, T. (2008). The progressive elaboration of a multinational harmonised database for freight transit through Alps & Pyrenees. Presented at the 8th International Conference on Transport Survey Methods, Annecy, France.

McKinnon, A. C. (2007). Synchronised auditing of truck utilisation and energy efficiency — A review of the British Government's transport KPI Programme. In: Presented at the World Conference on Transportation Research, Berkeley, California. Available at: http://www.greenlogistics.org/PageView.aspx?id = 150

McKinnon, A. C., & Leonardi, J. (2009). The Collection of Long-Distance Road Freight Data in Europe. In: P. Bonnel, M. Lee-Gosselin, J. Zmud & J.-L. Madre (Eds), *Transport survey methods: Keeping up with a changing world.* Bingley, UK: Emerald.

Menge, J., & Lenz, B. (2008). Company survey approach. Paper presented at the COST355 final conference, Annecy, France.

Pérez-Martinez, P. (2008). The vehicle approach for freight road transport energy and environmental analysis in Spain. Contributed Paper presented at the 8th International Conference on Transport Survey Methods, Annecy, France.

SESP-Ministère de l'Écologie, de l'Energie, du Développement durable et de l'Aménagement du territoire. (2008). *Enquête transport routier de marchandises (TRM) and SiTRAM (all modes)*, Paris. Available at: http://www.statistiques.equipement.gouv.fr/rubrique.php3?id_rubrique = 33

US Census Bureau. (2003). *Vehicle inventory and use survey.* Available at: http://www.census.gov/svsd/www/vius/products.html

US Census Bureau. (2009). *2007 commodity flow survey advance data files*, Service Sector Statistics Division, Commodity Flow Branch. Available at: http://www.census.gov/svsd/www/cfsmain.html

Chapter 19

Identifying and Reconciling the Data Needs of Public Transit Planning, Marketing and Performance Measurement

Gerd Sammer

Abstract

More than ever before, public transit must compete in the transport market. This competition is, on the one hand, against steadily increasing car traffic; and on the other hand, between public transit operators. This, in turn, leads to new demands regarding the type, content and quality of data needed for planning and management. Frequently, traditional travel behaviour surveys do not provide sufficiently accurate and detailed information about public transit demand. To plan public transit, frequently a precise description of all trip stages, including the first and the last mile, is necessary. To achieve this, an adaptation of the traditional survey methods is necessary. In many countries, public transit associations have been established to integrate services offered by individual public transit operators with the help of through-ticketing and a coordination of lines and timetables into what looks, to the user, like a single system. To distribute revenue among the operators involved, detailed surveys of passengers are needed. Measuring the quality of public transit service and surveying customer satisfaction are new tasks. Such data are the basis for quality assurance and are essential for gaining and keeping customers of the public transit system. New technologies such as the Global Positioning System, automated passenger counts and Smart Card Payment Systems offer new possibilities to collect data more efficiently and cost-effectively. This article covers essential aspects of surveys and the collection of data that are crucial for the planning and management of public transit; it points to state-of-the-art methods and offers potential solutions.

Transport Survey Methods

ISBN: 978-1-84855-844-1

19.1. Introduction

This resource paper focuses on innovative and efficient methods for sampling passengers, collecting data, analysing travel patterns and using administrative data for public transit planning and marketing as well as for travel demand model development. Many public transit systems have grown into multi-modal systems that include bus, shuttle, light rail, commuter rail, car-sharing and para-transit services. Public transit operators are interested in collecting empirical data about their systems to get a better insight into the mobility market (e.g. mode of access to and egress from trip stages for the first and last mile, the potential of additional and loss of demand). Questions of interest include: Who uses public transit data? How do people use them? When do they use them? How might usage be improved and enhanced, and what might cause it to be impaired? How can public transit markets best be segmented or visualised? New tasks are gaining in importance, such as the distribution of revenues among the operators of an integrated transport system, which generate new requests for data. These questions require the collection of data about attitudes, opinions, preferences, customer satisfaction and behaviour from users and non-users of the public transit system (Strambi, Trépanier, & Cherrington, 2009). Some of these data may be captured through surveys; other data may be administrative ones or ITS data.

This paper sets out to document, as well as possible, the level of knowledge from an international point of view. Naturally this tremendous challenge is difficult to meet. One crucial problem is the fact that much has been written about travel behaviour surveys in general, but the specific questions that are essential for planning and marketing purposes, as well as for the development of public transit models for scientific research, have not been as extensively addressed by academics. It seems that, in their daily business, public transit operators have little time to contribute to academic publications about the problems of collecting, analysing and using relevant public transit data and the solution of such problems. This statement is supported by the fact that there are comprehensive project reports and so-called "grey literature" regarding this subject.

The paper is structured as follows: the first section presents an overview of the essential activities within the framework of public transit (henceforth, p.t.) that have specific needs and users in terms of survey data. In the succeeding section the problems associated with data collection and data analysis are dealt with in detail, and a valid and representative method of modelling travel behaviour for p.t. is presented. The subsequent sections touch on important requirements for p.t. operation, including revenue distribution among p.t. operators, measuring the quality of p.t., the use of new technologies to survey p.t. passengers, and attitudinal surveys for p.t. The current level of knowledge, open questions and possible solutions are discussed, with the objective of achieving an optimal method of quality assurance. The terms used for this discussion are those suggested for a standardised and generally used terminology for transport surveys (Sammer, 2006).

19.2. Target Activities of Public Transit Planning with Specific Data Needs

The number and kind of tasks and requirements regarding p.t. have increased markedly over the last few years. On the one hand, p.t. has to face stronger competition in the transport market — not only from car transport but also from other p.t. operators. Reasons for this development are the increasing motorisation in many countries and the end of the monopolistic position of many p.t. operators. On the other hand, p.t. has to take on new tasks, for example providing a minimum amount of p.t. for those sections of the population without a car, offering a transport system which is resilient against congestion in conurbations and contributing significantly to the solution of environmental and climate problems. These new tasks demand new activities in the planning and management of p.t. and, for these additional tasks, high-quality data and information are needed (see Table 19.1, first and second column). Examples of new tasks are the performance-related distribution of revenue among p.t. operators participating in a p.t. association (in Europe this important task has so far not been dealt with in a satisfactory way) or measuring the quality of service offered by individual p.t. operators in order to provide customer-oriented quality assurance.

The type and number of users of p.t. data has also grown in the last decade. They range from transport authorities, p.t. operators and transport decision-makers, transport associations and different stakeholders to transport planners, modellers and researchers (see Table 19.1, third column). All of these user groups have different requirements regarding the accuracy of the data that influence survey needs, but cost restrictions are always a major problem when trying to provide the required data quality. Table 19.1 provides an overview of the target activities for the planning and operation of p.t., specific data needs and the essential user groups discussed in the paper.

19.3. Counting Public Transit Passengers

How to ascertain, with the required level of precision, the demand for p.t. is a crucial question. On the one hand, valid information about the demand for p.t. is important in order to document the importance of p.t. compared to other modes of transport, that is non-motorised (foot, bike) and motorised (car driver and car passenger) private transport. On the other hand, data are needed for traffic forecasts and the planning of p.t. in competition to other modes of transport. A number of publications point out that different data sources lead to results that differ considerably (Axhausen, Köll, & Bader, 1998; Brög & Erl, 2004a, 2004b; Follmer & Kunert, 2004; Campbell & Leach, 2004; Sammer, Fallast, Lamminger, Röschel, & Schwaninger, 1990). In the United States it could be observed that census data were up to 40% below the passenger data collected by p.t. operators. The same is true for the "National Household Travel Survey 2001", which shows that p.t. was only used

Table 19.1: Target activities of p.t. planning and operation, their specific data needs and user groups.

Target activity for p.t. planning and operation	Specific data needs	Users of data
Transport modelling for operational and infrastructure decision	Detailed information about trip stages and intermodality	Transport authorities, modellers, researchers
Counting p.t. ridership correctly	Accurate figures of p.t. ridership and compatibility of different data resources	Transport authorities, p.t. operators, transport decision-makers and stakeholders
Providing adequate p.t. service for mobility-impaired people	Definition and identification of these specific groups and their needs for p.t.	Transport authorities, transport decision makers, transport planners, researchers
Revenue distribution to operators	Accurate allocation of passengers mileage of p.t. operators over the year, error estimation of sample	Transport associations, transport authorities, p.t. operators
Measurement of service quality of p.t.	Usability for improvement of service quality and marketing	p.t. operators, transport associations
Benchmarking of p.t. operators	Comparability of data collected for benchmarking indicators	p.t. operators
Measurement of customer satisfaction for p.t.	Identification of potential of new customers and loss of current customers of p.t. for marketing	p.t operators, transport associations, transport authorities
Decision-making process for p.t. measures	Identification of stakeholders' attitudes to identify barriers and drivers of the decision-making process	Transport authorities, transport planners
Marketing and information campaigns for p.t.	Perceptive customer satisfaction data for different target groups, relevant for marketing	p.t. operators, transport associations
Automation of data collection for user counts and travel behaviour	Highly accurate and reliable data covering the whole year; cost-effective data management	Transport associations, transport authorities, p.t. operators, researchers

for 1.6% of all trips. A similar phenomenon can be observed in Europe: local and national travel behaviour surveys show considerably lower numbers of passengers than the official passenger statistics published by p.t. operators. What are the reasons behind these discrepancies and which figures are correct? To find an answer to these questions it is essential to analyse the survey techniques used by the various data sources. Which survey technique is most accurate? What should be done to improve the accuracy of current survey techniques?

By and large the following data sources are available:

- *Passenger surveys of p.t. operators*
  These passenger surveys try to ascertain the number of passengers transported and provide two important metrics: the number of passengers transported per year and passenger mileage per year. In general the number of tickets sold is used to calculate these data. In the case of season ticket holders (commuters, students, etc.) a certain frequency of use is assumed on the basis of a theoretical maximum frequency of use. Comparisons with behaviour surveys (Axhausen et al., 1998; Dähne & Reinhold, 2008) show that the real frequency of use is considerably overestimated — by as much as one third. One needs to bear in mind that the number of passengers transported cannot be compared with the frequency of trips mentioned in behaviour surveys, because in passenger counts the repeated use of modes of p.t. for one single trip is added up. The miles covered per passenger provide an approximation for the distance between boarding and alighting stops but cannot be compared with the results of a behaviour survey which describes the distance from door to door. Differences may also be due to the definition of the modes of transport whenever a "dominant mode of transport" is defined (see Section 19.4.1). Some p.t. operators count boarding and alighting passengers automatically. Depending on the concrete measurement method it is possible to assign passengers transported more or less accurately to individual p.t. transport vehicles. In many European countries the passenger surveys conducted by p.t. operators are used for official transport statistics and for the calculation of the revenues of subsidised tickets for school-children, students and pensioners, etc. This may also cause a tendency of biased but desired overestimation. Some questions have to be addressed: What are the most promising techniques and tools to improve the current, relatively inaccurate measurement of passenger trip data? Is the use of passenger surveys to collect behavioural data the best technique?
- *Statistical surveys (Censuses)*
  Many countries use statistical surveys to gather information about commuters, school-children and students. Few of these surveys are censuses (full population surveys); most are sample surveys. None of these surveys collect information about the use of different modes of transport for one specific day but ask for the "normally used" mode of transport to get to work or school. This leaves room for interpretation as far as the "normally used" is concerned which often leads to some bias. Quite a few statistical surveys are household surveys. If the response rate to certain questions is low, the lack of responses leads to some bias and non-representative results regarding the use of p.t.; this is clearly the case in the United

States (Campbell & Leach, 2004). The analysis of the rate of non-response shows that typical non-respondents are also typically users of p.t.: tenants of flats, single-person households, minorities, low-income people (Word, 1997; APTA, 1992). In Europe the problem of non-response is less severe, since people are obliged to participate in a census. Comparisons of censuses and behaviour surveys focusing on one specific date show that an expansion of figures from censuses for the trip purpose commuter/student–transport leads to an overestimation of the importance of p.t. compared to the results of behaviour surveys. There is an obvious explanation: the expansion of the census data to obtain figures for a whole year does not take into account holidays, sick leave, public holidays or days when the "normally used" mode of transport is not used for whatever reason, because no information is available about these days. But this bias can be solved by weighting with automatic counts over the whole year (see Section 19.7).

- *Travel behaviour surveys of households*
  There are a number of indications that travel behaviour surveys among households do not capture the demand for transport and particularly p.t. in an accurate way — figures are too low. The Austrian national travel behaviour survey of households of 1983 produced figures for p.t. trips, which were 48% lower than reality; for motorised private transport the bias was somewhat smaller (35%) (Sammer et al., 1990). In Germany, a discussion among academics focuses upon the question of whether the national household survey of 2002 assessed p.t. in a correct way (Brög & Erl, 2004a, 2004b; Follmer & Kunert, 2004). The national survey "Mobilität in Deutschland 2002" showed that the share of p.t. in the modal split was 8%; comparable results of local transport surveys point to a share of 12%.

Key reasons for the under-representation of p.t. in behaviour surveys are:

- the problem of non-response because the respondent feels overwhelmed, for example to fill in detailed questions about used p.t. lines (item non-response);
- a lack of effort and awareness by some organisations carrying out surveys to increase the response rate (unit non-response; there is a clear evidence that non-respondents use p.t. more than respondents);
- the low importance of p.t. compared to the use of cars from the point of view of some users;
- an inappropriate weighting of the data (failing to take into account the non-response problem in a proper way) and insufficient quality control at the interview level;
- another important element when comparing household travel survey with p.t. counting has to be mentioned: it has to be considered if the compared population is identical.

One central problem, which has not yet been solved in a satisfactory way, is the expansion of the results from travel behaviour surveys which are based on only a few specific dates to obtain estimates for the annual demand (Pfeiffer & Schmidt, 2006). This problem does not only concern the p.t., it exists also for the other modes.

19.4. Specific Requirements of Public Transit Planning for Behavioural Travel Surveys

To develop p.t. systems, on the one hand as an environmentally friendly alternative to motorised private transport particularly in conurbations, and on the other hand to provide at least a minimum offer of transport for all those population groups that have no car, all planning activities require new information which cannot be provided by traditional travel behaviour surveys. For example, for the planning and modelling of the first and last mile of p.t. trips, additional detailed information about the access and egress mode is needed (Kite, 2009). Little is known about the gathering of this kind of detailed information especially of such high quality. Two crucial questions need to be asked. Firstly, which survey design can provide an optimal balance, that is avoid overwhelming the respondent while collecting a sufficient amount of high-quality data for the planning of p.t. at reasonable cost (see Section 19.5)? Secondly, how can one achieve a sufficient standardisation of survey variables and coding to achieve comparability? (Sammer, 2006).

19.4.1. Detailed Information about Trip Stages (Intermodality)

The planning of p.t. must take the whole trip from door to door into account if p.t. is intended to compete against other modes of transport, particularly cars. This means that travel behaviour surveys must provide more information than generally collected in traditional behaviour surveys, such as origin, destination of trips and intermediate destinations of trip stages, trip purpose, used mode(s) of transport for all trip stages (foot, bicycle, p.t. [bus, shuttle, light rail, tram, para-transit service], car driver and car passenger, taxi, car-sharing, etc.), length and duration of trip. In order to explain and model travel behaviour regarding p.t., it is essential to record the individual trip stages in detail, including the first and last mile, because the user assesses individual trip stages differently regarding the generalised costs. Trip stages are the subdivisions of a trip for which different modes of transport are used. A trip stage is defined by an origin or destination of a trip and an intermediate destination as well as two intermediate destinations where the mode of transport is changed, independent of which type of mode is used.

It is recommended to collect the following information as part of the survey of those trips which involve the use of p.t.:

1. differentiation of p.t. by mode of transport (bus, school bus, tram, railway, underground train, trolleybus, plane, ferry, para-transit, etc.) as well as line identification (number of line);
2. identification of all modes of transport used for one trip, ordered by sequence of use;
3. access to and egress from the mode of transport, differentiated by foot, bike, car driver, car passenger (park-and-ride, kiss-and-ride), motorbike, etc.;
4. length and duration of all trip stages;

5. stops to change (intermediate origins and destinations);
6. type of ticket, transportation cost and parking fees, etc.

When designing a survey one must ask whether the request of such detailed information might not reduce the willingness of the respondent to answer all questions within a behaviour survey. Experience is lacking in this regard. Conference proceedings of several years back provide some clear indication that the willingness to provide such additional information is comparatively low (Table 19.2, Sammer & Fallast, 1983). For example, the data in Table 19.2 indicate that the item non-response rate of written questionnaires is significant higher for detailed information of trip stages than for household, personal or trip items. Whereas the item non-response for the latter ones ranges from 0.1% to 5%, it ranges from 15% to 28% for the investigated trip stage item. This leads to the question: What is the optimal length of a behavioural survey from the respondent point of willingness to respond all questions? The answer to this question is an optimisation task between the criteria of cost, quality (unit and item non-response rate) and data needs.

These results lead to the debateable consideration that a survey that is meant to provide such detailed information should be conducted in two stages. In a first step the traditional questions regarding travel behaviour are asked and, in a second interview a few days later (preferably by phone), detailed information about p.t. trips is collected. Experience shows that this process increases the rate of response significantly (DATELINE Consortium, 2003). The delay in time between first and second stage does not create any memory problem. The question occurs how the

Table 19.2: Item non-response effect for different types of written travel surveys and items (non-response rate in %, Sammer & Fallast, 1983).

Type of survey	Target population	Household item (%)	Personal item (%)	Trips item (%)	Reported distance for the trip stage from origin and destination to p.t. stop and car park (%)
Delivery and collection survey	Urban	4	1	0.1	20
Mail out, mail back household survey	Urban	5	2	0.2	17
Mail out, mail back household survey	Rural	3	2	0.1	28
On board p.t. survey	User of public transit	–	2	0.3	15

application of Global Positioning System (GPS) tools (Stopher, 2009; Wolf, 2009) can solve this problem in a proper way.

19.4.2. Coding the "Dominant Mode of Transport"

If different modes of transport are used for one trip with several trip stages, one has to define and code a "dominant mode of transport" for certain analysis and modelling purposes. This has consequences for the analyses and it must be taken into account for the interpretation of the results. There are generally two ways to do this:

(a) A classification on the basis of a hierarchical ranking based on the amount of flexibility from the users' point of view (low ranking = little flexibility, e.g. timetable, course). The mode of transport which has the crucial impact upon the modal choice is the "dominant mode of transport". A standardised ranking can be used, for example 1 = p.t., 2 = car driver, 3 = car passenger, 4 = bicycle, 5 = pedestrian; the mode of transport with the highest ranking is the "dominant mode of transport" (Köstenberger et al., 1985). There is also the option of basing dominant modal choice on the subjective decision of the respondent. Such a classification needs an in-depth survey of the users' motives of the mode choice, which involves a great effort. This classification has the advantage that it offers a bit of important information for the analysis of users' decisions about the modal choice for a certain trip, but it has the disadvantage that the modal split as a result of the survey has to be interpreted in a different way than is currently done.
(b) Defining the "dominant" mode of transport as whichever one is used for the longest trip stage.

To enable better comparison, there is a clear need for an international coding standard for the mode of trips and trip stages.

19.4.3. Survey of Mobility-Impaired People

One crucial function of p.t., which is actually gaining in importance, is the provision of a minimum standard of accessibility for mobility-impaired people. These are (Graf et al., 2007) people who own no car, such as children, youths, people without a driving license, members of low-income groups such as single parents, physically handicapped people, immigrants, illiterate persons, etc., and they have specific requirements for p.t.. In most countries, no reliable quantitative information is available about this group. That means that sampling and surveying is a major problem for this "hard to reach group" and there is a need for specific solutions (Riandey & Quaglia, 2009; Behrens, Freedman, & McGuckin, 2009). The size of this group as a percentage of the entire population is considerably higher in developing countries (estimated at above 40% in highly developed European). Mobility-impaired people are also worth considering from the point of view of rising energy prices, because there is an increasing risk that, for reasons of low income, the availability of cars might decrease. At the moment little is known about the specific mobility needs and mobility behaviour of this group, and the same is true for its identification and categorisation. However, the needs of physically handicapped people are fairly well-researched as far as the physical design of public modes of transport and stops is concerned (Kohaupt & Schulz, 2007); in some countries such needs are dealt with in national regulations (e.g. The Scottish Government, 2006).

A representative survey of the needs and the travel behaviour of mobility-impaired people is complex compared to traditional travel behaviour surveys. The following problems need to be addressed: identification of this group of people, sampling, dealing with the problem of low response, survey of the specific mobility needs and requirements for p.t. service (which differ considerably depending on the degree of mobility-impairment) as well as the survey data expansion since the population of inference is unknown due to a lack of a specific definition. The development of suitable survey techniques needs to be researched, too. Given current experience, one can state the following:

- Questions about mobility-impaired people in traditional travel behaviour surveys does not provide the necessary results in high quality. This is mainly due to sample problems and the risk of overloading the survey questionnaire.
- Good results were obtained with a survey in two stages: the first stage should be a screening to identify the target group and the second stage should consist of face-to-face interviews in households. This second interview can be used to get detailed information about stages of p.t. trips, etc. (Sammer & Röschel, 1999).
- In both stages, the problem of non-response and the issue of expansion should be taken into account to gain valid results. It has to be mentioned that there is evidence that non-respondents tend to be public-transport users (Campbell & Leach, 2004).

19.5. Revenue Distribution among Public Transit Operators

The question of distributing the revenue among the operators participating in an integrated transport system that tends to be organised as a kind of p.t. association is of particular interest in Europe, where such integrated transport systems exist in many regions (Kummererow & Spichal, 2006). To the user, such an integrated transport system looks like a uniform operator for a whole region (e.g. Berlin, Greater London, Paris, Vienna), a medium-sized town (e.g. Bern, Switzerland; Freiburg, Germany) or a federal state (e.g. Styria in Austria). Integrated transport systems offer standardised ticket prices, through-ticketing, coordinated timetables for all lines integrated in the system and a standardised information and quality-assurance system. The revenue distribution among the participating operators is either based on services offered, such as bus, train and seat-mileage of a certain defined quality, or on passenger mileage. Most integrated transport systems prefer the second method, because it offers transport operators an incentive to provide customer-friendly services in their area (Laschinsky & Meißner, 2007). At the moment it is still difficult to gather valid information about passenger mileage classified by type of ticket and transport operator. The following methods are possible or in use (Sammer, Röschel, & Saurugger, 2007):

- *Registration of the tickets of all boarding and alighting passengers at every stop*
  So far no system is in use which permits a complete survey of all boarding and alighting passengers for all modes of transport (rail, underground system, trams and buses). Some operators manage to record such data partly at railway and underground stations. The fast development of a contact-free registration technology for tickets will make this a useable method in future. But issues of efficient data processing, error rates and cost efficiency of this technology still need to be discussed.
- *Approximate recording of passenger mileage via an automatic counting of boarding and alighting passengers at all stops*
  Such systems are currently tested in the market. Problems of these systems are the fact that a split by type of ticket is possible only with new types of smart card recording systems (Chapleau, Trépanier, & Chu, 2009). The processing of all collected data is time-consuming and costly; thus the question arises whether the recording precision is good enough and the systems are cost-efficient. For revenue distribution purposes an estimation procedure based on the number of tickets sold by types and an estimation algorithm is required; the procedure operates either according to the principle of minimum information gain or maximum entropy and is used also for the estimation of origin–destination matrices based on automatic passenger counts (Wilson, 1974; Snickers & Weibull, 1977; Sammer & Zelle, 1982; Friedrich, Nökel, & Mott, 2000).
- *Interviews of a sample of passengers regarding the mode of transport used and the routes used in the p.t. system; expansion on the basis of the number of boarding and alighting passengers at all stops*

This is the method which is currently most frequently used. There are a number of unresolved problems regarding the sampling, the expansion and the accuracy of the results. These issues are dealt with in the following sections.

19.5.1. Distribution of Revenue on the Basis of Sample Interviews in Public Transit Vehicles and Passenger-Counting

Currently this method is mainly used in Germany and Austria (Engelhardt, Bock, Spichal, & Paul, 2005; Sammer et al., 2007; Friedrich et al., 2000). Regarding the survey technique, the following questions need to be addressed:

- *Interviews of passengers in vehicles and sampling*
  Interviewing passengers has the objective of recording the stops where the passenger got on and off, the route (including all stops to change) and the operator of the used p.t. vehicles for all relevant trips, broken down by type of ticket. Interviews are

conducted in the vehicles, which means that of all passengers boarding at a certain stop a representative sample has to be taken; these individuals are then interviewed. A representative sampling is very difficult to obtain during rush hours. Normally every vehicle is divided into different seat and standing room categories to simplify the sampling by concentrating on the selection of relevant categories (Verkehrsverbund Steiermark, 2004; Sammer et al., 2007). No studies about the representative validity of this kind of sampling and the problem of non-response are known; this seems to be a blank spot on the research map. It has to be mentioned that on-board interviews raise the problem of multiple counting of trips; this must be taken into account by appropriate questions and in the follow-up procedure.

- *Expansion*
  The expansion is done on the basis of the number of boarding and alighting passengers and some additional information about the types of tickets sold. Either a sample of passengers is counted or automated counting devices are used. Because of the cost involved, counting a sample population and sample interviews are only done for one specific day per year. The expansion is done with the help of an estimation method, as mentioned above (Wilson, 1974; Snickers & Weibull, 1977; Sammer & Zelle, 1982; Sammer & Fallast, 1996; Heeren, Kamlah, & Schwabe, 2004). One central, unsolved problem of the expansion are the fluctuations or the variance of the routes and the origin–destination relation during the year, due to weekends, the holiday season, spare time activities, seasons and the weather, days of illness, etc. Currently there is no reliable information available, because it would be too costly to conduct such surveys the whole year round. Some expansion methods use profiles of transport volume over time taken from passenger counts as an approximation (Heeren et al., 2004; Keppeler & Schulze, 2004).
- *Precision of the expansion for the whole year*
  Revenues are distributed on the basis of the expansion of results for the whole year. The expansion methods documented in literature offer no estimates for the exactness of the surveys or methods to estimate the sampling error. They do not take the variations of the origin–destination relationship and/or the routes during the year into account (Pfeiffer & Schmidt, 2006). This issue is addressed in a study (Sammer et al., 2007), and the error potential depending on probable variations over the year is estimated. A relative error for the distribution of revenues among the different operators between 5% and 25% is calculated with an error probability of 5%. Since this kind of result is definitely not satisfactory, a considerable amount of research is still required in this field.

19.6. Measuring the Quality of Public Transit

Measuring the quality of p.t. service is gaining in importance since p.t. operators, who held a monopoly in most countries in the last century, now have to compete against each other and against other modes of transport in most markets. Measuring quality has a number of objectives. It is meant to provide data for the benchmarking of p.t. operators (Allgeier & Hemmersbach, 2007; Mulley et al., 2000; UITP, 2002), for marketing strategies (Waerden van der & Timmersmans, 2008), to win new customers and retain existing customers, to train staff, to plan customer-friendly infrastructure and service, etc. Figure 19.1 provides an overview of the essential quality characteristics of p.t. that are currently used, and suitable measurement methods. Measuring the quality of the infrastructure and the operation (network density, timetable density, accessibility, etc.) is a technical issue and is not dealt with in detail in this paper (see (1) in Figure 19.1). The other quality characteristics are covered under the heading of survey techniques and are dealt with in detail on the following pages (see (2)–(5) in Figure 19.1). Very different methods with considerable difference in value are available to measure the quality of p.t. service (Anreiter & Schaafkamp, 2006; Theißen, 2005); they are all based more or less on European Standard EN 13816, Service Quality in Public Passenger Transport (European Committee for Standardisation, 2001; Österreichisches Normungsinstitut, 2002).

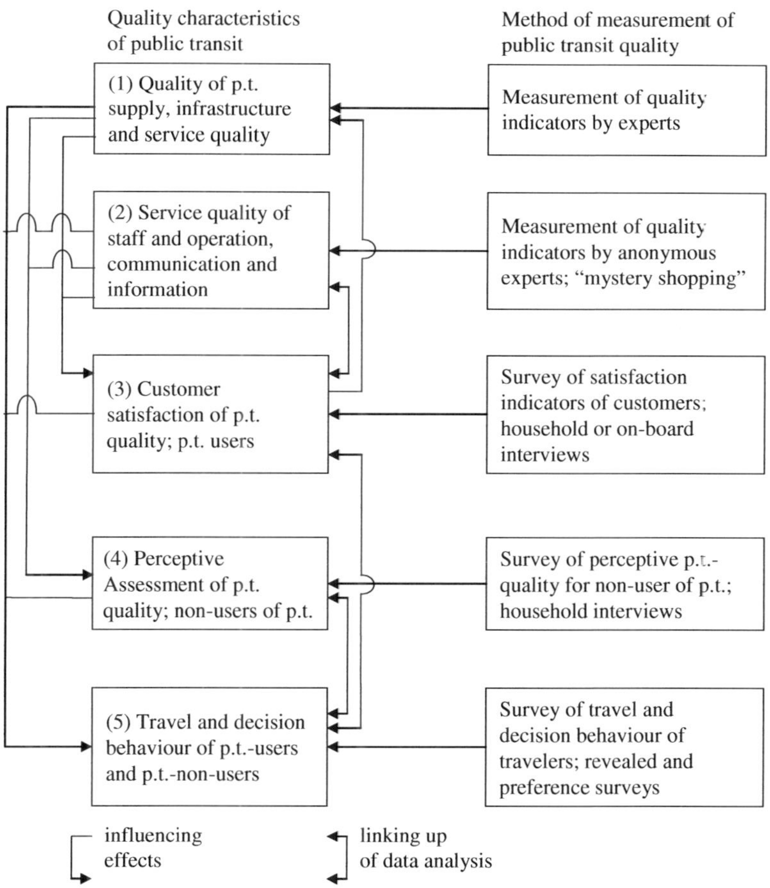

Figure 19.1: Quality characteristics and methods of measurement of public transit quality (overview).

Figure 19.1 shows the key groups of quality characteristics of p.t. and the instruments and characteristics used for ascertaining them and their interconnection. One major quality characteristic relates to the available p.t. supply, infrastructure and service quality of the network and the timetable (see (1) in Figure 19.1). These quality characteristics can be ascertained by experts. The quality of the operation is defined by the service quality of staff, operation, communication and information (see (2) in Figure 19.1). An appropriate data collection procedure is "mystery shopping". The above-mentioned groups of quality characteristics have a considerable impact on customer satisfaction (see (3) in Figure 19.1). The latter can be ascertained by carrying out either household surveys or on-board surveys. Non-users provide their perceptive assessment of p.t. quality, which often deviates strongly from reality and may be based on experience in the distant past (see (4) in Figure 19.1). Their opinion is extremely important, above all as a basis to win new passengers. Non p.t. users are usually interviewed at home. Assessing travel and decision behaviour of users and non-users is an essential basis for developing marketing measures aimed both at p.t. user "retention" and winning of new p.t. users (see (5) in Figure 19.1). Decision behaviour is influenced by all four groups of quality characteristics mentioned above. Suitable survey instruments are Revealed surveys and Stated Preference surveys.

In the following section essential questions are addressed for the survey of data for different methods of quality measurement of p.t..

19.6.1. Service Quality of Staff and Operation

This kind of quality characteristics has the objective to record the everyday quality of the service of a p.t. operator in a standardised way (see (2) in Figure 19.1). The quality of service depends on the level of education and training of the operator's staff and their willingness to offer good service. The quality of service is defined by the interaction between customers and staff (friendliness of the staff; quality of the information provided by drivers, customer centres and call centres; help provided to boarding and alighting physically handicapped passengers; professional abilities of the staff; etc.) but also by the quality of the vehicle operation as such, for example the cleanliness of the vehicles and stops, the functional quality of the announcements at stops, the reliability of vehicles calling at stops, etc. A standardised questionnaire (e.g. with precisely defined questions to ask the conductor, ticket collector or information clerks) and a standardised sampling procedure (which vehicle of which route is to be covered where and when) are necessary to record the service quality of all p.t. operators in a comparable way. The interviews should be done by interviewers who are not known to the p.t. operator and at an unknown but clearly defined date. This kind of survey is called "Mystery Shopping" (Lüttenberg, Person, & Felscher-Suhr, 2004). Occasionally this kind of quality test is carried out by the p.t. operator's own staff, which leads to the risk of bias (Lindenberg, Prelle, Keese, Th., & Uhlenheit, 2007). No analysis about the quality, comparability or validity of this method was found in literature. Therefore, it seems to make sense to use a validation survey whenever this method is redesigned, based on parallel interviews of the same vehicle and the same staff to obtain an estimate for the precision of the survey. The survey quality is usually depicted with the help of a one-dimensional scale with points or marks to be allocated. It is possible to do some weighting and aggregation to obtain a quality index. "Mystery Shopping" is frequently used for benchmarking purposes or to adjust the payment for the services of p.t. operators by using bonuses or deductions (Sammer, Klementschitz, & Roider, 2001). It is also used to improve the quality of services by training the staff. It has been noted that oftentimes the experts' perception of the quality of p.t. differs considerably from the perception of the passengers (Vanhanen & Kurri, 2007).

19.6.2. Customer Satisfaction with Public Transit

The measurement of user and non-user satisfaction with p.t. has gained in importance over the last few years (see (3) and (4) in Figure 19.1). The liberalisation of p.t. has increased the request for data as a basis of strategic planning. On the one hand, the quality of service and the operation as a whole should be improved by scrutiny; on the other hand, such data are important for marketing and promotion. For the existing and potential user of p.t., subjectively perceived customer satisfaction is crucial for the choice of mode of transport. The survey, measuring and analysis of customer satisfaction with p.t. serve the objective of improving the quality of information and service. This, in turn, will help to retain current users of p.t. and to attract potential new users. There is a vast variety of methods currently in use, and they differ considerably regarding the quality of the results. Important elements of the measurement of customer satisfaction are addressed in the following sections.

19.6.2.1. Target Respondents for Customer Satisfaction Surveys The currently used survey types fall into one of the two categories: (a) those that focus upon the users of p.t. systems in order to collect information about their travel behaviour and their satisfaction (Georgi, Wüst-Rocktäschel, Ch., & Heller, 2006; Krietemeyer & Wergles, 2006; Crusins & Pajonk, 2007) and (b) those that include all transport users, even those people who currently do not use p.t. (Brög & Kahn, 2002; Socialdata, 2006; Brög & Ferber, 2000) in order to retain existing customers and to attract new customers.

19.6.2.2. Feature-Based Versus Event-Based Methods to Survey Customer Satisfaction Most customer satisfaction surveys use a method that concentrates on a range of quality features. Users and non-users of p.t. are asked about their satisfaction in general, without addressing the experience of one specific trip. The results reflect overall expectations on the basis of general perceptions. This survey technique has the advantage that it is easy to handle for both interviewer and respondent, who are asked to react to a predefined item battery of quality features.

There is the disadvantage that, due to the pre-selection of features, important aspects from the respondent's point of view may be missing. A general, comparatively superficial impression is ascertained. On the other hand, real p.t. trips on a specific day are the focus of an event-based survey. In reaction to open questions the respondents comment upon their experience and give reasons for their satisfaction or dissatisfaction with specific trips. The respondents mention those reasons that are really "important" to them (Socialdata, 2006; Brög & Kahn, 2002). This kind of survey can only be conducted with actual users of p.t. and is quite demanding for interviewer and respondent alike. Since specific p.t. trips are covered, it is possible to differentiate by line, p.t. operator, time of day, etc. This survey technique makes it possible to gather information about customer defections. A combination of the two survey methods makes it possible to collect data about users and non-users of p.t. (Isfort, 2004).

19.6.3. Customer Satisfaction Scales

The scales used to ascertain customer satisfaction have an important impact upon the results of the surveys.

- The ÖPNV-Kundenbarometer (customer satisfaction scale for short distance p.t.) used by the German market research agency TNS Infratest (Krietemeyer & Wergles, 2006) is based on a scale similar to German school marks (1 = completely satisfied, 2 = highly satisfied, 3 = satisfied, 4 = less satisfied, 5 = unsatisfied). Results are published as average of all marks by all respondents. The asymmetric scale (four positive and one negative value) leads to a biased result which tends to be more positive than the real assessment of the situation.
- Symmetric scales with an uneven number of values, such as the five-point scale: highly satisfied, satisfied, neither satisfied nor unsatisfied, unsatisfied, highly unsatisfied, frequently induce respondents to opt for the medium value, which can distort the results. Such results tend to be presented as percentage shares of satisfied and unsatisfied respondents, with those sitting on the fence being counted as half satisfied and half unsatisfied (Crusins & Pajonk, 2007).
- There are also symmetric scales in use with an even number of values, such as the four-point scale: highly satisfied, satisfied, unsatisfied and highly unsatisfied. Results are either presented as percentage shares to indicate the degree of satisfaction or as a satisfaction index based on the formula.

$$\text{Satisfaction index [\%]} = \text{share of satisfied respondents [\%]} - \text{share of unsatisfied respondents [\%]}.$$

Such a satisfaction index can range from + 100% to − 100% (Socialdata, 2006). This kind of symmetric scale has the advantage that the respondent is forced to take an unambiguous position.

- Scales should have no more than four to six points to make the difference between the points clear (Vanhanen & Kurri, 2007).

Scales to measure satisfaction are generally problematic, and no convincing solution has been found so far. Whenever the raw data are transformed into a satisfaction index, neutral answers such as "no comment" or "neither/nor" are taken into account in very different ways. This is particularly important if open questions are asked regarding the respondent's satisfaction as is commonly the case in event-based surveys. What does it mean if one feature is not mentioned at all: is it unimportant, is the respondent reasonably satisfied with it or does he or she have no opinion about it?

19.6.4. The Overall Index of Customer Satisfaction

It is essential for various purposes to determine an overall index; for example, to calculate bonuses to or deductions from the respective shares whenever revenues are distributed among p.t. operators (Sträuli, ter Hofte, & Scheidegger, 2006; Sammer et al., 2001). There are quite a few possible methods to determine an overall index. Transport users can be asked directly about their overall satisfaction, individual satisfaction characteristics can be weighted by experts and then aggregated, the users of p.t. can be asked about the relative importance of individual characteristics or the same weight can be used for all characteristics (technique of Multi-Criterial Analysis and Analytical Hierarchy Process; Wikipedia, 2008). Experience shows that it is preferable to use a weighting of individual characteristics based on some indirect questioning of the target group if one wants to generate a customer-oriented satisfaction index.

19.6.5. Direct and Indirect Determination of the Importance of Individual Service Features

To plan the improvement of service quality and marketing measures of p.t., one needs to know the subjective importance of service features from the customer's point of view. There are basically two methods to ascertain this subjective importance (Sarikaya et al., 2004): one can ask customers directly about the importance of features. An evaluation of individual features usually leads to the result that nearly all features are highly important, that is one can observe "exaggerated expectations". This phenomenon is caused by the fact that a directly asked question raises the awareness of the respondent for the addressed issue. This effect can be reduced by asking customers to rank the features or by allowing them to distribute only a limited number of points depending on the importance of features, but it cannot be avoided totally. Indirect questioning about the importance of service features is an alternative method. This can be done in different ways.

19.6.5.1. Correlation of Satisfaction Indices and Metrics of Travel Behaviour Suitable for this method are metrics for customer retention/loyalty (Sarikaya et al., 2004), the probability of customers defecting from p.t. or of p.t. gaining customers from other modes of transport (Brög & Kahn, 2002; Socialdata, 2006). By using a two-dimensional model with the variables "feature-specific customer satisfaction" and "potential of winning customers" or "potential customer defection", one obtains a clear impression of the service features which are particularly important or which features urgently need to be improved (Figures 19.2 and 19.3). In both figures the axes depict the respective mean of the two variables. To determine the potential of customers who might defect from p.t., the reasons mentioned by the transport users that speak against or for the use of an alternative mode of transport for a specific trip are analysed. All current p.t. users are considered non-captive users of the p.t. mode who have no objective or subjective reasons not to use an alternative mode of transport. They are the potential number of customers who might defect from p.t.. In Figure 19.2, for example, clear evidence is quite obvious that the quality feature "cleanliness in vehicles" (see satisfaction feature (8)) has a low satisfaction value and a high-risk potential of loss of p.t. passengers to other modes. This result indicates an urgent call for action by the p.t operator.

To determine the potential of customers who might be attracted by p.t., the reasons mentioned by the users of alternative modes of transport (car, bike, walking on foot), which speak against or for the use of p.t. transport for a specific trip are analysed. All those current non-users of p.t. are considered non-captive users of

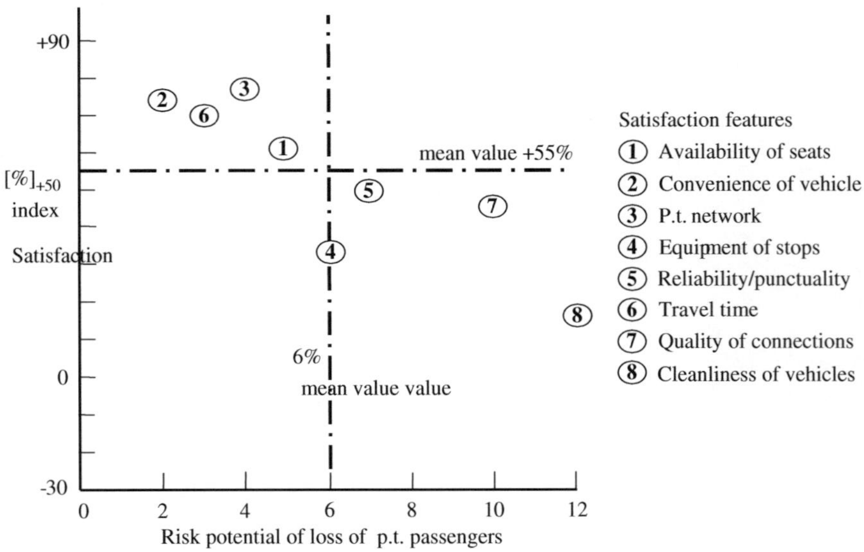

Figure 19.2: Importance–performance matrix of customers' satisfaction characteristics, satisfaction-index and risk potential of loss of p.t. passengers to other modes (Brög & Kahn, 2002).

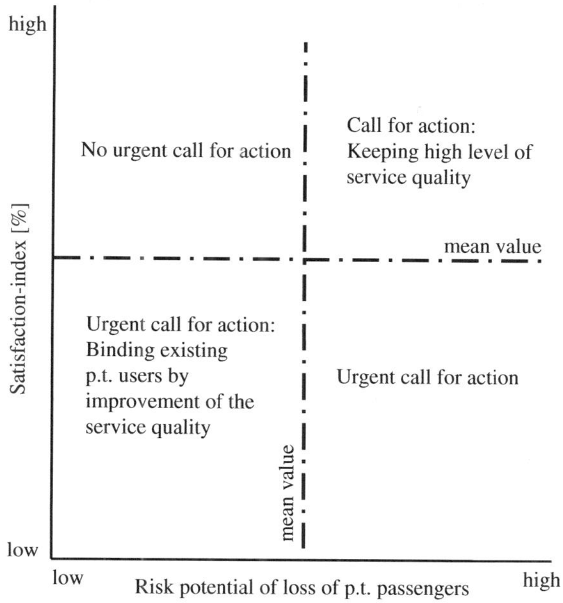

Figure 19.3: Derived need for action from the importance–performance matrix of customers' satisfaction characteristics (Brög & Kahn, 2002; Socialdata, 2006).

alternative modes of transport, who have no objective or subjective reasons not to use p.t. They are the potential number of customers who can be attracted.

The following eight categories of reasons, which speak for or against the use of a certain mode of transport, are important: objective availability (e.g. availability of a car), constraints (e.g. heavy luggage, bad weather), accessibility of the offer (e.g. timetable), information about the mode of transport, travel time, travel cost, travel comfort and subjective behavioural disposition (subjective willingness to use the mode of transport). To determine the potential, an in-depth, interactive interview technique based on a situational approach is used (Brög & Erl, 1996) to check the reasons for or against the use of alternative modes for all concrete realised trips by the users and non-users of p.t.

19.6.5.2. Indirect Determination with the Help of the Stated Preference Technique The stated preference technique is a very interesting approach that is not currently used for the issue in question. Various scenarios using different quality features of the analysed modes are presented to the respondent and he is asked about his hypothetical choice decision of these alternative modes of transport (Sammer, 2003). The respondent makes his modal choice decision based on different mode characteristics of each scenario and he expresses his satisfaction indirectly by his mode choice. The analysis results in the calculation of coefficients of the utility function for a choice function (e.g. a Logit function) where the variables presents quality features, travel time and costs, etc., and the coefficients represent the weights of influence or importance for the feature variables.

Both ways to determine the importance of quality features in an indirect way are comparatively time-consuming and costly. The indirect determination means a quantum leap regarding the usability and practical implementation of the results of a customer satisfaction analysis in order to improve the service quality and the marketing approach.

19.7. Use of New Technologies to Survey Public Transit Passengers

Currently a whole range of new technologies are being developed or put into use that are bound to change the survey methods used in connection with p.t.. In the following sections some of these technologies are discussed, along with their potential for use with p.t..

19.7.1. Automatic User Count

It is essential for the planning of p.t. operators and traffic authorities to get regular, reliable transport data covering the whole year. Automatic counting devices at stops and at the doors of vehicles are an excellent approach. Thus the number of boarding and alighting passengers per stop and line can be measured more or less accurately, as well as the time and the number of passengers in a vehicle. So far, two problems have been identified. Firstly, the precision of this method, which is either based on measurements by light barriers or the weight on the steps at the door of the vehicle, is not satisfactory, especially for peak times. The devices need to be repeatedly checked and calibrated, requiring great effort. Secondly, the transfer and processing of the data requires a well thought-out and standardised system of data management, which is rarely available at present. The big advantage of this method is the provision of counts for each stop the whole year round, which can be used for the extension of results from sample surveys and for the estimation of origin–destination matrices from stop to stop (Friedrich et al., 2000).

19.7.2. Global Positioning Systems (GPS) and Handheld Digital Computers for Ride Check Surveys

The availability of passenger data for p.t. systems is particularly unsatisfactory in developing countries. Manual or automatic counting has the disadvantage that the stops are badly documented. The combination of GPS, "handheld computers" based on BlackBerry technology and a GIS database has proved its effectiveness for counting samples of boarding and alighting passengers (Dondo & Rivett, 2004; Mokonyama & Schnackenberg, 2006). A further advantage is the quality of the counting, which is difficult to falsify (Bogner & Flügel, 2006).

19.7.3. Use of a Smart Card Payment System

So far only a few countries have seen the broad-based introduction of this technology. One of these countries is Finland (Keppeler, Kröpel, Sinisalo, & Sirkiä, 2006); smart cards have also been introduced in Quebec in Canada (Chapleau et al., 2009). In addition to the number of boarding and alighting passengers at each stop per line and the time of the data collection, these systems offer the following information for each passenger: identification of the stop where he/she got on/off or changed, the ticket type and the trip fare. This permits researchers to draw origin–destination matrices from stop to stop for the whole year. When using this system, season tickets are not recorded by all systems. However, soon this will be possible by using a contact-free registration of tickets. Provided that a suitable data management system is developed, this technology will permit an accurate distribution of the revenues to operators who participate in an integrated transport system. Thanks to the identification of recurring behaviour patterns, it is possible to identify individual passengers in an anonymous way (Tseytin, Hofmann, O'Mahony, & Lyons, 2006). A very promising project was carried out in Montreal with the use of a combination of contactless smart card automatic fare collection systems, integrated automatic vehicle location and of geographic information system (Chapleau et al., 2009). Thus these data can provide interesting behaviour metrics for planning purposes, particularly regarding the long-term origin–destination pattern of passengers from stop to stop. However, it must be mentioned that this method provides no information about the access and egress mode of p.t. users and the travel patterns of non-users of p.t., which is an essential disadvantage. The accuracy of the data still needs to be determined.

19.7.4. Using Electronic Passenger Information Systems to Survey Travel Behaviour

The introduction of electronic passenger information systems, which are used via the Internet, mobile phones or fixed telephone networks, offers a new data source that is worth analysing. The initial analysis shows that there is a big systematic difference between the search for connections and actual p.t. trips (Börger, Keppeler, & Lankowsky, 2006; Becker, Müer, & Wulff, 2004). While the relative profile of all queries for a weekday corresponds quite well to the actual trips, only one fifth of the trips on weekdays from Monday to Friday are covered by queries. For weekends this figure is close to 50%, because many weekend trips are only occasional trips and information about timetables or lines is more frequently needed.

19.8. Attitudinal Surveys for Public Transit

To successfully implement a transport policy that is p.t.-oriented, information about the attitude of all stakeholders is helpful and necessary. On the basis of such information, the political decision-making process is easier to plan (Guidemaps

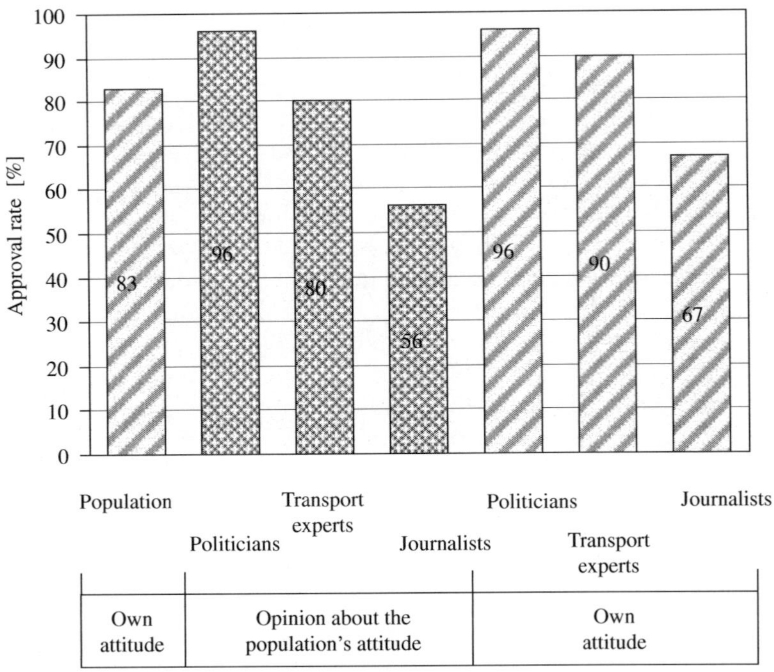

Figure 19.4: Attitudes of various stakeholders towards the local public transit policy, approval rate to the following statement: "In the event of conflict public transit has priority over car traffic" (Sammer et al., 1993).

Consortium, 2004). Therefore, the attitudes of people involved are ascertained for certain types of decision making (Sammer, Röschel, & Wernsperger, 1993). Target groups for attitudinal surveys are, on the one hand, the population as a whole, and on the other hand, decision-makers of all political parties, transport experts, stakeholders and journalists as opinion leaders. In the past it not only proved useful to survey the personal opinion of these specific groups but also their opinion about the attitude of the general public. Thus it is possible to reveal differences between the actual attitude of the population and its perceived attitude. An example is shown in Figure 19.4. Such information is important for a successful decision-making process. A clear and comprehensible phrasing of the questions regarding their attitude and a sufficient response rate of the political decision-makers who have a low willingness to respond (in our experience) is essential for the quality of the survey. A symmetric scale with an even number of no more than six points has proved very useful to ascertain attitudes.

19.9. Conclusion

An analysis shows that p.t. requires certain, specific data that go far beyond the usually surveyed travel behaviour data that are needed and collected for transport planning. In particular, information about customer satisfaction, customer retention and the acquisition of new customers is of considerable economic importance for p.t. systems. But more detailed information about trip stages and mode chains than currently obtained in attitudinal surveys is also needed for p.t. modelling. The question of data quality is important for the use of survey data for the distribution of revenues among p.t. operators; this needs to be researched further.

References

Allgeier, J., & Hemmersbach, A. (2007). Benchmarking und zwischenbetrieblicher Vergleich (Benchmarking and inter-company comparison). In: *Der Nahverkehr, Öffentlicher Personenverkehr in Stadt und Region*, Heft 6, pp. 63–65, offizielles Organ des Verbandes Deutscher Verkehrsunternehmen, Alba Fachverlag, Düsseldorf.

Anreiter, W., & Schaafkamp, Ch. (2006). Systematische Qualitätsmessung (Systematic measurement of p.t. quality). In: Der Nahverkehr, Öffentlicher Personenverkehr in Stadt und Region, Heft 11, pp. 27–33, offizielles Organ des Verbandes Deutscher Verkehrsunternehmen, Alba Fachverlag, Düsseldorf.

APTA. (1992). *Americans in transit, a profile of public transit passengers*. Washington, DC.

Axhausen, K. W., Köll, H., & Bader, M. (1998). Public transport usage intensity of season ticket holders in the City of Innsbruck, report to the Innsbrucker Verkehrsbetriebe, Innsbruck and Ampass.

Becker, J., Müer, A., & Wulff, S. (2004). Fahrgastinformation über Handy und das Mobilitätsverhalten (P.t.passenger.information via mobile phones and travel behaviour). In: Der Nahverkehr, Öffentlicher Personenverkehr in Stadt und Region, Heft 12, pp. 58–62, offizielles Organ des Verbandes Deutscher Verkehrsunternehmen, Alba Fachverlag, Düsseldorf.

Behrens, R., Freedman, M., & McGuckin, N. (2009). The challenges of surveying "hard to reach groups": Synthesis of a workshop. In: P. Bonnel, M. Lee-Gosselin, J. Zmud & J.-L. Madre (Eds), *Transport survey methods: Keeping up with a changing world*. Bingley, UK: Emerald.

Bogner, W., & Flügel, F. (2006). Automatisierte Erhebungssysteme für Fahrgasterhebungen (Automatic systems to survey p.t. passengers). In: Der Nahverkehr, Öffentlicher Personenverkehr in Stadt und Region, Heft 11, pp. 60–61, offizielles Organ des Verbandes Deutscher Verkehrsunternehmen, Alba Fachverlag, Düsseldorf.

Brög, W., & Erl, E. (1996). *Can daily mobility be reduced or transferred to other modes?* Round table 102, European Conference of Ministers of Transport, Economic Research Centre, Report No. CEMT/RE/TR(96)1.

Brög, W., & Erl, E. (2004a). Neue Studie Mobilität in Deutschland 2002. Oder: Der Vorhang zu und alle Fragen offen (A new study of mobility behaviour in Germany; The curtain comes down and all questions remain unanswered). In: mobilogisch — Ökologie. Politik. Bewegung, 02/2004.

Brög, W., & Erl, E. (2004b). Mobilität in Deutschland 2002 — Klärende Worte zur Klärung (Mobility in Germany 2002 — Some clear words to clarify the issue). In: mobilogisch — Ökologie. Politik. Bewegung, 04/2004.

Brög, W., & Ferber, G. (2000). Brennpunkt Kundenorientierung. Chancen und Probleme der Kundenzufriedenheitsuntersuchung (Focus on customer-orientation. Possibilities and problems of customer satisfaction analysis). Konferenzbericht für den Kongress Nahverkehr 2010, 2 to 3 März 2000, Linz.

Brög, W., & Kahn, Th. (2002). Kundenzufriedenheit im Spannungsfeld zwischen Kundenbindung und Kundengewinnung (Customer satisfaction seen in the context of retaining existing and gaining new customers). Vortrag zur Fachtagung "Messung der Dienstleistungsqualität im ÖPNV" der VDV-Akademie, 26/27 Juni 2002, Bonn.

Börger, St., Keppeler, B., & Lankowsky, Ch. (2006). Elektronische Fahrplaninformation: Analyse des Nutzerverhaltens (Electronic timetable information: Analysis of its usage). In: Der Nahverkehr, Öffentlicher Personenverkehr in Stadt und Region, Heft 3, pp. 50–56, offizielles Organ des Verbandes Deutscher Verkehrsunternehmen, Alba Fachverlag, Düsseldorf.

Campbell, S., & Leach, P. (2004). Counting transit so that transit counts. Prepared by Trans Management Inc., a Washington, D.C., transportation consulting firm, for the American Public Transportation Association, Washington, D.C.

Chapleau, M., Trépanier, M., & Chu, K. K. (2009). The ultimate survey for transit planning: Complete information with smart card data and GIS. Presented at the 8th international conference on transport survey methods, Annecy, France.

Crusins, J., & Pajonk, E. (2007). Brandenburger Verkehrsunternehmen stellen sich dem Fahrgasturteil (Public transport operators in Brandenburg face the verdict of p.t. passengers). In: Der Nahverkehr, offizielles Organ des Verbandes Deutscher Verkehrsunternehmen, Heft 4, pp. 56–59, Alba Fachverlag, Düsseldorf.

Dähne, O., & Reinhold, T. (2008). Monatliche Mobilitätsprofile vollständig ermitteln (Determination of monthly profiles of mobility). In: Internationales Verkehrswesen 3/2008, pp. 68–72, DVV Media Group GmbH, Hamburg.

DATELINE Consortium. (2003). Final report of DATELINE, Deliverable D6, DATELINE (Design and application of a travel survey for European long-distance trips, based on an International Network of Expertise). 5th Framework Programme, European Commission-DG TREN, Brussels.

Dondo, C., & Rivett, U. (2004). Representing ride check survey data in a GIS, the case study of Cape Town, South Africa. GIS-Development, March, available at: www.gisdevelopment.net.

Engelhardt, A., Bock, S., Spichal, M., & Paul, H. (2005). Einnahmenaufteilung: So geht es (Revenue distribution. That's how it works). In: Der Nahverkehr, Öffentlicher Personenverkehr in Stadt und Region, Heft 1–2, pp. 56–59, offizielles Organ des Verbandes Deutscher Verkehrsunternehmen, Alba Fachverlag, Düsseldorf.

European Committee for Standardisation. (2001). European Standard 13816, Service Quality in Public Passenger Transport, Brussels.

Friedrich, M., Nökel, K., & Mott, P. (2000). Keeping passenger surveys up-to-date: A fuzzy approach. Transportation Research Records, No. 1735, Washington, D.C., pp. 35–42.

Follmer, R., & Kunert, U. (2004). Mobilität in Deutschland 2002 — Einige Worte zur Klärung (Mobility behaviour in Germany 2002 — Some clarifying words). In: mobilogisch — Ökologie. Politik. Bewegung, 04/2004.

Georgi, A., Wüst-Rocktäschel, Ch., & Heller, B. (2006). Was Fahrgäste wirklich wollen (What passengers really want). In: Der Nahverkehr, offizielles Organ des Verbandes Deutscher Verkehrsunternehmen, Heft 3, pp. 57–61, Alba Fachverlag, Düsseldorf.

Grafl W., Meschik, M., Meth, D., Neumann, A., Sammer, G., Unbehaun, W., et al. (2007). ÉGALITÉ — Ein gleichberechtigter Alltag im telematikgestützten Verkehrsgeschehen (Everyday with equal rights for telematic-aided transport). Forschungsprojekt im Rahmen des Programms I2-Intelligente Infrastruktur des Bundesministerium für Verkehr, Innovation und Technologie, Wien.

Guidemaps Consortium.Kelly, J., Jones, P., Barta, F., Hössinger, R., Witte, A., et al. (2004). *Successful transport decision-making handbook* (Vols. 1 and 2). London: PTRC.

Heeren, M., Kamlah, D., & Schwabe, S. (2004). Fahrgastzahlen präziser erheben (Collecting the number of passengers more accurately). In: Der Nahverkehr, Öffentlicher Personenverkehr. in Stadt und Region, Heft 10, pp. 32–37, offizielles Organ des Verbandes Deutscher Verkehrsunternehmen, Alba Fachverlag, Düsseldorf.

Isfort A. (2004). Was erwartet der Fahrgast? (What does the public transport passenger expect?). In: Der Nahverkehr, Öffentlicher Personenverkehr in Stadt und Region, Heft 7–8, pp. 7–10, offizielles Organ des Verbandes Deutscher Verkehrsunternehmen, Alba Fachverlag, Düsseldorf.

Keppeler, B., Kröpel, St., Sinisalo, K., & Sirkiä, A. (2006). Nutzung des E-Ticketings für die Analyse der Verkehrsnachfrage (Using electronic ticketing to analyse travel demand). In: Der Nahverkehr, Öffentlicher Personenverkehr in Stadt und Region, Heft 6, pp. 54–61, offizielles Organ des Verbandes Deutscher Verkehrsunternehmen, Alba Fachverlag, Düsseldorf.

Keppeler, B., & Schulze, M. (2004). Repräsentative Hochrechnung von Stichprobenbefragungen (expanding-up of sample interviews). In: Der Nahverkehr, Öffentlicher Personenverkehr in Stadt und Region, Heft 11, pp. 42–48, offizielles Organ des Verbandes Deutscher Verkehrsunternehmen, Alba Fachverlag, Düsseldorf.

Kite Consortium, STRATA Karlsruhe, University of Karlsruhe, INRETS Paris, ETH Zuerich, BOKU Vienna, TIS Lisbon, CDV Bruno, Inovaplan Ettlingen, IMOB Dieperbeck. (2009). A knowledge base for intermodal passenger travel in Europe. Research Project of the Sixth Framework Programme, Priority 6.2, Sustainable Surface Transport FP6-2005-TREN-4, founded by the European Commission, Brussels.

Kohaupt, B., & Schulz, A. (2007). Unbehinderte Mobilität in Hessen (Unimpeded mobility in Hesse). In: Der Nahverkehr, Öffentlicher Personenverkehr in Stadt und Region, Heft 6, pp. 24–27, offizielles Organ des Verbandes Deutscher Verkehrsunternehmen, Alba Fachverlag, Düsseldorf.

Köstenberger, H., Kriebernegg, H., Sammer, G., Ecker, H., Fallast, K., Sadila, A., et al. (1985). Nahverkehrskonzept Zentralraum Salzburg — Verkehrsuntersuchung 1982 — Bericht (Masterplan for local transport in the Salzburg region — Transport Study 1982 — Report) commissioned by the Regional Government and the City of Salzburg, p. 34.

Krietemeyer, H., & Wergles, K. (2006). Kundenzufriedenheit aktuell messen (Measuring customer satisfaction on time). In: Der Nahverkehr, Öffentlicher Personenverkehr in Stadt und Region, Heft 7–8, pp. 28–34, offizielles Organ des Verbandes Deutscher Verkehrsunternehmen, Alba Fachverlag, Düsseldorf.

Kummererow, H., & Spichal, M. (2006). Neue Tarifkonzepte und die Frage der sachgerechten Einnahmenaufteilung (New concepts for trip fares and the question of an appropriate revenue distribution). In: Öffentlicher Personenverkehr in Stadt und Region, Heft 7–8, pp. 35–37, offizielles Organ des Verbandes Deutscher Verkehrsunternehmen, Alba Fachverlag, Düsseldorf.

Laschinsky, N., & Meißner, St. (2007). Einnahmenaufteilung in den Verbünden gestalten und verhandeln (Designing and negotiating the distribution of revenues in integrated public transport systems). In: Der Nahverkehr, Öffentlicher Personenverkehr in Stadt und Region, Heft 6, pp. 28–31, offizielles Organ des Verbandes Deutscher Verkehrsunternehmen, Alba Fachverlag, Düsseldorf.

Lindenberg, W., Prelle, M., Keese, Th., & Uhlenheit, A. (2007). Üstra nimmt Qualitätstest erfolgreich selbst in die Hand (Üstra [Transport operator of Hannover] runs its own successful quality tests). In: Der Nahverkehr, Öffentlicher Personenverkehr in Stadt und Region, Heft 3, pp. 54–57, offizielles Organ des Verbandes Deutscher Verkehrsunternehmen, Alba Fachverlag, Düsseldorf.
Lüttenberg, R., Person, D., & Felscher-Suhr, U. (2004). Dienstleistungsqualität gezielt ermitteln und verbessern (Targeted investigation and improvement of service quality). In: Der Nahverkehr, Öffentlicher Personenverkehr in Stadt und Region, Heft 7–8, pp. 56–59, offizielles Organ des Verbandes Deutscher Verkehrsunternehmen, Alba Fachverlag, Düsseldorf.
Mokonyama, M., & Schnackenberg, E. (2006). Harnessing innovation in passenger transport research in Africa. SATC 2006: Southern African Transport Conference, p. 9.
Mulley, C., Mageean, J., Nelson, J. D., Bonetti, A., Marciani, M., Cossu, P., et al. (2000). EQUIP, Extending the quality in public transport, transport research. Fourth Framework Programme urban transport DG VII-76, Brussels.
Österreichisches Normungsinstitut. (2002). EN13816 Transport Logistik und Dienstleistungen — Öffentlicher Personennahverkehr, Fachnormenausschuss 228, Wien.
Pfeiffer, M., & Schmidt, J. (2006). Auswertung von Fahrgastbefragungen-Punkt-und Intervallschätzungen bei komplexen Stichprobendesigns (Analysis of passenger interviews — point and interval estimates in the case of complex sampling designs). In: Der Nahverkehr, Öffentlicher Personenverkehr in Stadt und Region, Heft 5, offizielles Organ des Verbandes Deutscher Verkehrsunternehmen, Alba Fachverlag, Düsseldorf.
Riandey, B., & Quaglia, M. (2009). Surveying hard to reach groups. In: P. Bonnel, M. Lee-Gosselin, J. Zmud & J.-L. Madre (Eds), *Transport survey methods: Keeping up with a changing world*. Bingley, UK: Emerald.
Sammer, G. (2003). Ensuring quality in stated response surveys. In: P. Stopher & P. Jones (Eds), *Transport survey quality and innovation* (pp. 365–375). Oxford: Pergamon.
Sammer, G. (2006). Processing, analysis and archiving of travel survey data. In: P. Stopher & C. Stecher (Eds), *Survey methods, quality and future directions* (pp. 240–270). Oxford: Elsevier.
Sammer, G., & Fallast, K. (1983). Effects of various population groups and of return methods on the return of questionnaires and the quality of answers in large scale travel surveys. In: *Proceedings of new survey methods in transport, 2nd international conference*. Australia 1983, Hungerford Hill, Pokolbin N.S.W., 12th–16th September 1983 (p. 6).
Sammer, G., & Fallast, K. (1996). A consistent simultaneous data weighting process for traffic behaviour. Presented at the 4th International Conference on Survey Methods in Transport, Oxford, UK.
Sammer, G., Fallast, K., Lamminger, R., Röschel, G., & Schwaninger, T. (1990). Mobilität in Österreich 1983–2011 (Mobility in Austria 1983–2011), Editor: Österreichischer Automobil-, Motorrad- und Touring Club, Wien.
Sammer, G., & Röschel, G. (1999). Altern in der Steiermark–Lust oder Last?, Verkehr und Mobilität, Mobilität älterer Menschen in der Steiermark (Getting older in Styria — Fun or burden?, Transport and mobility, mobility of elderly people in Styria). In: Seniorenreport Steiermark, Hg. Gerald Schöpfer, Schriftenreihe der Arbeitsgemeinschaft für Wirtschaft und Sozialgeschichte, Graz.
Sammer, G., Röschel, G., & Wernsperger, F. (1993). Die verkehrspolitische Einstellung zum öffentlichen Verkehr in Klagenfurt-Meinungen und Fakten, Attraktivierungs- und Überprüfungsprogramm des ÖPNV in Klagenfurt 1992/93 (The transport political attitudes

to public transport in Klagenfurt — opinions and facts, a programme to check public transport in Klagenfurt and to make it more attractive, 1992/93). Im Auftrag der Klagenfurter Stadtwerke, Verkehrsbetriebe, Graz.

Sammer G., Klementschitz, R., & Roider, O. (2001). Ö-QUIS — Öffentlicher Busverkehr — Qualitätssicherung bei Leistungsbestellungen im Land Oberösterreich und im Land Salzburg, Teil 1 und 2: Entwicklung eines Bewertungssystems (Ö-QUIS Public bus-transit — Quality assurance when ordering transit services in the regions Upper Austria and Salzburg, Part 1 and 2, Development of an assessment system). Forschungsprojekt i.A. Amt der Salzburger Landesregierung Abteilung 6 und Amt der Oberösterreichischen Landesregierung Baudienstzentralabteilung — Verkehrskoordination, Institut für Verkehrswesen Universität für Bodenkultur, Wien.

Sammer, G., & Zelle, K. (1982). A method for updating existing data bases of an origin-destination traffic matrix. In: *Proceedings of PTRC — summer annual meeting 1982*, Warwick.

Sammer, G., Röschel, G., & Saurugger, V. (2007). Methode zur Erhebung der Aufteilung von Verkehrsverbundeinnahmen (Method to determine the distribution of revenues within integrated transport systems). Im Auftrag der ÖBB-Personenverkehrs AG und der ÖBB-Postbus GmbH, Wien.

Sarikaya, M., Dörkes, Ch., Donath, Th., & Schulte, K. (2004). Wie sehen Bürger ihren ÖPNV? (How do citizens see their p.t.?). In: Der Nahverkehr, offizielles Organ des Verbandes Deutscher Verkehrsunternehmen, Heft 4, pp. 20–24, Alba Fachverlag, Düsseldorf.

Snickers, F., & Weibull, J. W. (1977). A minimum information principle — Theory and practice. *Regional Science and Urban Economics*, *7*, 137–168.

Socialdata. (2006). ÖPNV im Verkehrsverbund Ost-Region-Vertiefende Analyse — Kundenzufriedenheit (Public transport within the integrated transport system in the Vienna Region-in-depth-analysis — Customer satisfaction). Im Auftrag des Verkehrsverbundes Ost-Region.

Stopher, P. R. (2009). Collecting and processing data from mobile technologies. In: P. Bonnel, M. Lee-Gosselin, J. Zmud & J.-L. Madre (Eds), *Transport survey methods: Keeping up with a changing world.* Bingley, UK: Emerald.

Strambi, O., Trépanier, M., & Cherrington, L. (2009). Data for public transit planning, marketing and model development: Synthesis of a workshop. In: P. Bonnel, M. Lee-Gosselin, J. Zmud & J.-L. Madre (Eds), *Transport survey methods: Keeping up with a changing world.* Bingley, UK: Emerald.

Sträuli, C., ter Hofte, K., & Scheidegger, D. (2006). Qualitätslenkung über den Markt (Steering quality through the market). In: Der Nahverkehr, offizielles Organ des Verbandes Deutscher Verkehrsunternehmen, Heft 11, pp. 62–65, Alba Fachverlag, Düsseldorf.

Theißen, A. (2005). Salzburger Verkehrsverbund testet Qualität flächendeckend (The Public Transport Association of Salzburg tests public transport quality for the whole region). In: Der Nahverkehr, Öffentlicher Personenverkehr in Stadt und Region, Heft 11, pp. 64–67, offizielles Organ des Verbandes Deutscher Verkehrsunternehmen, Alba Fachverlag, Düsseldorf.

The Scottish Government. (2006). Improved public transport for disabled people. Vol. I — Report, Scottish Executive Social Research. Available at: http://www.scotland.gov.uk/Publications/2006/05/16145134/0

Tseytin, G., Hofmann, M., O'Mahony, M., & Lyons, D. (2006). Tracing individual public transport customers from an anonymous transaction database. *Journal of Public Transportation*, *9*(4), 47–60.

UITP. (2002). *Marketing as an instrument in greater client satisfaction and better benefits*. A UITP Position Paper, Internal Association of Public Transport.

Vanhanen, K., & Kurri, J. (2007). *Quality factors in public transport*. Report of the Transport Research Programme JOTU; WSP LT Consultant Ltd. and the Finish Local Transport Association, Finland.

Verkehrsverbund Steiermark. (2004). Fahrgasterhebung 2004 — Handbuch für Interviewer (Public transport passenger survey 2004 — Handbook for interviewers). Graz.

Waerden van der, P., & Timmersmans, H. (2008). Data collection approaches for public transport marketing: Some examples. Presented at the 8th international conference on transport survey methods, Annecy, France.

Wikipedia, the free encyclopedia. (2008). Analytic hierarchy process. Available at: http://de.wikipedia.org/wiki/Analytic_Hierarchy_Process, 11th April 2008.

Wilson, A. G. (1974). *Entropy in urban and regional modelling*. London: Pion.

Wolf, J. (2009). Mobile technologies: Synthesis of a workshop. In: P. Bonnel, M. Lee-Gosselin, J. Zmud & J.-L. Madre (Eds), *Transport survey methods: Keeping up with a changing world*. Bingley, UK: Emerald.

Word, D. (1997). Who responds: 1990 Census, Commerce Department, Bureau of Census.

Chapter 20

Data for Public Transit Planning, Marketing and Model Development: Synthesis of a Workshop

Orlando Strambi, Martin Trépanier and Linda Cherrington

20.1. Introduction

Public transit has a central role in the move towards sustainable transportation systems and improved quality of life, particularly in large urban areas of developed and developing countries. Changes are required in public transit infrastructure, service and travel behaviour to reach the ambitious goals of efficiently improving accessibility for all sectors of society and reducing the impact of transportation on the environment. To plan and deliver quality and affordable public transit, adequate data are needed for the following purposes:

- to plan for a better system, from the perspectives of users, operators and society;
- to model travel behaviour in order to understand the needs of users of public transit and
- to market the system, by telling people — users and non-users alike — that public transit may match travel needs and will promote societal goals.

The workshop drew from the knowledge and experience of participants representing 10 different countries who discussed the main issues regarding methods for data collection in public transit. The workshop addressed two questions: what is the state-of-the-practice in data collection for public transit planning, marketing and model development; and what are the relevant strategic needs for research in these areas?

Significant contributions to the discussion came from a comprehensive resource paper by Sammer (2009) and two contributed papers by Chapleau, Trépanier, and

Transport Survey Methods

ISBN: 978-1-84855-844-1

Chu (2008) and van der Waerden, Timmermans, and Berenos (2008). A synthesis of the main conclusions of the lively exchange of ideas occurring in the workshop is presented in the following sections.

20.2. Changing Scene for Public Transit

The workshop participants started by characterising the changing scene in the public transit industry and identifying the data requirements arising from the current and prospective context. The following points deserve particular attention:

- Public transit is becoming increasingly multi-modal, multi-agency and multi-operator, adding to the complexity of obtaining complete and compatible data. As noted by Sammer (2009), collecting data about each stage of a transit trip increases in importance, for reasons as diverse as understanding and modelling mode/route choice and determining the distribution of revenue among multiple operators.
- Transit agencies and operators have an increasing focus on business practices; in some cases, a market-like situation prevails, reinforcing the need for a business approach. This requires data to develop better performance measures, to understand user needs and consumer satisfaction and to benchmark performance against peer transit agencies.
- Public policy in several regions is focusing on increasing the use of public transit as an important element for attaining sustainable transportation and urban quality of life. Better information is required about the impact of public transit on travel patterns in general and on the environment as well.
- Several workshop participants raised the issue of concern for personal security related to the use of public transit. This is an important problem to be tackled as part of effective marketing to attract and retain public transit users.

Some trends relating to data collection in the context of public transit were also identified. In sample surveys requiring interviews, non-response is an increasing problem. In this regard, minimising bias is a crucial requirement in the design of sampling methodologies, in the validation of survey responses and in the selection of analysis methods.

On the other hand, the introduction of new technologies to support the operation and management of public transit systems also makes available a wealth of data for planning, modelling and marketing (Strathman et al., 2008). Notwithstanding the availability of data that can be collected through new technology, some important challenges lie ahead. First, even basic information such as the number of passengers is still subject to measurement problems (Sammer, 2009). Second, new methodologies and analysis techniques are to be developed to process the usually very large quantity of data and to obtain meaningful results. Third, although abundant, these data do not yet provide a complete coverage of the range of information required for public transit planning, modelling and marketing.

The workshop participants also acknowledged that in many parts of the world, public transit is still a single-mode, low-technology system, sometimes operated by a multiplicity of informal operators, working under no structured organisation. While this condition makes data requirements no less important, there are additional challenges and complexities to data-collection efforts.

20.3. Data Needs and Data Sources

Two distinct perspectives on data for public transit were considered by the workshop participants:

- *data uses*, including planning, modelling, marketing, providing user information and support to operation and management and
- *data users*, including public transit agencies, operators (and their associations), planners, market specialists, researchers, modellers, decision makers, stakeholders and transit users.

This approach helped to identify a number of data types and match them to a particular perspective. Some types contemplate objective data, in the sense that they reflect measurable supply characteristics or usage levels; other types are considered subjective data, often related to perceptions and attitudes concerning public transit. The following data needs were highlighted by the workshop participants:

- ridership and revenue data, at the route level, by user category and time period;
- information about trip stages and intermodality;
- data documenting the needs of specific groups of the population, particularly persons with disabilities;
- indicators of service quality (objectively and subjectively measured);
- measures of consumer satisfaction (subjectively measured);
- data for peer benchmarking (requires comparable data among systems);
- data for allocation of passenger miles (and revenues) by mode or type of service, or by service provider;
- information to identify market potential (additional uses, new users, lost users) and
- data from surveys of public attitude towards public transit.

Considering the variety of data needs, it is natural to expect that distinct data sources should be required. Three basic categories of data sources were identified by the workshop participants and labelled *administrative*, *passive* and *active*. Administrative data concerns network infrastructure and service, as well as basic operational data. Examples of administrative data are:

- routes and schedules;
- intramodal and intermodal terminals, stations and stops;
- vehicle fleet characteristics;

- operators and
- costs and revenues.

Data obtained through intelligent transportation systems (ITS) were labelled passive, since the data flow from normal operation, without active intervention into the transit system to collect further data. Under the general label of ITS, a wide array of technologies (and acronyms) is being adopted by the public transit sector, such as automated fare collection (AFC), automatic passenger counters (APC), automatic vehicle location (AVL), advanced traveller information systems (ATIS), video cameras (and associated image processing capabilities) and others. Although the main concern for implementing these systems is to support management and operation, the systems are also an important source of information about public transit users. Notably, most applications of ITS automatically provide some contribution of operating and ridership data with a space and time stamps.

Active data-collection efforts encompass data gathering by surveys of users, non-users and other stakeholders. Possibilities may range from on-board to household; face-to-face to Internet; paper-and-pencil to global positioning system (GPS); revealed or stated preference and attitudinal surveys.

Other sources complement the data requirements of public transit planning, modelling and marketing. These include land use and census data, usually in the form of geographic information system (GIS) files. Compatibility with these more general information sources has to be considered from the inception of specific public transit databases.

In closing this section, it should be noted that several questions were raised during the workshop concerning methodological, technical and practical difficulties with ITS data:

- Which segment of users do they represent?
- How to identify, reach and sample those under-represented or not represented?
- How to assure the validity of automatically collected data?
- How to deal with very large amounts of data?
- How to obtain trip and user data not collected by automatic means?

In spite of the massive amount of data that can be potentially obtained from administrative and automatic (passive) means, sample surveys (active data collection) will remain an important component of data collection in public transit to address specific issues of interest.

20.4. Focused Surveys

Where ITS systems are deployed, large amounts of data about operation and ridership are available, usually at a low marginal cost. This will allow surveys to be more focused on specific themes and population groups, opening some interesting possibilities in this regard.

Small sample household surveys may focus on characteristics of interest about public transit use and the user. Alternatively, an in-depth look at public transit could be feasible through a sub-sample of larger conventional household surveys. The environment of a (small) sample household survey allows the collection of information about public transit at a higher level of spatial resolution. The spatial level of detail required for the analysis of public transit demand is not adequately met by zone sizes conventionally used for strategic modelling (and data collection). Another possible advantage of household surveys over on-board surveys is that potential users of public transit can be screened among non-users; understanding their attitudes and behaviour may be a key to targeting marketing efforts.

The need to collect data on usage patterns for multi-modal systems requires the conduct of origin–boarding–alighting–destination (OBAD) surveys. These surveys are usually conducted on-board a single mode, but data are collected about the use of other modes on the same trip, from origin to destination. Transfers between vehicles or modes can be investigated, aiming to reduce the cost of time for transfers on multi-modal trips.

Marketing actions, service planning and decision making also have to rely on information collected about specific groups in the population. These range from surveys of consumer satisfaction to more structured data collection about consumer preferences, also including surveys about attitudes of the general public and decision makers towards public transit. Specific surveys are required to address the special needs of those with some form of mobility restriction or impairment. Surveys are also used to identify new markets for public transit, such as the creation of premium services, which may also address the need to reduce car dependency and use.

Focused surveys are instrumental in modelling user behaviour, targeting market segments and identifying relevant service improvements to retain current users, to attract new users and to strengthen the revenue/cost ratio of operators.

20.5. Strategic Research Needs

As noted by Sammer (2009), scientific literature about data collection and analysis methods specifically aimed at public transit does not abound. Nevertheless, the growing importance of public transit as a means towards sustainable transportation and the potential of ITS-generated data set the scene for new fields of investigation. The workshop discussions reported in previous sections raise some opportunities for research in methods to optimise the use of both automated and social data for public transit planning.

The need to cope with decreasing response rates for surveys in general presents a challenge, but also an opportunity in the case of public transit. While the large amount of data produced by new technology can provide a more complete representation of system use and users, survey techniques, when required, will justify additional research to address problems related to non-response. Table 20.1 shows a comparison of results of two different types of surveys conducted in two regions in

Table 20.1: Characteristics obtained from two different surveys in The Netherlands.

Characteristic	Level	Apeldoorn (written questionnaire)		North-Holland (Internet-based questionnaire)	
		Frequency	Percentage	Frequency	Percentage
Response rate			13.5		14.1
Gender	Male	343	50.7	144	22.7
	Female	328	48.5	491	77.3
	Unknown	5	0.7	–	–
Age	35 and younger	129	19.1	413	65.0
	Between 35 and 56	294	43.5	216	34.0
	56 and older	219	32.4	4	0.6
	Unknown	34	5.0	2	0.3
Public transit card	Yes	144	21.3	83	13.1
	No	532	78.7	552	86.9
Education	Low	163	24.1	91	14.3
	Medium	230	34.0	275	43.3
	High	273	40.4	269	42.4
Car availability	No car	58	8.6	28	4.4
	One car	399	59.0	271	42.7
	More than one car	219	32.4	336	52.9
Public transit use (general)	Rarely	86	12.7	155	24.4
	Sometimes	403	59.6	344	54.2
	Often	179	26.5	136	21.4
	Unknown	8	1.2	–	–

Source: van der Waerden et al. (2008).

the Netherlands, with the purpose of constructing a map of potential demand for public transit (van der Waerden et al., 2008). As discussed by the authors, the results clearly indicate a bias related to survey methodology, with considerably different sample composition in each case. These difficulties can be tackled in two ways: the improvement of data-collection methods aiming to increase response rates, or, on the other hand, the development of analytical techniques to minimise the bias resulting from non-responses. While this is a general problem, in the particular case of public transit a combination of survey data with automatically generated data can be a promising research avenue.

Where ITS technologies are in place in public transit networks, the data produced should be meaningfully combined with administrative data (system characteristics), land-use and census data, in addition to specific tailored survey data. Data fusion methods and techniques, with the objective of combining different data sources to produce desired information, as exemplified by Chapleau et al. (2008), have a large potential for public transit. Results can provide a fuller description of the use of public transit systems and deeper insights on travel behaviour (see, for example,

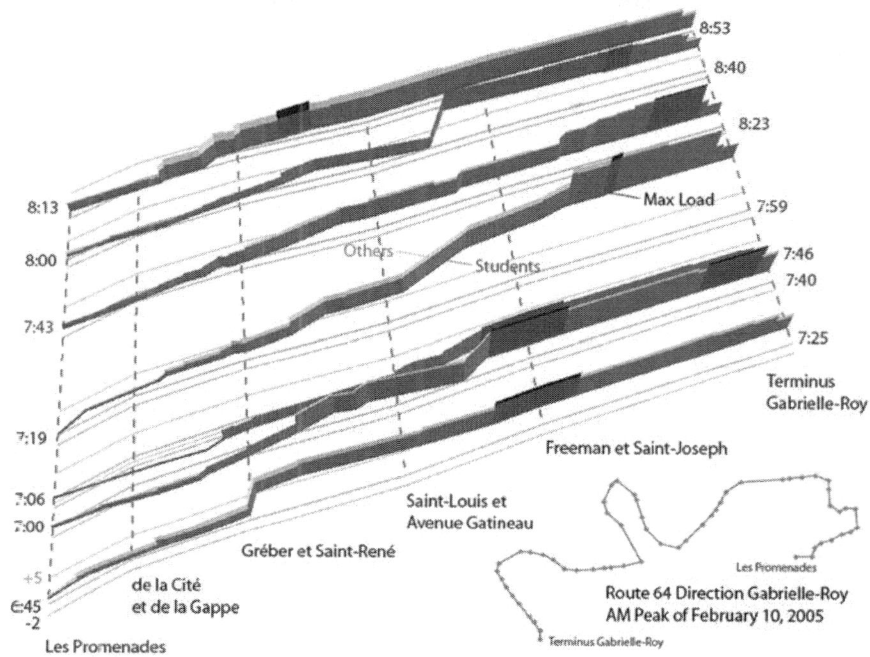

Figure 20.1: Synthesis of time–space diagrams and load profiles of a single route derived from smart card automated fare-collection system data (reprinted with permission).

Figure 20.1). The availability of multi-day data (from fare-collection systems) also provides the opportunity to enrich behavioural analysis and to identify latent demand from current users. There is thus a need to investigate the application of ITS for public transit, especially on-board passenger counting systems, fare-collection systems and fleet management systems. At least the following elements must be examined in this regard: type of technologies, structure of collected data and compatibility of collected data with other systems.

Finding new potential markets is a key to the financial sustainability of public transit systems. Household surveys provide relevant information to address the data requirements for such analyses. However, the high cost and non-response associated with conventional household surveys call for innovative approaches. The use of the Internet as a tool to identify potential demand for public transit, as proposed by van der Waerden et al. (2008), presents an interesting alternative. However, research is needed to identify methodologies for control of sample results to minimise bias, as noted above.

As indicated in the previous section, surveys (household surveys in particular) can focus on topics of specific interest for public transit. In countries where the use of public transit is low, sampling methods should be designed to guarantee that minimum required information on the use of transit modes is obtained. In all cases, procedures should be developed to obtain and record the information related to public transit use at higher levels of spatial resolution, as well as methods to analyse the information at this level of detail.

20.6. Conclusions and Recommendations

Whether it is for system planning, modelling travel behaviour or promoting the use of public transit, the collection of suitable data is still not a single task. The issues reported in this workshop can be synthesised in three major areas that require research focus in the next several years:

- *Development of ITS for public transit.* The technologies related to surveillance, management, control and reporting of public transit operations are evolving very rapidly and are increasingly adopted in many systems. These devices collect a lot of data, and there is a need to influence the design of ITS in order to optimise their use in the public transit field, looking in particular at the issue of data compatibility. There is a also a need to identify what distinguishes transit ITS from other ITS technologies.
- *Improved techniques for survey data collection.* Active data collection is increasingly hard to conduct, especially for surveys. Research has to cope with concerns about data quality (response, sampling and validation). From the perspective of public transit, household surveys should include specific questions on route use, transfer points, form of payment and other disaggregate information. Research is needed to compare the accuracy of household survey data about use of public transit with the results of on-board passenger surveys. The size of sample needed to ensure the validity of household surveys as a contribution to public transit planning is not known. In addition, privacy concerns are arising with the development of new technologies.
- *Data integration and harmonisation.* There is a need to better integrate the data coming from all ITS devices, AFC, APC, AVL and other sources. Data may exist but may not be available in a form usable by public transit planners and researchers. Data must also be collected from external ITS devices (such as those on road networks) and other sources such as census, land use and transportation infrastructure inventories, etc. Public transit must be seen as a part of the whole transportation network, and partnerships must be developed with researchers working with other modes of transportation.

Finally, the business practices of public transit operators may need to be changed or reviewed to better assess the new context in which they are inserted. To achieve this task, organisational changes may be required, in association with the new data needs.

Funding for better public transit, in particular for the introduction of ITS technologies with the potential for data recovery, must compete with the demands of other transportation modes, including roads infrastructure. The importance of public transit in the move towards sustainable transportation systems and improved quality of life will hopefully be a consideration when choices must be made concerning the public transit system.

Acknowledgements

The workshop participants were Bruno Allard, Canada; Daniel Bergeron, Canada; Robert Chapleau, Canada; Linda Cherrington, USA; Itzhak Ditzian, Israel; Regine Gerike, Germany; Jane Gould, USA; Dana Gruschwitz, Germany; Alban Igwe, Nigeria; Kim McDonough, Canada; Marcela Munizaga, Chile; Tim Raimond, Australia; Jean-Louis Routhier, France; Gerd Sammer, Austria; Orlando Strambi, Brazil; Maisa Tobias, Brazil and Martin Trépanier, Canada.

References

Chapleau, R., Trépanier, M., & Chu, K. K. (2008). The ultimate survey for transit planning: Complete information with smart card data and GIS. Presented at 8th International Conference on Transport Survey Methods, Annecy, France, May 2008.

Sammer, G. (2009). Identifying and reconciling the data needs of public transit planning, marketing and performance measurement. In: P. Bonnel, M. Lee-Gosselin, J. Zmud & J.-L. Madre (Eds), *Transport survey methods: Keeping up with a changing world*. Bingley, UK: Emerald.

Strathman, J. G., Kimpel, T. J., Broach, J., Wachana, P., Coffel, K., Callas, S., Elliot, B., & Elmore-Yalch, R. (2008). *TCRP Report 126: Leveraging ITS data for transit market research: A practitioner's guidebook*. Washington, DC: Transportation Research Board.

van der Waerden, P., Timmermans, H., & Berenos, M. (2008). Data collection approaches for public transit marketing: Some examples. Presented at 8th International Conference on Transport Survey Methods, Annecy, France.

PART V

TECHNOLOGY APPLICATIONS

Chapter 21

Collecting and Processing Data from Mobile Technologies

Peter R. Stopher

Abstract

In the recent past, mobile technologies that track the movement of people, freight and vehicles have evolved rapidly. The major categories of such technologies are reviewed and a number of attributes for classification are proposed. The willingness of people to engage in such technologically based surveys and the reported biases in the make-up of the sample obtained are reviewed. Lessons are drawn about the nature of the samples that can be achieved and the representativeness of such samples is discussed. Data processing is addressed, particularly in terms of the processing requirements for logged data, where additional travel characteristics required for travel analysis may need to be imputed. Another issue explored is the reliability of data entered by respondents in interactive devices and concerns that may arise in processing data collected in real time for prompting or interrogating respondents. Differences, in relation to the data user, between data from mobile devices and data from conventional self-report surveys are discussed. Potentials that may exist for changes in modelling from using such data are explored. Conclusions are drawn about the usefulness and limitations of mobile technologies to collect and process data. The extent to which such mobile technologies may be used in future, either to supplement or replace conventional methods of data collection, is discussed along with the readiness of the technology for today and the advances that may be expected in the short and medium term from this form of technology.

Transport Survey Methods

ISBN: 978-1-84855-844-1

21.1. Introduction

In each of the past three conferences in this series, there has been something included in the conference on the topic of mobile technologies. In the 1997 conference, the topic of mobile technologies being applied to travel activity measurement was sufficiently new that a workshop was not devoted to it, but there was, instead, a paper on the topic in a session on 'New Technologies' (Murakami, Wagner, & Neumeister, 2000). By the next conference in 2001, a workshop was devoted to the topic of 'Using Technology to Improve Transport Survey Quality', with two papers that were published on the topic, and a workshop summary (Wermuth, Sommer, & Kreitz, 2003; Wolf, Loechl, Thompson, & Arce, 2003; Murakami, Morris, & Arce, 2003). In the following conference in 2004, there were two workshops devoted to the topic, one on Web-based and one on non-Web-based technologies, the latter being one that was primarily orientated to mobile technologies, again with two published papers and a workshop summary (Wolf, 2006; Kracht, 2006; Lee-Gosselin & Harvey, 2006). In this conference, there are three workshops that dealt with some aspect of mobile technologies. This progression from one paper in a paper session in 1997 to three workshops in 2008 clearly illustrates the rapidity of the evolution and adoption of mobile technologies for measuring travel behaviour. In this book, there are concerns being raised about non-response in surveys using mobile technologies, the design of surveys using some form of mobile technology and this chapter which focuses on the collection and processing of the data from such surveys.

Principally, there are two technologies that have been the focus of most of the work on developing mobile tracking technologies to inform on people's travel: global positioning system (GPS)-based devices and mobile telephone-based devices, generally using GSM (Global System for Mobile communications). The former use 24 satellites currently orbiting the earth at a high level and compute position from triangulation to at least 3 or 4 in-view satellites, whilst the latter use triangulation and signal strength from base stations to determine position. These two technologies can, themselves, be combined or incorporated into various secondary technological devices. For example, GPS receivers can be used as stand-alone devices, or can be incorporated into a personal data assistant (PDA), a portable computer or a mobile telephone (Murakami & Wagner, 1999; Doherty, Noël, Lee-Gosselin, Sirois, & Ueno, 2001; Kochan, Bellemans, Janssens, & Wets, 2006). Likewise, GSM can be used as a stand-alone device. The Personal Handy-Phone System was developed and tested in Japan to provide positioning information to within ± 60 m and has been used as the underlying technology for the P-POINTS system (Personal Positioning Information System), which uses Pico cells to improve accuracy (Shibuya, Nakatsugawa, Kubota, & Ogawa, 2000). GSM can also be incorporated into a PDA or other communications devices, and may be combined with a GPS device (e.g. Ohmori, Nakazato, Sasaki, Nishii, & Harata, 2006; Hato, 2006; Itsubo & Hato, 2005).

At present, the principle difference between GPS and GSM devices is the accuracy of positioning and the availability of signals for positioning. GPS devices generally still require a clear view of the sky, although increasing sensitivities are permitting these devices to obtain progressively more reliable position information when there is no clear view, and even when some amount of structural material lies between the device and the sky. Nevertheless, they still do not record in tunnels, and experience some problems in urban canyons — locations where the device is surrounded by nearby tall buildings. However, GPS devices will usually work equally as well in rural areas as in urban areas, and are relatively unaffected by the density of development in the vicinity. So far as accuracy is concerned, GPS devices today are usually able to provide position information to within ± 5 m or less. GSM devices, on the other hand, will provide the most accurate positioning in dense urban areas, where base stations are closely spaced, and will provide little or no position in remote areas where base stations either do not exist or are extremely widely spaced. GSM devices will provide position in tunnels and other locations where GPS struggles or is unable to provide a position, including inside buildings. However, even in dense urban areas, position information from GSM devices is usually not much better than ± 40 m, and may be in excess of 100 m even within suburban areas (Varshavsky et al., 2006). For this reason, when it was mandated in the United States that mobile telephones should be able to report the position of the person using a mobile phone to dial emergency services, most telephone manufacturers added a GPS chip to the mobile telephone, so that more accurate positioning would be possible. Apart from these basic features of GPS and GSM positioning, there are a number of attributes that could usefully be used to classify mobile technologies and which are useful in considering the issue of collection and processing of the data from these devices. In the next section, these attributes and classifications are put forward.

There is another new positioning system that is emerging at present and which could potentially replace GPS and/or GSM positioning in due course. This is Wi-Fi positioning. Recently, there have been releases of software that uses Wi-Fi to locate positions. The device that can be used for this is a PC, laptop, Tablet PC, smart phone or RFID tag. The more densely an area is populated with Wi-Fi signals, the more accurate is the positioning. In a recent paper (Cheng, Chawathe, LaMarca, & Krumm, 2005), it was found that position could be fixed to within 13–40 m, depending on the characteristics of the local environment. One of the advantages of Wi-Fi location over GPS is that it works well indoors and does not require line of sight to the sky, and is not subject to the reflection of signals that is a problem for GPS in dense urban areas. However, the technology has not yet received any sort of widespread implementation, so is somewhat a futuristic concept at this time. It is noted here for completeness in discussing mobile positioning technologies, but is not dealt with further, since there are no known applications of it in transport research to date.

Position can also be determined using a radio frequency identification (RFID) tag. These tags can be passive or active. Passive tags have no battery and are provided with energy from a reader or writer, such as a cash register or a mobile telephone, which incorporates the RFID tag. The passive tag does not appear to have a potential use for locating travellers. An active tag has its own power source and has a much longer communication range than the passive tag. It is costly, at present, so has not received serious consideration as a locational mobile technology. However, since a tag ID can be extracted from the tag signal, there is a possibility to use it to determine the location of a user (Yamada et al., 2005). However, this technology does not appear at this time to be a serious contender.

For a more comprehensive overview of mobile technologies that may be applicable to transport and travel surveys, the reader is referred to a report by Wolf, Bonsall, Oliveira, Leary, and Lee (2006). This report provides quite a comprehensive review of what was available in 2005–2006, although it should be noted that significant advances, at least in GPS, have occurred since the publication of that report.

21.2. Classifying Mobile Technologies

As used in current practice, there are a number of attributes that differ in the devices used. These attributes will have greater or lesser effects on the collection and processing of data. In this section, the focus is on devices used specifically for tracking person movements. There are, of course, an increasing number of situations in which data are collected automatically, such as from mobile telephones, credit cards, fare payment cards etc., from all of which data could be used and developed for tracking purposes, and for some of which data may already be commercially available or available to government agencies. These are not considered further in this section.

21.2.1. In-Vehicle or Portable

The first attribute suggested here is whether the device is designed for in-vehicle use alone, or is a portable device to be carried by a person. In the case of an in-vehicle device, it is not necessary for the device to have its own power source, because it can be connected to the vehicle battery. Further, the bulk and weight of the device is relatively unimportant, and installation can be either a simple plug in to the accessory slot in the vehicle, or can be a more permanent installation in the boot of the car or elsewhere. The in-vehicle GeoLogger®, developed by GeoStats, is an excellent example of an in-vehicle device and has been used extensively in the United States and elsewhere. The paper by Bradley, Wolf, and Bricka (2005) describes and shows this device. This in-vehicle device represented a major breakthrough in the development of usable GPS devices for a major household survey, following the Lexington proof-of-concept equipment (Wagner, 1997).

In-vehicle devices can also be set up to capture other information from the vehicle, such as speed, acceleration, engine condition, etc. It is also relatively easy to add to an in-vehicle device some form of dead-reckoning secondary device, such as a gyroscope or inertial system that will provide position information when the primary technology (GPS or GSM) is compromised. A good example of a GPS in-vehicle device that taps into other information is the one used by Georgia Tech in the

Commute Atlanta project (Jun, Ogle, & Guensler, 2007). Laptop computers and other devices may also be included in this category, especially when equipped with either or both GPS and GSM technology to provide positioning information.

Portable devices, on the other hand, are required to be small and lightweight, must incorporate their own power supply and are generally not able to incorporate other functionality, unless it is very small in size and light in weight and can be placed inside a common housing. For portable devices, the size and weight and the duration of battery power have been key issues in their development. Early devices, such as the bicycle device developed in the Netherlands (Draijer, Kalfs, & Perdok, 2000), were bulky and heavy, with the first one weighing in at a rather substantial 2 kg (including batteries). Since that development, however, personal, portable devices have become substantially smaller and lighter in weight, and battery life issues, which dogged most of the early designs — requiring either a battery with sufficient bulk and weight to compromise somewhat the portability of the device, or having a much smaller and lighter battery that did not provide adequate battery life — have largely disappeared as power requirements of the GPS devices have diminished and battery technology has improved.

GSM devices have largely been adaptations of conventional mobile telephones, with size, weight and battery life attributes that are similar to the mobile telephone, although requirements to transmit position data frequently, or to perform other functionality have been reported to reduce battery life significantly. Other portable devices include PDAs equipped with a GSM or GPS chip that allows positioning data to be obtained. These are usually constrained to the available dimensions of current PDAs on the market, together with the addition of, possibly, a GPS receiver.

21.2.2. Interactive or Passive

Devices can be interactive, in which case the user is expected to enter data that will be recorded along with the position records. This normally requires combination of the GPS or GSM device with a PDA, laptop computer or other data entry device. The first GPS devices to be tested in the Lexington proof-of-concept project coupled a PDA with a GPS receiver and logger (Wagner, 1997). Respondents were asked to enter such information as the identity of the driver of the vehicle (this was an in-vehicle device), the purpose of the trip and the identity of other persons accompanying the driver. Similarly, work on the CHASE method involved the development of an electronic questionnaire that was loaded into a PDA to supplement position data collected by GPS (Doherty et al., 2001). There are many other examples of interactive devices, but most have in common the idea of linking a PDA or laptop computer to a tracking device.

Alternatively, devices can be passive, meaning that respondents/users are required simply to carry the device with them or place the device in a vehicle, possibly also being required to turn the device on and maybe turn it off. No data entry is required of the respondent. Passive devices may also include either firmware that turns the device off and on, depending on detection of movement or GPS signals, or a response device that detects movement, such as a vibration sensor, which turns the device off and on.

While not a feature of the device itself, another aspect of the interactive versus passive classification comes from the manner in which data are collected. Using a passive device, the data may be collected and then processed without further reference to the user, or the user may be asked to respond to a prompted recall survey (Stopher, Bullock, & Horst, 2002; Stopher, Collins, & Bullock, 2004; Lee-Gosselin, Doherty, & Papinski, 2006; Tsui & Shalaby, 2006; Wolf, 2006; Stopher, FitzGerald, & Xu, 2007b; *inter alia*). In a prompted recall survey, respondents are provided with maps showing the travel recorded by the GPS and are then asked to respond to a series of questions about the travel, including such information as the mode(s) of travel used, the purpose of each trip, the identity and number of accompanying persons and possibly costs related to each trip. In the evolution of these methods, increasing use is being made of the Internet as the means to both provide the maps and obtain the input information.

21.2.3. Real-Time or Logged Data

Another attribute that may vary between devices is whether data are logged or are transmitted in real time. Given GPRS and other facilities, the technology clearly exists to provide position data in real time. However, there are rather restricted needs for real-time data in most applications. Clearly, in studying travel behaviour of people or vehicles, real-time data would only be needed if there were also a need to be able to communicate with the respondent to ask questions about a particular activity happening at a particular moment in time. To this author's knowledge, such situations have largely not arisen so far, although there has been discussion in various forums of the potential to do this. At the same time, real-time tracking of individuals could be seen as a major invasion of privacy and could lead to rather poor response rates. Given such concerns and the relatively high costs of implementing real-time data collection in many countries around the world, real-time data collection is not a major priority. There are examples, however, of studies that have used this.

Almost all applications in North America, the United Kingdom and Australia have used devices that log data. This is because of both the lack of justification for real-time data, and the cost of collecting it. However, this also then requires that the GPS or GSM device have a rather large memory capacity, which has resulted in off-the-shelf devices generally not being applicable, because they are not designed with large internal logging memories. In a recent search for small GPS devices that are available at retail, we were unable to find anything that could log more than about 1000 data points, which is not adequate for even one day of second-by-second tracking of individual travel behaviour. An average of about an hour of travel per day would produce 3600 second-by-second data points at a minimum assuming that the device is off whenever the person is not moving. Most devices currently in use are logging devices that are, to a greater or lesser extent, custom built for the purpose of tracking individuals or vehicles. Memory requirements for effective logging devices are a minimum of about 4 MB of memory dedicated to recording position information. For one day of data, this is much more than what is required, even for someone who travels as part of their work, but it does ensure that even outlier travellers are recorded completely, and then allows for multi-day data, which is one of the major advantages of using GPS to collect travel data.

21.3. Likely Sample Characteristics for a Survey Using Mobile Technologies

While there have been a significant number of GPS surveys conducted, there are relatively few reports about who responds and who does not respond to a GPS survey, perhaps because, while this factor may be important to analyse, time and resources have often not been available to undertake such an analysis. The first instance of a report on the characteristics of respondents versus non-respondents seems to be a paper by Hawkins and Stopher (2004). They concluded that there were distinct biases between those who responded to the GPS and the conventional travel survey, versus those who were unwilling to take the GPS devices. Hawkins and Stopher (2004) found that there were significant differences on 6 of 12 demographic attributes examined. The following were found to be significantly less likely to accept GPS devices:

1. Country of origin: non-English–speaking respondents;
2. Family structure: couples with older children;
3. Students: those in secondary education;
4. Low income earners;
5. People without a driver's license; and
6. Household size: those with large households.

However, in this study, the sample size was small (48 households), so the findings of this study may not be reliable.

In subsequent work by Stopher and others, different results have been found, mainly in comparisons of sociodemographic data to census data. Stopher, Clifford, and Montes (2008a) provide comparisons of a GPS panel recruited in 2005 and run through three waves (one wave in 2005 and two waves in 2006). In this panel of nominally 50 households, there was an underrepresentation of one-person households and an overrepresentation of four-person households. Unlike the earlier work of Hawkins and Stopher, there was not an underrepresentation of larger households. There was an underrepresentation of zero car owning households, which may be allied to the low levels of one-person households, and could also partially be due to changes in the region between 2001 and 2006. There were more children and more full-time students in the GPS sample than in the census data, which again is allied to the low level of one-person households. There was also a difference in the number of full-time workers per household, but this is very likely a result of the difference in time between the census and the GPS panel, since there were significant changes in unemployment rates over the period.

In two other studies, as yet unpublished, Stopher and others have found further information on those who respond to a GPS survey compared to those who do not. In a validation survey for the Victoria Integrated Survey of Travel and Activities (VISTA), no significant differences were found between households that accepted GPS devices and households that did not, except on length of residence. In this case, households that accepted GPS devices tended to have resided at their addresses longer than those who refused the GPS devices. No other significant differences were found. In a much larger panel of 200 households, it was found that one-person households were underrepresented, while four-person households were overrepresented, leading to a slightly higher than average household size, number of adults and number of children per household. There was also a substantial underrepresentation of non-car-owning households and overrepresentation of households with two or more cars (Stopher et al., 2008a).

In none of the studies discussed thus far were incentives used. The protocols were somewhat different in the studies. In Hawkins and Stopher (2004) and the VISTA study, households were first recruited for a standard household travel survey by interviewers knocking at the door of the household, and were subsequently asked if they would also be willing to undertake a GPS survey. If they were willing, then the GPS device was provided at that time. In the other studies discussed here, the protocol was to telephone the households, following the mailing of a pre-notification letter that explained the purposes of the survey, and recruit the household into a panel of GPS households. If the household agreed to participate, then the GPS devices were sent to the household by courier and subsequently collected by courier. Households were aware from the outset that they would be asked to repeat this process at either a 6- or 12-month interval.

Bradley et al. (2005) report on an analysis of who responded to a GPS validation survey conducted with the Kansas City Regional Household Travel Survey in 2004. In this survey, where the diary survey was conducted using telephone recruitment, mailing of diaries and CATI retrieval of data, the authors found that those who participated in the GPS component of the survey were more likely to have more workers per vehicle and higher incomes, while older households and those with fewer vehicles were less likely to participate. However, they also found that the strongest determinant of participation in the GPS survey was the number of diary vehicle trips per adult, with households making more trips per adult being more willing to participate in the GPS survey. This survey used in-vehicle GPS devices, rather than the personal portable devices of the Australian surveys.

It appears likely, from the limited data reported on GPS and similar surveys, that the sociodemographics of a GPS sample will not be markedly different from those of a conventional travel survey. As happens in conventional travel surveys, various concerns by individuals who live alone is likely to lead to underrepresentation of one-person households. On the other hand, the evidence to date seems to suggest that large households, which are often underrepresented in conventional diary-based surveys, may not be underrepresented in GPS and similar surveys, especially those that use passive devices. The probable reason for this is the size of the task required of a large household, when diaries are to be completed by or on behalf of each family member; by contrast, the requirement simply to carry a GPS device for each member (over a specified age) of a large household is likely not seen as burdensome. Hence, such households are more likely to be willing to undertake the survey. It does appear that GPS surveys will find it similarly difficult to persuade non-car-owning households to participate. Generally, conventional surveys have found this to be a problem, and the data available so far suggests that there is likely to be a similar problem in GPS surveys. This is more likely attributed to the survey being a transport survey than to being a function of the GPS device.

One other issue relating to the characteristics of respondents relates to the ages of those who are tracked. In surveys that use only in-vehicle devices, there is no direct issue with the ages of respondents. However, in surveys using portable devices, there is an issue relating to the ages of those who are asked to participate. In the surveys already mentioned in Australia, as a result of ethics considerations, principally, the ages of respondents have been limited to those over the age of 14 years. There are concerns that asking respondents much younger than this to carry GPS devices could be liable to issues relating to whether or not the devices would be carried at all times, whether devices would be 'borrowed' by other children and whether or not devices would be subject to serious damage during their use by younger children. One recent study of schoolchildren, however, had success in having children aged 8–11 years old carry portable GPS devices (Mackett, Brown, Gong, Kitazawa, & Paskins, 2007). In this case, the device was a wristwatch-style device, which the child carried on his or her wrist. Possibly one of the major factors in the children keeping the devices to themselves was that sampling for this study was school-based, so that there would have been a significant number of children in the specific years and schools (two schools were the subject of the paper), so that the tendency for children to lend their devices to others was probably minimised. In contrast, in an urban study using a small random sample, equipping children of this age range with GPS devices could well lead to a situation in which most participating children would be the only ones in their school and year with a GPS device to carry, leading to much curiosity, probable experimentation and 'loaning' of the device to close friends and others. This is an issue that will require further research to determine the age limit on use of the devices. However, the issue here is clearly in contrast to conventional surveys, where traditionally everyone over the age of five years was asked to complete a diary or have one completed for them by a responsible adult, and more recent trends, where diary surveys have extended to everyone, regardless of age.

21.3.1. Recruitment and Response Rates

There are some difficulties in considering response rates and recruitment. The vast majority of GPS surveys reported to date are validation surveys that have been undertaken in conjunction with a conventional household travel survey. As such, it is normally the case that respondents are first recruited to the household travel survey and are subsequently asked if they will also participate in a GPS survey, where attempts are made to ensure that no one drops out of the diary survey because they are asked to undertake the GPS survey. There are only limited instances so far of 'pure' GPS surveys. Clearly, response rates to the validation surveys will be influenced by the fact that the household has already been asked to commit to undertaking a diary survey, which often represents a significant respondent burden. There is also an important issue concerning the use of incentives, and the fact that receptiveness to incentives may vary from country to country.

In terms of response rates, the record seems to be somewhat mixed. While a response rate was not reported, the proof-of-concept test in Lexington, Kentucky in 1996 used a $50 incentive per household (Wagner, 1997). The literature is rather quiet on the response rates achieved. In Australia, for recruitment into three-wave panels, Stopher et al. (2008a) reported a response rate of around 28% to a telephone recruitment to undertake a GPS panel survey involving taking a GPS device for one month every six months. In a three-year, annual panel survey with GPS, an initial recruitment rate of 34% was achieved, where the goal was to recruit 150 households. A total of 167 households agreed to undertake the panel survey, but 15 (9%) of those households did not comply with the requested task, and 151 completed the GPS task. Neither of these panels were offered any incentives. This is a somewhat better recruitment rate to that experienced in other surveys in Australia, using telephone recruitment, but for a one-off, cross-sectional survey. In those surveys, the response rate is generally in the range of 20–25% for a similar recruitment method. In general, the recruitment in subsequent waves, intended to make up attrition in the panel, produced very similar response rates in the range of 25–35%. Anecdotally, it has been reported that the French National Survey, which is currently in the field and has a GPS component, was able to get about one-third of respondents to agree to take GPS devices (Murakami, 2008). Similarly, in Victoria, Australia, in the validation study for VISTA, about one-third of the households recruited to undertake travel diaries were also willing to take GPS devices (98 households out of 306). Of the 306 households recruited to complete travel diaries, 233 actually provided usable data, while 76 out of 98 households recruited for the GPS component provided some usable data. Thus, the completion rate for the diaries was around 76%, which is almost identical to the 77% completion rate for the GPS survey component.

At this point, with the information available, it appears that recruitment rates into a GPS survey, if done alone, are similar to recruitment rates into a more conventional survey, such as a paper and pencil or CATI survey. It appears that willingness to do both a conventional travel survey and a GPS survey is closer to about one-third of the sample that is initially recruited for the diary survey, while completion rates are around 75% for both types of survey. However, reports on the actual response rates achieved are difficult to uncover.

Evidence is not available on any technology other than GPS for the response rates and the nature of the sample collected. It is likely that there is considerable information available in many of the GPS surveys that have been conducted that has simply not been analysed or has not been reported to date. However, information that has been obtained to date relates to two different scenarios. The only cross-sectional GPS surveys that have been undertaken for which response rates and the make-up of the sample are reported are surveys that have been undertaken to validate a more conventional survey using some form of travel diary. In these cases, respondents are asked to take a GPS device *in addition to* completing a standard travel diary survey. Most of these surveys have been undertaken using a mailed CATI survey, with one case of a face-to-face survey (Hawkins & Stopher, 2004) and one case of a face-to-face recruitment for a self-administered survey to be collected by interviewer (VISTA). In all cases where a report has been provided of the recruitment rate for these cross-sectional surveys, the GPS survey has been accepted by about one-third of the sample that has agreed to undertake the diary survey. It is not possible, from this, to estimate the likely response rate to a stand-alone, cross-sectional GPS survey.

In the case of use of a GPS device in a panel survey, the recruitment rates have been found to be on the order of 25–35% (Stopher et al., 2008a). In contrast to a much simpler odometer panel, where the retention rate for a panel survey that was repeated on a three to four month cycle was around 90%, the retention rate for the GPS panels was found to be closer to 75% for an annual or six-monthly survey. Overall, this suggests that the recruitment rate for a GPS survey would be comparable to a conventional diary survey. In terms of retention in a panel, this also appears to produce figures similar to what would be expected in a conventional panel. Other transport panels, such as the German Mobility Panel, have shown about a 30% attrition rate for an annual survey (Zumkeller, Madre, Chlond, & Armoogum, 2006).

21.4. Processing the Data

In this section, the focus is on data from GPS devices, for which there are an increasing number of references in the literature. Processing data from GSM applications presents greater difficulty and literature is not as readily available. Key among the points for mobile telephone data are the fact that position is recorded much less frequently than with dedicated GPS devices, and that position is not already recorded in latitude–longitude terms, but must first be geocoded. Hence, the processing software will be more complex than that for GPS devices, which are discussed in detail in this section.

21.4.1. Removing Invalid Data Points

Most surveys using mobile technologies involve the collection of a large quantity of data. Frequently, the data range from second-by-second recording of position to recording position every 5 or 10 s. If a device records continuously for a 24-h day, this can result in as many as 86,000 records for a single day of travel. Multi-day data will, of course, result in much larger numbers of data points than this. The author has routinely collected 7 days of data from most respondents, with some surveys being for as many as 28 days. The amount of data from each GPS device can, thus, be very large indeed. Given that the average person actually travels for around 70 min per day, the first task is to sort out those records that represent valid travel and those points that relate to staying in a fixed position. With GPS in particular, even a stationary device will record apparent movement, simply as a result of the imprecision of the position fix. This results in the position 'wandering' usually over distances of up to 30 m or more. In addition, many GPS receivers will also record data points even when the signal is compromised, such as by having too few satellites in view, or by having the satellites in view aligned in such a way as to result in very imprecise position location.

Hence, the first task that must be undertaken is to remove invalid data points and those data points that represent no movement. Almost all processing software developed to date for GPS processing use two items commonly included in the NMEA data stream to sort out valid and invalid data points: the horizontal dilution of precision (HDOP) and the number of satellites in view. If fewer than four satellites are in view, this generally will result in an unreliable position location. Most GPS devices have built-in firmware that interpolates position after the number of satellites in view decreases below four. This position interpolation may continue for a number of seconds, and usually assumes that speed and heading remain constant from the moment before the number of satellites in view decreased below four. While such interpolation may be useful when a GPS device is used as a navigation device, it is not particularly useful for transport modelling and planning uses of the data, and such interpolated points need to be removed. Similarly, a high value of HDOP indicates that the satellites in view are aligned more or less in a straight line or are dispersed around the horizon, either of which conditions gives rise to an unreliable position. There is less agreement on what value of HDOP represents an unreliable position. In the author's work, a value of 5 is considered to be the maximum acceptable, and any value greater than 5 is assumed to be unreliable, and so is removed (Stopher, Jiang, & FitzGerald, 2005).

21.4.2. Identifying Trip Ends, Vehicle Stops and Activities

Included in this section are two different processing approaches, which depend on whether the device is an in-vehicle device or a personal device. In the event that the device is an in-vehicle device, then trip ends are not actually identified. Rather, the analysis in this step will be to identify the car stage ends, where a stage is a continuous movement with one mode, including any stops purely for traffic control purposes. Hence, the purpose of this initial step of analysis with in-vehicle devices is to identify stage ends, which may include stopping the car to drop someone off or pick someone up, to order and pick up items at a drive-through facility, to stop to post a letter or package, or to drive through or stop to use an automatic teller machine or similar facility, as well as times when the vehicle is parked and the occupants of the vehicle leave the vehicle on foot and spend a significant time in an activity out of the car. In the event that the device is a personal device, then this step is to identify trip ends. There are some specific differences between these, which are discussed further in the succeeding paragraphs.

An important issue that has not received much attention to date in the literature is that of defining what constitutes a trip or a stop for the purposes of processing data obtained by means of mobile technologies. In traditional transport planning, for example, a trip to pick up or drop-off a passenger is typically considered to be a trip, especially because the location of such pick-ups and drop-offs may not be on the chosen path between the prior trip origin and the following trip destination, and because such activities also result in a change of vehicle occupancy. On the other hand, traditional transport planning has largely ignored such stops as posting a letter, buying a cup of coffee or a newspaper, etc., especially when such activities are accomplished without significant route deviation. The question needs to be addressed as to whether such activities represent activities that should be recorded as separate activities from the travel, so that a consistent definition of trip purposes can be achieved. As is discussed in the following paragraphs, activities in mobile technology surveys are most often defined by the time that elapses for the activity, in contrast to the implicit definition in transport modelling and planning, which is that an activity is significant to travel only if it involves a routing decision or destination choice. Unfortunately, this latter definition of an activity of significance does not lend itself to ready definition from the data available from a passive mobile technology. It is probably advisable to code all definable trips from the mobile technology device, and leave decisions on which trips to 'link' together to the analyst who may use the resulting data.

Once the data representing no movement and invalid points have been removed, the remaining data usually require significant processing to be useful. If the mobile technology is an interactive device, in which, among other things, respondents enter some indication of the beginning and ending of each trip or stage, then the next step is a relatively trivial one: dividing the data into individual trips and stops. In the case of an interactive device, there is usually some form of respondent-initiated time stamping that fixes the time of start and end of the trip, and it is necessary simply to correlate this with the position records, and thereby define the beginning and ending of the trip. Similarly, for in-vehicle devices, especially where the power to the device is stopped by turning off the ignition of the vehicle, the beginning and ending of vehicle trips may be quite evident from the data. In such cases, it will usually be indicated by the cessation of recorded points, followed by a re-starting of recording. On the other hand, some devices that are used in vehicles have their own battery power and use the vehicle power supply only to recharge the battery. In such cases, the times when the ignition is turned on and off will not, of themselves, provide any identification in the record to note the beginning and ending of a trip. This will also be true when the accessory slot is continuously powered. Of course, even where this activity can be used to define the start and end of a trip, not all trip ends of interest necessarily involve turning off the vehicle engine.

When passive devices are used, however, this requires the use of some software procedure to identify the beginning and ending of trips. From the literature, it appears as though a period of no movement (dwell time) of anywhere from 30 s to 2 min has been used to define a stop (Schönfelder, Axhausen, Antille, & Bierlaire, 2002; Chung & Shalaby, 2004; Wolf, Schönfelder, Samaga, Oliveira, & Axhausen, 2004; Forrest & Pearson, 2005; Bricka & Bhat, 2006; Li & Shalaby, 2008; Bohte & Maat, 2008; Stopher, FitzGerald, & Zhang, 2008b). The problem with setting the dwell time to a low value of, say, 30 s, is that this may result in a number of stops for traffic lights, stop signs or other manoeuvres being identified as trip ends. On the other hand, if the time interval of no movement is set to, say, 2 min, this can result in failing to identify as trip ends, locations where the activity to be performed takes less than 2 min to accomplish. Purchasing fast food, picking up or dropping off a passenger, posting a letter in a roadside post box and other similar activities are all likely to take less than 2 min to accomplish, with no movement of the vehicle or the person. As noted in the earlier discussion concerning what constitutes a trip, the problem here is that some of the activities that can be accomplished in a very short time may define a destination choice and a route choice, while other activities that may take longer than the time interval used to define a trip end may not have such a defining role in travel behaviour. So far as the author has been able to ascertain, notwithstanding this potential problem with what defines a trip end, the majority of software applications developed to date use a simple specification of the elapsed time of an activity, with the majority of these being in the range of 90–120 s. However, Bricka and Bhat (2006) report that, in Austin, TX, two different definitions were used for the length of time of no movement — 45 or 120 s. With the shorter dwell time, the trip-underreporting rate was found to be 31%, whereas it dropped to 12% when the longer dwell time was used. So far, this appears to be the only documented case of attempting to use different dwell times to define when a trip ends.

One aspect of GPS data that does not seem to have been taken into the processing so far, but which could potentially lead to some important insights in identifying trips or stages, as well as assisting in subsequent imputation efforts, is information on multiple observations of essentially the same trip. For example, it is conceivable that such data would allow more ready identification of the situation where a person drops a child at school on the way to work on some days and drives directly to work on other days, where the drop-off of the child may not take the prescribed 2 min, nor involve a significant deviation in travel, but may be identifiable by small changes in travel time, scheduling of the trip and routing among multiple days of data.

In most GPS surveys, although probably not with GSM surveys, another step that needs to be undertaken at around this point in the processing is that of converting times and dates in the data stream from UTC values to local values. In some instances, the devices are able to report the local time and date in the NMEA data stream, while in other cases, this must be calculated using appropriate software. For one-day data, this is principally necessary only so that the actual clock time when activities occur can be retained in the data, and to check that all activities are within the time frame of the study. In the case of multi-day data, however, this becomes significantly more important to allow the data to be subdivided by day and trip records to be compiled on a daily basis.

21.4.3. Producing Maps and Undertaking Prompted Recall Surveys

Usually, at this point in the processing, maps of the travel and trip ends can be produced. These maps may be used for visual checking of the processing to date, or may be used as a part of a prompted recall survey. The visual checks at this point will identify several situations that may require correction to the data. For example, one of the problems with GPS devices is that of a cold start in recording position, meaning that the device does not begin recording valid positions at the outset of a new trip. This will show up in the data as a gap between the ending of one trip and the beginning of the next. It may also appear as a start of the first trip of the day at some distance from home, whereas it did, in fact, originate at home. There may also appear to be a gap at some point during the day, where for one reason or another, an entire trip was not recorded. This may result from an extended cold start problem with the device, or from the respondent forgetting to turn the device on, or from other problems with complete data.

On the other hand, some stops that are shown to be identified through the rules used for the minimum duration of an activity could result in identifying a spurious stop. Such stops may be evident from the fact that the location of the stop is actually on the road network, and occurs close to or at an intersection or other location where an extended traffic stop may have occurred. At other times, it may be evident that an identified trip is not a trip, but is excessive position wandering from a stationary device. This will show up as a straight-line trip that is not on any transport network.

In the event that a prompted recall survey is used, the maps produced at this stage may be provided as they are to respondents, or obvious edits may be undertaken, and modified maps sent out to respondents. In some instances, the maps may be made a part of an Internet-based survey, while in others they may be mailed to respondents. In either case, there will also be some form of questionnaire provided that will ask specific questions about each trip shown. In work originally pioneered at ITLS (Stopher et al., 2004), live editing of the map may be made possible through the Internet (Li & Shalaby, 2008). In other applications, the respondent simply writes down details of what changes should be made. In fully passive surveys, where no prompted recall is employed, the maps will be edited based on a set of rules that are applied to the trips, such as identifying the spurious stops, missed trips, spurious trips etc. The appropriate files are amended and the data then used for the remaining processing steps.

A potential problem in this area is that of map matching. To a large degree, identifying trip ends alone does not require map matching capabilities, although this is necessary for the production of sensible maps of the GPS traces. If any further processing beyond determining the geographic trace of the GPS record is required, then map matching becomes a necessary next step. In addition, there can be problems with different GPS devices with respect to deriving such measures as travel distance for trips, because of the problems of lateral movement of position from the GPS trace. GPS devices are by no means 100% accurate, and it is not uncommon to find that the individual data points recorded wander away from and then back to the actual route alignment. The distance added by such inaccuracies in position location, especially when position is recorded every second, can add considerable amounts to the estimated travel distance (Ogle, Guensler, Bachman, Koutsak, & Wolf, 2002; Stopher, Bullock, & Jiang, 2002). These are difficult to resolve if the underlying GIS information on the transport network is inaccurate, which is often the case (Marchal, Hackney, & Axhausen, 2006; Quddus, Ochieng, & Noland, 2007). Not only are there errors in the street networks themselves, but good records for bicycle, walking and even rail trips are often difficult to obtain, so that correction of the GPS trace to the actual alignment on the ground is not possible.

Commercially available GIS maps are often a problem. However, in this author's experience, when doing work under contract to state governments in Australia, there are often good GIS sources available from government sources that provide not only the streets and bus and rail lines, but also the locations of bus stops. Such files are, however, often not available commercially as yet. Regardless, the accuracy of these GIS sources is often unknown and could compromise the results of the GPS processing to degrees that are not known.

21.4.4. *Imputing Mode and Purpose*

For in-vehicle surveys, there is no issue of identifying mode, and only trip purpose remains to be defined, along with vehicle occupancy. Using an interactive device, both purpose and occupancy will normally be among the attributes requested, in which case the processing of this information is again relatively trivial. If a passive device is used, then either a prompted recall survey will follow and identify purpose, or the procedures discussed later in this subsection will need to be employed.

Similarly, for an interactive personal device, or a passive device with a prompted recall survey following, the identification of mode and purpose, along with members of the travelling party (number and household or non-household membership) will usually be a part of the survey, and identification of mode and purpose is again comparatively trivial. In an interactive device, these details would normally be requested, while they will also be part of a prompted recall survey for a passive device.

The interesting and more challenging issue comes when a passive device is used with no prompted recall survey. This is the preferred method for conducting a GPS survey by the author of this paper. To deal with this situation, there are three steps necessary. First, it is essential to obtain as accurate and detailed a GIS as possible for all modes of transport. This should include a detailed road network, detailed bus networks (with bus stop locations if possible), rail networks and networks for any other forms of public transport, such as subway, ferry etc. In addition, networks for walk and bicycle should, if possible, be obtained. Second, in the author's surveys, respondents are asked to provide the workplace addresses for each workplace that is routinely used by any member of the household, the addresses for each educational establishment attended by any member of the household and the two most frequently visited grocery stores. In addition to these addresses, the home address will, of course, be known, because it is required for sending out the mobile technology devices and other survey materials. Third, to identify trip purpose, a GIS of land uses is preferred, supplemented by geocoding of most educational establishments, shopping centres and health-related complexes (hospitals, clinics etc.). The geocoded locations are generally important because respondents will often not know the precise address of some of the non-home locations requested, and it is necessary to have a good geocoded base list of locations, from which the locations reported by respondents can be identified.

Using these input data, it is then possible to develop software algorithms that can define the mode and purpose. The earliest effort at identifying trip purpose from GPS traces appears to be the work of Wolf, Guensler, and Bachman (2001), which pioneered the way for purpose definition. Subsequent work by Schönfelder and Samaga (2003) in Europe further developed the procedures to define trip purposes from GPS data. The idea of imputing trip purpose from land-use data was further explored in a Bayesian approach by Scuderi and Clifton (2005). In a research project aimed at examining the extent of physical activity in travel, Oliviera et al. (2006) explored the identification of mode of travel using information from both GPS and an accelerometer.

In software applications at ITLS (Stopher, Clifford, Zhang, & FitzGerald, 2007a), both mode and purpose are identified through a hierarchical process. For mode of travel, the hierarchy begins with walking trips, which are identified based on maximum speed and acceleration. Following this, off-network public transport trips are identified, using the appropriate network information. The next mode to be identified is bus, which is identified through maximum speed, acceleration, beginning and ending of the trip on a bus route, and also, when the GPS device permits this, recording of stops at points that correspond to intermediate bus stops. The remaining trips will be car and bicycle. Bicycle trips have proved to be the most difficult to pin down, and the ITLS procedure uses a probability score that is built up from bicycle ownership, origin of the trip being at home or at a location previously identified as the destination of a bicycle trip, maximum speed and acceleration. The remaining trips are assumed to be by car. If the destination of the previous trip was not served by car, then it is assumed that the trip is by car passenger. Otherwise, all trips are assumed at this point to be car driver. It has been proposed that we introduce further tests to determine if there are other household members with an identical trip by time, route, and origin and destination location, which would allow identification of additional car passenger trips. However, the programming for this has not yet been developed. Schüssler and Axhausen (2008) have developed an automatic approach to identifying mode, which was necessitated by the need to process around 30,000 person days of GPS traces. As they note, and further supporting the experience of Stopher et al. (2007a), certain modes remain difficult to identify clearly, especially bicycle and car and, in the work of Schüssler and Axhausen (2008), long distance train and car.

For trip purpose, many trips can be classified based on the information collected on the workplaces, educational establishments and shopping locations. Recalling that home is already known, also, this allows identification of most home-work trips, many home-education trips and the most frequent home-shop trips. Also, by noting if neither end of the trip is at home, we can identify a broad group of non-home-based trips. In instances where the respondent has been unable to provide an address, for example, someone who works at multiple locations throughout each day, or someone who simply does not know the street address of their workplace, recourse to the geocoded list of locations will often again produce information that will allow a trip purpose to be assigned. As with identifying trips, some decision is needed at this point as to the level of detail required for the identification of trip purpose. Many travel-demand models use relatively aggregate trip purposes, such as home-based work, home-based education, home-based other, non-home-based work and non-home-based other. In this case, except when the addresses were not provided by the respondents, there is already more than sufficient information to categorise every trip into one of the trip purposes. Even where the address is missing, some simple heuristics, based on frequency of visiting the site during the data collection period and duration of the visit, may identify a workplace for which an address was not provided.

Where a more detailed classification of trip purpose is desired, the GIS of land-use would be used. In this case, again, a combination of the frequency of visits, the duration of the visit and the type of land-use of the parcel visited will allow disaggregation of the home-based other category into home-based shopping, home-based social and recreational and home-based medical purposes. Trips for the purpose of eating a meal can also be identified to a large extent. Problems tend to arise in multi-purpose land-use parcels, such as a shopping centre, which may contain shops and stores, medical facilities, cinemas, and services such as banking, legal, real estate and insurance. In such cases, the duration and frequency of the visit may again provide some clues, although it is likely that the identification of purpose will not be as accurate as one might desire.

Work has been done on using artificial intelligence (AI) programming to identify both mode and purpose. However, there are relatively few reports in the literature of such applications to date. It would appear that relying solely on AI programming might not provide the best solution, but rather a combination of the heuristic processed described in this paper with AI to refine those situations where there remains some doubt as to the actual mode and purpose. To proceed along this path, it is necessary to have a substantial number of records in which the actual mode and purpose is known, as well as having the normal data associated with a mobile technology survey only. Such data are often available from the validation exercises that have been rather common in the past few years. It is to be hoped that the opportunity will be taken to improve the programming for mode and purpose for fully passive systems.

21.4.5. Respondent-Entered Data and Real-Time Data

As noted earlier, when the mobile technology involves an interactive device, the respondent is asked to input certain information as travel proceeds. While this may seem to be an optimal arrangement, combining the accuracy of positioning data and time from a mobile technology with respondent input of data items that cannot be recorded, it relies on two things: respondent accuracy in providing the inputs, and respondent memory to provide the inputs at the appropriate time. Both of these may prove to be both burdensome to the respondent and of poorer quality than would be desired. Certainly, based on past experience with self-report surveys, it has to be expected that people will forget to enter data for some trips, and will also report incorrect details from time to time. So far as current literature goes, there are few reports about these potential errors, so it is somewhat unknown to what extent this is a problem. It would seem that some research into this issue is desirable to indicate whether or not there is sufficient reliability in self-reported data in a PDA or other device, and whether the added burden of entering such data has significant impacts on the response rates and conscientious use of the mobile technology device.

A second issue of concern is the application of mobile technologies to real-time data collection. If the data are to be used to prompt respondents or to interrogate them during travel, several issues arise. The first is likely to be a much more serious concern with privacy issues of using mobile technologies. In the use of logging devices, it is this author's belief that much of the reason that privacy has not arisen as a major response issue is that there is a significant separation of time between when the data are actually gathered and when the survey agency processes the data. By the time that a prompted recall survey is undertaken, there may have been a lapse of days if not weeks from when the data were collected. This will often serve to allay fears and concerns related to privacy. On the other hand, if the mobile technology allows real-time prompting or questioning, while a person is engaged in travel, privacy concerns are likely to be significant and substantial.

A further issue that arises here is the interpretation of the data that are being transmitted in real time. In this author's experience, the time required to process data from a GPS device and to be reasonably sure of what one is looking at, in terms of travel, makes it is unlikely that data collected in real time can be readily interpreted to support real-time questioning of the respondent. Again, perhaps because real-time collection of data has been rather limited to date, we have not found any literature discussing this issue, so it becomes an issue to be investigated.

21.5. Differences between Data from Mobile Technologies and Conventional Travel Surveys

The intent of this section is to address briefly the differences between data from mobile technologies and data from conventional travel surveys, from the viewpoint of considering whether or not mobile technology data could replace conventional surveys. It is not concerned with the validation efforts of using GPS with a conventional travel survey, which is work that is extensively written up and discussed elsewhere (Wolf et al., 2003; Wolf, 2006; Bricka & Bhat, 2006), although it is worth noting in this respect that most of the comparisons that have been made directly between GPS data and conventional survey data relate to comparisons with self-report data, and only the work of Stopher et al. (2007b) relates so far to comparisons to an interview survey. In this respect, looking at the potential for mobile technology data to replace conventional diary data and be used for travel demand modelling, the major differences between conventional travel survey data and data from mobile technologies can be summarised as follows.

Mobile technology data provide much greater accuracy about the locations of origins and destinations than self-report surveys can, especially for the many addresses that people do not know well enough to report. They also provide very accurate information on the start and end times of travel, and provide data that has not generally been available from conventional surveys on the detailed route travelled by the respondent. The total travel time and distance are also provided with much greater accuracy than can be obtained from conventional surveys. Because of the low burden, especially of fully passive devices, it is also possible to collect multiple days of travel data from a random sample of households within any study area. The additional richness of data from such multi-day surveys has yet to be fully explored, although there are indications of the additional insights that can be gained from such data (Stopher et al., 2008a). However, there is extensive experience with analysing results of conventional multi-day surveys from self-report procedures, as opposed to mobile technology surveys (Pas, 1986; Hanson & Huff, 1988; Pas & Sundar, 1995; Axhausen, Zimmermann, Schonfelder, Rindsfuser, & Haupt, 2002; Pendyala, 2003; *inter alia*). On the other hand, mobile technologies generally cannot provide data on mode of travel, vehicle occupancy or purpose of travel. Such data must be collected either by an associated interactive device, a prompted recall survey or the use of a combination of heuristic and AI programming. Mobile technologies are also unable to collect data on travel costs experienced by travellers, and also provide no data on alternatives to the actual travel performed, nor information about perceptions, attitudes and preferences of travellers.

For the purposes of conventional travel modelling, which has been traditionally one of the major uses of data from travel surveys, the lack of mode, purpose, travel costs and party size constitute the major barriers to replacing conventional travel diaries with mobile technology surveys, especially using fully passive devices. To the extent that it can be shown that programming can provide mode, purpose and other details that are required for modelling to a level of accuracy that meets or exceeds that of conventional surveys, then data derived from mobile technology may be able to replace conventional methods of surveying. However, to do this, it will be necessary to undertake controlled experiments in which people complete both conventional surveys and mobile technology surveys, and in which the data from both are fully processed to provide the requisite data for model estimation. Models should then be estimated from each type of data, and issues explored of the extent to which the two data sets produce similar models, and also of the extent to which further data repair is required of either type of data. In addition, more in-depth analysis needs to be done of the resulting trip records from both the mobile technology and the travel diary.

As an example, in a study recently conducted by the author in Victoria, Australia, a more in-depth analysis than usual was undertaken on the differences between trips from a conventional travel diary (albeit one of considerable sophistication) and a GPS survey. In this survey, the GPS data were collected for an entire week, with the diary day being one of the days within that week. It was found that there were instances when trips reported in the diary that did not have a match in the GPS data for that day appeared on another day in the same week as the diary day. There were other instances in which the trips reported in the diary bore no resemblance to any trip recorded by the GPS throughout the entire week. Further exploration of these situations, including in-depth interviews with the people undertaking the dual measurement, may provide important data about the reliability of conventional diaries compared to mobile technology.

21.5.1. Sample Size

Potentially, one of the major areas in which mobile technology data may offer significant gains over conventional diary data is in the area of sample size. In a recent paper, Stopher, Kockelman, Greaves, & Clifford (2008c) showed that sample sizes could be reduced dramatically as a result of the ability to collect multi-day data, with samples dropping to as small as 20–30% of the sample sizes required from one-day surveys, based on sampling error considerations. Of course, sampling error is not the only issue that must be considered. Variability in destinations travelled to is always an issue in conventional modelling of trip distribution. Reducing the sample size for error purposes may provide too little data to allow good destination choice models to be developed. In this case, a larger sample size may be needed, with fewer days of data, to achieve the best modelling outcomes. However, there is no experience to date with using multi-day data for model estimation, so that this is as yet an issue that is open for new research and exploration.

A clear advantage of data from mobile technologies is the rather clear identification of tours that result from the data. An illustration is provided in Figure 21.1. This shows two tours, one of which consists of trips 1, 2 and 3, and the second of which consists of trips 4 and 5. The duration of the tours, the location and duration of the stops on the tours, are all readily identifiable from the processed data from the GPS survey. This provides a valuable input for the development and estimation of tour-based models.

21.5.2. Description of the Regional Transport System

Apart from modelling, one of the other major uses of data from conventional travel surveys is to provide a description of the current situation in a region, with respect to

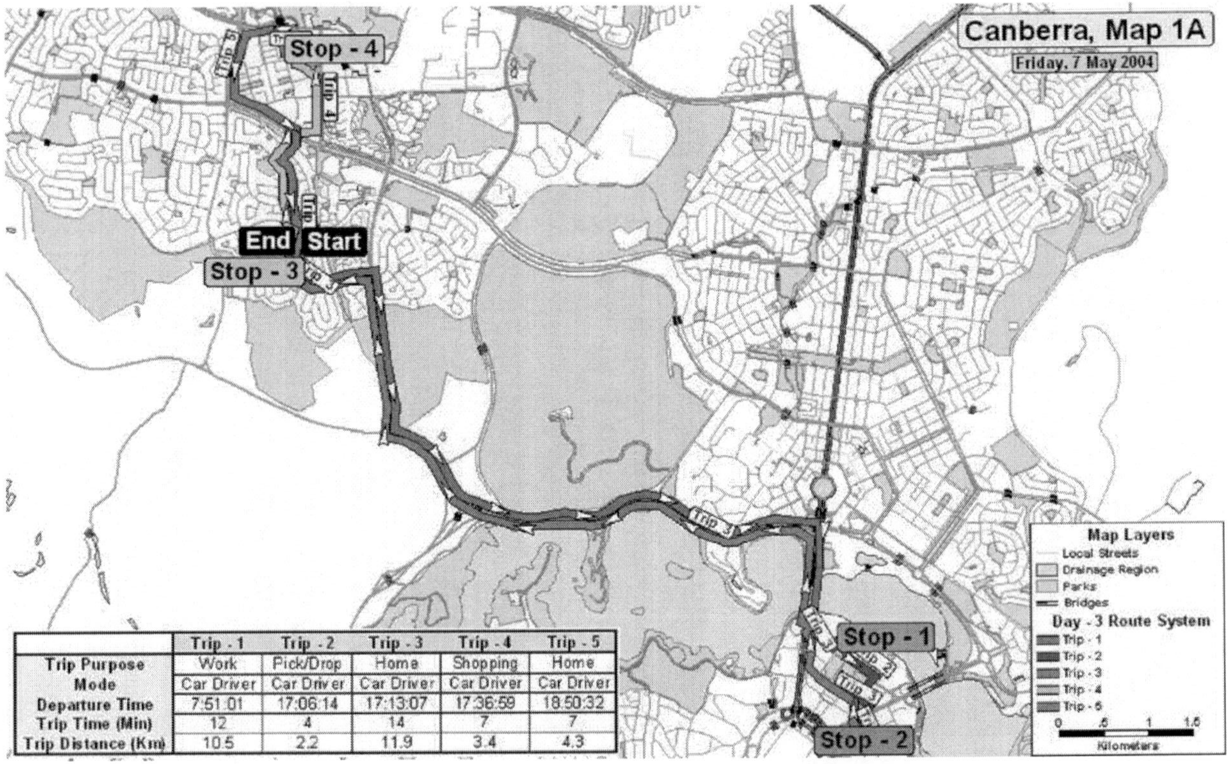

Figure 21.1: Tour identification from a GPS survey.

performance of the transport systems. There seems to be no reason why a survey using mobile technologies would not provide at least as good a description of the performance of the transport systems as can be obtained from aggregating data from a conventional self-report survey (Quiroga, 2000; Quiroga, Perez, & Venglar, 2002; Ogle et al., 2002; Hackney, Bernard, Bindra, & Axhausen, 2007). Perhaps in contrast to the conventional survey, it is likely that mobile technology surveys can provide a more accurate idea of the extent of congestion in the region. Figure 21.2 shows a speed-coded representation of travel from a GPS survey. This clearly shows the location and extent of congestion on a roadway at a specific time of day. If data are collected from 1000 to 2000 households over an entire region, the potential to map data of this type for many roadway facilities, and also to determine the performance levels of public transport systems, is clear. This would represent much superior information to what is obtained from conventional travel surveys.

21.5.3. Data Imputation

One further issue that seems pertinent to explore is that of data imputation. In conventional diary surveys, it is often necessary to impute data that are missing from the original survey form. People sometimes refuse to provide certain data, do not know the answer sought by the interviewer or simply omit data that are required. As a result, some level of data imputation is necessary to allow modelling to proceed from the data set. Usually, imputation is done either by inference from other data provided by the respondent, or using hot-deck imputation (Wilmot & Shivananjappa, 2003; Han & Polak, 2003). Imputation that has to do with sociodemographic data will be just as likely to be required in a survey using mobile technologies as it is in a conventional travel survey, because both types of survey are likely to use a similar method to collect sociodemographic data, which are likely to be subject to similar errors of omission or incorrect information.

Of more importance in comparing methods is the imputation that may be required in a conventional survey for trip characteristics, and what may be required in a mobile technology survey also for trip characteristics. It is not uncommon for people to omit details about trip purposes, or to misidentify the purpose of a trip, and also for details of the destination (or origin) of a trip to be omitted. To be able to use the full set of trips reported by a respondent, such data will usually be imputed, using inference where possible, or some form of logical deduction. However, one of the problems with this type of imputation is that it is not generally amenable to hot-deck imputation, and there are no checks that can be made on the correctness of the inferences and logical deductions made.

In the case of mobile technology data, especially data from a GPS device, imputation is also required to deal with cold start problems (discussed earlier) and data lost in urban canyons and other locations where the signals from satellites are interrupted to such an extent as to result in a gap in a trip. With the possible exception of the first trip made at the beginning of the first day, imputation for cold

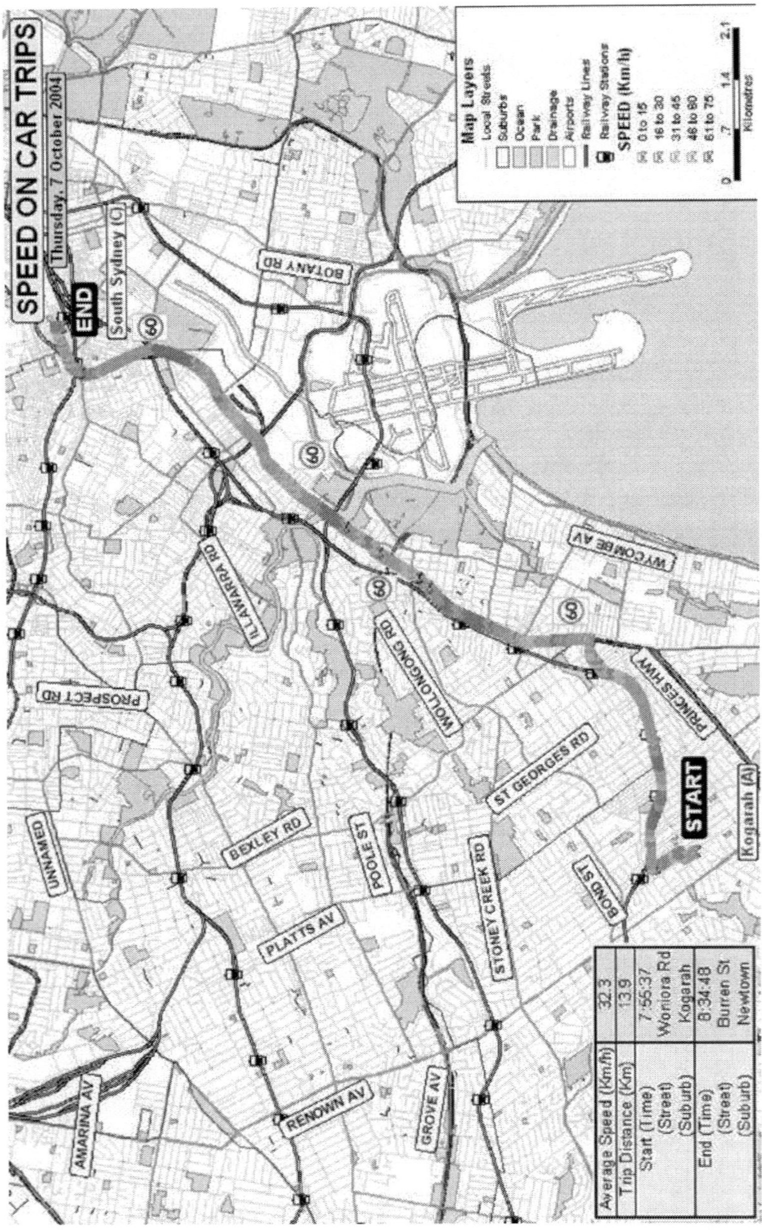

Figure 21.2: Speed-coded representation of travel on a roadway facility.

starts is relatively simple. The location of the origin of the trip must be the same as the destination of the preceding trip. It is also possible to develop some simple rules (e.g. Stopher et al., 2008b) by which the travel time and speed can be estimated for the missed segment of the trip, and a repair made. With multi-day data, and a trip that is repeated from day to day, it may even prove possible to use a form of hot-deck imputation, by imputing the details from the same trip made on another day of the week, where a cold start problem did not occur. Similarly, rules can be set up to handle the situation where a trip segment is missed between the origin and destination, by using information from a GIS network, and the speeds immediately prior to and immediately following the missed segment (Stopher et al., 2008b). The reliability of these imputations has not been checked to the knowledge of this author, but could be by amassing sufficient data to check the details on such missed trip segments. Furthermore, in current work at ITLS, data are also collected on the odometer readings of the household vehicles, which allows some checking that imputed distances are being estimated correctly in the aggregate.

21.6. Conclusions

The record seems clear that data collected using mobile technologies (principally but not limited to GPS) provide valuable information for transport analysts. Principal uses of such technology to date have been in the areas of validation of traditional diary surveys and evaluation over time, using panels, of travel behaviour changes targeted by a specific policy. Perhaps the major question to be answered at this point is whether or not one or more of these technologies — GPS, GSM, RFID, Wi-Fi or combinations thereof — are ready to replace conventional travel surveys. While possibly the first attempt at undertaking a GPS-only survey is shortly to be undertaken in the United States (by the state of Ohio), to date, no reports have been found of a complete substitution of mobile technology for conventional travel surveys. Probably the major reason that this has not been attempted to date is that the technology is only just now nearing the state where it is ready for deployment in a stand-alone mobile technology survey and the processing algorithms and data analysis procedures are still under final development. In this author's opinion, mobile technologies are probably now ready for proof-of-concept testing as a stand-alone travel survey mechanism and for subsequent use to attempt to develop travel-demand models and urban travel descriptions. The fact that no serious efforts have yet been made to build demand models from mobile technology data is probably a further element that must be tested before there is more widespread use of these technologies.

From the standpoint of the sample, it appears that the samples from mobile technology surveys will be slightly less biased than those from conventional surveys. While similar biases are likely to exist relating to one-person households and non-car-owning households, there appears to be some evidence that there will be less bias against large households. On the other hand, limitations of the use of mobile technologies by age may be an issue for such data. Research is needed to determine the extent to which demand models and other uses of the data will be limited by the lack of data on children under a certain age. There also would appear to be room for research on the minimum age at which the devices can be used, and there may be a need to demonstrate that there are no ethical concerns in having children under the age of 14 carrying such devices. Sampling rates and compliance with task appear likely to be comparable to similar recruitment methods for traditional surveys, and may possibly be a little higher.

New methods of analysis may also need to be explored to make maximum use of the data provided by mobile technologies, in that it is so easy to obtain extensive multi-day data, and that such data may provide insights into travel behaviour patterns that have not been available from conventional one-day surveys. At the same time, analysts will require access to standard software (which does not yet exist) to process and analyse the vast amounts of data of which mobile technologies are capable of producing. Currently, the software for processing is under development by a number of researchers, but there is no standard in the methods used, the heuristics applied, or the use of AI or fuzzy logic programming to enhance the results. One or more standard pieces of software will need to be developed to permit the devices to be used more widely.

The principal advantages are obviously the much greater accuracy in time and location that can be provided compared to what people are able to report in conventional surveys, the information available on detailed route choices (which has never been available from conventional surveys), the ability to produce multi-day data without significantly increasing respondent burden and the ability to map tour patterns and determine activity durations much more accurately than from conventional survey methods. As shown by the validation checks against conventional diary surveys, a more complete set of trips are recorded by mobile technologies and correct timing is also provided compared to diaries. On the negative side, vehicle occupancy and travelling party size are not determined from a mobile technology survey unless an interactive device is used. For passive devices, which offer the lowest respondent burden, it appears that software to provide mode and purpose may be capable of providing as good information as is customarily obtained from traditional diary surveys, where there are always a percentage of records that provide mode and/or purpose that are either incorrect, missing or impossible. Again, however, comparative analysis of these results has not been undertaken to date, and this would be a useful exercise to demonstrate the level of usefulness of mobile technology data.

In the short term, it would appear likely that further improvements are probable in GPS technology, in particular, especially with the increasing popularity of in-vehicle navigation devices, which will likely lead to increased sensitivity, even better handling of urban canyon problems and other related issues and also further reductions in power requirements, leading to longer battery life for portable devices. At the same time, the potential addition of a dead-reckoning technology, which could pick up when satellites are no longer in view and signals from them cannot be received, may also become more practical. This could be achieved through some form of miniaturised gyroscope, or other motion detector that is also able to determine heading and speed of movement. It also seems likely that most developments in the short term are likely to take place in GPS applications. As the European Galileo system comes on line (which is expected, currently, by 2013), GPS may be able to provide much greater positioning accuracy and be less subject to signal drop out.

In the medium term, combination systems, especially using GSM and GPS as a result of the requirements in the United States to be able to locate emergency calls from mobile phones, are likely to emerge as an increasingly good alternative to combined GPS and dead-reckoning systems. However, whether positional accuracy from GSM will improve much beyond the present is an unknown. Certainly, there are no indications currently that most developed countries will increase significantly the density of mobile relay stations, so that accuracy is not likely to be improved in urban areas. Increased coverage outside urban areas may occur for GSM, although in such areas, GPS is well able to provide a continuing level of high accuracy, so the impetus for improving positional accuracy of GSM may simply not arise commercially. In the medium term, it also seems likely that Wi-Fi positioning may become more of a contender to replace both GPS and GSM. However, the accuracy of positioning by Wi-Fi will have to improve significantly to provide a real alternative to GPS. Current positioning of 13–40 m is comparable to GSM, but much poorer than is achieved by GPS, especially using the latest chipsets available.

Acknowledgements

I would like to acknowledge the many constructive suggestions I received for this paper from the chair of the workshop session, Jean Wolf, and the discussant, Kay Axhausen. Their comments provided me with the means to make many improvements to the paper. Its remaining inadequacies and errors are entirely my responsibility.

References

Axhausen, K. W., Zimmermann, A., Schonfelder, S., Rindsfuser, G., & Haupt, T. (2002). Observing the rhythms of daily life: A six-week travel diary. *Transportation*, *29*, 95–124.

Bohte, W., & Maat, K. (2008). Deriving and validating trip destinations and modes for multiday GPS-Based travel surveys: Application in the Netherlands. TRB 87th Annual Meeting Compendium of Papers DVD, Transportation Research Board, Washington, DC.

Bradley, M., Wolf, J., & Bricka, S. (2005). Using GPS data to investigate and adjust for household diary data non-response. Paper presented to the 10th Transportation Planning Applications Conference, Portland, OR, April.

Bricka, S., & Bhat, C. (2006). Comparative analysis of global positioning system-based and travel survey-based data. *Transportation Research Record*, *1972*, 9–20.

Cheng, Y.-C., Chawathe, Y., LaMarca, A., & Krumm, J. (2005). Accuracy characterization for metropolitan-scale Wi-Fi localization. *Intel Research*, available at: http://www.placelab.org/publications/pubs/IRS-TR-05-003.pdf, accessed 25 January, 2008.

Chung, E., & Shalaby, A. S. (2004). Development of a trip reconstruction tool to identify traveled links and used modes for GPS-based personal travel surveys. Paper presented to the 83rd Annual Meeting of the Transportation Research Board, January.

Doherty, S. T., Noël, N., Lee-Gosselin, M. E. H., Sirois, C., & Ueno, M. (2001). Moving beyond observed outcomes: Integrating global positioning systems and interactive computer-based travel behavior surveys. *Transportation Research Circular No. E-C026*, March, pp. 449–466. Transportation Research Board, National Research Council, Washington, DC.

Draijer, G., Kalfs, N., & Perdok, J. (2000). Global positioning system as data collection method for travel research. *Transportation Research Record, 1719*, 147–153.

Forrest, T., & Pearson, D. (2005). Comparison of trip determination methods in household travel surveys enhanced by a global positioning system. *Transportation Research Record, 1917*, 63–71.

Hackney, J. K., Bernard, M., Bindra, S., & Axhausen, K. W. (2007). Predicting road system speeds using spatial structure variables and network characteristics. *Journal of Geographical Systems, 9*(4), 397–417.

Han, X. L., & Polak, J. W. (2003). Imputation with non-ignorable missing values: A stochastic approach. In: K. W. Axhausen, J. L. Madre, J. W. Polak & P. Toint (Eds), *Capturing long distance travel* (pp. 172–187). Baldock: Research Science Press.

Hanson, S., & Huff, J. (1988). Systematic variability in repetitious travel. *Transportation, 15*, 111–135.

Hato, E. (2006). Development of MoALs (mobile activity loggers supported by GPS mobile phones) for travel behavior analysis. Paper presented at the 85th Annual Meeting of the Transportation Research Board, Washington, DC, January.

Hawkins, R. & Stopher, P. R. (2004). Collecting data with GPS: Those who reject, and those who receive. Paper presented at the 27th Australasian Transport Research Forum, Adelaide, September.

Itsubo, S., & Hato, E. (2005). A study of the effectiveness of a household travel diary using GPS-equipped cell phones and a WEB diary through a comparative study with a paper-based travel survey. Paper presented at the 85th Annual Meeting of the Transportation Research Board, Washington, DC, January.

Jun, J., Ogle, J., & Guensler, R. (2007). Relationships between crash risk and temporal-spatial travel distances using GPS instrumented vehicles. *Transportation Research Record*, National Academy of Sciences, Washington, DC.

Kochan, B., Bellemans, T., Janssens, D., & Wets, G. (2006). Dynamic activity-travel diary data collection using a GPS-enabled personal digital assistant. Paper presented at the Innovations in Travel Modelling Conference, May 21–23, Austin, TX.

Kracht, M. (2006). Using combined GPS and GSM tracking information for interactive electronic questionnaires. In: P. R. Stopher & C. Stecher (Eds), *Travel survey methods: Quality and future directions* (pp. 545–560). Amsterdam: Elsevier.

Lee-Gosselin, M. E. H., Doherty, S. T., & Papinski, D. (2006). Internet-based prompted recall diary with automated GPS activity-trip detection: System design. TRB 85th Annual Meeting Compendium of Papers CD-ROM, Transportation Research Board, Washington, DC.

Lee-Gosselin, M. E. H., & Harvey, A. S. (2006). Non-web technologies. In: P. R. Stopher & C. Stecher (Eds), *Travel survey methods: Quality and future directions* (pp. 561–568). Amsterdam: Elsevier.

Li, Z., & Shalaby, A. (2008). Web-based GIS system for prompted recall of GPS-assisted personal travel surveys: System development and experimental study. TRB 87th Annual Meeting Compendium of Papers DVD, Transportation Research Board, Washington, DC.

Mackett, R., Brown, B., Gong, Y., Kitazawa, K., & Paskins, J. (2007). Children's independent movement in the local environment. *Built Environment*, *33*(4), 454–468.

Marchal, F., Hackney, J. K., & Axhausen, K. W. (2006). Efficient map-matching of large GPS data sets — Tests on speed monitoring experiment in Zurich. *Transportation Research Record*, *1935*, 93–100.

Murakami, E. (2008). Travel survey methods new technologies subcommittee minutes. Transportation Research Board Meeting, January.

Murakami, E., Morris, J., & Arce, C. (2003). Using technology to improve transport survey quality. In: P. R. Stopher & P. M. Jones (Eds), *Transport survey quality and innovation* (pp. 499–506). Oxford, UK: Pergamon Press.

Murakami, E., & Wagner, D. P. (1999). Can using global positioning system (GPS) improve trip reporting?. *Transportation Research C*, *7*, 149–165.

Murakami, E., Wagner, D. P., & Neumeister, D. M. (2000). Using global positioning systems and personal digital assistants for personal travel surveys in the United States. *Transportation Research Circular E-C008*, pp. III-B/1–III-B/21, Transportation Research Board, Washington, DC.

Ogle, J., Guensler, R., Bachman, W., Koutsak, M., & Wolf, J. (2002). Accuracy of global positioning system for determining driver performance parameters. *Transportation Research Record*, *1818*, 12–24.

Ohmori, N., Nakazato, M., Sasaki, K., Nishii, K., & Harata, N. (2006). Activity diary surveys using GPS mobile phones and PDA. Paper presented to the 85th Annual Meeting of the Transportation Research Board, Washington, DC, January.

Oliviera, M., Troped, P., Wolf, J., Matthews, C., Cromley, E., & Melly, S. (2006). Mode and activity identification using GPS and accelerometer data. Paper presented at the 85th Annual Meeting of the Transportation Research Board, January, Washington, DC.

Pas, E. I. (1986). Multiday samples, parameter estimation precision and data collection costs for least squares regression trip-generation models. *Environment and Planning A*, *18*, 73–87.

Pas, E. I., & Sundar, S. (1995). Intra-personal variability in daily urban travel behavior: Some additional evidence. *Transportation*, *22*, 135–150.

Pendyala. (2003). Measuring day-to-day variability in travel behavior using GPS data, available at: http://www.fhwa.dot.gov/ohim/gps/travel_char.htm, accessed 13/06/07

Quddus, M., Ochieng, W., & Noland, R. (2007). Current map-matching algorithms for transport applications: State of the art and future research directions. *Transportation Research C*, *15*, 313–328.

Quiroga, C. A. (2000). Performance measures and data requirements for congestion management systems. *Transportation Research Part C*, *8*, 287–306.

Quiroga, C. M. Perez, & Venglar, S. (2002). Tool for measuring travel time and delay on arterial corridors. In: K. C. P. Wang, S. Madanat, S. Nambisan, & G. Spring (Eds), *Applications of advanced technologies in transportation: Proceedings of the 7th International Conference*. Reston, VA: American Society of Civil Engineers.

Schönfelder, S., Axhausen, K. W., Antille, N., & Bierlaire, M. (2002). Exploring the potentials of automatically collected GPS data for travel behaviour analysis — A Swedish data source, in GI-Technologien für Verkehr und Logistik, Institut für Geoinformatik, Universität Münster, No. 13, pp. 155–179.

Schönfelder, S., & Samaga, U. (2003). Where do you want to go today? — More observations on daily mobility. Paper presented to the 3rd Swiss Transport Research Conference, Ascona, March.

Schüssler, N., & Axhausen, K. W. (2008). Identifying trips and activities and their characteristics from GPS raw data without further information. Paper presented at the 8th International Conference on Survey Methods in Transport: Harmonisation and Data Comparability, Annecy, May.

Scuderi, M., & Clifton, K. (2005). *Bayesian approaches to learning from data: Using NHTS data for the analysis of land use and travel behavior*. Washington, DC: Bureau of Transportation Statistics, US Department of Transportation.

Shibuya, A., Nakatsugawa, M., Kubota, S., & Ogawa, T. (2000). A high-accuracy pedestrian positioning information system using pico cell techniques. 52nd Vehicular Technology Conference, Boston, MA, pp. 496–500.

Stopher, P., Collins, A., & Bullock, P. (2004). GPS surveys and the internet. Paper presented at the 27th Australasian Transport Research Forum, Adelaide, September.

Stopher, P., Clifford, E., Zhang, J., & FitzGerald, C. (2007a). Deducing mode and purpose from GPS data. Paper presented to the Transportation Planning Applications Conference of the Transportation Research Board, Daytona Beach, Florida, May.

Stopher, P., FitzGerald, C., & Xu, M. (2007b). Assessing the accuracy of the Sydney household travel survey with GPS. *Transportation*, *34*(6), 723–741.

Stopher, P., Jiang, Q., & FitzGerald, C. (2005). Processing GPS data from travel surveys. Paper presented to the Second International Colloquium on the Behavioural Foundations of Integrated Land-Use and Transportation Models: Frameworks, Models, and Applications, Toronto, June.

Stopher, P. R., Bullock, P., & Horst, F. (2002). Exploring the use of passive GPS devices to measure travel. In: K. C. P. Wang, S. Medanat, S. Nambisan, & G. Spring (Eds), *Proceedings of the 7th International Conference on Applications of Advanced Technologies to Transportation* (pp. 959–967). Reston VA: ASCE.

Stopher, P. R., Bullock, P., & Jiang, Q. (2002). GPS, GIS and personal travel surveys: An exercise in visualisation. Paper presented to the 25th Australasian Transport Research Forum, Canberra, ACT, October.

Stopher, P. R., Clifford, E., & Montes, M. A. (2008a). Variability of travel over multiple days: Analysis of three panel waves. Paper presented to the 87th Annual Meeting of the Transportation Research Board, Washington, DC, January.

Stopher, P. R., FitzGerald, C., & Zhang, J. (2008b). In search of a GPS device for measuring person travel, Special Issue of *Transportation Research C*, on Emerging Commercial Technologies, *16*, 350–369.

Stopher, P. R., Kockelman, K., Greaves, S. P., & Clifford, E. (2008c). Reducing burden and sample sizes in multi-day household travel surveys. *Transportation Research Record*, Transportation Research Board, Washington, DC.

Tsui, S. Y. A., & Shalaby, A. S. (2006). *Enhanced system for link and mode identification for personal travel surveys based on global positioning systems. Transportation Research Record*, *1972*, 38–45, Transportation Research Board, Washington, DC.

Varshavsky, A., Chen, M. Y., De Lara, E., Froelich, J., Haehnel, D., Hightower, J., et al. (2006). Are GSM phones the solution for localization? 7th IEEE Workshop on Mobile Computing Systems and Applications, Washington, DC.

Wagner, D. P. (1997). Report: Lexington area travel data collection test: GPS for personal travel surveys. Final Report for OHIM, OTA, and FHWA.

Wermuth, M., Sommer, C., & Kreitz, M. (2003). Impact of new technologies in travel surveys. In: P. R. Stopher & P. M. Jones (Eds), *Transport survey quality and innovation* (pp. 455–481). Oxford, UK: Pergamon Press.

Wilmot, C., & Shivananjappa, S. (2003). Comparison of Hot-deck and neural network imputation. In: P. R. Stopher & P. M. Jones (Eds), *Transport survey quality and innovation* (pp. 543–554). Oxford, UK: Pergamon Press.

Wolf, J. (2006). Application of new technologies in travel surveys. In: P. R. Stopher & C. Stecher (Eds), *Travel survey methods: Quality and future directions* (pp. 531–544). Amsterdam: Elsevier.

Wolf, J., Bonsall, P., Oliveira, M., Leary, L., & Lee, M. (2006). Review of the potential role of "new technologies" in the National Travel Survey: Report of work conducted in Work package 1, Report to the Department for Transport, UK, available at: www.dft.gov.uk/pgr/statistics/datatablespublications/personal/methodology/ntsreports/ntsreviewtechnologies, accessed 04/02/08.

Wolf, J., Guensler, R., & Bachman, W. (2001). Elimination of the travel diary: An experiment to derive trip purpose from GPS travel data. *Transportation Research Record, 1768*, 124–134.

Wolf, J., Loechl, M., Thompson, M., & Arce, C. (2003). Trip rate analysis in GPS-enhanced personal travel surveys. In: P. R. Stopher & P. M. Jones (Eds), *Transport survey quality and innovation* (pp. 483–498). Oxford, UK: Pergamon Press.

Wolf, J., Schönfelder, S., Samaga, U., Oliveira, M., & Axhausen, K. W. (2004). 80 Weeks of GPS traces: Approaches to enriching trip information. *Transportation Research Record, 1870*, 46–54.

Yamada, I. S. Shiotsu, Itasaki, A., Inano, S., Yasaki, K., & Takenaka, M. (2005). Secure RFID tag system. *Proceedings of the UbiComp 2005 Workshop on UbiComp Privacy*.

Zumkeller, D., Madre, J.-L., Chlond, B., & Armoogum, J. (2006). Panel surveys. In: P.R. Stopher & C. Stecher (Eds), *Travel survey methods: Quality and future directions* (pp 363–398). Amsterdam: Elsevier.

Chapter 22

Mobile Technologies: Synthesis of a Workshop

Jean Wolf

22.1. Participation and Scope

Over the past decade, increasing attention has been given to the application of mobile technologies for the collection and processing of travel survey data at the ISCTSC conference. In fact, two workshops at the 2008 conference were dedicated to this topic (the other one focused on non-response challenges in GPS-based surveys), and many of the other workshop reports mentioned mobile technologies as a possible solution for data-collection challenges facing their survey areas, such as studies of freight movement, tourism, and hard-to-reach groups. This increasing attention reflects the growing opportunities for the application of mobile technologies in travel surveys to address issues with respondent burden, sample biases, breadth of details collected, and the accuracy of key travel data elements (such as origin and destination locations, trip times, travel durations and distances, and routes taken) that are ongoing challenges with more traditional survey methods.

This workshop was well attended, with 27 participants representing many countries. About half of the participants had some experience with mobile technologies. It was agreed from the start that the workshop would indeed focus on data collection and processing using mobile technologies, would avoid lengthy discussions on non-response issues since this topic was already addressed by another workshop, and would limit discussion on broader study design topics. To allow adequate time for a thorough discussion on data collection and processing, it was also decided to focus the workshop on Global Positioning System (GPS) and mobile telephone technologies, although recognition of other possible mobile technologies was given. This decision was based on the existing body of evidence (i.e., existing travel survey research) already established with these more promising and seemingly ubiquitous technologies.

Transport Survey Methods

ISBN: 978-1-84855-844-1

22.2. Presented Papers

The workshop resource paper (Stopher, 2009) provided a comprehensive and in-depth review of GPS data-collection and processing methods and challenges. This resource paper also touched on mobile phone technologies (specifically GSM or Global System for Mobile communication phones) to the extent that relevant research results were available. To augment this wealth of information, three other papers were presented which provided detailed descriptions of different types of mobile technology applications in travel surveys. The first of these reported on two mobile phone case studies conducted in South Africa that collected nearest cell phone tower locations of study participants on five-minute intervals (Krygsman, Nel, & de Jong, 2008). The second paper (Bohte & Maat, 2008) reported on a large-scale, multiday (1104 participants for a one-week study duration) GPS study conducted in the Netherlands that used a Web-based interface to confirm travel details. The third paper (Schüssler & Axhausen, 2008) provided details on a large-scale GPS data processing effort performed in Switzerland on a dataset that had been previously collected for a different purpose (4882 participants with an average of 6.6 days of GPS data). Finally, to further assist the workshop in focusing on core issues, Barbara Noble gave a brief presentation as the workshop discussant.

22.3. State of the Art — Devices and Data Collection

GPS studies are shifting from vehicle-based to person-based deployments as device technologies continue to evolve, with device sizes continuing to shrink while storage and power capacity continue to increase. Furthermore, where there may have only been a few custom GPS data loggers available for household travel surveys just a few years ago, there are now more than 30 versions of battery-operated GPS data loggers available on the market today. Person-based solutions are desirable to collect detailed data on all modes of travel and mirror traditional travel survey study units (i.e., all persons in a household, all adults in a household, etc.).

However, even though technology options now support person-based studies, there are still advantages to vehicle-based methods that should be considered. These advantages include a reliable, constant power supply, limited respondent burden (i.e., no need to carry or recharge), and a limited chance of devices being switched inadvertently between persons in the household. Given these tradeoffs, it is possible that some study objectives (especially those focused specifically on vehicle travel or on populations that are highly vehicle dependent) could be better met with a vehicle-based study.

There have been many GPS-enhanced travel surveys conducted in the past decade, but most of these have included both GPS and diary data collection. The trend is moving away from this combined method (which actually increases respondent burden) and toward GPS-only studies (with or without a prompted recall feedback/confirmation stage). This evolution matches the adoption (or acceptance) rate of

GPS technology solutions for travel survey data collection as these technologies prove themselves to be a reliable source of travel data.

Methods for equipment deployment vary from country to county, with interview methods often driving the most cost-effective option chosen. For example, in countries where face-to-face interviews are common (e.g., Australia, the United Kingdom, and France), equipment is delivered during the initial recruitment visit; in which household information is collected and travel diaries or logs are provided. In countries where computer-assisted telephone interviews (CATI) are common (such as the United States), equipment is often delivered by a study-specific courier service, a commercial shipping service (such as United Parcel Service or Federal Express) or the US Postal Service.

Viable alternatives to GPS technology do exist and are being evaluated. Mobile phones are one obvious alternative, given their widespread use. Location calculations based on phone communication towers are becoming a standard method for estimating travel times on congested corridors; this same approach can be used for estimating travel patterns in a region, as reported by Krygsman et al. (2008). Other feasible technology solutions include Bluetooth, WiFi, RFID, and smartcards (see Stopher, 2009). These alternatives to GPS each offer varying levels of detail, accuracy, cost, geographic coverage, and population coverage. Generally speaking, GPS technology is the most accurate for outdoor travel monitoring with no limit to coverage area. Of course, mobile phones with GPS are also an option and are considered to be another version of GPS technology for the purpose of this report; several studies in Japan have explored and reported on this option (see Hato, 2006, for an example).

22.4. State of the Art — Data Processing

Since the first GPS-enhanced travel survey conducted in Austin, TX in 1997, ongoing software and algorithm development efforts have resulted in fairly reliable detection of key travel diary elements based on successful trip end identification (with associated trip origins and destination locations, start and end times, and travel routes and distances defined). Much work has also been completed focusing on algorithms to define travel mode and trip purpose; these variables are more challenging in that they are not directly obtained from the GPS data (i.e., they must be derived). Work has also begun to derive other trip-related details, such as number of household travel companions and parking lot use/parking fees paid.

Most software developed to date has used a rule-based approach, with more recent research shifting toward the use of fuzzy logic. The use of geographic information system (GIS) datasets (e.g., road networks, transit networks, points of interest, and land use details) is important to improve the ability and accuracy of these algorithms.

There have been varying levels of validation performed to date on GPS processing software, with a general lack of ground truth datasets available to perform this critical task. Prompted–recall interfaces have been developed and implemented in some GPS-enhanced studies to obtain participant clarification or confirmation of detected and derived travel details, but participant burden using this method may be an issue depending on the method of feedback (e.g., paper, Internet, CATI, and CAPI methods have been used to date).

Although mobile phone technologies seem to be an obvious method for collecting travel data, only a handful of studies have been conducted using custom computer-assisted personal interviewing (CAPI) software loaded on phones with GPS as a prompted recall tool. Likewise, only a few studies have been conducted on the usefulness of tower-based location data collected at intermittent or infrequent intervals for estimating travel patterns.

22.5. Issues — Devices and Data Collection

There was a range of issues identified with mobile technology devices and data collection. Here is a summary of those discussed in the workshop.

22.5.1. *Device Design*

With respect to device design and functionality, the presence and design of the on/off switch is a critical factor impacting the success of data collection. If the button is easy (i.e., too easy) to press, data logging can stop unintentionally. Other device elements such as voices or flashing lights can be helpful to some participants and possibly confusing or alarming to others.

22.5.2. *Device Functionality*

Given the multitude of GPS logging devices available on the market, it is logical that these devices would support different types of logging rules. Some logging rules (such as a speed filter that omits the logging of zero point speeds) are desirable for travel surveys to conserve storage capacity and download times. Not all devices support the same rules in the same manner. Furthermore, firmware issues with some devices may not retain user-selected settings. Any device selected for a study should be tested thoroughly through a range of real-world survey conditions in a pilot or pre-test phase prior to large-scale deployment.

22.5.3. *Device Output*

Even though all GPS receivers are capable of generating standard NMEA (National Marine Electronics Association) GPS message formats, the logging component of many devices parses these messages to select only those fields considered essential by the manufacturer. Some devices support various output formats (such as .txt, .csv, .nmea, and .gpx), but there is often little or no flexibility in selecting additional desired fields. The minimum set of fields required for generating travel survey data includes date, time, latitude, and longitude. Speed and heading are also desirable fields; if these are not provided by the device, however, they can be derived. Other GPS fields that can be used as indicators of positional accuracy include the number of satellites used in the position calculation and horizontal dilution of precision (HDOP, a measure of satellite dispersion in view). These latter fields are useful, although algorithms to detect bad or potentially bad points that rely on the primary data elements are still necessary.

22.5.4. *Recruitment of and Communication with Participants*

In the person-based GPS studies conducted to date, many have established a cutoff age for participants below which equipment is not provided. There are both practical and ethical considerations for defining an appropriate age threshold for GPS deployment. Like other survey methods, the subpopulation likely to use GPS devices may not fully represent the targeted population. Methods for recruitment and communicating with GPS study participants are needed to address and minimize this bias as well as to encourage compliance with equipment usage rules for the duration of the study period.

22.5.5. *Device Deployment and Retention*

Beyond the cost of the devices (which continue to decline rapidly, from \$1000 USD or greater just a few years ago to less than \$100 USD now), the cost to deploy and collect the devices may still be a significant portion of a GPS study budget. In face-to-face interviews, this cost is actually absorbed in the cost of the home visits before and after the assigned travel period. Studies using mail out or courier-type delivery methods have proven susceptible to significant equipment loss, at least in the United States, as households who initially agree to participate by phone become uninterested and/or too busy to participate or return the equipment. Even repeated attempts by phone, mail, and personal visits to these households do not necessarily result in the collection of these devices.

22.5.6. *Missing or Messy Data*

Although passive mobile technologies offer great opportunities for the highly accurate collection of multiday travel data, the use of these technologies is still dependent on participants to use the equipment as instructed. Furthermore, technology devices are not immune to component failure that can also result in missing or messy data.

22.5.7. Data Retention

Although data retention is actually the final step of a GPS study, it can also be considered as the final stage of data collection as well. Multiday GPS datasets contain highly accurate location and time information on travel patterns for all instrumented household members or vehicles. Once these data are collected and processed, there are ethical and legal concerns related to the security of the data and access by others. Rules for archiving and accessing these datasets must be established, implemented, and monitored. Encryption methods may be needed to provide security for both data storage and transfers.

22.5.7.1. Mobile phone technologies Given the millions and millions of mobile phones in use worldwide, it certainly seems that these personal devices should be leveraged as personal travel monitors. There are still many issues to be addressed before this will be feasible for large-scale samples, however. These issues include the costs for obtaining reliable and usable data from mobile phone network providers (in the case of tower-based location tracking) and the practicality of large-scale location-enabled mobile phone studies (including commercial and device application considerations).

22.6. Issues — Data Processing

There is no commercially available software that would allow a planning agency to purchase its own GPS equipment and to conduct its own study. The primary reason for this is that the software developed to date is still evolving rapidly, with no standard methods proven yet as the best for detecting trips and trip details, and for deriving other travel details. Many researchers believe that we are still in a discovery phase. Evaluations of software methods using advanced techniques such as artificial intelligence and fuzzy logic have only just begun.

Ground truth data for calibrating and validating data processing algorithms is essential, but can be costly and/or challenging to acquire. Prompted recall does offer some potential in this area, although in the end this method still relies on the accuracy of respondent reporting and may introduce additional respondent burden. It is possible that some direct observation may be needed in this area.

22.7. Research Needs — Devices and Data Collection

Although the general consensus among workshop attendees was that small, self-powered GPS logging devices are now available for person-based GPS travel surveys, it was also agreed that continued improvements are desired in several areas, including form factor, simplicity/ease of use, power capacity, and storage capacity.

Research is needed to compare deployment methods with respect to study purpose, cost, equipment loss, and response rates. Given the desire for large-scale

GPS-based surveys, further research is warranted in this typically high cost, potentially high loss, and area.

The paper presented on using cell phone tower location traces definitely stirred interest at the workshop. It was agreed that continued research in this area is needed. Most workshop participants were also aware that the continued worldwide rollout of GPS-enabled mobile phones presents new opportunities and issues, as mentioned previously. A feasibility study is needed to evaluate deployment, support, and costs of a population-based GPS-enabled mobile phone sample using custom software either for passive logging and transfer or for a prompted recall user interface that also supports logging and transfer.

22.8. Research Needs — Data Processing

Given the desire to use mobile technologies to collect detailed travel survey data, it is logical that many of the same issues that face traditional travel surveys apply to technology-based surveys — including the need for standard data elements and formats, and standard data processing algorithms and software. Workshop attendees who were considering how to implement their own mobile technology study asked if there were any data processing software products available — either for a fee or free. There was quite a bit of deliberation over whether the state of the research was ready to standardize data processing algorithms and software. In the end, it was agreed that there was still significant work left to do on the software development side and that standardizing the data elements and formats available from mobile technologies was the logical and necessary first step.

Having said that, there has been substantial research performed to date on several key mobile technology (primarily GPS) data processing steps, including trip end identification, travel mode assignment, and trip purpose determination. It was agreed that a comparative analysis of the strengths and weaknesses of the various methods used to determine these trip details would be valuable. Insufficient research has been conducted to impute other trip level details, such as the identification of travel companions and/or the number of companions on the trip. The question also was raised about the necessity of these variables. Workshop attendees agreed that data processing algorithms were needed that could work with or without GIS data. Finally, the development and provision of validation datasets for testing data processing was identified as a key research need.

22.9. Related Research Needs

Some surveys are using a mix of traditional and advanced survey modes, such as travel diaries, CATI, and Internet-based travel reporting in an effort to obtain better response rates from targeted subgroups. Mobile technology options may soon join the suite of "reporting" methods offered in travel surveys. Research is needed on the various challenges and biased inherent in these different methods, with a tradeoff analysis performed to assess the benefits of increased participation versus the biases introduced by different methods.

Research is also needed to fully evaluate and quantify the tradeoffs (e.g., in respondent burden, data quality and accuracy, sample bias, study costs) between the different mobile technology options available (including GPS with CATI, GPS with some form of prompted recall, and GPS-only methods). This same evaluation should include mobile phone technologies (with and without GPS).

Finally, after a comprehensive discussion of mobile technologies for collecting and processing travel survey data, it was clear that some of the traditional travel diary data elements can be collected or imputed reliably with some level of effort, while others will prove to be challenging or simply unattainable. A thorough review of travel demand modeling data needs seems appropriate to determine how existing models can be adapted to meet new data sources such as GPS and mobile phones — or if new models are needed to accommodate the strengths (and weaknesses) of these new sources. Other modeling paradigms should also be examined, including the number of travel days to be collected (which can easily be increased with passive mobile technologies), the number of persons per household sampled (one person per household may be sufficient and would reduce study costs), and the minimum age of participation (i.e., is the collection of travel data for children under the age of 16 necessary).

Acknowledgements

The workshop participants were Tracy Anderson, United Kingdom; Patrick Bonnel, France; Andrew Clark, Canada; Kelly Clifton, United States; Eric Cornelis, Belgium; Sean Doherty, Canada; Robert Follmer, Germany; Martin Kagerbauer, Germany; Nina Karasmaa, Finland; Thipphaphone Khenmingmongkhon, France; Stephan Krygsman, South Africa; Nick Lanyon, United Kingdom; Kees Maat, the Netherlands; Roger Mackett, United Kingdom; Steve McFeely, Ireland; Fabrice Marchal, France; Philippe Marchal, France; Michael Meschik, Austria; Barbara Noble, United Kingdom; Juan de Dios Ortúzar, Chile; Matthew Roorda, Canada; Daniel Sauter, Switzerland; Nadine Schüssler, Switzerland; Peter Stopher, Australia; Susan Swain, United States; Liva Vågane, Norway; and Shuning Yuan, France.

References

Bohte, W., & Maat, K. (2008). Deriving and validating trip destinations and modes for multi-day GPS-based travel surveys: A large-scale application in the Netherlands. Presented at the 8th International Conference on Transport Survey Methods, Annecy, France.

Hato, E. (2006). Development of MoALs (mobile activity loggers supported by GPS mobile phones) for travel behavior analysis. Presented at the 85th Annual Meeting of the Transportation Research Board, Washington, DC.

Krygsman, S., Nel, J., & de Jong, T. (2008). The use of cellphone technology in activity and travel data collection in developing countries. Presented at the 8th International Conference on Transport Survey Methods, Annecy, France.

Schüssler, N., & Axhausen, K. (2008). Identifying trips and activities and their characteristics from GPS raw data without further information. Presented at the 8th International Conference on Transport Survey Methods, Annecy, France.

Stopher, P. R. (2009). Collecting and processing data from mobile technologies. In: P. Bonnel, M. Lee-Gosselin, J. Zmud & J.-L. Madre (Eds), *Transport survey methods: Keeping up with a changing world.* Bingley, UK: Emerald.

Chapter 23

What Is Different About Non-Response in GPS-Aided Surveys?

Stacey Bricka

Abstract

Over the past decade, transportation researchers have leveraged global positioning system (GPS) technology to improve the accuracy and increase the depth of spatial and temporal details obtained through household travel surveys. While earlier studies used GPS as a supplement to traditional household travel survey methods, measuring the accuracy of trips reported (Wolf et al., 2006), studies are now underway to identify the methods and tools that will allow us to do away with paper diaries entirely and simply rely on GPS to obtain trip details. This paper finds that while GPS clearly helps to improve participation among some groups, it decreases participation among others. Thus, it should be considered a tool in the household travel survey toolbox and not "the" solution to non-response issues in household travel surveys.

23.1. Introduction

The desire to use GPS for travel surveys stems from the ability to glean from the data streams more precise travel times and distances, accurate route information, interim stop locations and vehicle operation characteristics (Doherty, Papinski, & Lee-Gosselin, 2006; Kalfs & Saris, 1997; Madre et al., 2008; Swann & Stopher, 2008). In addition, recent technological advances have led to decreased equipment costs and increased prominence in the consumer market (Kracht, 2006).

Adding to the context, recent advances in activity-based travel demand modelling require more detailed travel behaviour details, both about the trips themselves as well

Transport Survey Methods

ISBN: 978-1-84855-844-1

as about variations in travel and household interactions, for longer periods of time than the traditional 24 h period. These increased demands being placed on traditional household survey participants are being met with resistance, as evidenced by declining response rates and corresponding increases in collection costs. GPS data collection is thought to decrease respondent burden through the passive collection of travel details (Lee-Gosselin, Doherty, & Shalaby, 2006). Trip information is subsequently identified in the GPS data streams through algorithms or follow-up prompted recall surveys (Wolf, Bonsall, Oliveira, Leary, & Lee, 2006).

Clearly, passive collection of trip details is less onerous than completing paper travel diaries and can be deployed for longer periods of time, at levels of "little or no respondent burden" (Doherty et al., 2006; Kalfs & Saris, 1997; Wolf et al., 2006). However, few studies examine the deployment of this technology within the household survey context from the perspective of the respondent, the logistics of deployment or with regard to the representativeness of the resultant data sets. Lee-Gosselin et al. (2006) recognize the problems of equipment deployment and recovery, as well as privacy concerns, which, when considered in tandem with factors that influence the decision to participate in these surveys and those that influence consumer acceptance of technology, contribute to unit non-response bias.[1] In addition, while the newer GPS units are smaller with lighter power supplies, respondents must still remember to carry them around and charge them (Doherty et al., 2006; Hawkins & Stopher, 2004; Wolf et al., 2006). However, units are not carried around all the time, due to respondent-cited reasons such as a "lack of time" (Draijer, Kalfs, & Perdok, 2000) and the "busyness of life" (Swann & Stopher, 2008). This creates item non-response bias in the resultant data sets.

The purpose of this paper is to investigate the non-response challenges associated with GPS-based surveys, focusing primarily on the identification of non-response levels at the unit level, in order to determine if the data files ultimately obtained by GPS are statistically different from those obtained through traditional survey methods with regard to the demographic characteristics of the survey participants. Unit non-response (or the representativeness of the households that agree to participate) is hypothesized to be a function of (1) traditional travel survey non-response factors and (2) technology acceptance levels, and will be identified in this paper using quantitative techniques. Details regarding item non-response (how completely the households perform the survey task) will be explored at a qualitative level since this is a relatively new area of research.

This paper is organized as follows: following this first section (Introduction) is an international review of the use of GPS in travel surveys in the second section. Next, in the third section, a literature review is presented that focuses on (1) factors influencing non-response, both from the travel survey literature as well as from

---

1. Throughout this paper, the references to unit and item non-response are consistent with that of accepted survey literature where, in the context of GPS studies, item non-response refers to the lack of data capture by the GPS unit due to respondents not carrying the units as instructed, not charging the units on a regular basis, or unit malfunction.

relevant information technology and marketing studies and (2) prior research specifically investigating unit non-response in GPS-based surveys. The fourth section presents an evaluation of unit non-response in two recent U.S. travel surveys in order to identify whether the composition of the GPS data files differs from those collected through traditional means, followed by a qualitative discussion of item non-response in the fifth section. The paper concludes with recommendations for mitigating non-response in GPS-based surveys.

23.2. GPS in Travel Surveys

The purpose of this section is to provide an international synthesis of GPS applications in travel surveys. The details in this section were drawn from published literature, as supplemented by direct inquiries to colleagues in various countries and input obtained at the workshop associated with this topic that was held as part of the 8th International Conference on Transport Survey Methods. It is limited to those documents written in English and focuses primarily on papers submitted for consideration to the Transportation Research Board annual meetings from 2001 to the present.

23.2.1. United States

The use of GPS in regional travel surveys throughout the United States began with the Lexington (Kentucky) proof of concept study in 1996 and continues today, with deployment in more than 16 regional studies. Texas has had the most deployments, with a 10% sample of GPS households in six regional travel surveys since 1997. Other regions with GPS deployments in travel surveys include Los Angeles; Pittsburgh; St. Louis; Kansas City; Reno; Portland; Seattle; Chicago; Washington, DC; and Baltimore (Bricka & Bhat, 2006). In addition, both California and Ohio have deployed GPS as part of their statewide surveys. GPS sample sizes ranged from 46 households (Pittsburgh) to a projected 1600 households in a Washington, DC/ Baltimore study (NuStats Travel Survey Archives). Except for the Portland proof of concept test and the recent study in Chicago, both of which deployed wearable GPS devices, all remaining U.S. studies were conducted using in-vehicle units that store the data within the unit with data downloads in between deployments.

The application of GPS to regional travel surveys in the United States has largely been for the purpose of trip rate auditing — identifying the magnitude and characteristics associated with trip underreporting. The range for levels of trip underreporting includes 5–7% in the most recent studies (Kansas City and Reno) and 81% in a 2001 study (Laredo, Texas). In general, levels of trip underreporting have decreased as research into trip underreporting correlates has been used to improve survey methods (Bricka & Bhat, 2006). In addition, algorithms for identifying trips in the GPS data streams have been strengthened with the use of prompted recall surveys. Specifically, earlier studies simply computed missing trips as the difference between trips detected in GPS versus those reported in CATI. Later studies excluded from the GPS-detected trips those trips that respondents were told *not* to report in their logs, such as trips outside the study area or for commercial purposes, and, when prompted recall follow-up surveys were employed, where respondents indicated that GPS-detected trips were actually traffic manoeuvres or movements within parking lots.

One additional U.S. project is the Commute Atlanta project. It is important to note that this is a GPS-based instrument vehicle monitoring study, not a household travel survey. A three-year project utilizing a random sample of 500 vehicles across 260 representative households, this is the longest-running continuous GPS collection effort to date (Elango, Guensler, & Ogle, 2007; Ogle, Guensler, & Elango, 2005). Although the burden associated with an extended deployment has not specifically been evaluated to date, Ogle et al. (2005) reported difficulties in recruiting low-income households into the sample.

23.2.2. Canada

Based on a review of international efforts, the Halifax STAR project enjoys the designation of the largest GPS sample within the context of a household travel survey project undertaken to date. The project design calls for 2200 surveys of completed diary and GPS data, and, as of January 2008, data has been obtained from 1400 households. The wearable GPS devices used in this study transmit the data to a central database as the travel takes place. The retrieval interview phase of the household travel interview includes collection of the diary data, immediately followed by a prompted recall survey using the previously transmitted data. Information is not available about the type of sample used in this study or any indications of non-response bias (Harvey, 2008).

The use of GPS to collect travel details in Canada began in 1995 and, with the exception of the STAR project, has largely been focused on methods development and post-processing techniques using data collected from small-scale pilots and other applications using non-random samples (Doherty et al., 2006; Lee-Gosselin et al. (2006); Lee-Gosselin, 2008). One example is a study of diabetic patients, which included wearable GPS, accelerometers, heart rate and blood glucose monitoring, and a prompted recall diary administered with a BlackBerry. Among a non-random sample of patients from a rehabilitation facility, the majority were elderly patients who were not at all intimidated by the technology (Doherty, 2008).

23.2.3. South America

GPS has not been used within the context of any household travel surveys to date in South America. However, GPS technology has been employed extensively to document level-of-service data for all modes of travel (such as travel-time data by auto and transit) (Strambi & Garrido, 2006).

23.2.4. Europe

The most visible GPS application in Europe is the large-scale national travel survey being conducted in France. The anticipated GPS sample size is 1500 households out of a national sample of 17,000 households, with a two-week deployment of the GPS units. While the national survey employs a random sample, the GPS units will be provided only to those participants who volunteer to carry them. This is similar to the "opt in" practice in the United States, where eligible households are asked if they would like to participate in the GPS sample (Madre 2007; Madre et al., 2008). Also planned for this project is a GPS non-response study.

According to Wolf, Schonfelder, Samaga, Oliveira, and Axhausen (2004), the Ratt Fart project in Sweden was a traffic safety project that equipped hundreds of cars in three Swedish cities with GPS for a two-year period. The units provided drivers with feedback if speeding (Wolf et al., 2004). Information was not available regarding whether this was a random sample or if there was any type of non-response bias noted.

In addition to these larger studies, European colleagues have conducted pilot tests of GPS within the context of household travel surveys and other large-scale GPS studies not linked to travel surveys. Most were focused on the methods and processing, as with our Canadian colleagues. Documentation regarding non-response issues was unavailable.

Several European countries are in the planning stages of their next national surveys, however, and most plan to include some type of GPS component. These include the following:

- The United Kingdom is in the planning stages for adding GPS to their national travel survey, for validation purposes. They are designing a GPS pilot that will collect diary and GPS data for seven days. The goal is to equip 100 households.
- Germany is conducting a national travel survey as well and has included a GPS pilot in two regions. The GPS is being deployed for two to four weeks, for the purpose of obtaining additional trip data.
- Norway's next travel survey is being planned for 2009. It will be a large-scale survey of 15,000–20,000 persons. A GPS subsample is being considered for validation purposes.
- Austria is planning its next national travel survey as well. A GPS subsample of approximately 100 households is under consideration for two to three regions. The current thought is that the survey itself would be for one to two days, while the GPS would be deployed for about a week.

23.2.5. Asia

GPS studies in Japan have largely focused on the use of GPS-equipped cellular phones within small-scale, non-random samples. This includes the MoALS (mobile activity logger), which was tested by 31 non-random recruits. The MoALS unit was interactive, with participants logging departure times, transfers of mode, arrival time and location information with a corresponding Web diary that feeds back the reported travel information for confirmation and obtains other trip details. The advantage to the cell-phone GPS technology is the instantaneous transmission of data as it is obtained (Itsubo & Hato, 2006).

Other Japanese studies include an 18-probe person survey with deployments ranging from 5 to 70 days and non-random samples of 10 to 384 persons to evaluate information provision, measure levels of service and identify route choice (Asakura & Hato, 2006). Because non-random samples were used, non-response bias was not addressed as part of this research.

Israel is currently conducting a household travel survey that includes 2000 households recording travel for 48 h. A GPS subsample calls for equipping the most frequent driver in 77 households with GPS. In Jerusalem, a GPS-only study is in the design stages.

23.2.6. *Australia*

Based on available publications, Australian colleagues have made the strongest advances in the transition to GPS-derived trip tables within the context of household travel surveys. Advances here have largely been in support of household travel surveys designed to measure reduction in vehicle kilometres travelled associated with the national TravelSmart programme. For example, small-scale ($n = 200$) GPS-based travel studies have been conducted in conjunction with odometer surveys within a panel survey framework in Victoria and South Australia (Stopher, Swann, & FitzGerald, 2007). Other GPS-based household travel surveys have been conducted in New South Wales and Sydney (Stopher, Xu, & Fitzgerald, 2006). These surveys all used random samples, and non-response was studied (as discussed in Section 23.4).

23.2.7. *Summary*

In sum, based on available English-language publications, the use of GPS in household travel surveys is becoming more commonplace throughout the world. There has a been a shift in focus from the use of GPS for trip auditing to the use of GPS for building the trip files, based largely on the ability to obtain more accurate trip details, particularly reported times, route choice and interim stop detection. While large-sample traditional household travel surveys are commonplace, the first round of large, random GPS-based travel surveys is now underway. As this data becomes available, they will provide a rich source of information on the issue of non-response in GPS-based travel surveys, among other topics such as the study of route choice or time-of-day travel patterns.

23.3. Literature Review

The study of non-response within the context of household travel surveys enjoys a strong history of discussion at these international travel survey conferences, as have discussions on how to mitigate that non-response. The purpose of this section is to briefly summarize the key demographic characteristics associated with non-response in household travel surveys, as well as demographics associated with technology acceptance, in support of the underlying hypothesis for this research: that non-response in GPS-based travel surveys is a function of the non-response associated with traditional travel surveys and the characteristics of those that do not quickly embrace new technologies. Supporting literature is summarized in three categories: non-response in travel surveys, technology acceptance and, for the few sources available, specific inquiries into non-response in GPS-based surveys. An international perspective is provided where possible.

23.3.1. Non-Response in Travel Surveys

A review of unit non-response in travel surveys suggests that it can be characterized by both demographic as well as travel behaviour characteristics. As indicated in Table 23.1, the most common demographic characteristics associated with non-response in traditional household surveys include being low income, being young

Table 23.1: Non-respondents in travel surveys.

	Low income	Young	Lower education	Large households	Minority	Urban dwellers	Low travel levels	Non-public transport user
German Mobility Panel[a]	X	X	X				X	
Sweden's TSU92[b]							X	X
International Review[c]							X	
Zurich Social Networks[d]		X	X					X
US Summary[e]	X	X		X	X			
TMIP Report[f]		X	X		X	X		
Australian Study[g]							X	

[a]Kuhnimhof, Chlond, and Zumkeller (2006).
[b]Forsman, Gustafsson, and Vadeby (2007).
[c]Bonnel, Armoogum, and Madre (2007).
[d]Frei and Axhausen (2007).
[e]Bricka (2006).
[f]Nonresponse in household travel surveys.
[g]Richardson, Ampt, and Meyburg (1996).

(under age 25), having a lower education level, being a part of a larger household, being of minority descent and living in an urban area. Travel characteristics typically associated with non-respondents include having lower-than-average trip levels and not using public transit.

23.3.2. Technology Acceptance

In addition to non-response factors, participation in GPS-based travel surveys is influenced by factors associated with levels of technology acceptance. The marketing research and information technology literature indicates that those who are most likely to accept technology have the following characteristics:

- male (de Blaeij, Dijst, & Farag, 2006)
- young (de Blaeij et al., 2006; Im, Bayus, & Mason, 2003; Macklin, 2008)
- highly educated (de Blaeij et al., 2006; Im et al., 2003; Macklin, 2008; Vijayasarathy, 2004)
- higher income (de Blaeij et al., 2006; Im et al., 2003; Macklin, 2008)

In addition, Vijayasarathy (2004) was the only one to find acceptance higher for those who were married and middle aged.

Aside from demographic characteristics, characteristics about the technology itself influence acceptance levels. According to the Technology Acceptance Model, acceptance of a new technology is a function of the perceived usefulness of the technology and its ease of use (Childers, Carr, Peck, & Carson, 2001; Goeke & Faley 2007).

Within the context of GPS-based travel surveys, amenability to participate, therefore, is expected to depend on how the technology is introduced (for what purpose and how easy it will be to use), as well as the characteristic of the respondent, with non-respondents expected to be older, less educated and having a lower income.

23.3.3. Non-Response in GPS-Based Surveys

To date, three studies have specifically evaluated non-response in GPS studies. These include one study from Australia and two studies from the United States (both using data from the 2004 Kansas City Regional Household Travel Survey). The purpose of this section is to summarize the findings from these studies.

In Australia, Hawkins and Stopher (2004) investigated non-response among a subsample of GPS participants in the Greater Sydney Metropolitan Region's household travel survey. In this study, wearables and in-vehicle units were offered to 89 households, of which 48 agreed to participate and 41 declined. At the larger survey sample level, non-response bias was minimal, particularly with regard to age, sex, household size and dwelling type. In the GPS subsample, statistically different participation rates existed with regard to county of birth (foreign-born less likely), household type (households other than couples-only or couples with children less likely), educational attainment (lower educated less likely), income (low-income less likely), licence status (non-licensed less likely) and household size (households other than two-person households less likely).

In the United States, the Kansas City Regional Household Travel Survey included a GPS subsample of 228 households. Eligible households from the main survey sample were offered the opportunity to participate in the GPS study. Eligibility criteria included: (1) the household owned at least one vehicle, (2) all household vehicles had a functioning cigarette lighter or 12-volt power adapter (to the best of the owner's knowledge) and (3) the household indicated an interest in participating (NuStats, 2004). Of the 3049 participating households, 1640 indicated an interest in participating in GPS and 228 ultimately provided GPS and CATI data. The GPS households were larger than the general survey population households, owned more vehicles (by definition since zero-vehicle households were excluded) and had higher incomes, on average (NuStats, 2004).

A second study (Bradley, Wolf, & Bricka, 2005) conducted a more in-depth analysis of the Kansas City GPS sample using a logistic regression. The study found that households with more workers per vehicle and higher incomes tended to be more willing to participate, while older respondents and households with fewer vehicles were less willing to participate. In terms of travel characteristics, the strongest explanatory variable was the number of diary vehicle driver trips per household adult.

23.3.4. Expectations

As we move from traditional travel surveys with trip details obtained through paper diaries to GPS-based trip files, the question is how representative the GPS-based files will be, given known levels of unit non-response associated with household travel surveys and known biases towards technology. By identifying the presence and magnitude of non-response in the GPS-based files, practitioners can identify methodological and procedural improvements to mitigate that non-response as we move forward with this alternate data collection approach.

The traditional non-response characteristics include the demographic characteristics of being low income, having lower levels of education, being of minority descent, being young, being a part of a large household and living in an urban area. Travel characteristics include having low travel levels and not using public transit. For most household travel surveys, non-response is present in at least one, if not all, of these areas (although most studies employ design and process techniques to reduce it, or mitigate it post-collection through the use of weighting and adjustment factors).

The attributes associated with those less likely to be accepting of technology are low income, with a lower education level and older. As a result, it is expected that GPS-based trip files will differ from those obtained through traditional methods by higher-than-"normal" levels of non-response for the low-income households and those with a lower education level. However, the use of technology may mitigate the non-response among younger respondents. This will be tested in the next section.

23.4. Unit Non-Response Analysis

The main hypothesis being tested in this study is that non-response in a GPS-based subsample of a larger travel survey effort is a function of the non-response existing in the main sample and the factors that influence technology acceptance. The expectation is that the characteristics of participants in the GPS subsample will mirror those of the main sample, with higher levels of income and educational attainment non-response. However, the non-response associated with being younger is expected to diminish, as younger adults are expected to be more accepting of technology. Ultimately, it is expected that the combined influences of demographic characteristics and technology acceptance factors will result in the unweighted GPS-based trip file being different in composition from the unweighted trip file collected through traditional methods.

This analysis relies on two survey data sets, both of which include a traditional household travel survey effort with a smaller GPS subsample. The first study is the Kansas City Household Travel Survey that was conducted in spring 2004 under the sponsorship of the Mid-America Regional Council and the Kansas and Missouri Departments of Transportation. As part of the Kansas City survey, complete demographic and travel behaviour characteristics of 3049 randomly sampled households were obtained, including details about 32,011 trips for 7570 household members. The GPS component of the study involved equipping the vehicles of 294 households with GPS equipment to record all vehicle travel during the assigned travel period. Of the 294 households, both CATI and GPS data are available for 228 households (NuStats). The GPS subsample in this study was formed through offering the GPS option to eligible households (an "opt in" approach).

The second study is the Washington, DC/Baltimore regional household travel survey, which as of January 2008 was in the final stages of data collection. Sponsored by the Metropolitan Washington Council of Governments and the Baltimore Regional Council of Governments, this survey is comprised of a traditional household travel survey of 15,000 households, with a subsample of 1600 households in the GPS component. The interim data file used for this analysis includes demographic and travel behaviour characteristics for 10,836 households, of which 850 participated in the GPS subsample. The GPS subsample in this study was formed through a two-stage process. First, households were randomly assigned to a GPS or non-GPS condition in the sample. During recruitment, those in the GPS condition were not offered the opportunity to participate but simply introduced to the study as being one where they keep a log for 24 h and use the GPS in-vehicle units for 72 h. Respondents could "opt out" of the GPS portion of the study, but non-GPS condition households were never offered the opportunity to "opt in." By including this survey in the analysis, it will allow for a general comparison of the impact of an "opt in" versus "opt out" design for future studies. It also provides for an alternative GPS approach to how most surveys are conducted in the United States and other countries.

The analysis of each sample is conducted in two stages: first, a descriptive comparison of the GPS sample to the main sample will be undertaken, benchmarking both against census. The purpose of this comparison is to identify the extent to which non-response bias, as reflected by deviation in each sample from the general population characteristics, is present. Second, using the approach from Hawkins and Stopher, a chi-square test of the key demographic variables is undertaken, to determine if statistical differences between the composition of the main and GPS samples exist. Finally, the differences in non-response associated with each approach (opt in vs. opt out) for identifying the GPS samples will be compared and discussed.

The results of this analysis are reflective of travel surveys in which households are asked to both keep travel logs and use the in-vehicle GPS devices for a short period of time. For wearable GPS studies, longer deployment periods and GPS-only travel surveys, these non-response issues are likely to change as a reflection of variations in respondent burden.

23.4.1. Kansas City GPS Sample

A summary of the demographic characteristics of households is shown in Table 23.2. This includes those that were recruited into the study, those that form the main sample (3049 households), the 228 households that comprise the GPS sample and census data for the region.[2]

As indicated in the table, non-response for the main sample (defined as lower proportions than what is present in the census data) is characterized as follows:[3]

- There is slight non-response for households in the urban area (2% lower than census).
- Low-income participation is 6% below census levels.
- Minority participation is about 5% lower than census.
- Hispanic participation is 2% lower than census.
- The proportion of young adults (aged 18–24) is 3% lower than census proportions.
- With regard to educational attainment, considering only those aged 25 + , those with a high school education or less are underrepresented by about 12%.

In comparison, the non-response for the GPS subsample includes the following:

2. Ideally, this analysis and those in the future should also include a comparison of the characteristics of those who "opted in" to the study. In this analysis, the data was no longer available.
3. While most travel surveys have non-response among larger households and those living in urban areas, those variables were used as controls during data collection and thus levels are consistent with census.

Table 23.2: Kansas City non-response.

Characteristic	Recruited households (%)	Main sample (%)	GPS sample (%)	Census data (%)
Household size				
1	27.5	27.6	17.5	27.4
2	32.9	34.2	41.2	32.9
3	16.2	16.1	21.5	16.2
4 +	23.5	22.1	19.8	23.5
Household vehicles				
0	7.4	4.6	0.0	7.4
1	33.9	31.8	27.6	33.9
2	41.7	44.8	57.9	41.7
3 +	17.0	18.8	14.5	17.0
Geography				
Urban	20.6	18.2	12.3	20.6
Suburban 1st ring	26.0	26.2	19.7	26.0
Remainder	53.4	55.5	68.0	53.4
Household income				
<$15k	9.6	8.1	3.6	12.2
$15k–<$25k	9.7	9.3	5.0	11.3
$25k–<$50k	29.8	29.9	27.2	30.1
$50k–<$100k	36.1	38.0	46.2	33.6
$100k +	13.7	14.6	18.1	12.8
Income refusals	5.5	5.5	3.4	–
Minority status				
Minority	N/A	15.1	5.7	20.1
Non-minority	N/A	84.9	94.3	79.9
Hispanic origin				
Yes	N/A	3.2	2.2	5.5
No	N/A	96.8	97.8	94.5
Age				
Child (under 18)	N/A	27.3	28.6	26.6
Young adult (18–24)	N/A	5.3	2.6	8.6
Adult (25–64)	N/A	53.5	58.3	53.6
Older adult (65 +)	N/A	13.9	10.4	11.3

Table 23.2: (*Continued*)

Characteristic	Recruited households (%)	Main sample (%)	GPS sample (%)	Census data (%)
Educational attainment (age 25 +)				
≤High school	N/A	28.6	22.3	40.6
Some college/ technical	N/A	28.2	30.9	30.0
College	N/A	27.1	27.3	19.5
Post-college	N/A	16.2	19.5	9.9

Sources: Kansas City Regional Household Travel Survey (unweighted) and Census 2000; includes the 4001 recruited households, the 3049 regional households that completed the general travel survey (and comprise the "main sample" and 228 households that participated in the GPS component.

- Larger households are underrepresented by 4%.
- There is an increase in non-response for urban dwellers from 2% in the general sample to 8% in the GPS sample. In addition, those living in the first suburban ring are underrepresented by 6% in the GPS sample.
- Low-income participation is now 15% below census levels (it was 6% lower for the full sample).
- Minority participation is now 15% lower than census (it was 5% lower for the full sample).
- Hispanic participation is 3% lower than census (a 1% increase in under-representation).
- The proportion of young adults (aged 18–24) is 6% below census proportions (double the rate of underrepresentation in the full sample). In addition, the proportion of elderly, which were overrepresented in the main sample, is now 1% underrepresented in the GPS sample.
- With regard to educational attainment, considering only age 25 + , those with a high school education or less are now underrepresented by 18% (a figure that was 12% in the full sample).

In reviewing the GPS sample non-response, the growth in non-response among low-income households, those with lower education levels and the elderly is consistent with the literature regarding technology acceptance (these groups are less likely to accept new technology).

A descriptive review of the data with regard to unit non-response suggests differences in the demographic composition of the GPS sample as compared to the main survey sample. However, this provides only a cursory review of the differences. A stronger technique to test for differences between the two samples is the chi-square statistic at the 0.05 level of significance (replicating the approach used in Hawkins & Stopher, 2004). Table 23.3 shows the key variables tested, the chi-square degrees of freedom, the calculated and critical levels of chi-square and the result (reject or do

Table 23.3: Kansas City sample differences.

	df	X^2 calculated	X^2 critical (0.05)	Result
Household size	3	17.665	7.82	Reject
Area	2	15.110	5.99	Reject
Income	6	22.563	12.59	Reject
Minority	1	16.967	3.84	Reject
Hispanic	1	0.769	3.84	Do not reject
Age	3	17.021	7.82	Reject
Education	3	7.329	7.82	Do not reject

not reject). The results of the chi-square testing indicate that the composition of the main and GPS samples are statistically different with regard to the distributions of household size, geography, income, presence of minorities and ages of the household members. The samples are the same with regard to the proportions of Hispanic households and educational levels of the adults in the households.

In sum, the differences between the two samples suggest that non-response issues present in traditional household travel survey samples do carry through to the GPS-based samples, and indeed become greater, particularly with regard to size, urban dwellers, low income, minority and the age of the household members. This is true for the "opt in" approach, where households are offered the opportunity to participate in GPS studies. In the next section, an alternative method for securing household participation in samples is evaluated — one in which households are randomly assigned to the GPS condition and the survey package (diary plus GPS) is presented as one option to the households (opt out).

23.4.2. *Washington, DC/Baltimore GPS Sample*

A summary of the demographic characteristics of households included in this interim sample is shown in Table 23.4. This includes those that comprise the main sample (10,836 households), the 850 households that comprise the GPS sample and census data for the Washington, DC/Baltimore urban area (recruitment data is not available). As indicated above, in this study, households were randomly assigned to the GPS condition. When recruited into the survey, they were told the survey task included both keeping a log for 24 h and having an in-vehicle unit for 72 h. Households that objected to the GPS unit could "opt out" of the GPS condition. However, non-GPS condition households could not "opt in" as they did in the Kansas City study.

In terms of non-response in the main sample, there is evidence of bias against large households (about 14% lower than census), low-income households (3%), minority households (11%), Hispanic households (5%) and children and young adults (9%) (note that these characteristics are unique to the interim file and not representative of the distribution of the final data set).

Table 23.4: DC/Baltimore non-response.

Characteristic	All households (%)	GPS sample (%)	Census data (%)
Household size			
1	33.0	17.5	26.4
2	37.6	41.2	30.9
3	13.3	21.5	17.0
4 –	16.1	19.7	25.7
Household vehicles			
0	8.6	0.0	10.8
1	36.2	27.6	34.3
2	37.6	57.9	37.3
3 +	17.6	14.5	17.7
Household income			
<$15k	5.9	3.6	5.5
$15k–<$50k	22.3	5.0	25.5
$50k–<$75k	19.3	27.2	20.9
$75k–<$100k	14.9	46.2	16.9
$100k–<$150k	23.4	46.2	18.5
$150k +	14.2	18.1	12.7
Income refusals	11.8	3.4	–
Minority status			
Minority	29.0	5.7	39.9
Non-minority	71.0	94.3	60.1
Hispanic origin			
Yes	3.2	2.2	8.8
No	96.8	97.8	91.2
Age			
Child (under 18)	19.9	22.7	25.3
Young adult (18–24)	4.8	4.8	8.7
Adult (25–64)	59.2	56.5	56.9
Older adult (65 +)	16.0	16.1	9.1

Sources: Interim Washington, DC/Baltimore Household Travel Survey (unweighted) and Census 2000; includes the 10,336 regional households that completed the general travel survey through December 2007 and 1008 households that participated in the GPS component to date.

In comparison, the GPS sample shows improvement in the proportion of larger households and proportions of children and young adults in the sample. This reflects mitigation of non-response through the key technology appeal to young adults. Also in line with the technology acceptance literature is the fact that the GPS sample non-response levels for low-income households, minorities and Hispanics further degrade as compared to the main sample.

Table 23.5: DC/Baltimore sample differences.

	df	X^2 calculated	X^2 critical (0.05)	Result
Household size	3	25.559	7.82	Reject
Income	5	67.776	11.07	Reject
Minority	1	19.687	3.84	Reject
Hispanic	1	4.964	3.84	Reject
Age	3	11.245	7.82	Reject

According to a chi-square test, the proportions of the GPS-based sample with regard to household size, income, minority status, Hispanic status and age are statistically different from the main sample files (as shown in Table 23.5). With regard to household size, the difference is largely attributed to the increased proportion of large households. For income, it is the large imbalance of too few low-income households and too many high-income households. For age, the increase in children in the GPS sample (correlated with the larger households) resulted in the statistical difference.

Again, the factors that influence technology acceptance can be seen in the decreases in low-income households and minority households, and the increase in the children's participation.

23.4.3. Summary

The purpose of this analysis was to identify levels of non-response in both the main and the GPS samples for two studies and to identify differences related to both technology acceptance and general participation levels in household travel surveys. In addition, two different approaches to the creation of GPS samples were considered: those in which households can "opt in" (they are invited to do the GPS along with their diaries) and those in which the households can "opt out" (they are introduced to the study as being diaries plus GPS, but if they object to the GPS, they are removed from that condition). Again, these findings are reflective of GPS studies conducted in tandem with a log-based household travel survey and with short deployment periods.

With both approaches, the distributions of the respondent characteristics are statistically different between the GPS and main samples. However, the "opt out" approach used in the DC study resulted in a higher proportion of large households with children in the sample (offsetting even the non-response introduced through traditional methods), an important group for travel behaviour studies. This suggests that the "opt in" approach is preferable to the "opt out" approach currently employed in most U.S. studies.

Regardless of the approach employed, the influence of technology acceptance factors can be seen. In Kansas City, there is a decline in the proportion of elderly, low-income and minority participants as one moved from the main to the GPS samples. For the DC study, there is a decline with regard to participation by low-income people and minorities, but an increase in younger participants in the sample. This suggests that as methods and processes for conducting a GPS-based trip collection study are identified, techniques to mitigate non-response associated with the traditional approach must be enhanced with particular attention to low-income, minority and elderly participants to provide a more representative sample.

23.5. Item Non-Response Insights

Item non-response is a measure of how completely the respondents participated in the survey. Discussed in some detail by Adler (2002), the use of GPS technology to obtain travel behaviour details is thought to reduce item non-response for household travel surveys, particularly with regard to documenting all trips made within the travel period.

As indicated earlier, the use of GPS to obtain more accurate trip details is seen by practitioners to be associated with "little or no respondent burden" (Wolf et al., 2006; Kalfs & Saris 1997; Doherty et al., 2006). Still in its infancy (but growing rapidly), the study of GPS deployment to date has focused primarily on the methods and processes needed to translate the data streams into trips and subsequent analyses. Very few studies to date have considered respondent burden with regard to GPS deployment, although some of the methodological studies have made passing reference to this issue:

- Lee-Gosselin et al. (2006) recognized the problems of equipment deployment and recovery, as well as privacy concerns.
- Doherty et al. (2006), Wolf et al. (2006) and Hawkins and Stopher (2004) all recognized that while the newer GPS units are smaller with longer-lasting power supplies, respondents must still remember to carry them around and charge them.
- As a result, units are not carried around all the time, due to respondent-cited reasons such as a "lack of time" (Draijer et al., 2000) and the "busyness of life" (Swann & Stopher, 2008).

The purpose of this section is to review the GPS process to assess the level of respondent burden as compared to the traditional household travel survey burden, and provide a review of the one study available that specifically addressed this issue with respondents.

23.5.1. Respondent Burden

In a traditional household travel survey, respondent burden is defined as "the perceived 'difficulty', dissonance, or intrusion that individuals associate with a survey that they are being asked to do" (Ampt, 1997). It can be measured in terms of the representativeness of the households that agree to participate (unit non-response). It also can be measured with regard to how completely the households complete the survey task (item non-response).

Most research into the replacement of traditional travel survey methods with GPS cite improved collection of more details and reduced respondent burden as benefits of this technology (Wolf et al., 2006; Kalfs & Saris, 1997; Doherty et al., 2006). Guensler and Wolf (1999) lay the groundwork for the argument that GPS technology allows researchers to obtain more detailed information "without increasing the burden on the participant." However, none of the publications identified to date attempt to quantify the differences in respondent burden or show that it remains unchanged. This summary is an initial attempt to at least identify the differences in response burden.

As shown in Table 23.6, in a traditional survey (Traditional Survey) where household members keep travel diaries for a specific travel period, the burden is typically associated with time on the telephone, reviewing mail (if advance letters are used), keeping a log, reporting results over the telephone and coordinating with other household members (in the case of the main household respondent). If the traditional survey uses both diaries and GPS (Traditional + GPS), the burden is actually increased, because of the presence of a GPS unit. Burden is highest for those studies that include the traditional diary recording and GPS deployment as well as a prompted recall survey (where the GPS data is "fed back" to the respondent and he/she is asked to fill in "blanks" in the data with regard to missed trips, trip purpose, etc.) (Traditional + GPS + Prompted Recall). If the survey uses only GPS to collect trip data (GPS Only), the level of burden may decrease some, but this would be offset if a prompted recall survey is conducted (GPS + Prompted Recall). An attempt to illustrate this is shown in Table 23.6. Note that the actual level of burden varies for each stage, while this example treats each stage with the same weight. Thus, the focus here is on how many stages are present for both traditional and GPS-based travel surveys.

Clearly, the reduction in respondent burden occurs for studies where GPS is used to obtain the travel details, with subsequent imputation of trip details as the trip file is created (GPS Only). What is interesting is that using GPS to obtain trip details followed by a prompted recall survey (GPS + Prompted Recall) effort has about the same level of burden as the traditional survey (Traditional Survey), while studies that employ both traditional methods and GPS, with or without a prompted recall survey follow-up ([Traditional + GPS] and [Traditional + GPS + Prompted Recall]) have almost twice the level of respondent burden. This can result in the unit not being used as planned (so not all trip data is collected by GPS), units being refused (non-increased non-response bias) and respondents forgoing activities, as detailed in the next section.

23.5.2. Respondent Perspective

Only one paper to date directly addresses the GPS burden issue faced by respondents with actual interviews of respondents. As part of an effort to reduce respondent

Table 23.6: Levels of respondent burden.

Survey stage	Traditional survey (1)	Traditional + GPS (2)	Traditional + GPS + prompted recall (3)	GPS + prompted recall (4)	GPS only (5)
Recruitment call	X	X	X	X	X
Provision of diaries	X	X	X		
Provision of GPS units		X	X	X	X
Record travel (manual)	X	X	X		
GPS maintenance		X	X	X	X
Telephone retrieval	X	X	X		
Equipment return		X	X	X	X
Prompted recall survey			X	X	
Total	4	7	9	4	3

burden, improve data quality and capture more complete travel by GPS, Swann and Stopher (2008) conducted a series of focus groups with household travel survey participants who had actually used GPS devices during the survey effort. Four groups were conducted, with good and bad performers who carried GPS units for one-week and one-month periods. The findings from these groups include the following:[4]

- For optimal use, the GPS units should be similar (in terms of size, shape and battery length) to the types of devices that respondents are familiar with: mobile telephones, iPods and commercial GPS units, for example.
- The most common complaint was short battery life (particularly with regard to expectations based on the types of devices that respondents are used to).
- The level of understanding about the task varies across households, with those understanding technology better tending to ignore instructions and relying on their own knowledge with regard to what to do with the units. All households understood the importance of charging batteries nightly.
- Many participants reported keeping their units near their keys and cellular phones to remind them to carry it with them, or charging them in an "obvious and visible location."
- Participants who do not use the devices tend to be someone other than the primary household respondent.
- Most participants denied deliberately leaving the unit behind. However, those jogging or going out indicated they did so because the unit was too bulky. Others suggested that they would accidentally leave it behind due to the "busyness of life" or because it needed to be charged and they did not have time to wait.

These focus groups are the first published dialogues with GPS survey participants regarding respondent burden issues, and were enlightening with regard to the intra-household organization in response to the survey request (who kept the survey going within the household, for example). They also show the importance of the GPS technology matching other known and widely accepted technology (cell phones and iPods), in terms of both proper "upkeep" (charging nightly, ensuring signal reception, etc.) as well as reducing item non-response (less likely to leave the units behind).

23.6. Conclusions

The purpose of this paper was to investigate and identify the non-response challenges of GPS-based travel surveys. The underlying hypothesis was that non-response in a GPS-based trip file is a function of the unit non-response existing in the general

4. The details come directly from Swann and Stopher.

sample as well as bias that is related to technology acceptance. The issue of non-response in traditional household travel surveys has been explored in-depth in the literature base, as well as at these conferences. The issue of non-response in GPS-based travel surveys has received little attention to date, but is likely to become of greater importance as the industry transitions from traditional to GPS data collection techniques and as the analysis of large-scale random-sample GPS studies currently being conducted in Canada, France, Israel and the United States begins.

23.6.1. Unit Non-Response

Prior studies suggest that non-response in the general travel survey sample is a function of income, age, education levels, household size, race/ethnicity and the density in which subjects live. The information technology and marketing literature points to income, education levels, age and gender as factors that influence technology acceptance. This analysis captures both elements, with the non-response in GPS-based subsamples increasing for low-income, lower educated and minority households. Clearly, efforts to mitigate non-response in GPS studies must match or increase those in place for mitigating non-response in the traditional household surveys. This includes using an "opt out" approach for building the GPS subsample.

Given that the study of non-response in GPS-based studies is not well documented, some standards or commonly accepted reporting indicators are needed. These include the type of sample (random or choice based), the recruitment method (part of a traditional study or a GPS-only study? Opt in or opt out?) and recruitment rates, completion rates, use of incentives and any other design factors that influence the level of non-response in the GPS data. Reporting of these elements was not available for the studies reviewed in this paper, but is essential for these and the upcoming GPS studies in order to develop standards and identify best practices with regard to GPS study designs that minimize unit non-response.

23.6.2. Measuring Respondent Burden

Given the characteristics of the non-responders, the mitigation of non-response for GPS-based surveys clearly must be at the forefront of the survey design and process specifications, making the effort as easy as possible for the respondents to participate. In addition, the literature is clear that the units themselves must "mimic" widespread technology (Swann & Stopher, 2008), and the introduction of the technology portion of the study to respondents must clearly convey that the units are easy to use and useful (Childers et al., 2001; Goeke & Faley, 2007). Accompanying materials should be targeted to lower literacy levels. Although not directly discussed, the use of incentives to help balance the "cost" of participating in the survey should be strongly considered.

The measurement of respondent burden is a difficult task. This is sometimes considered in units of time — how many minutes/hours did it take the participating household to move through the survey process? Clearly GPS reduces the amount of time to record travel in a log, but the battery life and portability of the unit are factors that may increase respondent burden and offset the "time savings." One approach for measuring respondent burden may be that developed by Axhausen (2007) to estimate response rates prior to fielding a survey. This approach entails reviewing the survey instrument and process and assigning them summing points based on question type and action (Axhausen, 2007). In this case, the need to charge the unit daily would incur a point penalty, as would the need to participate in a prompted recall survey. The points are summed, and per Axhausen's example, higher point surveys are associated with lower participation rates. Making the process and tasks simpler will help to maximize participation rates and possibly overcome the identified non-response bias.

23.6.3. *Item Non-Response*

As the survey design and processes are being developed, it is important to also consider the factors influencing item non-response: you can get the household to agree to take the units, but will they use them as instructed? Again, here the equipment design is critical: smaller, slimmer, less bulky units that have longer battery lives are preferred (Swann & Stopher, 2008). Respondents should be reminded to keep the units near their keys or phones, so that they remember to take the units with them on all trips. An important note is that as we mitigate unit non-response and have more low-income, low-education and ethnic minority GPS samples, similar qualitative research is needed to ensure that these design and process improvements work for all GPS participants, not just the technophiles.

As with unit non-response, it would be useful to document how the study is introduced to respondents, what instructions were provided to aid in obtaining complete and accurate travel details via GPS and how to handle possible unit malfunctions. These details will help in identifying the factors that influence (both to increase as well as to decrease) the levels of non-response within the GPS data.

23.6.4. *Summary*

In conclusion, this is an exciting time for the transportation industry. As technology advances, so do our models, our data requirements and our options for ensuring high-quality and increasingly detailed data obtained for longer periods of time and with greater accuracy. The GPS technology "fits the bill," but can we identify the appropriate techniques to mitigate non-response and ensure representative samples? That is the challenge for this workshop and our colleagues.

Acknowledgements

Special thanks to those who provided information and publications regarding the use of GPS in specific travel surveys worldwide, including Yasuo Asakura, Kay Axhausen, Sean Doherty, Andy Harvey, Martin Lee-Gosselin, Elaine Murakami, Juan de Dios Ortúzar, Peter Stopher, Orlando Strambi, Jean Wolf and Johanna Zmud. In addition, thanks to Charles Gorugantula of the Mid-America Regional Council and Bob Griffiths of the Metropolitan Washington Council of Governments for permission to use the data from their studies in this analysis, and Heather Contrino and Stephen Greaves for their reviews and input on the paper content. Finally, the author wishes to thank the two anonymous reviewers for their comments and suggestions to strengthen this effort. Any errors or omissions remain the responsibility of the author.

The research described in this paper has been funded wholly or in part by the United States Environmental Protection Agency (EPA) under the Science to Achieve Results (STAR) Graduate Fellowship Program. EPA has not officially endorsed this publication, and the views expressed herein may not reflect the views of the EPA.

References

Adler, T. (2002). Reducing the effects of item nonresponse in transport surveys. In: P. Jones & P. Stopher (Eds), *Transport survey quality and innovation*. Oxford: Elsevier.

Ampt, E. (1997). Understanding the people we survey. Paper read at Transport Surveys: Raising the Standard, Grainau, Germany.

Asakura, Y., & Hato, E. (2006). Tracking individual travel behaviour using mobile phones: Recent technological development. *11th International Conference on Travel Behaviour Research*. Kyoto: IATBR.

Axhausen, K. W. (2007). *Predicting response rate: A natural experiment*. Zurich: Institute for Transport Planning and Systems, Swiss Federal Institute of Technology.

Bonnel, P., Armoogum, J., & Madre, J. -L. (2007). National transport surveys: What can we learn from international comparisons? Paper read at TRB 86th Annual Meeting, Washington, DC.

Bradley, M., Wolf, J., & Bricka, S. (2005). Using GPS data to investigate and adjust for household diary data non-response. *TRB 10th Planning Applications Conference*. Portland, OR: Transportation Research Board.

Bricka, S. (2006). Design considerations to mitigate non-response in regional household travel surveys. Paper read at 10th National Tools of the Trade, Nashville, TN.

Bricka, S., & Bhat, C. (2006). A comparative analysis of GPS-based and travel survey-based data. *Transportation Research Record*, *1972*, 9–20.

Childers, T., Carr, C., Peck, J., & Carson, S. (2001). Hedonic and utilitarian motivations for online retail shopping behavior. *Journal of Retailing*, *77*(4), 511–535.

de Blaeij, A. T., Dijst, M., & Farag, S. (2006). Effects of data collection methods in travel behaviour surveys: Comparing an Internet and a mail sample. Paper read at TRB 86th Annual Meeting, Washington, DC.

Doherty, S. (2008). GPS studies in Canada. Waterloo, Ontario, Canada.

Doherty, S., Papinski, D., & Lee-Gosselin, M. (2006). An internet-based prompted recall diary with automated GPS activity-trip detection. Paper read at TRB 85th Annual Meeting, Washington, DC.

Draijer, G., Kalfs, N., & Perdok, J. (2000). Global positioning system as data collection method for travel research. *Transportation Research Record, 1719*, 147–153.

Elango, V. V., Guensler, R., & Ogle, J. (2007). Day-to-day travel variability in the commute Atlanta study. *TRB 86th Annual Meeting*. Washington, DC: Transportation Research Board.

Forsman, A., Gustafsson, S., & Vadeby, A. (2007). The impact of nonresponse and weighting in a Swedish travel survey. Paper read at TRB 86th Annual Meeting, Washington, DC.

Frei, A., & Axhausen, K. W. (2007). Size and structure of social network geographies. *Arbeitsberichte Verkehrs und Raumplanung, 439*.

Goeke, R., & Faley, R. (2007). Leveraging the flexibility of your data warehouse. *Communications of the ACM, 50*(10), 107–111.

Guensler, R., & Wolf, J. (1999). Development of a handheld electronic travel diary for monitoring individual tripmaking behavior. *78th Annual Meeting of the Transportation Research Board*, January. Transportation Research Board, Washington, DC.

Harvey, A.S. (2008). GPS studies in Canada. Halifax, Nova Scotia.

Hawkins, R., & Stopher, P. (2004). *Collecting data with GPS: Those who reject, and those who receive*. Edited by Institute of Transport Studies. Sydney: University of Sydney.

Im, S., Bayus, B., & Mason, C. (2003). An empirical study of innate consumer innovativeness, personal characteristics, and new-product adoption behavior. *Journal of the Academy of Marketing Science, 31*(1), 61–73.

Itsubo, S., & Hato, E. (2006). A study of the effectiveness of a household travel survey using GPS-equipped cell phones and a WEB diary through a comparative study with a paper based travel survey. *TRB 85th Annual Meeting*. Washington, DC: Transportation Research Board.

Kalfs, N., & Saris, W. E. (1997). New data collection methods in travel surveys. In: D. F. Ettema & H. J. P. Timmermans (Eds), *Activity-based approaches to travel analysis*. Oxford: Pergamon.

Kracht, M. (2006). Using combined GPS and GSM tracking information for interactive electronic questionnaires. In: P. Stopher & C. Stecher (Eds), *Travel survey methods: Quality and future directions*. Oxford: Elsevier.

Kuhnimhof, T., Chlond, B., & Zumkeller, D. (2006). Nonresponse, selectivity, and data quality in travel surveys. Paper read at TRB 85th Annual Meeting, Washington, DC.

Lee-Gosselin, M. (2008). GPS studies in Canada. Quebec, Canada.

Lee-Gosselin, M., Doherty, S., & Shalaby, A. (2006). Personal data collection using mobile ICTs: Old wine in new bottles. *Second International Specialist Meeting on ICT*. The Netherlands.

Macklin, B. (2008, January 7). *US TV trends: The impact of DVRs, VOD, and the web*. eMarketer. Available at: http://www.emarketer.com.ezproxy.lib.utexas.edu/Reports/Viewer.aspx?code = emarketer_2000440&page = 2&xsrc = go_to_2_reportx

Madre, J.-L. (2007). *GPS subset in the French national travel survey*. Washington, DC: Transportation Research Board.

Madre, J.-L., Armoogum, J., Flavigny, P.-O., Hubert, J.-P., Marchal, P., & Yuan, S. (2008). Person-based GPS subset in the French national travel survey (ENTD 2007–2008). *Transportation Research Board Annual Meeting*. Washington, DC: Transportation Research Board.

Nonresponse in household travel surveys. Federal Highway Administration. Available at: http://tmip.fhwa.dot.gov/clearinghouse/docs/surveys/nonresponse/ch2.stm

NuStats. (2004). *Kansas City regional household travel survey: GPS study final report*. Austin, TX: Mid-America Regional Council.

Ogle, J., Guensler, R., & Elango, V. (2005). Commute Atlanta value pricing program: Recruitment methods and travel diary response rates. Paper read at TRB 84th Annual Meeting, Washington, DC.

Richardson, A. J., Ampt, E. S., & Meyburg, A. H. (1996). Non-response issues in household travel surveys. *New Concepts and Research Needs in Household Travel Surveys*. Irvine, CA: Transportation Research Board.

Stopher, P., Swann, N., & FitzGerald, C. (2007). Using an odometer and a GPS panel to evaluate travel behaviour changes. *TRB Transportation Applications*. Daytona Beach, FL: Transportation Research Board.

Stopher, P., Xu, M., & Fitzgerald, C. (2006). Assessing the accuracy of the Sydney household travel survey with GPS. *28th Australasian Transport Research Forum*.

Strambi, O., & Garrido, R. (2006). Processing, analyzing, and archiving standards. In: P. Stopher & C. Stecher (Eds), *Travel survey methods: Quality and future directions*. Oxford: Elsevier.

Swann, N., & Stopher, P. (2008). Evaluation of a GPS survey by means of focus groups. *TRB 87th Annual Meeting*. Washington, DC: Transportation Research Board.

Vijayasarathy, L. (2004). Predicting consumer intentions to use on-line shopping: The case for an augmented technology acceptance model. *Information and Management*, *41*(6), 747–762.

Wolf, J., Bonsall, P., Oliveira, M., Leary, L., & Lee, M. (2006). *Review the potential role of "new technologies" in the National Travel Survey*. London: Department for Transport.

Wolf, J., Schonfelder, S., Samaga, U., Oliveira, M., & Axhausen, K. W. (2004). Eighty weeks of global positioning system traces. *Transportation Research Record*, *1870*, 46–54.

Chapter 24

Nonresponse Challenges in GPS-Based Surveys: Synthesis of a Workshop

Heather Contrino

24.1. Introduction

Using GPS to capture observational travel data has increasingly become a viable data-collection approach for transportation organizations. While the technology for person- or vehicle-based passive trip data collection has been around for some time, GPS is a fairly new methodology in the context of travel behavior research. As GPS data collection still relies on various levels of human acceptance and follow-through (e.g., use, installation, data downloads, battery replacement, or charging), GPS data collection provides some unique quality, methods, and logistics challenges, especially when incorporating into large-scale travel behavior studies. The intent of this workshop was to focus on nonresponse challenges, specifically, in the conduct of GPS-based studies.

This workshop provided a forum for practitioners and researchers to exchange information on the characteristics of nonresponding persons and households in GPS-based surveys with an emphasis on current study design and techniques for mitigating nonresponse. Workshop participants shared experiences in implementing GPS studies and methods for dealing with nonresponse — defined as the failure to obtain participation from households or persons in the GPS component of the study. While the primary discussions and documentation in the workshop centered on unit nonresponse, designers and implementers of GPS-based travel surveys know that both unit and item nonresponse are prevalent and bring a unique set of response challenges. Sound approaches for identifying sources and remedies for both types of nonresponse are still being explored and developed by the transportation research community.

The workshop centered on the resource paper provided by Stacey Bricka, *Nonresponse Challenges in GPS-Based Surveys*. In the resource paper, Dr. Bricka

Transport Survey Methods

ISBN: 978-1-84855-844-1

demonstrates how the increasing independence of GPS-based travel surveys from the constructs of traditional paper diary household surveys is making the issue of measuring and reducing nonresponse increasingly more important for GPS studies. While the majority of GPS studies are still tied to traditional household travel survey methods, there is a growing focus in the research community on the independent application of GPS technology. Dr. Bricka contends that, while GPS-based studies improve response rates among some population groups, the current GPS survey methodology actually exacerbates nonresponse among other population groups.

Additional presentations on the topic of GPS for travel survey applications were provided by Philippe Marchal, *A Study of Non-Response in the GPS Sub-sample of the French National Travel Survey* and Jean Wolf, *Synthesis of and Statistics for Recent GPS-enhanced Travel Surveys*. The discussant for the workshop was Dr. Stephen Greaves (Marchal et al., 2008; Wolf and Lee. 2008).

Topics included in the scope of the workshop ranged from both unit and item nonresponse in the context of study design, sample design and selection, and post processing. Early on in the workshop, participants prioritized areas of discussion. As discussed in the following section, the workshop placed emphasis on three main areas:

1. An inventory of GPS-based surveys and key characteristics of their study design.
2. Sample selection and methodology that may introduce coverage bias and unit nonresponse.
3. Possible solutions in measuring and reducing nonresponse, including additional required research in this area.

24.2. State of the Practice

The workshop had the benefit of participation of representatives of 10 countries, including the United Kingdom, Germany, Israel, Australia, Austria, Switzerland, France, the United States, Canada, and Norway. All of these countries are either implementing or planning a GPS study of regional or national geography. In addition, there are several examples of metropolitan area based GPS studies, primarily in the United States. These last are not included individually in the following table.

As shown in Table 24.1, there is some variation in the methodologies and scope of GPS studies around the world. The data-collection period, sample size, geographic coverage, and household member/vehicle selection vary considerably. The most common, and perhaps most important, similarities across the international research community are the size of the GPS study and its position within the construct of the traditional household travel survey design.

Most countries are planning or have conducted limited GPS studies with small sample sizes. In addition, all of these studies have been conducted as subsamples to

Table 24.1: International GPS pilots and studies.

Country	Sample	Methods	Time period	Other
United Kingdom	TS subsample, 3 regions opt out, $n = 100$	In person, all household members, incentives	7 day	
Germany	TS subsample, 2 regions	In person, all household members	2–4 weeks	Prompted recall
Israel	TS subsample, opt in regional, $n = 77$	In person, most frequent driver, no incentives	48 h	Prompted recall
Australia	TS subsample, regional opt in, $n = 58$	In person/mail back, all household members, no incentives	1 week	Prenotification, correction factors
Austria	TS subsample, 2–3 regions, opt out, $n = 100$	All household members, no incentives, multimode	1 week	
Switzerland	No GPS TS: $n = 25{,}000$	1 day CATI, no incentives, household based	1 day TS	Planned for 2010?
France	TS subsample, opt in national, $N = 800$	In person, 1 hh member 18 + , no incentives	1 week	Advance letter 1 day prompted recall
United States	TS subsample, several MPO and regional studies, $n = 100$–1500	Methods vary	Varies (usually 2 days to 1 week)	
Canada	No GPS — but vehicle-based study, vehicle registration file, as frame, $n = 20{,}000$, national	Trip log, fuel log	Varies (averages 1 week)	Looking into GPS study
Norway	GPS subsample, travel survey $n = 15$–20 K, national	Telephone travel survey, person based, no incentives	1 day	Advance notification

the traditional household travel survey. This is important because it signifies that (1) we are in the very early stages of this methodology in the national travel context, and (2) for many reasons, GPS has not yet become a method in and of itself, integrated with the traditional household travel survey approach.

24.3. Nonresponse Challenges

Both unit and item nonresponse are challenges in all types of social science research, and GPS studies are no exception. Unit nonresponse refers to the absence of participation by the eligible population of interest. In the context of GPS studies, this can include full or partial lack of participation. Item nonresponse refers to missing data within participant records. This can be due to human error, nonparticipation, or technical issues in collecting and deciphering the travel data. While the latter is an important element of nonresponse in GPS studies (e.g., missing days, missing trips, missed stops that are difficult to identify and understand), the workshop's focus was unit nonresponse.

To examine nonresponse in GPS studies, we look first to the current knowledge on nonresponse in traditional household travel surveys. While GPS studies have unique participation challenges and attractions, most are conducted in concert with household travel surveys, a methodology for which much more nonresponse research and quantification has been conducted. In reviewing unit nonresponse in travel surveys, we find both demographic and travel behavioral impacts. The most common demographic characteristics associated with nonresponse in traditional household travel surveys include low income, age less than 25 years, lower level of education, larger households, single person households, minority households, and urban households (Bricka, 2008). Travel characteristics commonly associated with nonresponse in household travel surveys (some correlated to demographic characteristics) include infrequent or non-travelers, high frequency travelers, and households that do not use public transit (Bricka, 2008).

As noted by Bricka (2008), other factors, including the technology itself, influence acceptance levels in GPS studies. Therefore, the way the technology is presented (use, appearance, and logistics) is an important consideration in improving response rates.

The common characteristics of GPS studies and their known or potential impact on nonresponse bias include those described below.

24.3.1. Traditional Travel Survey Integration

One of the major drivers of nonresponse in GPS studies is their almost universal (in terms of study design) dependence on the household travel survey. It is rare that regional or national GPS studies are conducted as stand-alone data-collection efforts. On the contrary, with few exceptions, GPS data collection is incorporated into the existing household travel survey study design.

This is problematic for many reasons, one of which is the compounding of nonresponse due to the reliance on the traditional travel survey sample and recruitment. When GPS data collection is a "follow-on" to a travel survey, nonresponse can occur at several stages: at travel survey recruitment, at GPS recruitment, and at the time of trip data collection. As a result, GPS studies often have the greatest nonresponse levels with traditional travel survey nonresponse groups: large households, low-income households, younger households, and frequent travelers. This is compounded by nonresponse due to aversion to technology, additional respondent burden, and other unknown/not fully researched characteristics of nonresponders.

24.3.2. Integration of the Two Data-Collection Approaches

This integration of the two data-collection approaches (traditional household travel survey and GPS) not only inherently increases nonresponse for GPS studies (as a rule they are typically conducted as a follow-on with already recruited travel survey households) but also limits the ability of researchers to reduce the level of nonresponse bias. The two data-collection approaches are very different and a re-examination of methods may be in order. While there are many benefits to conducting GPS studies in tandem with traditional household travel surveys, there are risks and disadvantages to applying household travel survey methods to an entirely different data-collection activity.

24.3.3. Sample Frames

Another issue that may have an impact on both unit and item nonresponse is the sample frame. While some aspects of this deal with the issue of coverage bias, the frame used in GPS studies may not always facilitate maximizing representativeness and quality. For example, most vehicle-based GPS studies utilize the household from the traditional travel survey as the population of interest.

24.3.4. Opt-In versus Opt-Out

As GPS studies are almost exclusively conducted as follow-ons to household travel surveys, two approaches are commonly used. The first approach, opt-in, provides the respondent with more flexibility in the decision to participate in the GPS portion of the study. The opt-out approach assumes formally (through sample selection) and informally (through questionnaire wording) that the respondent will participate and must actively refuse to be excluded from the GPS portion of the study. The choice to participate in the opt-in approach, while inherent in most social science research, may make it too easy for persons or households to be nonresponders to GPS studies.

In addition, in both recruitment methods, there is little room for typical refusal conversion strategies.

24.3.5. *Person-Based versus Vehicle-Based*

There are large differences between person-based and vehicle-based GPS in recruitment, logistics, coverage, and nonresponse. Across the world, there is much variation both in the type of GPS data collection (person vs. vehicle) and in the number of people or vehicle per household included in the study. Many design characteristics can be different for these two types of studies including, but not limited to, the population of interest; the logistics of equipment shipping, maintenance, and installation; and the retrieval of data. As stated previously, a major research question for GPS studies is whether or not the traditional household travel survey method is the optimal method. The same can be said for person- versus vehicle-based GPS. Using the same methodology for both types of studies may not support the reduction of nonresponse and overall data quality.

24.4. Research Needs

As there is little documented work that identifies and quantifies nonresponse bias, the research areas regarding nonresponse in GPS surveys are numerous. With increasing nonresponse levels across all types of studies with the general public, there is a major need to rethink approaches to new technologies. Clearly technology has the ability to take people out of the equation (if not entirely, then partially), which potentially makes GPS, and technologies like it, a real opportunity in the travel survey methods arena.

Recommendations for first steps in examining nonresponse in GPS studies include building the body of literature on where and to what degree nonresponse bias is prevalent, and focusing on an evaluation of methods that are most effective. In this construct, the recommended research priorities include the following:

- *Defining nonresponse*: There is little agreement or documentation on the definition of nonresponse. While it is fairly straightforward to measure when a household or person refuses to participate, the prevalence of partial households and partial data create many gray areas.
- *Quantifying nonresponse*: There are some examples of attempts to measure nonresponse in GPS subsamples (France, the United Kingdom, and metro areas in the United States are examples), there is really very little documented work in this area. The typical methodology of conducting GPS as a travel survey subsample offers some advantages, as both demographic and trip information is available for nonresponders. However, little is known of households that opted out of the travel survey and little is known about GPS participants with missing data, days, or trips. Identifying nonresponders and quantifying the effect will help to improve the

methods, protocols, and recruitment efforts in these households. While France and the United Kingdom have done some work in this area, there is still very little information on the likely nonresponse groups for GPS.

- *Minimum nonresponse reporting standards*: As recently done with household travel surveys, minimum documentation and reporting standards need to be established for GPS studies. Documentation on the levels of nonresponse at various stages in the GPS study is difficult to obtain, even if it is available.
- *GPS methods:* Some surveys are using a mix of traditional and advanced survey modes, such as travel diaries, CATI, and Internet-based data collection. The common thread, however, is that GPS studies are typically conducted as a subsample to the traditional travel survey, mirroring their sample frame and data-collection methodology. Research is needed to look at GPS out of the travel survey context so that the optimal methods for gaining participation, collecting valid data, and reducing both unit and item nonresponse can be identified.
- *Other methods:* There is a great deal of variation across GPS studies in terms of the methods and protocols employed. This includes the data-collection period, population of interest (household vs. person based), eligible populations, use of incentives, and data quality enhancements such as prompted recall. Research is also needed to fully evaluate and quantify the tradeoffs (e.g., in respondent burden, data quality and accuracy, sample bias, study costs) between the different mobile technology options available (including GPS with CATI, GPS with some form of prompted recall, and GPS-only methods).

24.5. Conclusions

Using GPS to capture observational travel data has increasingly become a viable data-collection approach for transportation organizations. As GPS data collection still relies on various levels of human acceptance and follow-through (e.g., use, installation, data downloads, battery replacement, or charging), GPS data collection provides some unique quality, methods, and logistics challenges, especially when incorporating into large-scale travel behavior studies. While there is much variation in methods and protocols across GPS studies, the most significant common factor in all is that they are conducted as subsamples to the traditional household travel survey. This integration limits the ability of researchers to identify, quantify, and alleviate nonresponse in GPS studies. In addition, there is an absence of standards and documentation with regards to most aspects of GPS research, including nonresponse. Much work needs to be done to maximize the benefits of new technologies and build a body of literature on methods, bias, and protocols within the transportation community.

Acknowledgements

We thank all of the participants of the workshop: Tracy Anderson, United Kingdom; Jimmy Armoogum, France; Anthony Boucher, United Kingdom; Stacey Bricka,

United States; Heather Contrino, United States; Jon Martin Denstadli, Norway; Robert Follmer, Germany; Stephen Greaves (discussant), Australia; Prokop Jehlicka, Germany; Thipphaphone Khenmingmongkhon, France; Catherine Lawson, United States; Michelle Lee, United States; Robert Leore, Canada; Fabrice Marchal, France; Philippe Marchal, France; Rami Noah, Israel; Gerd Sammer, Austria; Nadine Schüssler, Switzerland; Yuan Shuning, France; Susan Swain, United States; and Jean Wolf, United States.

References

Bricka, S. (2008). Non-response challenges in GPS-based surveys. Presented at the 8th International Conference on Transport Survey Methods, Annecy, France.

Marchal, P., Roux, S., Yuan, S., Hubert, J.-P., Armoogum, J., Madre, J.-L., et al. (2008). A study of non-response in the GPS sub-sample of French National Travel Survey. Presented at the 8th International Conference on Transport Survey Methods, Annecy, France.

Wolf, J., & Lee, M. (2008). Synthesis of and statistics for recent GPS-enhanced travel surveys. Presented at the 8th International Conference on Transport Survey Methods, Annecy, France.

Chapter 25

Electronic Instrument Design and User Interfaces for Activity-Based Modeling

H. J. P. Timmermans and E. Hato

Abstract

In this chapter, we address the question if and how modern technology can be used to design questionnaires, diaries, web sites, and experiments to improve the validity of reliability of active data collection instruments. In particular, it discusses the history of computer-assisted activity diary data, reenactment sessions, stated preference methods, and interactive computer experiments with a special focus on the design of these instruments in terms of respondent support and user interfaces. Empirical evidence and experience suggests that although fascinating instruments may increase respondent motivation and involvement and therefore improve the reliability of the measurements, there is also the danger that respondents' answers are influenced by features of the electronic instrument that are not essential, reducing validity and reliability.

Abbreviations

GIS	geographic information system
GPS	global positioning system
GSM	global system for mobile communications
HTML	hypertext markup language
PAPI	paper-and-pencil interviewing
POI	point of interest
PDA	personal digital assistant
RFID	radio frequency identification

Transport Survey Methods

ISBN: 978-1-84855-844-1

SC	stated choice
TSL	Transport Simulation laboratory
VEDETT	vehicle embedded data acquisition enabling tracking and tracing

25.1. Introduction

Over the years, information needs of transportation researchers have evolved substantially due to new topics appearing on research and policy agendas. Traditionally, information was required about the numbers of trips by transport mode, origin–destination matrices, and traffic intensity on links of the transportation network that are disaggregated by population segments purpose and transport zones where relevant. Such data were used to monitor trends in travel demand and served as input to the well-known four-step travel demand forecasting models. These models were typically used to justify new investment in infrastructure or to predict the likely consequences of such new investment on traffic flows and transport demand characteristics. In addition to traffic counts, travel surveys were used to collect the required data.

Recently, data requirements have changed because transport demand management has become increasingly more important. In academia, this has resulted in a shift away from the four-step modeling approach, first to tour-based models and then to activity-based models. Activity-based models, in addition to data about trip generation, transport mode choice, destination choice and route choice, require data on activity participation, trip chaining, travel party involved, and timing and duration of activity episodes. Consequently, at least in academic research, traditional travel surveys, organized around trips, were gradually replaced by activity–travel and time-use diaries, organized around the sequence of activities or time episodes. Moreover, because many policies options such as road pricing were new, impact assessment and feasibility studies by definition could not be based on historical data. Consequently, the application of stated preference methods gained in popularity.

Changing information needs also reflected the tendency of increased complexity in modeling transport demand. For example, while the first generation of activity-based models remained close to trip- and tour-based models (basically adding activity participation), more advanced activity-based models incorporate household decisions, task allocation, substitution between physical and virtual travel, and travel arrangements. Recently, this list has been extended, as witnessed by research on social networks, information and communication technology, and dynamic activity-based models that examine life trajectories, annual regimes, and rescheduling behavior (see Arentze & Timmermans, 2007, for an overview). Moreover, there is evidence of theoretical enrichment, or at least diversification. Researchers have explored decision styles under uncertainty, attitudinal theories, (accumulated) prospect theory, theories of bounded rationality, computational process approaches, coevolutionary choice theory (e.g., Senbil & Kitamura, 2004; Avineri & Bovy, 2008; Polak, Hess, & Liu, 2008; Zhu & Timmermans, 2008; to name a few). Theoretical diversification has triggered the need for data capturing behavioral processes underlying observed travel patterns.

All these developments have led to increased respondent burden in a variety of ways. First, the need to explore the wider context has meant an increase in the *amount* of information that respondents need to process before they complete the requested task. Second, the *consistency* of the responses has become critical. The four-step transport demand models consisted of a set of independent sub-models and hence consistency of data across different choice facets was less relevant. In contrast, the quintessence of activity-based models concerns the interdependencies between choice facets, household members, time episodes, etc. The traditional survey, consisting of a set of simple, independent questions, has been gradually replaced and supplemented with detailed activity–travel diaries, social network generators, game-like interactive computer experiments, virtual reality reenactment protocols, etc.

This increased complexity in measurement instrument and data collection, together with the associated respondent burden, may lead to lower response rates. More importantly, it may cause less valid responses for a variety of reasons. First, even motivated respondents may become tired and/or bored at some stage and start adopting simplifying responses or skip detail. Second, respondents may not be able to maintain a consistent framework across different interrelated sets of questions. Third, more complex tasks require more cognitive effort. Due to cognitive limitations, respondents may not consider all manipulated information but rather may simplify these descriptions and respond on the basis of the information considered (Hess & Rose, 2008).

Modern technology may help to improve the validity and reliability of data collection. For example, to reduce missing information, computer software can prompt respondents if a particular question has not been answered. Similarly, computer software can detect highly unlikely response patterns and ask the respondent to verify the provided information. Interactive computer experiments can create a sense of involvement and may help better retrieving information from memory. Finally, modern technology such as global positioning system (GPS), cellular phones, radio frequency identification (RFID) tags, etc. may be used as an alternative to direct questioning methods.

In this context, the distinction between active and passive data collection is relevant. Passive data collection does not involve any direct questioning or involvement of respondents. Rather, respondents are observed and their behavior is recorded. The main advantage of passive data collection is its unobtrusiveness (unless approval is *a priori* required and respondents are aware of being tracked). Moreover, the accuracy of data collected through modern technology such as GPS, although not perfect, likely outperforms that of direct questioning. On the other hand, by definition, such technology only allows collecting data on observed travel behavior and hence is not appropriate for measuring preferences, decision processes, attitudes, and motivations. Active data collection, which involves direct questioning of respondents, will always remain necessary to collect such data.

This resource paper will be mainly concerned with active data collection. In particular, we will address the question of whether modern technology can be used to design questionnaires, diaries, web sites, and experiments to improve the validity and reliability of these data collection instruments. Due to a lack of systematic methodological research on this topic, this paper is largely based on personal experiences. The key issue is how to design an instrument to create excitement, enhance involvement and motivation, and reduce boredom, while making sure that technology does not become overwhelming. The discussion will focus on electronic instrument design and user interfaces. For each type of instrument, we will first briefly summarize its history and then discuss issues related to instrument design.

25.2. Behavioral Data

25.2.1. Computer-Assisted Activity Diaries

25.2.1.1. History Traditionally, activity diaries were collected using paper-and-pencil formats. Respondents were typically invited to record sequentially the activities they conducted and the travel involved across a designated day. To the best of our knowledge, the first computer-assisted tool for collecting activity–travel diary data was MAGIC (Ettema, Borgers & Timmermans, 1994). It was used for collecting three different kinds of data: (i) a plan for a daily activity–travel schedule, (ii) the actual implementation of the planned schedule, and (iii) adaptation of the schedule in response to external policies. The program was developed to collect data for estimating SMASH, an activity-based computational process model (Ettema, Borgers, & Timmermans, 2000). The focus of this model was not on the outcomes of travel decisions, but rather on the process of adding, deleting, and modifying activities in an agenda. Because such data cannot be easily collected using paper and pencil, dedicated software was written. It recorded the sequential actions of a respondent in developing and modifying a schedule. The same software could be used to record the implemented schedule.

The program was written in MS-DOS. Consequently, it was not very appealing in the sense that size and differentiation of fonts, color schemes, etc., were rather limited. Moreover, because it was text-based, graphics, and other visualizations could not be used. In fact, some routines had to be written in Assembler to quickly capture the data, update screens, and perform very basic operations on the computer screen.

Soon after, software development tools for Windows became much easier to use with the appearance of GUI building tools better integrated with development environments. This made software development much easier, and allowed the combination of text, graphics, and even video and animation. It led to a new generation of computer-assisted activity–travel diaries. CHASE (Doherty & Miller, 2000) has become the best known of these. Its functionality is similar to MAGIC, but it allowed for different time horizons in the planning of activities. Later (see e.g., Doherty, Lee-Gosselin, Burns, & Andrey, 2002), the system was also used to collect data about rescheduling.

New technology triggered elaborations of these tools. For example, CHASE-GIS (Kreitz & Doherty, 2002) can be viewed as a version of CHASE, augmented with a map. It combines the tabular weekly planner view of CHASE with a geographic zoomable geographic information system (GIS)-based vector map interface implemented within the activity–travel entry dialogue. The map view includes a map of the study region that serves both as a display tool and as an input tool for spatial data. All spatial information can be entered by clicking on the map. The shortest route is automatically generated and displayed. This route can be corrected by the user through the addition of further reference points along the route on the map. The route is then redrawn, including the new reference points, and the route length is recalculated. The program iCHASE (Lee, Doherty, Sabetiashraf, & McNally, 2000) is an Internet version of the CHASE software. REACT (Lee & McNally, 2001) is very similar to iCHASE. Both supported, albeit in a limited sense, Web applications. A similar tool, based on a hand-held device, is EX-ACT (Rindsfüser, Mühlhans, Doherty, & Beckmann, 2003).

Another example is VIRGIL, developed jointly between Hasselt and Eindhoven, has the same functionality. It was only used in small-scale projects for internal use, and therefore was never widely published (Janssens, Wets, De Beuckeleer, & Vanhoof, 2004). It did serve, however, as the basis for developing a personal digital assistant (PDA) system with reduced functionality. PARROTS (Bellemans, Kochan, Janssens, Wets, & Timmermans, 2007, 2008) uses GPS to automatically trace travel patterns. In addition, it allows respondents to provide additional information about their (planned and rescheduled) activity–travel behavior. Previous experiences with computer-assisted data collection tools, such as MAGIC and VIRGIL, and knowledge-based semi-automated correction software, such as SYLVIA (Arentze, Hofman, Kalfs, & Timmermans, 1999) were incorporated in the design of PARROTS, subject to limitations of the PDA in terms of memory, speed, and battery life. The most important feature of these tools is their ability to guide the data entry process and to assist the respondent in avoiding inconsistencies. The portability of PDA makes them especially appropriate for real-time data collection.

All these computer-assisted data collection tools were developed for academic purposes. Commercial software for an advanced Internet-based household travel diary survey instrument has been developed by the Resource Systems Group (Adler, Rimmer, & Carpenter, 2002). The instrument first invites respondents to provide information about their household and their vehicles. Next, they are requested to provide details of an activity–travel diary. Pretests of the instrument indicated that respondents prefer to begin listing activities and then providing details about travel between the listed activities. Based on this finding, activities are listed first, then location information for each activity is requested, and finally respondents are invited to provide details about the trips, connecting consecutive activities. Location data can be entered by using an interactive geocoding module that allows respondents to specify locations in terms of street address, nearest intersection, business name, map point, selection of external location, or a selection from a list of previously identified locations (Adler et al., 2002). In the case of address, intersection, and business name, respondents receive immediate feedback whether the information provided resulted in a successful identification. Ambiguous matches are presented to the respondent.

25.2.1.2. Prompted recall instruments In principle, the electronic instruments discussed in the previous subsection collect all information simultaneously and actively. The example of PARROTS is, however, also an example of combined active and passive data collection: route and duration information is collected using GPS, while information about the other choice facets is collected using the diary. The combination of different technology offers interesting new opportunities. In recent years, several prompted recall instruments have been developed and tested. They vary in their means of administration and the technology that is used for passive data collection. Interest in this approach stems from the desire to check GPS traces and augment activity–travel diaries (e.g., Wolf, Myers, & Arce, 2001; Wolf, Schonfelder, Samaga, Oliveia, & Axhausen, 2004). Figure 25.1 displays the main idea, underlying prompted recall instruments (Hato, 2006).

As discussed by Doherty, Papinski, and Lee-Gosselin (2006), several prompted recall interfaces and approaches can be distinguished. *Sequential methods* involve systematically querying subjects by phone or Internet for missing or supplemental travel attributes. Alternatively, the data can be collected simultaneously with tracking. For example, Murakami, Wagner, and Neumeister (2000) used GPS along with a hand-held computer to prompt for trip purpose and passenger information immediately prior to each auto trip. *Temporal/tabular prompted* recall methods involve a time-ordered display of GPS or cellular phone-detected activities and trips, including start and end times, travel times/modes and activity locations. *Spatial*

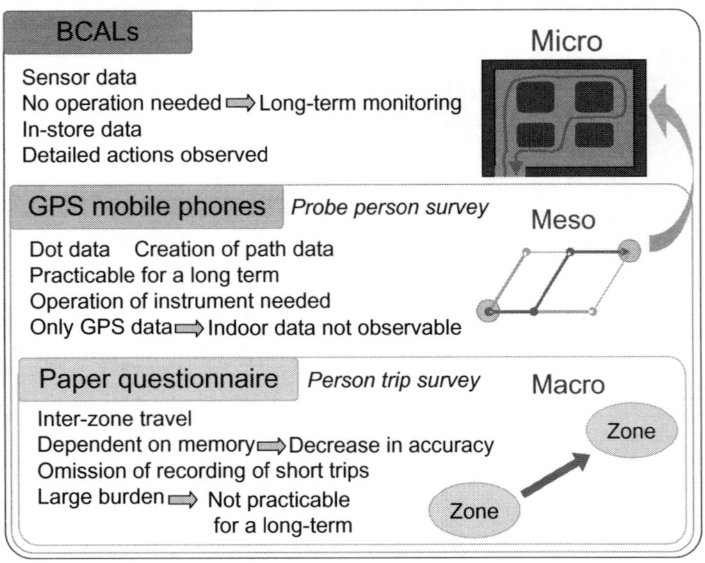

Figure 25.1: Basic principles underlying prompted recall instruments.

prompted recall methods involve use of a GIS or cellular phone to generate a map showing a person's routes, activity stops/location, and an array of text boxes or other map attributes depicting such items as mode, speed, location name, start/end times, trip and activity sequence, overlaid on the road and land-use network for context. Hato and Asakura (2001) used a web-based GIS, along with GPS mobile phones, for input of trip purpose in a travel–activity diary. Marca (2002) used a simple map generator coupled with an hypertext markup language (HTML) form for input of attributes such as destination name and involved persons for each trip segment. Stopher and Collins (2005) used both maps and a tabular (temporal) presentation of vehicle-based GPS-detected trips to prompt for passengers, trip purposes, and travel costs using a paper-and-pencil questionnaire and a map-based animation. Another recent example is Auld and Mohammadian (2009).

An early prototype was Hato and Asakura's (2001) probe person system. Space–time paths of travelers are automatically detected. Travel behavior information is extracted and identified using accelerometers, barometers, sound sensors, etc. Transport mode is identified based on acceleration. Elevation of a subject is identified using atmospheric pressure. This part of the system aims at automatically recognizing behavior by preparing sensor signal sequence data corresponding to behavior patterns. Survey methods using multiple sensors allow a long-term travel survey of high quality, while reducing respondent burden. However, because there are limits to such identification, a web diary supplements the data collection. The process is displayed in Figure 25.2 (Hato & Kitamura, 2008).

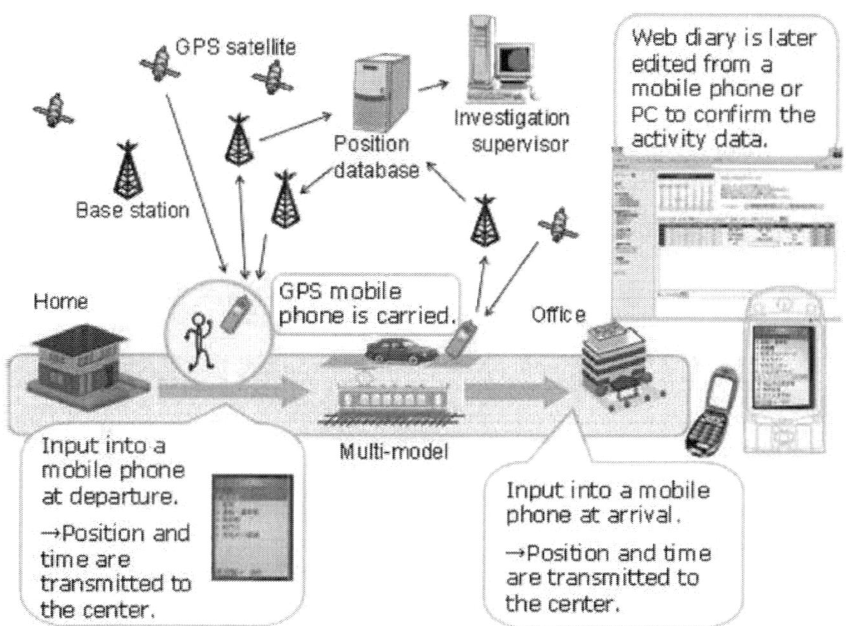

Figure 25.2: Outline of probe person systems.

This type of data collection is not necessarily restricted to GPS devices or GPS-enabled cellular phones. In the context of the FEATHERS project (Janssens, Wets, Timmermans, & Arentze, 2007), a VEDETT (vehicle embedded data acquisition enabling tracking and tracing) was tested. Its main functionality is logging vehicle parameters, but this functionality was extended with location data. The VEDETT is installed parallel to the controller area network (CAN) of the vehicle, allowing monitoring and logging virtually every electrical signal that passes through the CAN. These include traveled distance, vehicle speed, fuel consumption, chosen gear, engine revolutions per minute, engine temperature, and others. The VEDETT tool is also equipped with a GPS-based element and global system for mobile communications (GSM)-based communication technology. The logged data are transmitted to a central data collection point. These data are processed. Total distance is calculated from the mileage logged at the arrival and departure point of the trip. Average vehicle speed is calculated from the logged speed data. Coordinates of the departure and arrival point of the trip are translated into addresses, using reverse geocoding. A web site application prompts respondents to provide information about the motivations for all trips made by car, and the number of passengers in the car. To minimize burden, the web site allows respondents to designate frequently visited addresses as point of interest (POI), labels locations (e.g., "home," "work," "grandparents"). The system also has a learning capability: motivation and number of passengers are inferred from previous evidence and respondents only need to confirm the imputed information.

RFID tags (active and passive) represent another technology, one which is especially useful in small-scale environments. The accuracy of position information of a GPS is about 15 cm to 5 m, and hence there are limits to the spatial accuracy of GPS technology. RFID tags offer an alternative for such cases. Subjects can be asked to carry a card with an IC chip that transmits weak radio waves. Readers installed at specific locations can process these waves to record an individual's position. The accuracy of these measurements varies from 10 cm to 50 m, depending on the type of RFID tag. As with the GPS systems, respondents should check the recorded data and their interpretation, and add any additionally required information that cannot be captured with the tags.

Figure 25.3 gives an overview of a social experiment (Hato & Kitamura, 2008). A RFID tag-based survey system was implemented in a commercial mall in the central urban area of the Matsuyama metropolitan area. The system involved passive tag readers with an effective range of 10 cm to authenticate people who had to put a card over a reader. In addition, active tag readers with an effective range of 50 m were used to trace people.

The experiment was implemented for one month in which incentivizing information was provided to pedestrian visitors. The number of participants in the experiment was 260; the number of samples for only passive tags was 130, and the number of samples for active/passive tags was 130. The results of analysis show that the visiting frequency to the city center increased from 1.5 per week before the experiment started to an average of 2.5 per week after the experiment started. Moreover, the duration of stay in the city center increased from 100 min to 120 min,

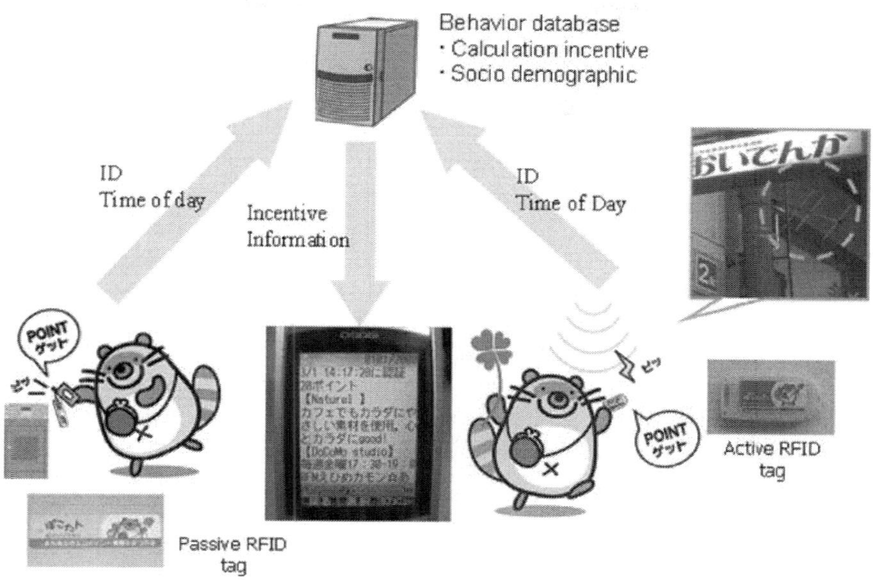

Figure 25.3: Incentive systems for pedestrians based on probe person systems.

while expenditure increased from 2000 yen to 2700 yen. In Eindhoven, Netherlands, a similar system was used to collect data on movement patterns in an office building (Tabak, de Vries, Dijkstra, & Jessurun, 2006). Although these examples may not constitute the core of activity analysis, they do demonstrate the potential of RFID tags.

25.2.1.3. Design and user interfaces This brief overview of computer-assisted activity–travel diary data collection indicates that progress has been made along three lines. First, these data collection instruments have become available for different platforms. The first systems were developed for laptops and thus could only be used in face-to-face interviews. Now, these instruments can also be accessed through the Internet, either by downloading software or directly online, through PDA's and cellular phones. Second, while the first system was text-based only, current systems include different visualizations. Third, while the first systems were integral data collection instruments, newer systems have made increasing use of mixed modes.

The instruments differ slightly in terms of the sequence of information that is collected. Most instruments trigger recall by the sequence of activities conducted. Subsequently, specific data about the activities and travel between consecutive activities is reported. Recalling trips would imply that memory is accessed in a more random order. By focusing on activities or time slots, memory is accessed in a more natural way, leading to a reduced chance of forgetting trips and activities.

The interfaces of the Windows-based instruments are very similar to each other: radio buttons, drop-down lists, text boxes, etc., are used for entering information.

25.2.1.4. Evaluation The superiority of computer-assisted instruments over traditional paper-and-pencil methods has been well documented in the literature (e.g., Arentze et al., 1999; Kreitz & Doherty, 2002). Nevertheless, computers represent a challenge to some respondents. Moreover, learning the software requires time. Furthermore, in many cases, the use of computers implies less flexibility. While a paper-and-pencil instrument can be completed at any time and place, computer-assisted instruments require access to a computer. Activating software does not only take time, but may also involve a higher mental threshold than simply completing a form. It often means that the requested information will probably only be provided at the end of the day or during other time slots when respondents have sufficient time to concentrate on the task, which may lower the reliability of the data. While paper-and-pencil formats can be completed in fragmented time slots, for example, during the commercial breaks of a favorite television show, this does not seem to be a good option for electronic instruments. Mobile technology may change this.

Paper-and-pencil instruments may also have some advantages in terms of layout and speed. Computer screens set constraints. In addition, routing causes a challenge and respondents get easily lost and frustrated when checking previous responses. Starting up the computer, logging in to the Internet, differential support of browsers, variable speeds, and screens, etc. cause Internet-based instruments to take longer to complete. Moreover, these instruments are more sensitive to interruptions and unexpected technical problems.

Thus, while computer-assisted activity–travel diary instruments improve data quality, they may lead to lower response rates either because the computer technology is not available or because they require a substantial amount of effort and tend to be slower. This hypothesis is supported by the study of Kreitz and Doherty (2002), who found that households without computers were less positive and indicated that the computer-assisted instrument required more effort than the paper-and-pencil interviewing (PAPI) method. Households, however, also indicated that the computer-based instrument was more interesting and easier to use, especially in case of long reporting periods, once respondents became acquainted with the software. This result supports our contention that learning software may take too much time if it is only used once.

Collecting location data offers a challenge in its own right. Kreitz and Doherty (2002) evaluated the use of CHASE-GIS. In particular, they examined the map-reading abilities of 178 persons with different socio-demographic characteristics, using the same map as the one used in CHASE-GIS. Subjects were asked to locate four locally well-known objects on the map. The time needed to find the locations and the geographical accuracy of the results was recorded. Findings indicated that only 43.1% of the recorded attempts were correct at the level of a block. However, if respondents knew the location, in only 3% of the cases did they fail to find it. It turned out that women generally needed more time than men to find a location on the map. The older and better educated a person and the higher their income, the faster the person located the locations. Those having a car at their disposal needed less time and achieved more precision than those with no motorized individual means of transport. Respondents who owned or used computers outperformed others in terms of faster and more accurate results. In general, these findings are consistent with a much larger literature on mental maps, mostly conducted in the 1970–1990s, in environmental psychology and behavioral geography (for a review, see Golledge & Timmermans, 1990).

Thus, in conclusion, a map will be a good interface for some respondents, but nevertheless it is advisable to offer multiple interfaces as respondents will have different knowledge about destinations: some know places by name, others can locate them on a map, and yet others may know the names of buildings, street addresses, or ZIP codes.

25.2.2. Reenactment Sessions

The computer-assisted data collection tools are typically designed to collect data about travel behavior. They differ primarily in terms of the guiding principle underlying the design of the instrument: some are based on a sequence of activities, while others are organized around the notion of a day planner. Regardless of these differences, all these systems rely heavily on context-independent memory recall: data collection instruments based on a sequence of activities assume that respondents can recall their behavior based on the sequence of activities they conducted, whereas day planners assume that retrieval is optimized by triggering the temporal sequence of activities.

Tan, Timmermans, and de Vries (2006a, 2006b) examined the feasibility of using virtual reality reenactment sessions. Such sessions may trigger different experiences, improving the reliability and validity of the recalled data. For example, virtual reality may trigger environmental cues, perceived atmosphere, and sense of a place or experience presence. It may be used to capture traveler's actions via "re-living the past," thus creating individual narratives of "what happened." Moreover, it may be of help in navigation and way finding, addressing the issue of spatial awareness.

25.2.2.1. Design and user interfaces Critical in the design of virtual reality-based interfaces is the need to maintain some perceptual similarity between the form of the representation and the environment being represented. Both realism (accuracy in terms of physical representation) and comprehensiveness (degree of completeness of the representation of all functions, environmental characteristics, etc.) are important in this regard. Based on these criteria, Tan et al. (2006b) developed a stereoscopic panoramic system. Panoramas are pictures captured by rotating a camera over 360 degrees. Their SPIN system was developed on the basic principles of stereoscopic vision. The SPIN Viewer has extended features to adjust for the pitch and the yaw at a specific focal point in the panorama to correct visual disturbances. The SPIN Render module handles the rendering of the stereo panoramas with a fixed field of view of 60 degrees vertically and 80 degrees horizontally (Figures 25.4 and 25.5).

Icons for performing actions with a mouse were transparently overlaid over the panorama in order not to obstruct viewing the environment. A radar icon displays

Figure 25.4: SPIN interface.

direction. A max speed icon indicates the maximum speed that is allowed for the current travel mode at the current location, considering traffic regulations. A weather icon shows actual weather conditions including temperature. A speed icon indicates the speed. It can be adjusted by clicking on the speed icon. Clicking on the travel mode icon pulls down the icons of the other transport modes. Time can be adjusted by opening a window that allows an increase or decrease of time in hours, minutes, and seconds.

An activity button opens a window from which an activity can be selected. Start time and duration are entered manually. Clicking an icon pops up a window with a log record of all actions that were performed. As is common practice in panoramic models, the arrow sign only shows up when there is a link. Navigation from one panorama to another is conducted by clicking on the arrow sign. When a "no entry" sign pops up, it indicates that with the current transport medium, the street cannot be entered. During loading, a loading bar is presented for a few seconds, informing the respondent about the time and distance traveled on a link.

Experience of congestion is mimicked by controlling the maximum allowable/possible speed: 0 km/h — standing still, heavy congestion; 25 km/h — slow movement, light congestion; 50 km/h — free flow. Entering a congestion mode by the user will adjust the travel speed of the current travel medium accordingly. The *Experience of time* is represented by displaying "current" time on the screen (HH:MM). This shows the time elapsed during travel or the conduct of an activity.

Stereoscopic Panoramic Interactive Navigation

About
This is a simulated environment of a limited part of the center of Eindhoven. It is composed of stereographic panoramas and stereoscopic glasses are needed to see a 3-D effect.In this virtual environment you can choose the path of travel from origin to destination. The path will be made up of interconnecting nodes. Nodes are at ***intersections*** of streets.

What you can do

- Link to another node; follow a link.
  A ***possible*** link is indicated when the mouse changes to an ***arrow***. To move to another node click on the ***right mouse button***.
  Tip: Place the mouse as close as possible along the path to go where you want to be and not directly on the target location.
- Activity: conduct activity.
  Click on . Follow the instruction in the dialog boxes.
- Change travel mode.
  E.g. from Walking  to Bike , etc.
  Each travel mode has its own characteristics and accessibilities.
- Speed: change the traveling speed.
  Click on the desired speed on icon 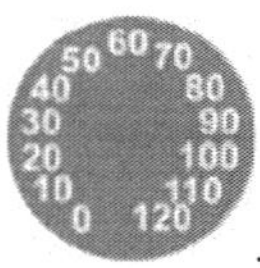

  . If you do not select your desired speed, a normal speed for the current mode is automatically used.
- Look around.
  Hold down the left button and move mouse left, right, up or, down.
- Time: changing the time.
  Click on digital clock. Enter a ***definitive*** time. Do this when you know the exact time that has elapsed for travel or conduct of an activity.
- Check Your Information Travelogue:
  Click on . A description of the events you have performed in the virtual environment is displayed for your verification. If you realized that some data is not correct or the sequence of events is not as you intended, you can go back by click the Back button . A Forward button  is also provided.

Only the 10 most recent actions are displayed. To see the whole history, use the Back and Forward buttons to scroll up and down.

Figure 25.5: Instructions to subjects: Stereoscopic Panoramic Interactive Navigation.

As to the treatment of time, it was practical to run the clock as the user travels to the locations of the activities using the chosen mode at an accelerated rate. At the intersections the clock is suspended. *Experience of weather* is captured by icons displaying the current situation at which the user executes a task. This is to override the static situation at the time the photograph was taken. Finally, *Experience of travel* is simulated by displaying the chosen travel mode with different icons for car, taxi, public bus, bike, and walking as chosen by user. The transfer of one mode to another is depicted in sequence accordingly.

25.2.2.2. Evaluation Tan et al. (2006a, 2006b) investigated the potential of virtual reality systems for data collection by comparing the validity of their stereoscopic, panoramic virtual reality instrument with the validity of a traditional paper-and-pencil diary in the context of pedestrians' activity schedules. They found mixed evidence. On the one hand, the virtual reality system performed better in reporting the number of stops and the number of activities in the schedules. On the other hand, most of the duration aspects were reported better by the traditional paper-and-pencil diary. These findings suggest that reenacting an activity agenda in virtual reality may help subjects to better retrieve the structural facets of their activity–travel diaries. Because the virtual reality system requires them to *reenact* the activity sequence very carefully it may be that memory is triggered more forcefully. The same property of virtual reality systems may also explain why the duration of activities was reported with less accuracy. While a traditional paper-and-pencil task requires subjects to estimate the duration of their activities, a virtual reality system involves compressed time. Cognitively, subjects thus first have to assess whether the time it takes to reenact activities and travel in virtual reality is a systematic function of real clock time. If it is, they need to mentally derive the constant of proportionality. If it is not, the cognitive effort involved is even more problematic as subjects must estimate duration as well. Thus, an individual's perception of time and the experience of the passage of time in virtual reality may trigger different cognitive processes. As a result, the reported durations may be less accurate.

Results obtained for route choice indicated that virtual reality performs worse than PAPI, using a map. The reason might be that a map provides an overview. In contrast, a virtual reality system does not provide that kind of overview. Hence, subjects may have more trouble recognizing how a particular node fits into the larger urban environment.

Thus, this study has provided some evidence that virtual reality may be another potentially relevant technology for collecting activity–travel data, especially for collecting data on the structural facets of activity–travel patterns, such as the number of stops and type and number of activities. It goes without saying that the technology used in this study cannot yet be readily applied to citywide data collection. However, considerable progress has been made in automating the process of creating panoramic views of cities. Soon such data will become available for many cities, at least in Europe.

25.2.3. Stated Preference Methods

25.2.3.1. History The data collection instruments reported in the previous section are concerned with actual travel behavior. Transportation researchers, however, have become increasingly more interested in collecting data about travelers' attitudes and preferences. To some extent, their data requirements can be easily included in any electronic data collection instrument. Other approaches, however, require more sophistication. Examples are the conjoint preference and choice methods, commonly called stated preference and choice methods in transportation research. These methods are more demanding in terms of instrument design and user interface because they rely on principles underlying the design of experiments.

25.2.3.2. Design and user interfaces Most conjoint analysis studies involved verbal descriptions of attribute profiles, although some researchers have used pictorial presentation. Klabbers and Timmermans (1999) developed a multimedia engine for stated choice (SC) and preference experiments that enables researchers to use varying presentation formats (textual, pictorial, auditory, and combinations of these). Pictorial presentation of attributes may lead to more reliable and valid measurement of utilities as pictures may improve the realism of the hypothetical situations.

One step further is the use of virtual reality techniques. What distinguishes virtual reality is the crucial role played by the subject, who is actively involved and not a passive observer. Dijkstra, van Leeuwen, and Timmermans (2003) developed their ICARUS system for interactive conjoint-based analysis in virtual reality of user satisfaction and decision making. The system depicts attribute profile descriptions in a three-dimensional virtual environment and allows subjects to interact with these profiles. A profile consists of a virtual environment model and dynamic virtual objects representing the attributes at different levels. Each level is a different state of the virtual object concerned.

Another, less extreme example can be found in Collins, Hess, and Rose (2007). The authors tried to mimic the decision-making process of a visitor to an online travel agent Site. Their approach included a large number of alternatives, based largely on real flight information, and allowed respondents to use extensive search and sort tools. Figure 25.6 shows the user interface that was used.

25.2.3.3. Evaluation Studies that examined the effects of using visualization have been largely confined to marketing. Much of this research has been based on the *dual-coding hypothesis*, which states that pictures tend to be processed simultaneously in an imagery system, whereas verbal representations are processed sequentially in an independent verbal system (Das, Kirby, & Jarman, 1975; Pick and Saltzman, 1978; Paivio, 1978; Finn, 1985). Green and Srinivasan (1978) suggested that, when faced with pictorial stimuli, respondents do not have to visualize potentially large quantities of information and therefore should be able to process a larger number of attributes. Differences in processing of pictorial and verbal information have been investigated by several authors with mixed results. For example, Holbrook and Moore (1981) in a study of sweaters found that pictorial representations evoked significantly more main effects than did verbal descriptions, but no significant difference in the number of attribute interactions was found. In a replication study on wrist watches, Domzal and Unger (1985) found sharply contrasting results. There was no evidence of any significant differences with respect to the number of main

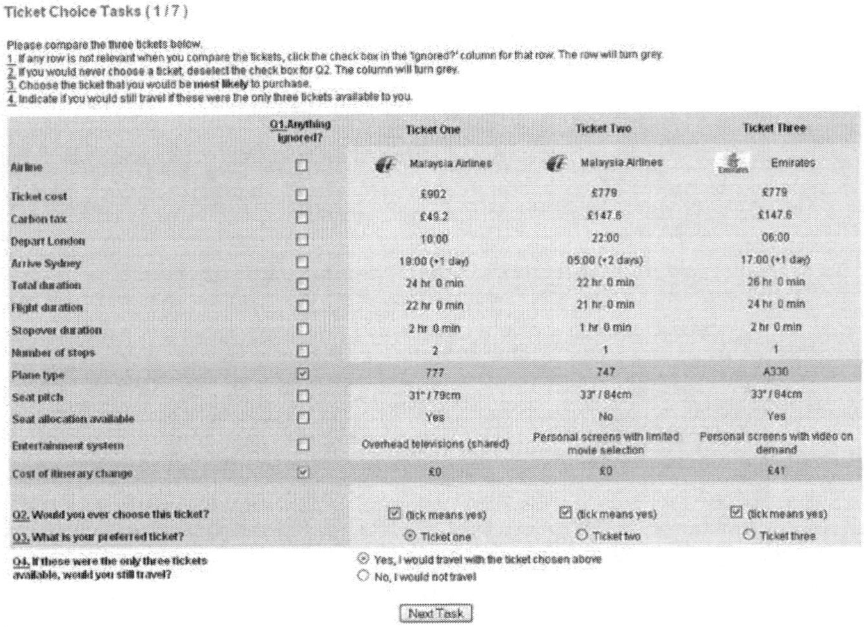
Ticket Choice Tasks (1 / 7)

Please compare the three tickets below.
1. If any row is not relevant when you compare the tickets, click the check box in the 'ignored?' column for that row. The row will turn grey.
2. If you would never choose a ticket, deselect the check box for Q2. The column will turn grey.
3. Choose the ticket that you would be **most likely** to purchase.
4. Indicate if you would still travel if these were the only three tickets available to you.

	Q1. Anything ignored?	Ticket One	Ticket Two	Ticket Three
Airline	☐	Malaysia Airlines	Malaysia Airlines	Emirates
Ticket cost	☐	£902	£779	£779
Carbon tax	☐	£49.2	£147.6	£147.6
Depart London	☐	10:00	22:00	06:00
Arrive Sydney	☐	18:00 (+1 day)	05:00 (+2 days)	17:00 (+1 day)
Total duration	☐	24 hr 0 min	22 hr 0 min	26 hr 0 min
Flight duration	☐	22 hr 0 min	21 hr 0 min	24 hr 0 min
Stopover duration	☐	2 hr 0 min	1 hr 0 min	2 hr 0 min
Number of stops	☐	2	1	1
Plane type	☑	777	747	A330
Seat pitch	☐	31"/79cm	33"/84cm	33"/84cm
Seat allocation available	☐	Yes	No	Yes
Entertainment system	☐	Overhead televisions (shared)	Personal screens with limited movie selection	Personal screens with video on demand
Cost of itinerary change	☑	£0	£0	£41
Q2. Would you ever choose this ticket?		☑ (tick means yes)	☑ (tick means yes)	☑ (tick means yes)
Q3. What is your preferred ticket?		◉ Ticket one	○ Ticket two	○ Ticket three

Q4. If these were the only three tickets available, would you still travel?
◉ Yes, I would travel with the ticket chosen above
○ No, I would not travel

Next Task

Figure 25.6: Stated preference task. *Source*: Collins et al. (2007).

effects occurring under both stimulus representation modes. Moreover, under the pictorial condition, significantly fewer interactions were found than under the verbal condition. Finally, they found evidence of task simplification. Smead, Wilcox, and Wilkes (1981) in a study of drip coffeemakers found that choices from verbal descriptions were perceived to be easier than choices from actual products, and found also differences with respect to relative attribute importance. Anderson (1987), studying preferences for clothing products in an industrial buying context, found the highest fit under the written product descriptions, while actual products without written descriptions resulted in the lowest fit. In contrast, Louviere, Schroeder, Louviere, and Woodworth (1987), studying preferences for state parks, found hardly any differences between the two representation modes with respect to mean part-worths.

It may well be that the dual-coding hypothesis is too simple. As suggested by the dual-loop theory (Rossiter and Percy, 1980) respondents may "translate" verbal information into pictures. Similarly, respondents may infer the benefits of features from pictures. Thus, it is also possible that, because the information in pictures is less structured, respondents may find the task more difficult and process fewer attributes (Zajonc, 1980). Conversely, according to the dual-loop theory, respondents may create mental images from verbal stimuli, which would result in simultaneous information processing even for the verbal mode.

One might argue that pictorial representations contribute to the degree of task realism. If pictures were used to show the design of the product and its attributes, respondents would take these representations into consideration during the conjoint task. If this results in a closer resemblance to actual purchase activities, the external validity of conjoint results will be enhanced. Vriens (1995) found support for such effects that design attributes have higher relative importance in the pictorial mode. However, he obtained a larger number of significant main attribute effects in the part-worth model when either mode was the second one administered, suggesting that respondents learned under either mode. For several dependent measures, superior results were found for the judgments of verbally represented stimuli, suggesting that verbal statements are easier to process.

More recently, the issue of how to best represent attribute levels in conjoint studies has also become an issue in housing and landscape research. It can be argued that, by visualizing the design of a house, subjects may be able to better articulate their preferences as they can better understand and appreciate the design options. On the other hand, visualization may also mean that much more information is provided, some of which may not be relevant to the measurement task. Hence, it is not known *a priori* whether different presentation styles will result in (a) different estimated housing preferences and attribute utilities, (b) equal utilities but with a different error variance, or (c) equal utilities and the same error variance.

Orzechowski, Arentze, Borgers, and Timmermans (2006) examined this issue. To provide empirical evidence on any of these three possibilities, two subsamples, assumed homogeneous, completed a conjoint choice experiment. One subsample was asked to complete a conjoint choice task in which only a verbal description of housing attribute levels was given. The other subsample completed a conjoint choice task that allowed them to inspect the housing design options in virtual reality. The results of the study indicated the lack of any significant difference between the two conjoint choice models in terms of internal and external validity. This suggests that the elicitation of housing preferences, at least in this study, is not strongly influenced by presentation style. Although not significant, the results of this study also suggested that the multimedia task had a lower error variance, which may imply that subjects gave fewer inconsistent or random responses. Thus, a multimedia presentation style may increase the reliability of the measurements. On the other hand, they found evidence of better face validity of the price attribute for the verbal description format.

In another study, Orzechowski et al. (2006) investigated the effects of preexperimental training, using virtual reality, on the validity and reliability of conjoint estimates of housing preference. The findings of this study indicated that the use of virtual reality helping subjects to learn about the nature of the attributes that are varied in the conjoint experiment significantly improves the validity of the results.

A study examining the impact of visualization in landscape design is Mambretti (2007). In a study on safety, she found that that different color schemes used in the visualization had a significant effect on the estimated effects of park attributes on judgments of safety, suggesting that the use of visualization may trigger different, unintended attribute processing, leading to less reliable results. Arentze, Borgers, Timmermans, and Del Mistro (2003) is the only study in transportation that we are aware of examining this issue. Their study involved SC experiment designed to evaluate the impacts of various choice task and respondent related factors on the potential validity of SC data in transportation research. More specifically, they were interested in the effect of using icons instead of verbal descriptions on the reliability of conjoint estimates for less-literate respondents. The choice of transport mode for work trips from a suburb to the Central Business District (CBD) of Pretoria (South Africa) was considered as a case. They found that adding pictorial material to a verbal description of attributes has neither an impact on error variance or on the measurement of attribute weights. The effort it takes to develop and present pictorial material, therefore, was not compensated by improved data quality. These results suggest that visualization cannot compensate for the efforts of task complexity and, hence, does not seem worth the extra effort involved.

25.2.4. Interactive Computer Experiments

25.2.4.1. History Transportation researchers have shown an interest in behavioral change. When data on changing behavior cannot be collected in traditional ways, interactive computer experiments may offer an alternative. In such interactive experiments, subjects indicate any change in their behavior in reaction to contextual change generated by the computer. Thus, while computer-assisted diaries are simply meant to be a recording instrument and computer-assisted conjoint measurements involve subjects expressing preferences for predesign attribute profiles, interactive computer experiments involve a series of related choices for changing situations generated by the computer. Some computer experiments remain close to diaries and conjoint tasks, for example, when subjects are invited to plan an activity–travel schedule under changing situations or when an experimental design is used for a rescheduling task. True interactivity occurs when the questions generated by the computer depend on a subject's previous answers. This is, for example, the case in learning experiments, another major application area of interactive computer experiments.

Computer-based travel simulator tools of varying degree of sophistication have a long tradition in transportation research (e.g., Adler, Recker, & McNally, 1993; Chen & Mahmassani, 1993; Adler & McNally, 1994; Koutsopoulos, Lotan, & Yang, 1994; Bonsall, Firmin, Anderson, Palmer, & Balmforth, 1997; Mahmassani & Liu, 1999; Bonsall & Palmer, 2004). They have been used especially for studying the impact of a variety of advanced travel demand measures in multimodal networks.

25.2.4.2. Design and user interfaces Aspects of design and user interfaces are discussed using the Transport Simulation laboratory (TSL) (Chorus et al., 2006) as an example. Travel simulators tend to have the following components: (i) a transportation network representation, (ii) a story line or narrative, (iii) response options. Key is the story line. It is required because of the relative complexity of the experiments. Because there is a tendency among subjects to grasp the purpose of the

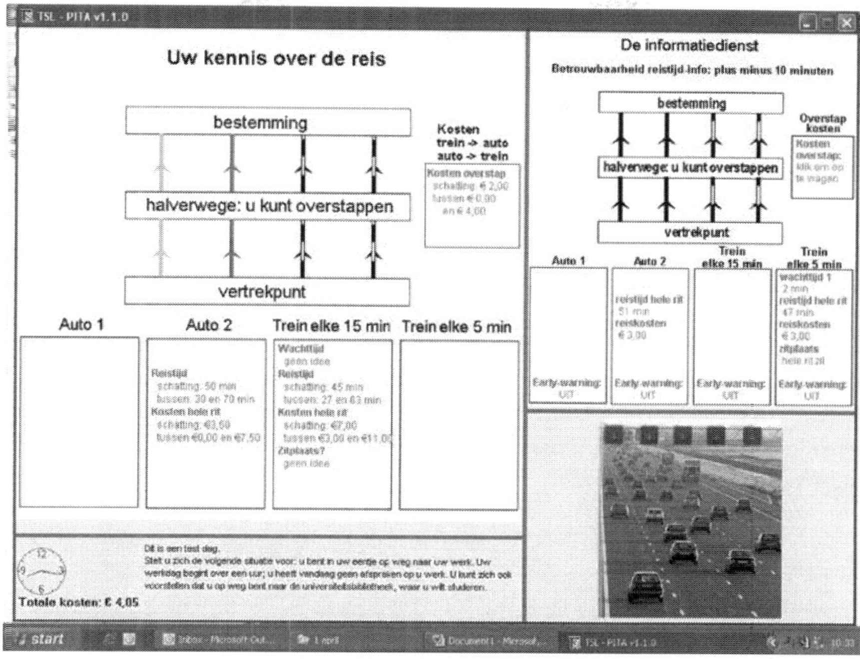

Figure 25.7: Screen plot of *TSL** (*in Dutch*). *Source*: Chorus et al. (2006).

experiment as quickly as possible, the challenge is to formulate aims and objectives and task specification concisely such as to trigger the right mindset. The representation of the network primarily serves as a visual aid of the description of the task. Finally, response options are needed to collect the required data.

Figure 25.7 shows the example of TSL, a travel simulator designed to collect data on the use of travel information. The computer screen consists of four parts: the lower left presents the trip context, the upper left shows the transport network, the upper right presents the information service, and lower right shows a visual aid. The main part of trip context consists of a story line describing trip purpose and possibly preferred arrival times. These story lines are presented at the beginning of each new trip through pop-up windows, after which they were located at the lower left side of the screen. Next to the story line, the trip context displays a clock, presenting accelerated time. It ticks away 1-min waiting time or in-vehicle time per actual second. In pre-trip conditions as well as during information acquisition and at the interchange point, the clock stands still. Finally, beneath the clock, a money counter registers the amount of money spent so far on a given trip, including both travel costs and costs of information acquisition. In this particular case, the transportation network was purely fictitious.

A simulated journey is started by clicking one of the arrows, and confirming the choice in a pop-up window that appears. Directly after confirmation, the subject is informed about actual costs of the alternative and whether or not they had a seat. As the trip commences, the clock starts ticking and the arrow that represents the chosen alternative incrementally turns red, indicating the amount of distance traveled thus far. At the interchange point, subjects are given the opportunity to adapt their route or mode. When interchanging between car and train, an interchange fee was charged. This fee is unknown in advance, although a best guess and a certainty interval were provided. At arrival, a brief trip summary is given, after which a new trip begins. Interactivity is achieved in that subjects can ask the system to provide (the price of) travel information and characteristics of a known alternative.

This travel simulator is just one example and the user interfaces will depend on the scope of the simulator. Simulator differ in terms of degree of realism, from instruments that break down the decision-making process based on some conceptual framework, to simulators that are close to reality with little explanation (e.g., driver simulators). They have in common, however, the fact that the researcher is in charge of the feedback mechanisms, the purpose being to examine user response patterns in a systematic way, and to motivate subjects by offering an interesting and fascinating environment and assisting them in completing quite complicated tasks.

25.2.4.3. Evaluation There is little systematic methodological research that we are aware of on the effect of design and user interfaces of such interactive computer experiments. Personal experience, however, suggests some potential problems. First, subjects tend to be overwhelmed or at least fascinated by the computer environment. Hence, they may view the interactive experiment as a game as opposed to a serious task. This suggests that researchers have to carefully explain the purpose of the experiment. Second, especially for first-time users, it takes time to master the necessary skills to operate the computer program. Researchers need to make sure that subjects know how to operate the instrument, allowing them to focus on the experimental task.

Comments made with regard to virtual reality systems also apply to these systems, which may or may not involve the use of virtual reality. Often, subjects need to come to the location where the system is available because of the required speed, operating system, and for explanation. It seems that, in the short run, such systems will be restricted to relatively small-scale specialized data collection efforts.

25.3. Conclusions and Discussion

Because the analysis and modeling of transport demand have substantially increased in complexity, collecting travel data has become more demanding. At the same time, response rates have dropped. While some commercial firms still report response rates of 70% and higher, reported response rates in the academic literature are more in the order of 10–30%. Increased complexity of data collection may further jeopardize the validity and reliability of data collection.

It seems that data quality depends on a number of factors. First, respondents need to be motivated to participate in the survey. As suggested by a substantial amount of research, once respondents are motivated, they are willing to spend more effort even if the data collection takes more time. Once motivated, a clear explanation of the task will in general improve the quality of the responses. Moreover, distractions should be minimized, so that respondents can focus on their tasks.

Innovative interfaces may help to improve data quality. A professionally designed, high quality, attractive interface may induce people to participate in the data collection and may motivate them. Visualization may help in better representing the attributes or choice alternatives of interest, enabling respondents to better understand the task. We have discussed some fascinating examples of the use of such technology: virtual reality, interfaces with Google Maps, etc. In other small-scale projects we have asked subjects to take pictures, which were automatically geocoded, could be included in blogs, and could be distributed to people included in their social networking web sites and mapped using Google Maps. For part of the world, 3D maps are becoming available, offering new opportunities. Interactive games could be developed as an alternative to or complement of stated preference methods to better understand how people react to a changing context in arranging their daily activities and associated travel.

However, as suggested by the scarce literature, visualization and attractive user interfaces may also cause problems. First, if the visualization contains elements, not relevant to the task, that trigger particular responses, measurements may be biased. In that case, respondents process consciously or unconsciously irrelevant attributes or shift their focus on attributes that are not relevant. Second, abundant visualization may trigger the idea that the data collection is more a game than a serious attempt to mimic actual decision making. In that case, respondents, in trying to serve the surveyor, may guess what kind of response is expected and react accordingly.

Thus, while visualization and attractive design and user interfaces may improve the motivation of respondents and help them to better understand the purpose of the data collection and the set of attributes and choice alternatives, there is also the potential problem of biased information processing. The limited available empirical evidence suggests this may be especially an issue if visual aspects are not relevant. The added simultaneous use of visual information does not seem to make much difference, or worse, leads to less valid and less reliable results.

References

Adler, J. L., & McNally, M. G. (1994). In-laboratory experiments to investigate driver behavior under advanced traveler information systems. *Transportation Research Part C*, *2*, 149–164.

Adler, J. L., Recker, W. W., & McNally, M. G. (1993). A conflict model and interactive simulator (FASTCARS) for predicting enroute driver behavior in response to real-time traffic condition information. *Transportation*, *20*, 83–106.

Adler, T., Rimmer, L., & Carpenter, D. (2002). Use of internet-based household travel diary survey instrument. *Transportation Research Record*, *1804*, 134–143.

Anderson, J. C. (1987). The effects of type of representation on judgements of new product acceptance. *Industrial Marketing and Purchasing*, *2*, 29–46.

Arentze, T. A., Borgers, A. W. J., Timmermans, H. J. P., & Del Mistro, R. (2003). Transport stated choice responses: Effects of task complexity, presentation format and literacy. *Transportation Research Part E, 39*, 229–244.

Arentze, T. A., Hofman, F., Kalfs, N., & Timmermans, H. J. P. (1999). System for logical verification and inference of activity (SYLVIA) diaries. *Transportation Research Record, 1660*, 156–163.

Arentze, T. A., & Timmermans, H. J. P. (2007). Modelling dynamics of activity-travel behaviour. *Proceedings 12th HKISTS Conference*, Hong Kong, China (CD-Rom, 30pp.).

Auld, J., & Mohammadian, K. (2009). GPS-based prompted recall activity travel surveying with automated data reduction. *Transportation Letters, 1*, 58–79.

Avineri, E., & Bovy, P. H. L. (2008). Parameter identification of prospect theory model for travel choice analysis. *Proceedings of the 87th Annual Meeting of the Transportation Research Board*, Washington, D.C. (CD-Rom, 16pp.).

Bellemans, T., Kochan, B., Janssens, D., Wets, G., & Timmermans, H. J. P. (2007). Paper-and-pencil versus personal digital assistant enabled surveys: A comparison. *Proceedings 12th HKSTS Conference,* Hong Kong, China (CD-Rom, 10pp.).

Bellemans, T., Kochan, B., Janssens, D., Wets, G., & Timmermans, H. J. P. (2008). In the field evaluation of the impact of a GPS-enabled personal digital assistant on activity-travel diary data quality. *Proceedings of the 87th Annual Meeting of the Transportation Research Board*, Washington, D.C. (CD-Rom, 15pp.).

Bonsall, P., Firmin, P., Anderson, M., Palmer, I., & Balmforth, P. (1997). Validating the results of a route choice simulator. *Transportation Research Part C, 5*, 371–387.

Bonsall, P., & Palmer, I. (2004). Modelling drivers' car parking behavior using data from a travel choice simulator. *Transportation Research Part C, 12*, 321–347.

Chen, P., & Mahmassani, H. S. (1993). A dynamic interactive simulator for the study of commuter behavior under real-time traffic information supply strategies. *Transportation Research Record, 1413*, 12–21.

Chorus, C., Molin, E. J. E., Wee, B., van, Arentze, T. A., Timmermans, H. J. P., & van Hoogendoorn, S. P. (2006). Observing the making of travel choices under uncertainty and information: Validation of a travel simulator. *Proceedings of the 85th Annual Meeting of the Transportation Research Board*, Washington, D.C. (CD-Rom, 18pp.).

Collins, A., Hess, S., & Rose, J. (2007). Stated preference survey design in air travel choice behaviour modelling. *Proceedings European Transport Conference*.

Das, J. P., Kirby, J. R., & Jarman, R. F. (1975). Simultaneous and successive synthesis. *Psychological Bulletin, 82*, 87–103.

Dijkstra, J., van Leeuwen, J., & Timmermans, H. J. P. (2003). Evaluating design alternatives using conjoint experiments in virtual reality. *Environment and Planning B: Planning and Design, 30*, 357–367.

Doherty, S. T., Lee-Gosselin, M., Burns, K., & Andrey, J. (2002). Household activity rescheduling in response to automobile reduction scenarios. *Transportation Research Record, 1807*, 174–182.

Doherty, S. T., & Miller, E. S. (2000). A computerized household activity scheduling survey. *Transportation, 27*, 75–97.

Doherty, S. T., Papinski, D., & Lee-Gosselin, M. (2006). An internet-based prompted recall diary with automated GPS activity-trip detection: System design. *Proceedings of the 85th Annual Meeting of the Transportation Research Board*, January 22–26, Washington, DC.

Domzal, T. J., & Unger, L. S. (1985). Judgements of verbal versus pictorial presentations of a product with functional and aesthetic features. *Advances in Consumer Research, 12*, 172–268.

Ettema, D., Borgers, A. W. J., & Timmermans, H. J. P. (2000). A simulation model of activity scheduling heuristics: An empirical test. *Geographical and Environmental Modelling*, *4*, 175–187.

Ettema, D. F., Borgers, A. W. J., & Timmermans, H. J. P. (1994). Using interactive computer experiments for identifying activity scheduling heuristics. Paper presented at the 7th International Conference on Travel Behavior, Valle Nevado, Santiago, Chile, June 13–16.

Finn, A. (1985). A theory of the consumer evaluation process for new product concepts. In: J. N. Seth (Ed.), *Research in consumer behavior* (pp. 35–65). Greenwhich, CT: JAI Press.

Golledge, R. G., & Timmermans, H. J. P. (1990). Applications of behavioural research on spatial problems I: Cognition. *Progress in Human Geography*, *14*, 57–99.

Green, P. E., & Srinivasan, V. (1978). Conjoint analysis in consumer research: Issues and outlook. *Journal of Consumer Research*, *5*, 103–123.

Hato, E. (2006). Development of MoALs (mobile activity loggers supported by GPS-phones) for travel behavior analysis. *Proceedings of the 85th Annual Meeting of the Transportation Research Board*, Washington, D.C. (CD-Rom, 19pp.).

Hato, E., & Asakura, Y. (2001). New approach for collection activity diary using mobile communication systems. *Proceedings of the 80th Annual Meeting of the Transportation Research Board*, Washington, D.C. (CD-Rom, 18pp.).

Hato, E., & Kitamura, R. (2008). Data-oriented travel behavior analysis based on probe person systems. *TRB Conference Proceedings 42, Innovations in Travel Demand Modeling: Summary of a Conference* (Vol. 2, pp. 187–196).

Hess, S., & Rose, J. M. (2008). Intra-respondent taste heterogeneity in instantaneous panel surveys. *Proceedings of the 87th Annual Meeting of the Transportation Research Board*, Washington, DC. (CD-Rom, 23pp.).

Holbrook, M. B., & Moore, W. L. (1981). Feature interactions in consumer judgements of verbal versus pictorial presentations. *Journal of Consumer Research*, *8*, 103–113.

Janssens, D., Wets, G., De Beuckeleer, E., & Vanhoof, K. (2004). Collecting activity-travel diary data by means of a new computer-assisted data collection tool. *Proceedings of the European Concurrent Engineering Conference (ECEC)*, April 19–21, Hasselt, Belgium.

Janssens, D., Wets, G., Timmermans, H. J. P., & Arentze, T. A. (2007). Modelling short-term dynamics in activity-travel patterns: Conceptual framework of the Feathers model. *Proceedings WCTR Conference*, Berkeley, USA (CD-Rom, 13pp.).

Klabbers, M. D., & Timmermans, H. J. P. (1999). Measuring tourism consumer behaviour using ESCAPE: A multimedia interview engine for stated choice and preference experiments. In: D. Buhalis & W. Schertler (Eds.), *Information and communication technologies in tourism 1999* (pp. 179–190). Springer Computer Science, Vienna.

Kreitz, M., & Doherty, S. T. (2002). Spatial behavioral data: Collection and use in activity scheduling models. *Transportation Research Record*, *1804*, 126–133.

Koutsopoulos, H. N., Lotan, T., & Yang, Q. (1994). A driving simulator and its application for modeling route choice in the presence of information. *Transportation Research Part C*, *2*, 91–107.

Lee, M., & McNally, M. G. (2001). Experiments with a computerized self-administrative activity survey. *Transportation Research Record*, *1748*, 125–131.

Lee, M. S., Doherty, S. T., Sabetiashraf, R., & McNally, M. (2000). ICHASE: An internet computerized household activity scheduling elicitor survey. *Proceedings of the 79th Annual Conference of the Transportation Research Board*, Washington, DC.

Louviere, J. J., Schroeder, C. H., Louviere, J. J., & Woodworth, G. G. (1987). Do the parameters of choice models depend on differences in stimulus presentation: Visual versus verbal presentation?. *Advances in Consumer Research*, *14*, 79–82.

Mahmassani, H. S., & Liu, Y. (1999). Dynamics of commuting decision behavior under advanced traveller information systems. *Transportation Research Part C*, *7*, 91–107.

Mambretti, I. (2007). *Urban parks between safety and aesthetics: Exploring urban green space using visualisation and conjoint analysis methods*. PhD dissertation, ETH, Zurich, Switzerland.

Marca, J. (2002). The design and implementation of an on-line travel and activity survey. Paper presented at The American Society of Civil Engineers, 7th International Conference on Applications of Advanced Technology in Transportation, Cambridge, MA.

Murakami, E., Wagner, D. P., & Neumeister, D. M. (2000). Using global positioning systems and personal digital assistants for personal travel surveys in the United States. Transport surveys: Raising the Standard, Transportation Research Circular E-008, TRB, National Research Council, Washington, DC. (pp. III-B/1–21).

Orzechowski, M. A., Arentze, T. A., Borgers, A. W. J., & Timmermans, H. J. P. (2006). Alternate methods of conjoint analysis for estimating housing preference functions: Effects of presentation style. *Journal of Housing and the Built Environment*, *20*, 349–362.

Paivio, A. (1978). A dual coding approach to perception and cognition. In: H. L. Pick & E. Saltzman (Eds), *Modes of perceiving and processing information* (pp. 19–29). Hillsdale, MI: Lawrence Erlbaum.

Pick, H. L., & Saltzman, E. (1978). *Modes of perceiving and processing information*. Hillsdale, MI: Lawrence Erlbaum.

Polak, J. W., Hess, S., & Liu, X. (2008). Characterising heterogeneity in attitudes to risk in expected utility models of mode and departure time choice. *Proceedings of the 87th Annual Meeting of the Transportation Research Board*, Washington, DC. (CD-Rom, 26pp.)

Rindsfüser, G., Mühlhans, H., Doherty, S. T., & Beckmann, K. J. (2003). Tracing the planning and execution of activities and their attributes — Design and application of hand-held scheduling process survey. *Proceedings IATBR Conference*, Lucerne, Switzerland (CD-Rom, 31pp.).

Rossiter, J. R., & Percy, L. (1980). Attitude change through visual imagery in advertising. *Journal of Advertising*, *9*, 27–38.

Senbil, M., & Kitamura, R. (2004). Reference points in commuter departure time choice: A prospect theoretic test of alternative. *Intelligent Transportation Systems Journal*, *8*, 19–31.

Smead, R. J., Wilcox, J. B., & Wilkes, R. E. (1981). How valid are product descriptions and protocols in choice experiments?. *Journal of Consumer Research*, *8*, 37–42.

Stopher, P., & Collins, A. (2005). Conducting a GPS prompted recall survey over the internet. *Proceedings of the 84th Annual Meeting of the Transportation Research Board*, Washington, DC.

Tabak, V., de Vries, B., Dijkstra, J., & Jessurun, A. J. (2006). User simulation model: Overview & validation — Capturing human behaviour in the built environment using RFID. In: J. P. van Leeuwen & H. J. P. Timmermans (Eds), *Progress in design & decision support systems in architecture and urban planning* (pp. 117–132). Eindhoven, the Netherlands: Eindhoven University of Technology.

Tan, A., Timmermans, H. J. P., & de Vries, B. (2006a). Route knowledge in complex environments: An analysis of pedestrian recall using stereoscopic panoramic interactive navigation. *Spatial Cognition and Computation*, *6*, 279–294.

Tan, A., Timmermans, H. J. P., & de Vries, B. (2006b). Using a stereo panoramic interactive navigation system to measure pedestrian activity scheduling behaviour: A test of validity. *Environment and Planning B*, *33*, 541–558.

Vriens, M. (1995). *Conjoint analysis in marketing*. Groningen: SOM.

Wolf, J., Myers, M., & Arce, C. (2001). Trip rate analysis in GPS-enhanced personal travel surveys. Paper presented at the 5th International Conference on Transport Survey Quality and Innovation, Kruger Park, South Africa.

Wolf, J., Schonfelder, S., Samaga, U., Oliveia, M., & Axhausen, K. W. (2004). Eighty weeks of global positioning system traces: Approaches to enriching trip information. *Transportation Research Record, 1870*, 46–54.

Zajonc, R. B. (1980). Feeling and thinking, preferences need to inferences. *American Psychologist, 35*, 151–175.

Zhu, W., & Timmermans, H. J. P. (2008). Bounded rationality cognitive process model of individual choice behavior incorporating heterogeneous choice heuristics, mental effort and risk attitude: Illustration to pedestrian go-home decisions. *Proceedings of the 87th Annual Meeting of the Transportation Research Board*, Washington DC. (CD-Rom, 19pp.).

Chapter 26

Electronic Instrument Design and User Interfaces: Synthesis of a Workshop

Elaine Murakami, Sharon O'Connor and Jane Gould

26.1. Introduction

The field of electronic user interfaces has seen a dramatic shift in the last 15 years. Timmermans and Hato in the preceding chapter provide a quick review including the use by Ettema et al. of MS-DOS for a computer-assisted tool for an activity diary called MAGIC in the early 1990s. Then, in 1995, the U.S. Department of Transportation Federal Highway Administration conducted a pilot project using GPS with a handheld Sony computer with a touch-screen interface. The proliferation of Microsoft Windows, Apple interfaces (both on personal computers and with iPhones), WiFi-enabled laptops for widespread Internet access, market penetration of mobile phones and, particularly, GPS-enabled mobile phones have made this a rapidly changing (and fun) environment. Electronic tools and advanced applications that travel survey researchers have used recently include:

Internet/Web
mobile phones (iPhones™, Blackberry™)
PDAs (Windows Mobile, iPod)
GPS and RFID
Web-based satellite imagery (Google Earth™)

These electronic tools have been used by survey respondents as well as trained interviewers in computer-assisted telephone interview (CATI) telephone centers.

While the available electronic tools and applications have changed and improved, the field of travel demand modeling has changed as well, with academia shifting away from the traditional four-step models to activity-based models that incorporate

Transport Survey Methods

ISBN: 978-1-84855-844-1

household decisions, task allocation, and substitution between physical and virtual travel. There is increasing use of activity/travel surveys to measure change in travel behavior in the context of environmental sustainability. The transportation community has increased the use of stated choice experiments, and new electronic tools have enabled these stated choice experiments to be dynamic and more complex.

While transportation analysts are demanding more complex data from respondents, response rates are dropping as respondents get pulled in more directions than ever: family, friends, work, social networking, volunteerism, and entertainment. Our challenge, therefore, is to design user interfaces that:

- pique respondent interest quickly
- motivate participation and involvement
- are easy and straightforward to answer
- reduce respondent burden as much as possible

To achieve these objectives is to both increase response rates and ensure data reliability.

26.2. Issues

26.2.1. Interface Design

Respondents don't read instructions.

It is the burden of the survey designer, not the respondent, to create a survey that is intuitive and user-friendly. Because there is a wide variation in respondents' familiarity with electronic devices, "one size does not fit all" and the survey researcher should, to the maximum extent feasible, customize the survey to the user. Characteristics such as cultural differences, ages, and languages should be considered in the design. For example, bold red text is considered negative in some cultures, and very positive in others, so the use of red text to display error messages may not be universally appropriate.

Electronic user interfaces can range from the almost nonexistent (such as a passive GPS data logger) to the very complex. For some GPS data loggers, the user interface is limited to whether or not the battery needs to be recharged and whether or not the device is receiving the necessary satellite signals. One GPS data logger designed by Peter Stopher includes a voice reminder and colored light indicator for both items. But for other surveys, there are many elements that need to be considered in the design.

For complex travel surveys, there exist a wide variety of tools. Lapietra and Pronello (2008) describe the many communication tools that can be incorporated into electronic instrumented surveys. "Language" is not limited to words, but may include audio, photos, video, graphics, symbols, and virtual reality/3D. Their work using graphic sliders also incorporated humorous photos to convey concepts as "crowded" or "spacious" for seat availability, but were not necessarily photos of a specific bus or rail car.

Timmermans and Hato (2009) reviewed examples of incorporating virtual reality by conveying ranges of congested travel using photos and video. They remind us that if a surveyor uses photos, he or she must be very careful about what is shown so that the respondents are not reacting to a characteristic that was unintentionally included (e.g., one photo is on a sunny day and one is dark and rainy; or one has green and another red but the survey is asking about bus size (number of seats)). Sometimes it may be better to use a black and white drawing to avoid unintended characteristics.

Web-based surveys can be designed to allow respondents to choose formats they prefer. For example, in identifying activity locations or trip destinations, respondents can choose between using "Yellow Pages" business name lookups, street names/intersection, or point/click on a map (Resource Systems Group, 1999). Some design decisions affect respondents' ability to simply read the survey — it should be kept in mind that small font sizes may be hard to read for certain individuals.

One of the benefits of recent improvements to Web-based surveys is the ability to incorporate responses to current travel into complex stated preference (SP) survey questions. Using familiar travel or other characteristics can increase a respondent's comfort level in the survey process (Lapietra & Pronello, 2008; Rose, Collins, & Hess, 2008). Similarly, fuzzy logic and string recognition can allow for open-ended responses instead of fixed choices (Horeni, Arentze, Dellaert, & Timmermans, 2008).

26.2.2. Recruitment and Motivation

Regardless of the technology used, recruiting and motivating participants continues to be problematic. Even when geographic data is obtained from GPS or other passive devices, there must be a level of trust between the respondent and the interviewer (or interview interface) so that behavioral details can be determined. Social exchange theory states that all human relationships are formed by the use of a subjective cost–benefit analysis and the comparison of alternatives (Dillman, 2007; Homans, 1958). For example, when a person perceives that the cost of an exercise, such as speaking with a phone interviewer or taking a Web survey, outweighs the perceived benefit, then the theory predicts that the person will choose to ignore or abandon the exercise. To successfully conduct any data collection study, researchers need to use procedures that ensure potential respondents see a clear benefit to completing the survey or household travel diary.

Issues discussed:

- How to design a Web interface that could work with National Travel Surveys.
- How to improve CATI systems, not just for self-reported surveys.
- Can nontraditional incentives (e.g., discounts on auto insurance) be a motivating factor for participation in highly detailed travel behavior studies?

26.2.3. Techno-Literacy Challenges

Familiarity and comfort with different technologies that can be used in activity/travel behavior research varies widely. There are clear age and income effects, and some technologies, such as mobile phones or Internet, simply may not be available in certain geographic locations.

Some groups are early adopters — college students are very often excited to use new computer-based technologies. Students at UCLA easily incorporated the use of Google Maps™ into a simple origin–destination survey (point on "home" and point on "school"). Similarly, Eiji Hato's work in Japan, documented in the preceding Timmermans and Hato chapter has volunteer respondents participate in complex surveys incorporating GPS location transmitted using a mobile phone, trip purposes using drop-down lists from the mobile phone screen, and user blogs to annotate daily travel and activities on a Web-based diary. His volunteer respondents are mobile-phone "literate"; other, less technical users would have much higher challenges taking part in this kind of study.

Sometimes we are forced to incorporate new technologies in our daily lives. One example given was a local government requiring people to pay their taxes using a Web-based payment system. Similarly, it is nearly impossible to get a paper ticket for airline travel. In the case of airline travel, there are now large populations who are familiar with online ticketing applications (OTA), so that designing an SP survey that has the look and feel of an existing OTA makes comprehension easier for respondents (Rose et al., 2008) (Figure 26.1).

However, in other communities, the access to computers and to Internet can be limited, or even close to zero. Older populations, impoverished households, and those with health or other disabilities are the most likely to have obstacles in using these new technologies, and it behooves the researcher to design different user interfaces that work best for different populations. But, even for populations who may not have much familiarity with new electronic tools, they may have a longer learning curve to use the interface, but they may enjoy participating in a novel experiment. For example, some field surveyors might use a touch-screen computer for a simple external station survey, or field surveyors might use a touch-screen computer for a visitor survey at an airport or historical landmark. A simple interface would not preclude someone without a computer or without Internet from successfully completing this kind of survey.

Another problem may be in the range in map-reading skills. Showing a GPS trace of trips from one (or many days) may be easy for some respondents to annotate, but it may be equally difficult and stressful for others who do not like to read maps, or have never looked at satellite imagery in Google Earth™, MapQuest™, or similar software.

There have been many improvements to incorporating GIS into activity/travel surveys. Previously, using GIS required people to install software onto their PC, which increased respondent burden. Today, Web-based GIS is much more user-friendly, and there have been advances in desktop GIS as well. Similarly, proprietary geographic files have incorporated business names and landmarks, resulting in improvements to

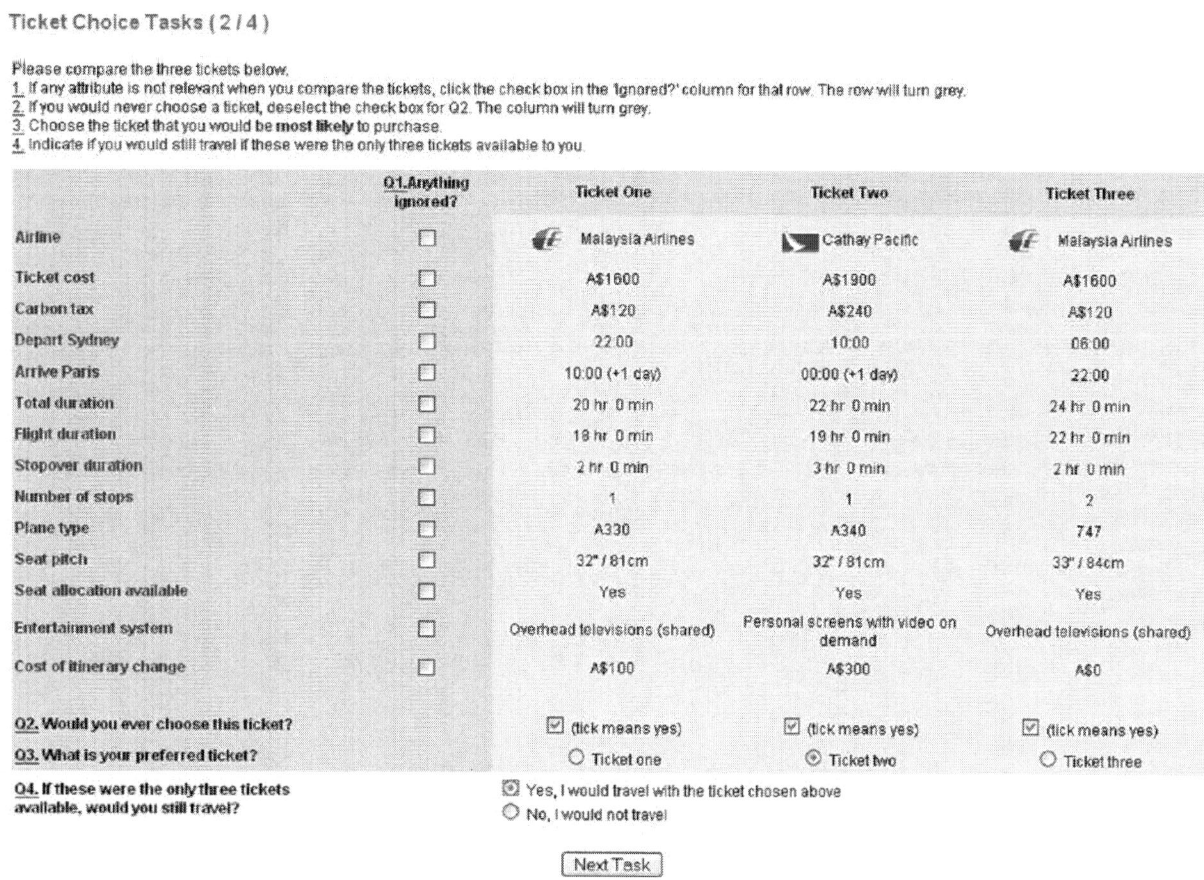
Ticket Choice Tasks (2 / 4)

Please compare the three tickets below.
1. If any attribute is not relevant when you compare the tickets, click the check box in the 'Ignored?' column for that row. The row will turn grey.
2. If you would never choose a ticket, deselect the check box for Q2. The column will turn grey.
3. Choose the ticket that you would be **most likely** to purchase.
4. Indicate if you would still travel if these were the only three tickets available to you.

	Q1.Anything ignored?	Ticket One	Ticket Two	Ticket Three
Airline	☐	Malaysia Airlines	Cathay Pacific	Malaysia Airlines
Ticket cost	☐	A$1600	A$1900	A$1600
Carbon tax	☐	A$120	A$240	A$120
Depart Sydney	☐	22:00	10:00	06:00
Arrive Paris	☐	10:00 (+1 day)	00:00 (+1 day)	22:00
Total duration	☐	20 hr 0 min	22 hr 0 min	24 hr 0 min
Flight duration	☐	18 hr 0 min	19 hr 0 min	22 hr 0 min
Stopover duration	☐	2 hr 0 min	3 hr 0 min	2 hr 0 min
Number of stops	☐	1	1	2
Plane type	☐	A330	A340	747
Seat pitch	☐	32" / 81cm	32" / 81cm	33" / 84cm
Seat allocation available	☐	Yes	Yes	Yes
Entertainment system	☐	Overhead televisions (shared)	Personal screens with video on demand	Overhead televisions (shared)
Cost of itinerary change	☐	A$100	A$300	A$0
Q2. Would you ever choose this ticket?		☑ (tick means yes)	☑ (tick means yes)	☑ (tick means yes)
Q3. What is your preferred ticket?		○ Ticket one	◉ Ticket two	○ Ticket three

Q4. If these were the only three tickets available, would you still travel?
◉ Yes, I would travel with the ticket chosen above
○ No, I would not travel

Next Task

Figure 26.1: Screenshot of OTA look-alike stated preference task (Rose et al., 2008, reprinted with permission).

integrating GIS with CATI surveys. Auld, Williams, Mohammadian, and Nelson (2009) at UIC have designed a Web-based survey which integrates GPS traces with Google Maps™ and asks respondents to annotate their travel including activity planning horizons and re-scheduling decisions (Figure 26.2).

Therefore, if different technologies are used to capture information from a range of populations, it becomes difficult to know how best to harmonize the results with the assurance that each survey is comparable.

26.2.4. Privacy and Ethics of Survey Research

Because passively collected GPS data allows such precise locations, the survey research community, mobile phone providers, and toll tag operators must make it clear to users that location information is being captured when using GPS-enabled cell phones or RFID toll tag transponders. Users must understand that there is a trade-off between privacy and location-detecting technology. It is one thing to want one's location known when calling for E-911 (emergency response) and another to get the location of the nearest ATM or specialty restaurant using a mobile phone or satellite navigation service.

As survey researchers, we should design systems where respondents have an option to "turn it off." As with today's examples, GPS data loggers and mobile phones, people can choose to leave the device at home, or turn it off, when they do not want to be tracked.

As survey researchers, it is our responsibility to maintain the confidentiality of the data and not misuse the data. In the long run, establishing trust of respondents contributes to increasing participant motivation, whether it is for a specific survey or surveys in general.

26.2.5. Web-Based Surveys

Several specific issues related to Web-based surveys were discussed. These included: (1) how to get representative e-mail addresses for a sample, (2) how to recruit respondents for a Web-based survey, and (3) potential benefits from a Web-based survey that are not possible from traditional survey methods.

E-mail address samples present particular issues. While there are some commercial sources for e-mail address lists, we are unsure of their completeness and how people with multiple e-mail addresses can be sampled appropriately. (Interestingly, surveyors used to ask people how many telephone lines they had.) In some situations, a business or a university, for example, may have a complete list of their staff or students who are identifiable, but finding a representative sample for a general-purpose survey can be nearly impossible. There are several commercial consumer panels, but recent work has documented different biases in different commercial panels.

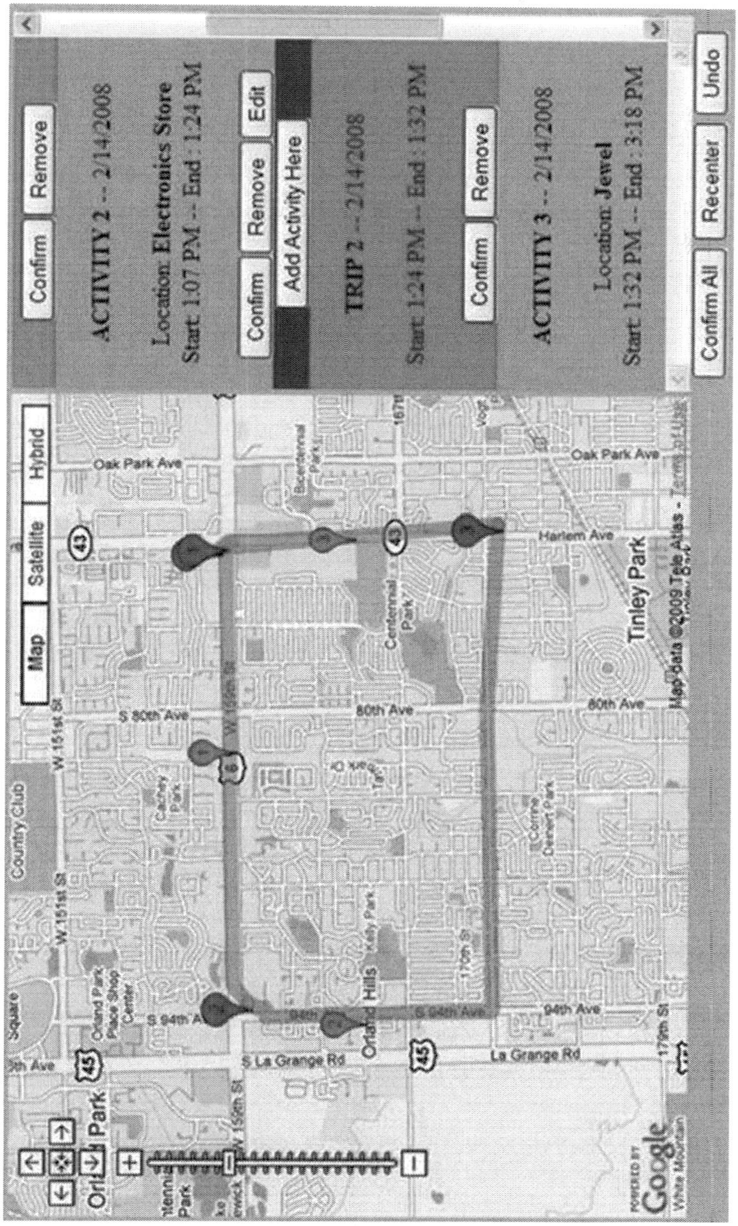

Figure 26.2: Example GPS trace in Google Maps™ (Auld et al., 2009).

There are still questions on how best to recruit respondents to a Web survey. If we recruit volunteers, what kinds of bias are introduced, and what methods should be used to adjust for that bias? We do not know if incentives can be used to improve representativeness, and reduce sample bias.

Web surveys offer the opportunity to capture GUI user metrics. For example, how long users take for each question and how much they switch back and forth between questions. Thus, there is potentially more information for the survey researcher to analyze and understand how respondents are completing their survey, beyond the actual responses.

26.3. Research Needs

In a perfect world, we would have many different tools to address the variety of respondents. For example, a very small, nonintrusive, passive system with GPS and other sensors to capture destination locations and modes would be ideal. This instrument should work "everywhere," that is, it should not be location-dependent. While there are some examples of implanted chips (pet and children IDs, and for automatic payment in bars), it is unlikely that this technology will be used in activity/travel behavior studies. But, the ubiquitous mobile phone, soon to be integrated with GPS on a widespread basis, may be close enough for activity tracking in time and space.

In a perfect world, we would also have integration between passive and active activity/travel behavior systems, along with integration with traffic operations data. For example, we would have GPS traces over many days, combined with respondents' motivations and purposes for travel. Traffic operations data includes both traditional data from traffic counting loops, noninvasive traffic counting and vehicle classification systems, video, and speed data from mobile phones.

How do we get to the perfect world from where we are today? Our priorities for research:

1. Develop interfaces that are congruent with the mental models of how respondents think about transportation issues and follow generally accepted/used Web design conventions.
2. Explore innovative means to recruit and motivate respondents.
   a. How to use blogs
   b. How to use familiar "shopping" Web designs as interface
   c. How to use virtual reality to assist with memory recall (e.g., showing people their GPS traces and street-level photos of their destinations)
   d. Understanding people who refuse to participate. What populations are we missing?
   e. Understanding inaccuracy in survey responses, understanding people who purposefully sabotage surveys.
3. Address whether respondents understand the trade-off between privacy and location-based services, including emergency response.

4. Develop algorithms to synthesize and process data streams into useable measures, for modeling and other needs.
5. Address how to harmonize data when using multiple survey modes.

26.4. Summary

While this workshop focused on electronic user interfaces, participants in the workshop were mindful of the use of the data after any travel survey was complete. One concern, especially tied to GPS data, is the massive amounts of data that can be collected, finding economic methods to transmit and store the data, and the ability to find open-source tools for processing the data into usable travel behavior data components. Processing GPS data also entails the ability to combine the data with GIS basefiles including highway and transit networks, including facility types, fares, and tolls. Another issue is data accessibility and the protection of individual confidentiality when such precise geographic information is captured.

Acknowledgments

The workshop participants were Bruno Allard, Canada; Andrew Collins, Australia; Daniel Ekwalla, Cameroon; Oliver Horeni, the Netherlands; Stephen Jones, United States; Nick Lanyon, United Kingdom; Solveig Meland, Norway; Michael Meschik, Austria; Virpi Pastinen, Finland; Dimitris Potoglou, United Kingdom; Cristina Pronello, Italy; Harry Timmermans, the Netherlands; Coralie Triadou, United Kingdom; and Liva Vågane, Norway.

References

Auld, J. A., Williams, C., Mohammadian, A., & Nelson, P. (2009). An automated GPS-based prompted recall survey with learning algorithms. *Transportation Letters: The International Journal of Transportation Research*, *1*(1), 59–79.

Dillman, D. A. (2007). *Mail and internet surveys: The tailored design method*. New York: Wiley.

Homans, G. C. (1958). Social behaviour as exchange. *American Journal of Sociology*, *63*, 597–606.

Horeni, O., Arentze, T. A., Dellaert, B., & Timmermans, H. J. P. (2008). Design of a computer-assisted instrument for measuring mental representations underlying activity-travel decisions. Presented at the 8th International Conference on Transport Survey Methods, Annecy, France.

Lapiettra, M., & Pronello, C. (2008). Experimental results from the test on the new stated preferences valuation method, the continuous attribute-based stated choice, in the web-based surveys. Presented at the 8th International Conference on Transport Survey Methods, Annecy, France.

Resource Systems Group. (1999). Computer-based intelligent travel survey system: CASI/internet travel diaries with interactive geocoding, available at: http://www.rsginc.com/pdf/RSG_SBIR_Report.pdf

Rose, J., Collins, A., & Hess, S. (2008). Search-based internet surveys: Airline stated choice. Presented at the 8th International Conference on Transport Survey Methods, Annecy, France.

Timmermans, H. J. P., & Hato, E. (2009). Electronic instrument design and user interfaces for activity-based modeling. In: P. Bonnel, M. Lee-Gosselin, J. Zmud & J.-L. Madre (Eds), *Transport survey methods: Keeping up with a changing world.* Bingley, UK: Emerald.

Chapter 27

Visualize This: Opportunities and Challenges for the Travel Survey Community

Catherine T. Lawson

Abstract

New computer-assisted techniques for visualizing data are evolving in a number of areas in transportation. For example, in engineering, 3D visualization and microsimulation techniques are being applied for the identification and evaluation of geometric and operational solutions for improving visually impaired pedestrian access to roundabouts and channelized turn lanes. For planning, visualization is being used for corridor analysis. Data visualization is being used as a tool for improving decision-making within transit agencies, as well as a tool for understanding truck trip generation on highways. Many of these new techniques take advantage of archived intelligent transportation systems (ITS) data. Examples of other innovative data sources include global positioning systems (GPS), geographic information systems (GIS), computer-aided design (CAD), and a variety of visualization tools available for use with travel survey data. As these various techniques and software applications move forward, consideration needs to be given to how the "lessons learned" from these applications can facilitate the use of data visualization techniques for travel survey data analysis and decision-making.

27.1. Introduction

The last half of this decade may well become known as the "viz age," during which visualization has transformed transportation research and practice. Called for explicitly in the SAFETEA-LU (Safe, Accountable, Flexible, Efficient Transportation

Transport Survey Methods

ISBN: 978-1-84855-844-1

Equity Act: A Legacy for Users) legislation,[1] visualization has captured the imagination of those developing techniques and tools to facilitate applications. Uses of visualization for transportation planning include visualization for public involvement, context-sensitive design, 3D computer-aided design and drafting (CADD) and project development. Exciting new opportunities are emerging for the use of visualization by the travel survey community. This community has the responsibility of conducting large-scale household surveys to better understand the transportation needs of the travelling public. These researchers are now experiencing a surge of applications of visualization, which will require strategic data stewardship to support and facilitate widespread use. The first section of this paper reviews examples of visualization applications used by various transportation-related fields. The second section will describe emerging uses of visualization techniques for the travel survey community. The final section will discuss issues facing the travel survey community as the opportunity to use visualization moves from research to practice.

27.2. What Is Visualization?

Visualization is not a new technology — or even a new term. The use of physical models for infrastructure planning (e.g. by-passes and interchanges) represented the notions of visualization, yet only provided a static supply-side view. Graphic artists and data analysts painstakingly produced visualizations prior to the age of computers. Tufte (1997) provides excellent guidelines on how visualization techniques can enhance the transmission of complex information to different audiences. He also warns of the misuse and potential for miscommunications, citing the space shuttle Challenger disaster.[2] He suggests that superior methods for displaying and analysing data are needed to produce truthful, credible and precise findings.

Visualization can be considered application dependent. Different techniques for the use of visualization vary from a simple drawing of an intersection, or a photosimulation illustrating current and proposed road alignments, to a 3D real-time simulation of a proposed toll plaza, depending on the context and the perceived need. The widespread use of visualization techniques is now possible due to advances in computing, both in hardware and in software. Software packages for transportation planning are capable of displaying both supply (the network) and demand (vehicle flows) at the same time.

For the transportation community, visualization is gaining popularity, yet, according to Hixon (2007), visualization remains misunderstood. He cites the fact

1. The SAFETEA-LU contains legislation addressing visualization techniques in plans and Metropolitan Transportation Improvement Projects (Manore, 2007a).
2. Tufte reviews the events prior to the explosion of the Challenger space shuttle and concludes the flawed decision to launch was due to poor visualizations, particularly in the ordering of the data and lack of clarity in cause and effect.

that no standards or guidelines are available to transportation professionals on the use and benefits of visualization. To date, decision makers have limited knowledge or understanding of the vast array of uses and tools that could bring them and the entire transportation community value. For the purposes of this paper, the travel survey community is a relative newcomer to the opportunities and challenges of visualization techniques, products and practices.

27.3. How Is Visualization Being Used in Transportation Planning?

Historically, visualization has been used in transportation planning in the form of graphic artwork. More recently, photosimulations made it possible to show how various alternative planning outcomes would look compared to an actual photograph. A growing interest in incorporating visualization techniques into many more areas of transportation planning spawned the Fifth International Visualization in Transportation Symposium and Workshop, held in Denver, Colorado, on October 22–26, 2006. This event brought together a wide range of visualization users (Transportation Research Board, 2006). The aim of the conference was to share experiences, primarily for project design and planning purposes. The recently published *Visualization for Project Development: NCHRP Synthesis 361* provides in-depth information on the progress and challenges for visualization (Hixon, 2006). Hixon (2007) sees visualization becoming "mainstream" in the near future as personnel trained in the use of a variety of visualization software packages join the workforce.

A powerful force behind the movement to increase the use of visualization is the 2006 FHWA interim guidance for implementing provisions of the SAFETEA-LU. This legislation mandates the use of visualization for transportation improvement plans and long-range statewide transportation plans. It also calls for the use of visualization "prior to the adoption of statewide and metropolitan transportation plans and metropolitan TIPs addressing SAFETEA-LU provisions." These mandates will prompt new, innovative approaches to using visualization in transportation (Hixon, 2007).

27.3.1. Computer-Aided Design and Drafting Applications

Computer software products, such as AutoCAD or MicroStation, have evolved from two-dimensional (2D) flat drawings to three-dimensional (3D) renderings of building and roadscapes. These visualizations make it possible to share a vision of a proposed project with technical staff and the public. Advancements in hardware (e.g. monitors and graphics cards) also improve the level of understanding of what a potential project would look like upon completion. For example, Figure 27.1 shows the features of a proposed road project.

Figure 27.1: A computer-aided rendering of a road project. *Source*: Charles (2006), reprinted with permission.

According to Walker (2007), the greatest impediment to use of visualization is the credibility factor. If the intended audience for a visualization presentation feels the renderings are unrealistic or purposely slanted for a particular outcome, credibility will suffer. When using visualization as a communication tool, special attention should be paid to following good practice. Care must be taken when renderings include vegetation and landscaping, traffic volumes, the field of view or people. He also recommends avoiding unnecessary technical stunts.

27.3.2. Animation and Simulation Software Applications

With advances in the computer gaming industry and in animation techniques in general, the transportation community was able to incorporate simulations into several important projects. An early example of computer-assisted visualization was the Border Wizard (see Figure 27.2). The purpose of the Border Wizard was to illustrate the trade-off between productivity and homeland security issues (Caldwell, 2002). The illustrations needed to demonstrate how freight might be penalized at national border crossings without a careful understanding of how to implement "post-9/11" measures for security. The intended audience was a diverse group of stakeholders involved in the movement of goods (i.e. truck drivers and shippers) and national security staff (i.e. those involved in customs, immigration and goods

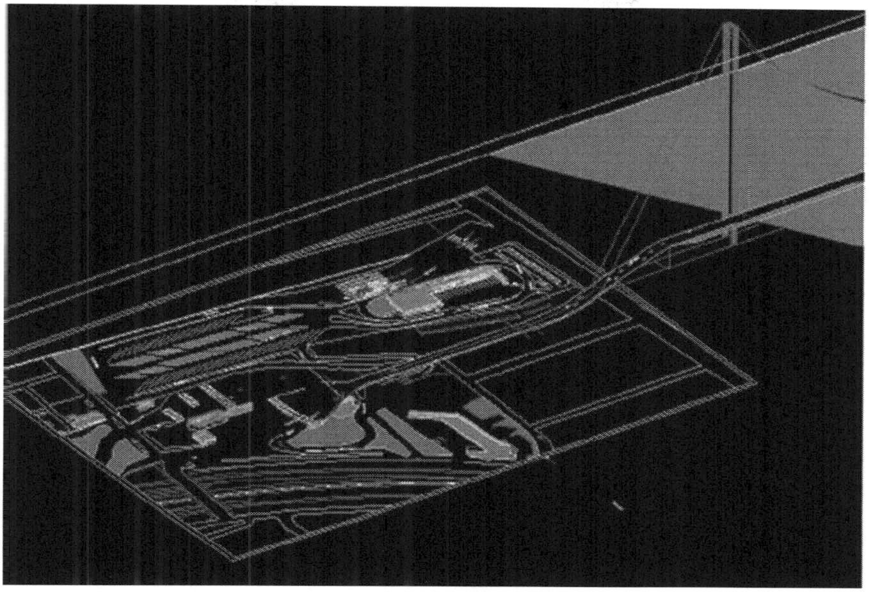

Figure 27.2: Border Wizard application showing animated truck processing. *Source*: Caldwell (2002), reprinted with permission.

inspection). Visualization was perhaps the only technique for bringing this group together to agree on just how the system would be developed and operated. Yet this early attempt is difficult to interpret and clearly primitive in its representations.

Another early example that brought attention to the usefulness of computerized visualization was TRANSIMS, a microsimulation modelling strategy that offered users an opportunity to "see" second-by-second movement of travellers in a 3D environment. Figures 27.3 and 27.4 illustrate the capabilities of the TRANSIMS visualizer. Although the intended purpose has yet to be accomplished (see Lawson, 2006), the value of the graphical interface provided a new awareness of the usefulness of visualization. At the same time, the visualizations are not particularly informative until the viewer understands the abstract nature of the representations of traffic volumes and congestion on the network. In many early attempts, the various features may have created more confusion than clarity.

Manore (2007b) cites other emerging applications, including the development of driving simulators; terrestrial, mobile and airborne LIDAR surveying systems; and temporal visualization using radio frequency identification (RFID) technologies. All of these promising uses will need a well-trained and computer-literate workforce. A number of universities are adding areas of study to their programmes to accommodate these future demands.

The Fifth International Visualization in Transportation Symposium included the opportunity for the transportation community to take advantage of the ever-expanding resource of archived ITS data and other high-end data production efforts (Transportation Research Board, 2006).

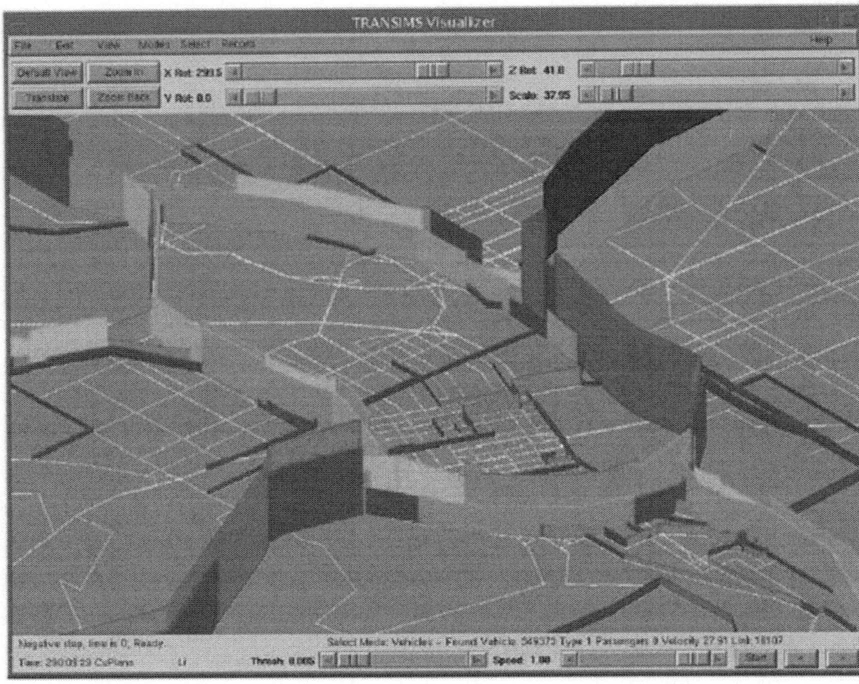

Figure 27.3: TRANSIMS visualization of traffic in Portland, Oregon. *Source*: Travel Model Improvement Program (TMIP) Graphic, provided courtesy of LANL, reprinted with permission.

Discussions at the session on Opportunities and Challenges in Transportation Data Visualization focused on the potential for making sense of large volumes of data. One area where considerable progress has been made is the use of traffic data (e.g. loop detector data). Examples of current websites for access to archived data from these systems include the Freeway Performance Monitoring System (PeMS) (https://pems.eecs.berkeley.edu/) and the Portland Oregon Regional Transportation Archive Listing (PORTAL) (http://portal.its.pdx.edu/). Claramont, Jiang, and Bargiela (2000) developed a framework for integrating, analysing and visualizing traffic data using GIS in the United Kingdom.

27.3.3. Data Visualization as a Tool for Transit Planning

A few pioneers in data analysis are able to visualize archived intelligent transportation systems (ITS) data from transit operations. For example, Kimpel (2007) provides an excellent discussion on the visualization of bus transit performance information. According to Kimpel (2007), transit agencies are faced with a set of internal problems including data collection, post-processing, integration

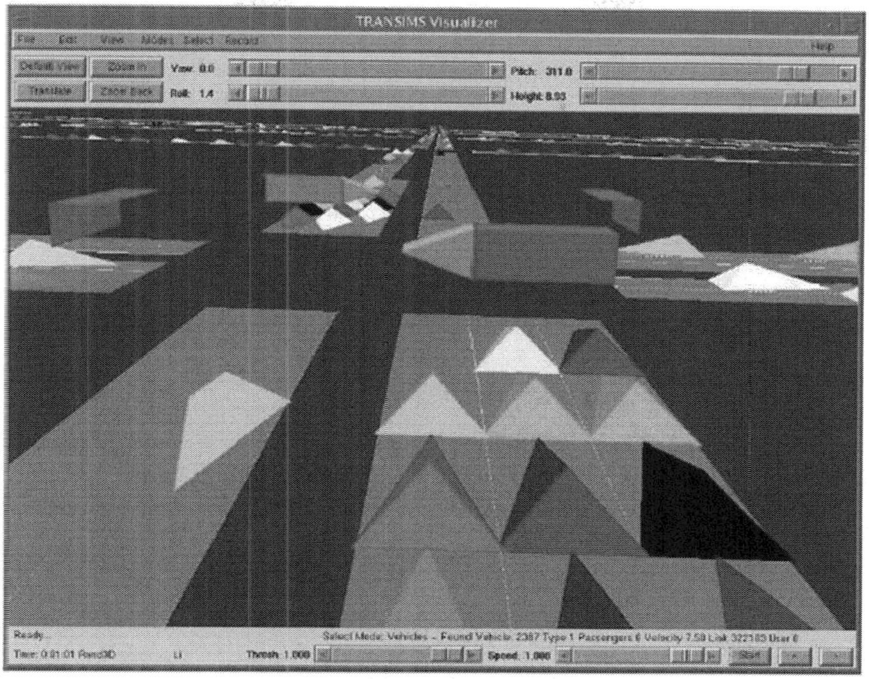

Figure 27.4: TRANSIMS visualization illustrating vehicle movement. *Source*: Travel Model Improvement Program (TMIP) Graphic, provided courtesy of LANL, reprinted with permission.

with back-end systems, limited budgets, inadequate training of in-house staff, inadequate fusion with existing geographic information systems (GIS), incompatible file formats and other operational issues.

A solution to these problems can be found by moving transit agencies to more general-purpose relational database management systems, GIS software and standard data models. In other words, all the issues preventing data visualization can and have been overcome by a few progressive agencies. As a result, it is possible for transit agencies to visualize performance data (see Figure 27.5). The major software facilitating this advance has been GIS (e.g. ESRI), using 2D and new 3D applications.

Tri-Met, the regional transit agency for the Portland, Oregon Metropolitan Area, has developed a number of custom software applications that make it possible for them to query and display a vast reservoir of archived automated passenger count (APC) and automated vehicle location (AVL) data streams. Figure 27.6 displays the high-demand locations within a regional context, as well as illustrating parts of the region that are underserved by transit. This technique could be enhanced by superimposing a map of the area to give the viewer a better sense of place.

To advance the use of data visualization, Kimpel (2007) suggests moving away from "one-off projects" and striving to incorporate longer-term performance-monitoring

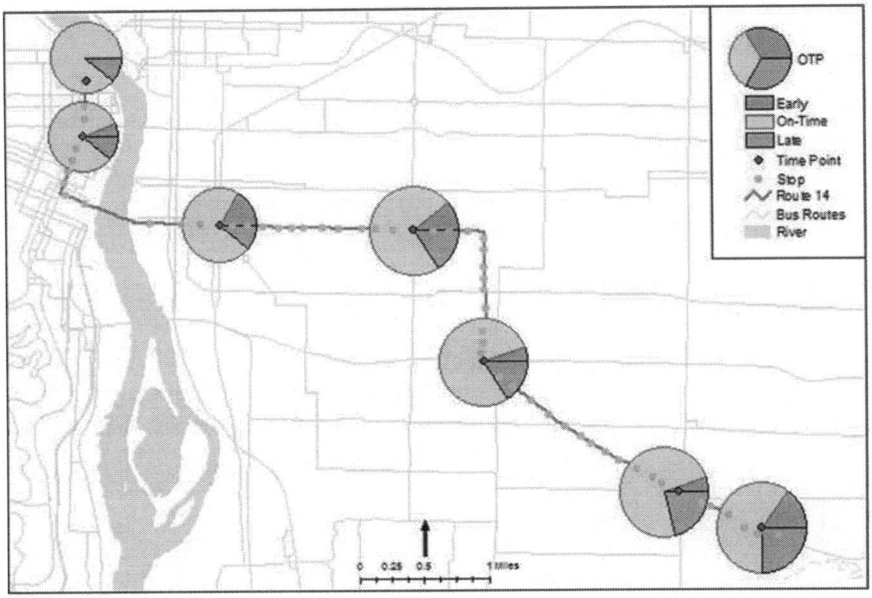

Figure 27.5: On-time performance visualization by stop location. *Source*: Kimpel (2007, p. 20), reprinted with permission.

Figure 27.6: On-time performance visualization by stop location. *Source*: Kimpel (2007, p. 27), reprinted with permission.

and evaluation programmes for transit operations. It is now possible to create information graphics "on-the fly" on an ongoing basis, using modern databases, GIS and statistical analysis software. These applications can serve day-to-day management needs of transit agencies, adding value with each glance at the archive ITS data and assisting long-term planning.

Transit data performance monitoring and transit data visualization require a level of data stewardship that treats data as an asset and leverages this asset to benefit the organization. Those agencies able to capture their data and reap the benefits of visualization need to provide guidance for promoting these capabilities to other transit operations.

27.3.4. Data Visualization as a Tool for Understanding Freight

Increases in the movement of goods, the subsequent increase in truck traffic and concerns about growing activity levels at ports and across international borders have made it necessary to gather new forms of freight data. Schmitt, Dronzek, and Rohter (2007) suggest the use of visualization to provide a better understanding of these activities. Early use of archived ITS data by the University of Washington produced new forms of graphical illustrations that provided levels of congestion over time along the length of I-5 corridor in the Seattle area (see Figure 27.7). The blue areas indicate extreme congestion by time of day (x-axis) by location (associated with the drawing of the road segment). The dynamic nature of the congestion apparent in these visualizations provides freight transportation planners with insights, both temporal and spatial.

Using GIS technologies, the New York State Department of Transportation (NYSDOT) funded a series of "proof-of-concept" studies to visualize not only the truck trip traffic, but also the employment landscape surrounding the road network.

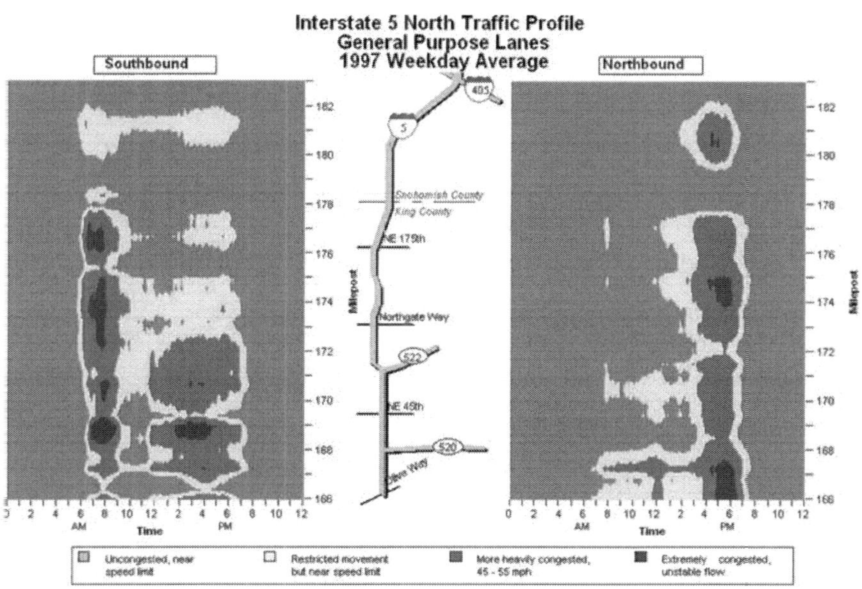

Figure 27.7: Visualization of traffic congestion over time and space. *Source*: Mark Hallenbeck, 1999 (personal communication, University of Washington/TRAC), reprinted with permission.

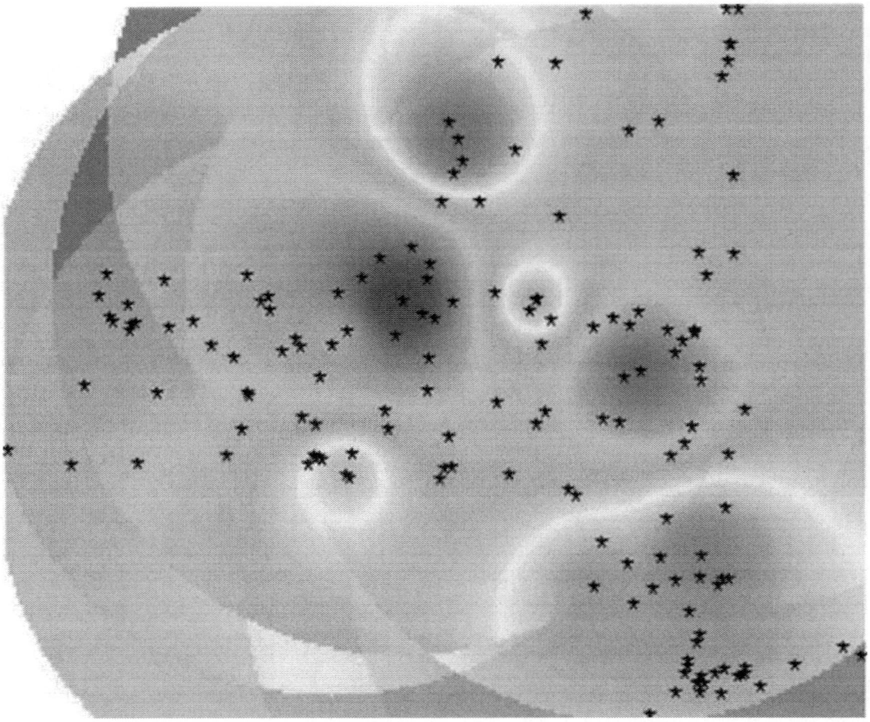

Figure 27.8: Visualization of speed variation in New York State. *Source*: Lawson and Jonnalagadda (2006), reprinted with permission.

The first experiment used a base map of continuous count data and weigh-in-motion (WIM) station locations (illustrated as stars in Figure 27.8) and archived WIM data containing a series of data elements, including the speed travelled and the weight of the axle passing over the WIM site. Figure 27.8 illustrates early attempts to merge archived ITS and GIS. The ITS data contains the speeds of trucks travelling on the network. The speeds recorded at each WIM station are processed and compared with all the other stations to produce a comparative scale illustrated with colours. The locations with the slowest speeds are indicated in blue and those with the fastest speeds in red. A colour ramp between these extremes is used to indicate variations. The stars form a rough approximation of the border of New York State.

Across the state, it is possible to "see" variation in truck speeds by any segment of time (e.g. daily, weekly, monthly averages, etc.). In Figure 27.8, the speeds are highest (in yellow, orange and red), relative to the other locations, in the corridor between Albany, New York and New York City and in Jefferson County. Data on weight of trucks can be displayed in a similar manner. For those individuals familiar with the trucking industry, the visualization technique is extremely helpful in illustrating freight behaviours with respect to weight and speeds. This is particularly useful in pinpointing congestion or identifying overweight trucks.

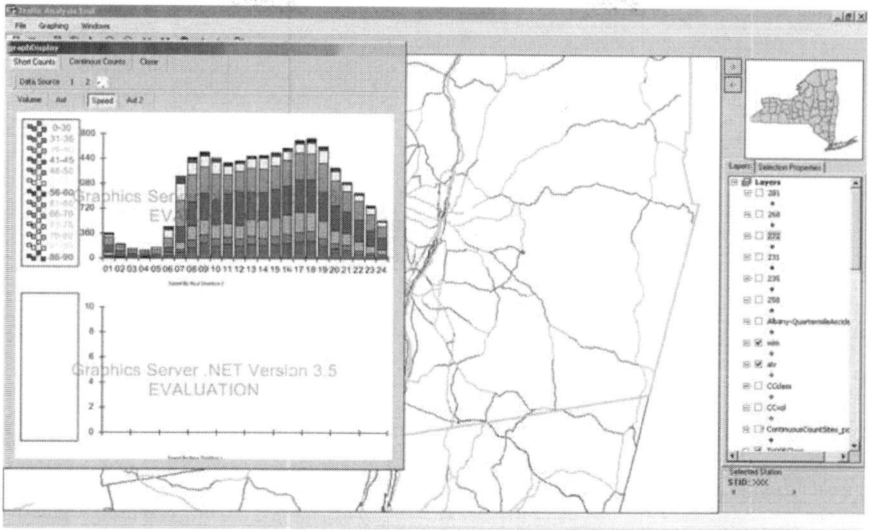

Figure 27.9: Traphic Tool visualization of short count speed data in New York. *Source*: Lawson and Muro (2007), reprinted with permission.

To continue the advancement of visualizations, ArcEngine, an ESRI developer tool, was used to develop software capable of using the available archived ITS data and displaying it at the direction of the user through a new graphic interface. For example, Figure 27.9 is a screen shot of the "Traphic Tool" software that makes it possible for an "unskilled" user to "click" on any road segment and instantly produce graphs of speeds, vehicle classes or other graphical interfaces within a modified GIS environment.

All of the features were developed to allow staff at NYSDOT to use visualization to improve their core functions and increase their access to datasets previously not available in a spatial context. The full value of these visualizations will depend on penetration into day-to-day tasks as the tool development team seeks to automate and customize easy access for users.

During the development phase, it was necessary for the NYSDOT staff to work closely with the computer science research team. To develop truly useful tools, the staff needed to "try out" the new features and give feedback to the developers, who then were able to make the changes and customize the tool to fit the needs of the user. The normal programme for research does not often allow for this type of interchange of ideas, as it requires the entire research project to be described in detail before the work begins.

The ability to adapt to advances in hardware and software, as well as to new ideas and relationships "discovered" during the development phase, is key to leveraging the development of visualization tools. The processing procedures depend on the functionality of the SQL server environment, making the vast quantities of archived data (e.g. second-by-second high-speed weigh-in-motion [HSWIM] data or global positioning systems [GPS] data) available to the user. The graphic interface also allows the user to "pop-out" smaller quantities of data through the "data source" button for additional analysis using SPSS or SAS.

27.3.5. *Four-Dimensional Representations for Dynamic Interactions*

According to Pack, Weisberg, and Bista (2007), 3D visualization works well for increasing the likelihood of early public acceptance, yet falls short for depicting dynamic interactions. Based on continued advancements in the computer gaming industry, the authors believe it is now possible to move from static 3D renderings to 4D visualization systems. The 4D visualization incorporates time as a fourth dimension, making it possible for the user to "fly through a virtual scene as it unfolds in real-time" (Howard, Weisberg, & Pack, 2008).

The Center for Advanced Transportation Technology (CATT Lab) at the University of Maryland is developing animations of road conditions with a variety of data streams, including point sensor data, incident data, incident response land closures, dynamic message signs, weather and other available data (Pack et al., 2007). In a similar effort to expand visualization techniques, Oak Ridge Laboratories have produced visualizations depicting aviation visualizations. The visualizations provide representations of flight proximities based on standard government safety guidelines and proposed flight plan adjustments (Howard et al., 2008). The authors note that what has yet to be determined is whether the benefits in other similar applications can be replicated in 4D applications. Such a study would provide direction for further developments using this visualization technology. Receiving feedback from users is important to this effort.

Recent advances in Web technologies make it possible for users to "create" their own visualizations of traffic incidents. According to Wongsuphsaswat, Pack, Filippova, VanDaniker, and Olea (2009), using a novel, Web-based, visual analytics tool called Fervor, users can interconnect various datasets as interactive maps, histograms and multi-featured visualizations. The ability to rank by feature and determine the strength of various relationships with relative ease should encourage the use of visualization and facilitate new understandings of relationships.

27.4. Opportunities for Incorporating Data Visualization into Travel Surveying Efforts

Travel surveys provide transportation planners and researchers with information on the trip-making behaviours of household members. Beginning in the 1950s, the need for aggressive urban transportation planning was fuelled by rapid population growth in urban areas, rapid growth of vehicle ownership and increasing movement of population to suburban locations (Pas, 1995). The first surveys collected travel data only of household members, with more recent efforts attempting to collect activity data or time-use data. These surveys collect data on all of the activities and locations of each activity over one or two days, for each member of a household. Household travel surveys are most often conducted by Metropolitan Planning Organizations (MPOs) or regional governments. (For more details, see Stopher & Greaves, 2007; for an international comparison of travel survey efforts, see Bonnel, Madre, & Armoogum, 2007).

27.4.1. Visualization of Flat Files Using GIS

In most cases, only flat files (e.g. SPSS or SAS) are created from the collected data and used to store the raw data elements. These files are used in statistical routines to estimate the parameters of transportation models. As GIS technologies increased in use, data was geocoded by home and stop locations. Early GIS attempts were hampered by low-quality address databases. A few efforts were successful, however, where high-quality GIS operations were able to geocode each location reported in the data. For example, in the Portland, Oregon, deployment, the data was geocoded and displayed in ArcView (see Figure 27.10).

The query tool in GIS allows for the analysis of the locations of individual activity types, socio-demographic characteristics and any combination on the survey data

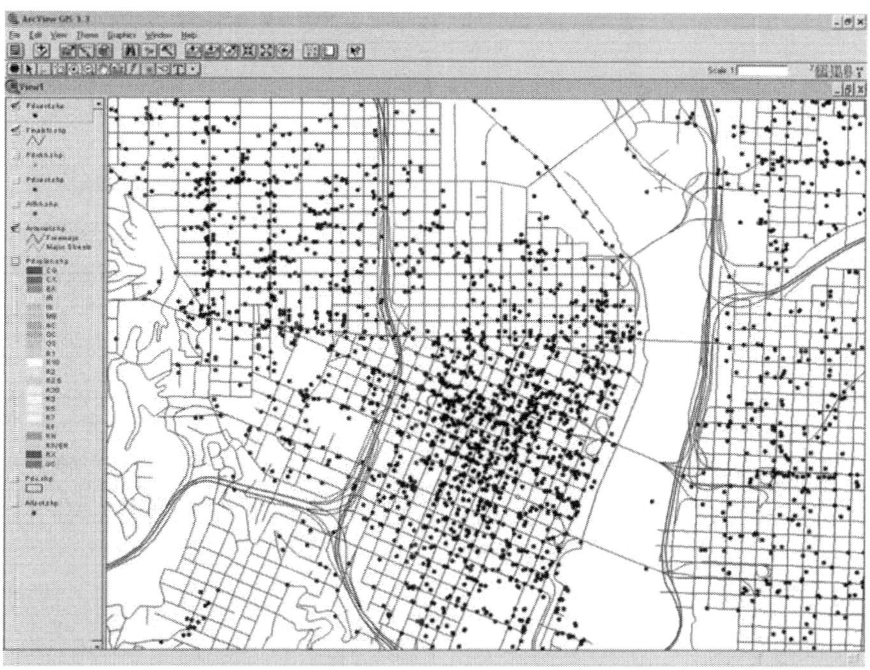

Figure 27.10: Example of Portland activity dataset geocoded activity locations. *Source*: Data provided by METRO, Portland, Oregon, 1998, reprinted with permission.

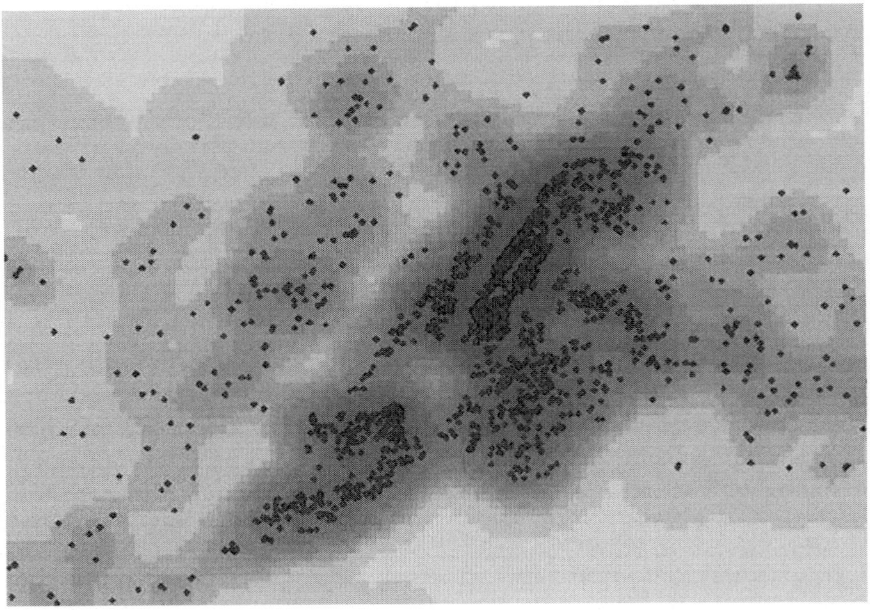

Figure 27.11: Density mapping of mis-reported transit trips in New York City region. *Source*: Data provided by NYMTC, 2007, reprinted with permission.

and other geospatial data. Geocoded data can also be used to analyse aspects of travel survey data. For example, a density analysis was conducted on the New York Metropolitan Transportation Council Regional Household Travel Survey collected in 1997/1998, to better understand the nature of the mis-reported transit trips (see Figure 27.11). While this visualization was quite helpful for the research team members, the visualization could be enhanced with an overlay of the Manhattan street grid to help anchor the concentration of mis-reported trips.

27.4.2. Visualization Technique for Space–Time Path Analysis

Kwan (2000) demonstrated the feasibility of visualizing activity travel data in GIS. Yu (2007) expands the notions of the space–time path concept using activity data. He has customized a procedure that makes it possible to connect all the trips taken by an individual in their sequential order, linked to each location where the activity took place, using GIS technologies. Figure 27.12 demonstrates a prototype system capable of generating space–time paths through temporal dynamic segmentation.

Chapleau and Morency (2005) used 3D GIS techniques to visualize the Montreal travel survey. Schönfelder and Axhausen (2004) developed a number of techniques for visualizing travel survey data, including the construction of a confidence ellipse.

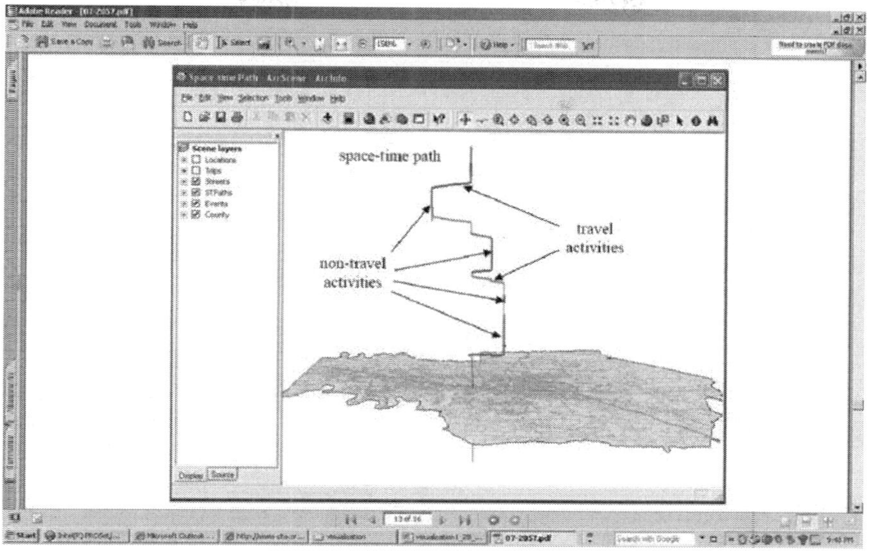

Figure 27.12: Creation of a process to visualize space–time paths. *Source*: Yu (2007, p. 59) reprinted with permission.

27.4.3. Incorporating Visualization Technologies into Travel Survey Deployments

Stopher and Greaves (2007) describe the entry of GPS technologies for the passive collection of person-based travel. The equipment is now available in a variety of small devices, including units that were reconfigured in response to the comments of users of the device. The feedback resulted in a GPS unit that is slimmer than a small cell phone, and has a longer battery life and a user-friendly "voice" that provides audible messages to help the user produce high-quality data (e.g. "low power" — it is time to recharge) (Swann & Stopher, 2008).

Most GPS loggers produce KML files that can be displayed in Google Earth®, and some are able to produce a variety of data formats, including CSV files that can be geocoded and used in GIS applications. Figure 27.13 is a mapping of a set of recorded points during a test of mode identification and spatial accuracy.

Hato (2007) has developed a procedure using GPS data from cell phones, RFID tags, photo logs and Web diaries in his "Probe Person System". Figure 27.14 illustrates the image created from aggregating pedestrian excursion data. The visualization technique provides day-to-day route choice variations. Other cutting-edge visualization experiments include the experimental Web interface being developed at the University of Illinois at Chicago that provides the GPS traveller with a set of spatial paths linked to questions about the activity undertaken at specific times and places (Mohammadian, 2007). Another example is being developed by researchers at the Massachusetts Institute of Technology (MIT) to visualize

Figure 27.13: Visualization of GPS experimental routings in Manhattan. *Source*: Lawson, Chen, and Gong (2008), reprinted with permission.

cell-phone data in Rome (see http://senseable.mit.edu/realtimerome). Reades, Calabrese, Sevtsuk, and Ratti (2007) provide insights into the visualization techniques and analyses developed with the data from the Rome project. In another study, Bar-Gera (2007) compares loop detector and cell-phone data using

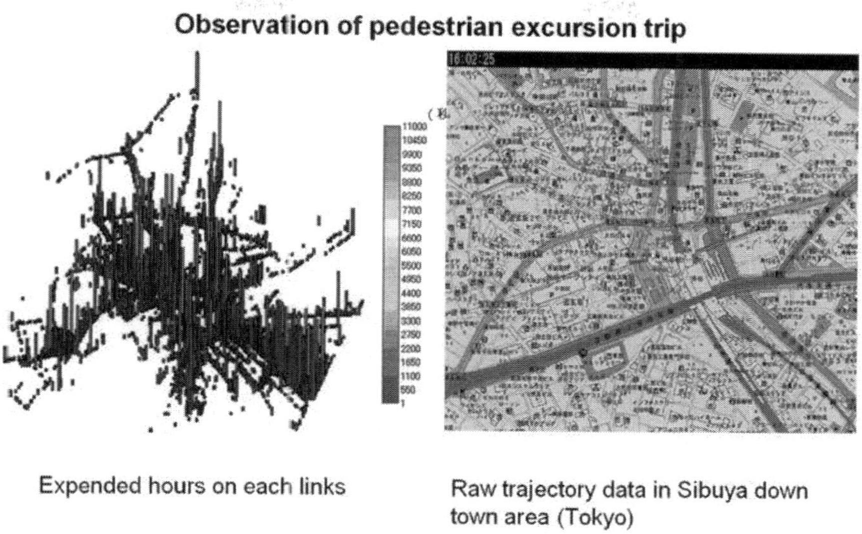

Figure 27.14: Visualizations produced using Probe Person technologies in Japan. *Source*: Hato (2007), reprinted with permission.

visualization techniques. Krygsman, de Jong, and Schmitz (2007) used GPS and Global Systems for Mobile Communications (GSM) to visualize daily urban rhythms.

27.5. Opportunities and Challenges for Travel Survey Community Visualization Efforts

As the opportunities for collecting GPS data expand, questions regarding appropriate data handling, archiving, access policy, confidentiality agreements and ownership remain unanswered. Sharp and Murakami (2002) expressed concern for the confidentially legislation, the Confidential Information Protection and Statistical Efficiency Act of 2002. This law mandates more stringent collection, protection and release of federal survey data procedures. Travel survey data collection efforts, such as the National Household Travel Survey (NHTS), could be adversely impacted. The consequences for the visualization of these data need to be determined.

Another threat to opportunities for visualization is the ability to archive travel survey data. Sharp and Murakami (2002) report that the Commute Atlanta project collected 365 days of one-second GPS data from 450 household vehicles. Within the language of the original contract, the parties inserted a "sunset clause" requiring the destruction of the data within six months of the end of the project.

Even with these threats, visualization has much to offer to the travel survey community. Whether individual travel diaries are collected with GPS technologies or with traditional surveying methods during the data collection phase of a deployment, the data can be visualized with MapQuest/Google technologies, GIS or other software innovations that would allow the survey participants to audit their own data. This improves the quality of the travel data for all future uses and allows the original data "provider" to play a role in the stewardship function of the data creation process. Consideration should also be given to understanding whether participants appropriately prepared through training with the GPS equipment would produce higher quality data.

The visualized data, whether on a person, household or aggregate level, will also improve the intra-industry tasks, improving data quality and productivity. For example, the original sampling frame can be geocoded and monitored during the initial contact phase and throughout the entire deployment (e.g. mapping contacts, refusals, completeness of survey instruments, return of GPS equipment, etc.). This "tool" would facilitate communications between consulting team members and MPO staff. Important tasks can be enhanced using visualization techniques including checking data quality, monitoring deployment progress and recognizing unexpected events (e.g. missing county data, high levels of incomplete surveys, etc.).

Travel survey data in general lacks an official programme for archiving data assets. Levinson and Zofza (2005) attempted to gather and archive existing household travel survey data documents and data with limited funding. New interest has been expressed in protecting these resources, perhaps following the archived ITS programme that has promoted standards and guidelines for long-term programme integrity (see http://www.fhwa.dot.gov/policy/ohpi/travel/adus.htm). In addition, a number of private sector firms offer their services for GPS data collection without a clear understanding of the future uses of these data. Visualization opportunity may not exist during the original contracting phase, making it difficult to include appropriate provisions. Individual consulting firms may not be motivated to archive data as they receive their revenue on a project-by-project basis, with little current incentive to archive for research and application development purposes.

Techniques to "smother" or make fuzzy sensitive locations can be accomplished using computerized operations. For example, home locations can be expanded to census tract or zip codes, while preserving the routing and specific stop patterns. This protects the GPS travel survey participants or traditional survey residence from exposure. The U.S. Census has a long history of developing techniques to reduce disclosure of cases. The credit card industry approaches privacy by allowing individuals to opt in or out of data-sharing programmes. Individual household members may be more willing to participate in a surveying/visualization effort if they can control the level of data release. Offering survey participants "provider-defined" release options could reduce concerns of the future use of private information.

Universities are on the forefront of the development of new visualization techniques. More funding to these innovative information "factories" with private and public sector partners should move the agenda forward. Table 27.1 provides a matrix of roles for the advancement of visualization for the travel survey community.

Table 27.1: Roles for the Advancement of Visualization for Travel Survey Community Members.

	Asset Acquisition/Data Collection	**Asset Preservation**
Person/ household	Must be convinced of value to participate Must be trained (self-trained) to answer surveys correctly/use equipment properly Can audit/review data paths (i.e. Google images)	Could be offered varying degrees of data release (similar to bank/credit cards selling customer information) — provider-defined release options
Private sector consulting	Advance capabilities and market shares with techniques acceptable to public Could capture entire value by saying data asset is private property of individual firm More productive use of time with visualization of "work in progress" (e.g. monitoring completeness of surveys across region)	No motivation to archive data if revenue is tied only to data collection and analysis phase Could make products out of lower order data releases Have superior knowledge regarding equipment and data formatting
Public sector agencies	More transparency and oversight of consultants with shared internal visualization of survey progress Need to require at least one version to be delivered in Open Source to make data visualization sustainable over time and across platforms Authorize new forms of dialogue-driven research to ensure feedback at the right time in the development process and from the ultimate users on a daily basis	Needs to set standard formats, but not standard on post-processing and output — these should be guidelines Provide funding for university partners and other research institutions to advance uses of archived data visualization

27.6. Discussion

Visualization has captured the imagination of the transportation planning community in general. The travel survey community members stand to make tremendous strides by integrating visualization techniques into nearly every facet of their activities. Visualization can be used in planning and implementing data collection, post-processing procedures and sharing results with technical staff and the public. To accomplish these gains, the travel survey community will need to address some specific action steps.

27.6.1. *Visualization for Travel Surveys Will Require Data Asset Management Strategies*

For visualization techniques to be used in practice for travel survey data, standards for data formats and procedures will be necessary. At the same time, it may be too early in the development process to establish generalized "best practices" for visualization outputs. A synthesis of current or proof-of-concept applications of visualization techniques needs to be produced and disseminated within the broad travel survey community for feedback and ongoing accumulation of "lesson learned" experiences. Incorporating a feedback loop in all aspects of the development and use of visualization appears to be an important element for success. By treating travel survey data as an asset, state DOTs and MPO can move forward with policies and procedures for good data stewardship practices. It is the quality of the data that will propel innovation with visualization techniques.

GPS data is produced on a second-by-second basis and can be saved as raw feeds or post-processed into any aggregate level desired, if the data is harvested and appropriately managed. Storage costs continue to decline, and many innovative uses of the data for visualization could be dependent on having retained the full set of second-by-second data streams. Software is needed to "tame" these vast data resources and make them easily accessible.

27.6.2. *Visualization Requires a New Form of Participatory Research*

The value of visualization is in the eyes of the beholder — in the case of travel surveys, there is value to be gained through the use of visualization for the internal processing and post-processing of the data. The new techniques require computer science solutions and know-how. Traditional request for proposals (RFPs) are not capable of anticipating potential visualization discoveries or unexpected environments such as rapid change in hardware, software and programming knowledge.

The best solution may be to scope the work as "experimental" with small phases and numerous dialogues (e.g. webinars or face-to-face weekly/bi-weekly meetings) as the work progresses so that the ultimate user can guide the computer science team towards meeting their needs in a timely manner. At the same time, the research will benefit from parallel advances. Through the exchange of ideas on both sides, gains can be made. These gains would not be possible in the normal course of research. The exchanges need to happen during the development stage of the software when most options and flexibility exist. The "dialogue" should include the equipment development community. Industry is able to produce small runs of experimental devices available for travel survey community to beta test.

27.6.3. Visualization Processes Should Be Developed in an Open Source Environment

Even the best ideas are at risk of software upgrades or hardware changes. By keeping at least one version of the work in Open Source, the agility, flexibility and sustainability over time and across platforms can be leveraged. If a single private firm captures all the value internally using public funding, there is a potential loss of future value. Open Source programming is often of higher quality because it has been examined by multiple developers using industry-wide best practices, rather than a single source operation, using their own internal standards of performance.

27.6.4. Visualization Will Need Continued Support from TRB and International Organizations

Finding a forum for the exchange of ideas and sharing of "lessons learned" is essential for rapid advancement and introduction of visualization to practice. Such events will reduce the risk of repeating previously learned lessons and can promote the sharing of travel survey community progress in "real-time".

For the travel survey community, the use of visualization depends on significant cooperation and coordination. Table 27.1 suggests some key roles across the travel survey community, including the survey participants (person or households), private consultants and public agencies. For example, visualization efforts will be enhanced if survey participants agree to provide high-quality data and are willing and able to audit these data. Concerns surrounding confidentiality of the data might be overcome if the participants are able to limit the use of their data (e.g. provider-defined release options similar to those used by the credit card industry for sharing information). While private sector firms can benefit from using visualization to enhance their own productivity, public sector agencies need to capture this value as well. It may be necessary to bring together representatives from each of these groups (e.g. potential participants, consulting firms and public agencies) to better understand how to move forward with the best interest of each group considered and the goal of successful visualizations accomplished as soon as possible.

27.6.5. Summary

Visualizations are being employed for a variety of uses in transportation. For planning infrastructure, technical experts are able to create 3D simulations to help decision makers and the public understand how the proposed improvement will function. Visualizations being developed for transit and freight planning rely on operations data being transformed through computerized and automated techniques using GIS and other spatial software.

The opportunity for travel survey community members to use visualization is uniquely different from these other uses because it requires the public, as survey participants, to be willing and able to provide the source data. These data can be collected through active methods (e.g. traditional travel surveys) or passive methods (e.g. using GPS to capture travel patterns) or a combination of these methods (e.g. modified surveys). Garnering participant cooperation, developing new equipment and techniques and providing guidance on how best to handle, analyse and preserve these data will encourage the use of visualizations to enhance our understanding of how people travel on a daily basis.

References

Bar-Gera, H. (2007). Evaluation of a cellular phone-based system for measurement of traffic speeds and travel times: A case study from Israel. *Transportation Research Part C, 15*, 380–391.

Bonnel, R., Madre, J. L., & Armoogum, J. (2007). National transport surveys: What can we learn from international comparisons? Paper presented at the Transportation Research Board Meetings, January 21–25, 2007, in Washington, DC.

Caldwell, H. (2002). Border wizard: Enhancing border security and productivity. Paper presented at the TRB Ports, Waterways, Freight & International Trade Conference, June 23–26, 2002, in Pittsburgh, PA.

Chapleau, R., & Morency, C. (2005). Dynamic spatial analysis of urban travel survey data using GIS. Paper presented at the ERSI User Conference in 2005. <http://www10.giscafe.com/link/display_detail.php?link_id = 13155>. Accessed March 10, 2008

Charles, L. H., III. (2006). *Visualization for project development: A synthesis of highway practice* (p. 13). Washington, D.C.: Transportation Research Board.

Claramont, C., Jiang, B., & Bargiela, A. (2000). A new framework for the integration, analysis and visualization of urban traffic data within geographic information system. *Transportation Research Part C, 8*, 167–184.

Hato, E. (2007). Behavioral data mining using auto ethnographical data collected through the probe person system. Paper presented at the Travel Survey Methods New Technology Subcommittee Meeting, January 23, 2007, in Washington, DC.

Hixon, C. (2006). *NCHRP synthesis 316: Visualization for project development* (http://onlinepubs.trb.org/onlinepubs/nchrp/nchrp_syn_361.pdf). Washington, DC: Transportation Research Board of the National Academies.

Hixon, C. (2007). Visualization issues for transportation agencies: Approaches and challenges. *TRNews, 252*, 11–14.

Howard, J., Weisberg, P., & Pack, M. (2008). Toward a four-dimensional, real-time AVL visualization system. Paper presented at the 2008 Annual Transportation Research Board (TRB) Meeting, January 13–17, 2008, in Washington, DC.

Kimpel, T. J. (2007). *Data visualization as a tool for improved decision making within transit agencies*. Final Technical Report TNW2006-14 prepared for Transportation Northwest (TransNow), University of Washington, Seattle, WA.

Krygsman, S., de Jong, T., & Schmitz, P. (2007). Capturing daily urban rhythms: The use of location aware technologies. Paper presented at the 10th International Conference on Computers in Urban Planning and Urban Management, July 11–13, 2007, in Iguassu Falls, Brazil.

Kwan, M. (2000). Interactive geovisualization of activity-travel patterns using three-dimensional geographic information systems: A methodological exploration with a large data set. *Transportation Research Part C*, *8*, 185–203.

Lawson, C., Chen, C., & Gong, H. (2008). GPS pilot project (ongoing research). Presented at the NYMTC Survey Workshop, January 10, 2008, in New York City, NY.

Lawson, C., & Jonnalagadda, S. (2006). *SWATH analysis: Using firm-level data to understand truck trips on New York state trade corridors*. Prepared for the Office of Policy and Strategy, New York State Department of Transportation, July, 2006.

Lawson, C., & Muro, A. (2007). *Visualization tool development: Using GIS and archived traffic data in a new environment of functionality*. Prepared for the Office of Policy and Strategy, New York State Department of Transportation.

Lawson, C. T. (2006). Microsimulation for urban transportation planning: Miracle or mirage?. *Journal of Urban Technology*, *13*(1), 55–80.

Levinson, D., & Zofza, E. (2005). *The metropolitan travel survey archive: A Case study in archiving*. Unpublished paper.

Manore, M. (2007a). It's about decisions: Improving transportation project development with visualization technologies. *TRNews*, *248*, 21–24.

Manore, M. (2007b). Visualization education and training. *TRNews*, *248*, 24–28.

Mohammadian, K. (2007). GPS-based activity survey video. A video clip produced of Joshua Auld's presentation at the University of Illinois, Chicago.

Pack, M. L., Weisberg, P., & Bista, S. (2007). Wide-area, four-dimensional, real-time, interactive transportation system visualization. Paper presented at the Annual Transportation Research Board Meetings, January 21–25, 2007, in Washington, DC.

Pas, E. (1995). The urban transportation planning process. In: S. Hanson (Ed.), *The geography of urban transportation* (2nd ed.) New York: The Guildford Press.

Reades, J., Calabrese, F., Sevtsuk, A., & Ratti, C. (2007). Cellular census: Explorations in urban data collection. *Pervasive Computing*, *6*(3), 30–38.

Schmitt, R., Dronzek, R., & Rchter, L. (2007). Visualization to solve problems in freight transportation. *TRNews*, *252*, 15–19.

Schönfelder, S., & Axhausen, K. W. (2004). Structure and innovation of human activity spaces, *Arbeitsberichte Verkehrs- und Raumplanung*, 258, IVT, ETH Zürich, Zürich.

Sharp, J., & Murakami, E. (2002). Travel survey methods and technologies resource paper. Paper presented at the National Household Travel Survey Conference, November 1–2, 2002, in Washington, DC.

Stopher, P. R., & Greaves, S. P. (2007). Household travel surveys: Where are we going?. *Transportation Research Part A*, *41*, 367–381.

Swann, N., & Stopher, P. (2008). Evaluation of a GPS survey by means of focus groups. Paper presented at the Annual Transportation Research Board (TRB) Meetings, January 13–17, 2008, in Washington, DC.

Transportation Research Board. (2006). Fifth International Visualization in Transportation Symposium and Workshop held in Denver, CO, October 22–26, 2006.

Tufte, E. R. (1997). *Visual and statistical thinking: Displays of evidence for making decisions.* Cheshire, CT: Graphics Press.
Walker, D. (2007). Visualization as common language for planning. *TRNews*, *252*, 7–10.
Wongsuphsaswat, K., Pack, M. L., Filippova, D., VanDaniker, M., & Olea, A. (2009). Visual analytics for transportation incident datasets. Paper presented at the Annual Transportation Research Board Meetings, January 11–15, 2009, in Washington, DC.
Yu, H. (2007). Visualizing and analyzing activities in an integrated space-time environment: Temporal geographic information system design and implementation. *Transportation Research Record: Journal of the Transportation Research Board*, *2024*, 54–62.

Chapter 28

Data Visualization Techniques: Synthesis of a Workshop

Michael Manore and Stephan Krygsman

28.1. Introduction

Workshop participants were introduced to innovative visualization techniques being used in transportation planning and explored the potential uses of visualization for improving travel survey data. It is argued that by improving and expanding visualization methods, transport researchers and practitioners will improve the quality of their data and more effectively convey the richness and insights of that data to policy makers and stakeholders. This would thereby support higher levels of communication, awareness, understanding, and decision-making. It is also envisioned that these tools and methods would improve the participation of travel survey respondents, also enhancing data quality and completeness. Interactive tools usable by policy makers or citizens are of special interest, as are innovative, information-rich, and aesthetically appealing graphs, charts, maps, and animated visualizations. They can be the results of data mining tools or statistical analysis, addressing either micro-data (individual records) or aggregate data. Making the data "come alive" is crucial for stakeholder discussion and participation. Examples used in this workshop focused on the problem definitions and visualization solutions involved rather than the technical work behind the data itself.

To introduce participants to the topic area of data visualization, Stephen Krygsman (discussant) made a brief presentation on the use of visualization in traditional transportation planning. He illustrated the value of visualization to facilitate communication among various stakeholders in any project through the display of a number of effective visualization examples. Krygsman contrasted successful applications of these tools with other examples that were poorly thought-out, limited, and/or confusing in their ability to convey any valuable information (Lawson, 2009; Chapleau, Morency, & Bourgeois, 2008; Lee, Wolf, Oliveira, & Kaiser, 2008).

Transport Survey Methods

ISBN: 978-1-84855-844-1

28.2. Developing an Adoption Strategy for Visualization

The workshop participants were tasked with identifying opportunities where the use of visualization could improve or enhance travel survey data collection, analysis, and preservation. It was determined that there were at least seven "stages" in the travel survey work flow where visualization could be introduced including: building the business case for travel surveys, data strategy development, establishing a sampling frame, recruitment, data collection, data quality, data analysis, and dissemination and preservation.

28.2.1. *Building the "Business Case" for Travel Surveys*

Visualization techniques could be employed to help define the audience for a proposed survey program. The use of integrated maps with special districts or attributes of a particular population of interest would help inform the agency responsible for the deployment. Visualization could also be used to help define the "problem" to be addressed in the travel surveying effort. Using visualization techniques to assess previous surveying efforts would help illustrate key facts or communicate previous results or issues. In building the business case, it is also necessary to ensure ethical standards are followed to reduce the possibility of improper use of the tools. For example, the parties need to be careful not to exaggerate certain factors that might skew the understanding of the issues.

Two key areas that will make visualization more informative will be the normalization (minimizing duplication) of travel survey data and further efforts to harmonize (bring together disparate data sources so they are more interoperable) the data across different deployments.

28.2.2. *Data Strategy Development*

Visualization techniques could be used to perform more graphics-based literature reviews, particularly where travel survey data have not been examined in this manner. It is anticipated that new understandings of what is known and not known will emerge. In some cases, the use of visualization might accelerate the process of moving from data to information, and from information to "knowledge." At the same time, there is a risk of reducing "thoroughness" and creating an environment of "visual overload." Since there is little experience in developing visual data strategies, there is also the increased risk of inappropriate uses of the data. To counteract this possibility, there is a need to develop guidelines regarding approach, tools, and methods.

28.2.3. *Establishing a Sampling Frame*

Visualization could contribute to the process of reviewing the "official" data, including special techniques for displaying population attributes and comparing those attributes to a particular sampling strategy. The use of visualization could ensure geospatial compatibility, making it easier to integrate datasets. It also could encourage "mixed mode" surveys by making it easier to identify areas with varying response schemes. To move forward on this aspect of using visualization for travel surveys, it will be necessary to secure funding to get existing data up-to-date. Such procedures and new methodologies will require the development of "visualization templates" to facilitate use and make adoption easier. The use of technology for goods movement data collection is likely to expand.

28.2.4. Recruitment

A good surveying effort requires a strong and successful recruitment strategy. Visualization techniques could provide an excellent medium to communicate the importance of the surveying project to potential survey participants. It is possible that this could even increase response rates if done effectively. In surveying efforts where new technologies (such as GPS) are offered to survey participants, the use of visualization could provide more effective training on the nature of the survey, and the use of the various tools, making it possible for participants to validate their own data and submit them properly.

Not enough is known about possible "biases" that might be introduced when using visualization techniques for recruitment. One concern that was voiced is whether respondents could more effectively see how their data impacts the results of the survey and "game" the data for particular results. To reduce this risk, guidelines on what visualization methods and tools would best support recruitment are needed, including research on how and to what extent visualization might influence response rates.

28.2.5. Data Collection

One important aspect during the data collection phase of a travel surveying effort is being able to monitor the progress of the effort. Visualization strategies will be very effective for not only making it possible to "see" the geographical location(s) of the survey effort, but also enabling researchers to monitor the quality of the data being gathered and detect trends, thereby exposing any poor quality data and other related problems. Such a strategy may also improve the CATI process.

Using visualization to monitor the "process and workflows" of a travel surveying deployment will make it possible to share this information among partners, including the deploying agency and outside consultants. By more effectively monitoring the products of the survey, it may also be possible to lessen the burdens on respondents. As previously mentioned, changing the nature of the work process may introduce biases in determining what constitutes quality data versus poor data. Research on the nature and extent of such biases will need to be undertaken.

28.2.6. Data Quality

The use of visualization techniques could make it possible to more effectively perform "real time" quality assessments. Post processing could be performed more effectively and with greater accuracy. A role of visualization is envisioned for data cleaning, inference, and imputation; visualization is also expected to add to the transparency of the data. Since the possibility exists that the visualization of data may influence perceptions of what constitutes quality data, research will need to be conducted to better understand this risk. It will also be important to consider research on how visualization influences decision-making at this stage. An important aspect during the data collection phase of a travel surveying effort is being able to maintain acceptable levels of quality. Knowing how visualization techniques affect the perceptions that these levels are being met will be key.

28.2.7. Data Analysis

More than any other stage, the use of visualization for data analysis has far-reaching possibilities. Visualization will be useful in extracting "patterns" that may have developed and help discover complex spatial–temporal relationships. Data fusion techniques incorporated in the visualization process could assist in identifying compatibilities between datasets and more effectively defining relationships (cause and effect) in the data. This process could easily be the "gateway" to opening possibilities for innovation. At the same time, it also increases the potential for misrepresentation and misinterpretation of results.

The development of "templates" for performing visualizations will help to defuse these problems with this new form of analysis. Research on the impacts of these new strategies will expand our knowledge on how visualization influences interpretation and analysis.

28.2.8. Dissemination and Preservation

Finally, visualization is envisioned to greatly assist in the promotion of more effective "data sustainability" by increasing the number and diversity of potential data users. Techniques could assist in "getting-to-the-knowledge" behind survey data and completing the circle of understanding by allowing professionals to more effectively prepare for future surveys using the stages described above.

Broadening the use of travel survey data could also increase concerns over privacy and ownership of the data. It will be necessary to develop tools and guidelines for using visualized data to better manage these risks.

28.3. Research Challenges

On the basis of the workshop participants' collective understanding of the potential benefits of incorporating visualization into travel surveying, the following perceived barriers and future needs were identified.

28.3.1. Perceived Barriers

At least in the early attempts to incorporate visualization, it may be difficult to attract the requisite talent and expertise to develop visualization strategies. Making the case for funding and presenting a clear understanding of the return-on-investment for making visualization a part of the travel survey process will be a challenge. In addition, it will be important to encourage a new level of education and training to effectively use visualization tools.

To date, there is lack of industry awareness in the travel survey community about visualization. Recognizing the role of ethics in the use of visualization techniques, transportation, in general, has not been an area of interest for technical "visualizers." And, even if the level of interest did exist, the current state of the available data from travel surveys is lacking in standards, formats, conditions, and levels of precision. Finally, visualization is envisioned to greatly assist in the promotion of more effective "data."

28.3.2. Needs Summary

To move the agenda of incorporating data visualization into travel survey deployments, there will need to be stable funding for reliable base-data resources, operating budgets for data maintenance and preservation, and the development and dissemination of appropriate visualization tools. As previously mentioned, there will need to be research on how visualization might improve response rates, particularly for hard-to-reach groups.

Research to identify and quantify the amount of "added value" that visualization can offer (return-on-investment) and on any new risks associated with this use (including biases and privacy thresholds). Research will be needed regarding "framing" in the context of traditional surveys and as part of stated preference and other types of surveys.

Through the application of tool development and advancements in computer science, it will be very feasible to develop templates and guidelines for implementing data visualization within the travel survey community. Integral to this aspect is the need to strive for harmonized, high-quality databases. By examining the educational and training needs, especially with respect to introducing computer science into the world of transportation in the context of travel surveys, progress could be very rapid.

One of the first steps would be to conduct a synthesis to establish the "state-of-the art" and better understand what is currently possible across the globe. The next step will be to implement the stages simultaneously wherever possible to increase the rate of available tools and methodologies.

Acknowledgements

The workshop participants were Jimmy Armoogum, France; Svetlana Beltyukova, United States; Christine Fox, United States; Stephen Greaves, Australia; Peter Jones, United Kingdom; Stephan Krygsman, South Africa; Catherine T. Lawson, United States; Michele Lee, United States; Michael Manore, United States; Catherine Morency, Canada; Elaine Murakami, United States; Virpi Pastinen, Finland; Angelika Schultz, Germany; and Chester Wilmot, United States.

References

Lawson, C. T. (2009). Visualize this: Opportunities and challenges for the travel survey community. In: P. Bonnel, M. Lee-Gosselin, J. Zmud & J.-L. Madre (Eds), *Transport survey methods: Keeping up with a changing world.* Bingley, UK: Emerald.

Chapleau, R., Morency, C., & Bourgeois, M. (2008). Simple and interactive spatial mobility analysis tool for data visualization. Presented at the 8th International Conference on Transport Survey Methods, Annecy, France.

Lee M., Wolf, J., Oliveira, M., & Kaiser, M. (2008). Data visualization in travel and physical activity studies. Presented at the 8th International Conference on Transport Survey Methods, Annecy, France.

PART VI

EMERGING/PERSISTENT SURVEY ISSUES AND DATA HARMONISATION

Chapter 29

Large-Scale Ongoing Mobility Surveys: The State of Practice

Elizabeth S. Ampt, Juan de Dios Ortúzar and Anthony J. Richardson

Abstract

Large-scale continuous mobility surveys have some advantages over less frequent (usually every 10 years), even larger-scale cross-sectional surveys; these advantages have been well documented in previous papers (Ampt & Ortúzar, 2004).

In this paper we first define what we mean by 'ongoing mobility surveys'. We then describe the state of practice in this context, briefly reviewing the state of affairs in all the cases that we are aware of. We then discuss some problems encountered in practice and offer ideas for improvement. In particular, we discuss a wide range of issues that are likely to act as barriers to a high quality and sustainable implementation and suggest approaches for improvement. Issues covered include sampling frames and sampling methods, survey methods, respondent burden, weighting processes and expansion, and the increased importance of developing and maintaining field staff motivation. We also touch briefly on the practical/political issue of securing ongoing funding. Throughout, we advance some thoughts to try and explain why this method has not gained wider acceptance, particularly in the Northern Hemisphere where there are more examples of travel surveys in general.

The paper also raises some ideas and issues about the way in which ongoing mobility surveys can best collect data for the environmental accounting of travel. Finally, we raise questions about the environmental impact of the survey methods themselves as a stimulus for further consideration.

Transport Survey Methods

ISBN: 978-1-84855-844-1

29.1. Introduction

This paper focuses on large-scale, ongoing, urban mobility surveys. Since there are various kinds of large-scale mobility surveys being carried out around the world, it is useful to commence by describing what we understand by this description. A large-scale, ongoing, urban mobility survey is typically commissioned for a particular city or region by either the city or regional authority responsible for transport planning and/or policy making. The aim of the survey is usually to describe what is currently happening in the realm of transport in the city or region and its immediate environs, and to use that data to plan or forecast what is likely to happen over the next years. By 'ongoing survey' we mean that it is carried out every day for the whole year, over several years. This definition leaves out data panels and national surveys (in Appendix 29.A.1 we list a series on national surveys conducted in the Northern Hemisphere[1]). It also leaves out cross-sectional surveys. Cross-sectional surveys repeat the same or similar surveys periodically across a city or country. On each occasion a sample of households is selected to represent the whole geographic area in question. The advantages and disadvantages are discussed in detail in Richardson, Ampt, and Meyburg (1995).

People choose ongoing surveys for several reasons. They provide data on temporal variability of travel — over seasons and over periods that can be considered different from 'normal' (e.g. periods of school holidays and other festive periods). Ongoing surveys also give an indication of changes in travel patterns in response to general changes in behaviour, for example responses to variations in cost of cars, public transport or petrol, modifications in land use and changes in social values (e.g. in response to climate change).

The core of an ongoing urban mobility survey is usually household based, with the sample selected periodically using the repeated cross-section method. This means that the same household is not being followed, as it would be in a panel survey,[2] but that new samples are chosen for each period, with some households possibly being selected again by chance[3].

The surveys we describe usually collect data on daily mobility only, and concentrate on travel within the targeted urban area, thereby excluding a deliberate focus on long-distance travel. This is because the uses of the data are focussed on the changes that happen and can be made in the urban area being studied. Most surveys we are considering gather one to two days of data.[4]

---

1. We are grateful to Prof. Dirk Zumkeller for having provided us with this list.
2. There is considerable literature on large-scale panel surveys, for example British Household Panel Survey Calibrated Time Use Data, 1994–2004. http://www.esds.ac.uk/findingData/snDescription.asp?sn = 5363.
3. This depends on what is considered a period; in Santiago, for example the sample is selected in advance for three years without repetition.
4. There is debate about the number of days that should be selected. In 1988, Pas (1988) suggested that two days of travel represents about 80% of travel variability, though more recent work (e.g., Mistro & Behrens, 2008) suggests that it is much less and implies that more days would be needed.

While the core surveys are household based and therefore focus on private travel, most cities also choose to have parallel (though not always ongoing) surveys of commercial vehicles and public transport, some traffic flow data and measurement of level of service, with some level of ongoing counts of all people and vehicles for all modes of travel. The survey data is usually used to produce trip matrices (by mode, purpose, time of day) and this involves combining data from several sources, most importantly with intercept surveys (also usually not ongoing). This is summarised in Ampt and Ortúzar (2004).

We have chosen to address this type of 'ongoing survey' in a research paper for this conference for several reasons. First, many of the issues raised are pertinent much more broadly to applications on a countrywide level. Second, we describe many of the problems encountered and, where possible, the solutions to enable others to avoid them in the future. Thirdly, we believe that this ongoing urban mobility survey is a very valuable way of achieving the targets of many city authorities; this paper should provide a useful basis for understanding the issues.

29.2. State of the Practice — What is Done

The main features of an 'ongoing survey' methodology are:

(a) Data collection should aim to cover every day of the year (not just typical days and periods). This allows the survey to:
- capture seasonal, weekly and daily variations in trip making;
- monitor the evolution of demand and of traveller responses to changes (foreseeable or otherwise) in the transport supply and movement capacity;
- achieve better training of the field personnel;
- lower operating costs (rather than paying by the hour, a fixed contract can be negotiated — however, this is not the most typical case);
- develop better and more expeditious quality control procedures, as fewer data items per day/week arrive at the office;
- obtain higher interest in the information, particularly by other potential users.

Furthermore, a continuous data collection system throughout the whole year allows intensification of the sampling rate in, for example newly developed areas or less-represented days of the week (such as weekends). These data simply receive different expansion factors.

(b) The data collection effort is kept running for several years, as described below, producing matrices and models at the end of each previously agreed period.

(c) The methodology uses two fundamental types of surveys to cover the most important objectives of the exercise: home interviews and intercept surveys; however, the latter are not always present — particularly if there is no special interest in producing trip matrices. Each type of survey is used to maximise the achievement of the study objectives, although it is not possible to look for a strict optimisation as neither the state of the art nor the state of practice yet allows for it.

(d) All the origin–destination (O–D) information is geocoded to allow for the future use of the data under any zoning system established. This does not detract from the need to design an appropriate hierarchical zoning system for the study, on the basis of fine zones, which can be aggregated into larger zones.
(e) Additional information such as traffic counts, bus and metro patronage by line, and level-of-service data for private and public transport can also be collected to validate and/or update models estimated with the household survey data.
(f) Simultaneously with the survey, computerised representations of the public and private transport networks are built or updated, level-of-service data by time period is gathered (e.g. using GPS devices), and land-use information is collected. All this is undertaken with the aim of eventually using the survey data to estimate strategic supply–demand equilibrium models of the study area transport system.
(g) Finally, matrices and models for the future can be updated taking advantage of this continuous data collection effort. Although it is possible to think about preparing partial trip matrices for specific purposes, the city matrices and models are normally updated not more often than every 12–18 months, depending on the type of city under scrutiny.

However, this ongoing nature also brings certain problems; for example it is more difficult to keep the field force motivated for long periods and thus a lot of personnel replacement and re-training is necessary. Similarly, it is not possible to keep a continuous level of advertising about the survey; this can have the effect of reducing respondent collaboration in periods without an advertising campaign.

29.3. Survey Methodologies

While the unifying feature of the surveys described in this paper is their ongoing nature, there have been a range of different methodologies used for the conduct of the surveys, depending on the demographics of the intended population, the history of travel surveys in the region, the available budget and the experience of those conducting the surveys. The following sections describe the methods employed in three different cities to highlight the possibilities as well as the experience gained in the design and conduct of ongoing surveys.

29.3.1. Self-Completion

Although the Santiago and Sydney surveys, discussed below, have adopted face-to-face interviews as their primary method of data collection, this is not the only way to conduct ongoing mobility surveys. This section of the paper will describe how the methods originally devised for the Victorian Activity and Travel Survey (VATS), which ran from 1994 to 2002 in Victoria, Australia, were modified over the years to arrive at the current incarnation, which is being used in its successor, the Victorian

Integrated Survey of Travel and Activity (VISTA). The VATS survey revealed two major issues with respect to the conduct of an ongoing mobility survey. The first related to the methodology itself, while the second related to its organisational administration over an extended time period.

The first methodological lesson pertained to the self-completion nature of the survey and, in particular, to the use of reminders; these were used because previous research (Wermuth, 1985) had shown that this could double the response rate achieved in mail-out/mail-back surveys. However, it had also been observed that the reported trip rates fell with increasing number of reminders, and hence it had been postulated (Brög & Meyburg, 1982) that non-respondents were more likely to have trip-making characteristics like those who respond late to the travel survey. They therefore assumed that non-respondents had lower trip rates than respondents. However, a more detailed analysis of the VATS data (Richardson, 2003) showed that the reducing response rate across the reminders was due to allowing respondents to use a later travel day, which introduced a degree of self-selection. In fact, the mobility of the non-respondents (when their travel day was specified by the interviewer) was more similar to the earliest respondents (who responded to the specified travel day) than it was to the late respondents. The lesson from this experience was that reminders should not be relied upon to increase response rates, since they inevitably introduced a degree of self-selection into choosing a travel day. If a self-completion questionnaire was to be used, then other methods would be required to increase the response rate.

The second lesson from VATS was concerned with the difficulty of keeping a qualified survey team together for a long period of time. One of the authors (Ampt) worked on VATS for three years (1992–1994) before moving. Another (Richardson) worked on VATS for seven years (1992–1998) and then also moved on. From 1998 to 2002, the survey was run by people not involved in the original design and with little experience in running a large-scale travel survey. In 2000, major changes were made to the design, with many untested new questions added. In addition, as the University demanded 'higher cost-effectiveness', several components were out-sourced to market research companies, who offered lower prices. Unfortunately, this resulted in lower quality data and a loss of control in administration and documentation. As a result, the past year of the data was never satisfactorily delivered to clients, and the survey was terminated after 2002. To this day, most analyses concentrate on the data from 1994 through 1999, before the design changes were introduced in 2000.

Following the cessation of VATS in 2002, several surveys have been designed and conducted by Richardson at The Urban Transport Institute (TUTI) using self-completion methods and building on the lessons learned in VATS, that is Perth (2002–2005), Brisbane (2003 and 2006), Sunshine Coast (2004 and 2007), Gold Coast (2004 and 2008), Melbourne (2004 and 2005), Auckland (2006) and Christchurch (2006). However, only one of these (Perth) has been a true 'ongoing' survey, in that it ran continuously for more than one year. However, the others have contributed to refining the methodology, which is now being implemented in VISTA (2007–2008), currently in its first year, but which may well be extended to run again as an 'ongoing' survey.

VISTA is still based on a self-completion questionnaire, but this is now incorporated in a substantially different overall methodology (to overcome many of the problems uncovered in VATS). The questionnaire is hand delivered to, and hand collected from, the survey households. This process is also supplemented by telephone motivational calls, telephone and postal reminders, and telephone clarification calls. The process consists of 11 essential steps:

1. *Selection of sample*: At the beginning of the survey period, a sample of households is drawn from the sample frame, which is a GIS database of properties within residential areas, but it could be a list of electricity connections or a list of residential addresses from a local government rates database. It is a cluster sample, based on a random sample of households within a stratified random sample of census collectors districts (CCD; groupings of approximately 250 residences, as used in the Australian census). Within each CCD, a cluster of 42 households is sampled (at an average rate of approximately 20% of residential addresses). Each week, those households designated for contact in that week are identified, and a list is made for each of the sub-regions accompanied by a GIS map showing the location of each address within the cluster.
2. *Address checking and pre-contact letter delivery*: On the Tuesday of each survey week, field staff in each of the CCDs checks for the existence and suitability of each listed address. At this stage, some will be classified as sample loss (i.e. no such address, address not findable, vacant block, business address, empty premises, etc.). Sample loss addresses are replaced by selecting another one from a pre-supplied random selection of replacement addresses. Other addresses will need to be corrected (e.g. for apartments, there may need to be correction or clarification of apartment numbers, where the apartment number is either missing or misrecorded in the sample frame). The field team is given strict instructions about what corrections they are allowed to make. In addition, field team is in contact with the fieldwork controller by mobile phone to clarify any problematic situations on the spot. Once the in-scope addresses have been established, field team delivers a pre-contact envelope (addressed to 'the householder') to each in-scope address (the letter is placed in the mailbox, with no contact being made with the householder at this time). The pre-contact envelope contains a pre-contact letter from the client and a survey brochure.
3. *Correction of sample addresses*: On the Wednesday of each week, the field corrections made on the previous Tuesday are entered into the sample address database, and a final list of in-scope addresses is produced for the questionnaire delivery. This is accompanied by an updated map, with any comments from field staff to enable the easy and correct location and identification of households for the delivery of questionnaires (e.g. big white house with poplar tree and red letterbox).
4. *Delivery of questionnaires*: On the following Saturday and Sunday, field staff (different from those who undertook the pre-contact letter delivery on the previous Tuesday) delivers the questionnaires in each of the CCDs. The questionnaires cover travel days from the next Monday through to the following

Sunday. Every attempt is made to personally contact a household member to deliver the questionnaire in person (since this has been found to significantly increase the response rate). Interviewers pay up to four visits to a household during the weekend to try to contact them (two attempts on Saturday and two attempts, if needed, on Sunday). If contact is made, the interviewers introduce themselves, briefly explain the nature of the survey and answer any questions about it.

If the household agrees to accept the questionnaires, then the interviewer asks how many people are likely to be in the household in the coming week, and then provides that many travel diaries (plus one) as part of the survey materials. They also inform the household when they will return to pick up the completed questionnaires, and where they might leave the questionnaire out for collection, if household members happen not to be at home. They are also asked for a contact phone number in case the field team needs to contact them during the survey. If contact cannot be made, a 'We-Missed-You' postcard is left at the household with an envelope containing all the survey materials (including six travel diaries to cover most household sizes). Information is also left about the pickup of the questionnaires the following week, and instructions for where to leave the completed questionnaires if the household does not expect to be home when they are collected. If a personal refusal is received at this stage, then the field team is instructed to first enquire whether the reason for the refusal is 'transport related' (e.g. I do not travel much, so why am I in the travel survey?). They then attempt to convert such refusals to an acceptance by stressing the importance of obtaining responses for all types of households. If the household cannot be converted, or if the reason for the refusal is not transport related, then the field person immediately asks two 'refusal questions': how many people are in the household and how many vehicles are in the household. This gives some information about refusals. These questions are asked anytime when a personal refusal is encountered.

5. *Motivational call*: On the day of each household's travel day, they are phoned by the survey office to ensure that they have received the survey materials, to answer any questions they might now have, and to remind them that today is their travel day. Phone numbers are obtained from households when survey packs are delivered to them, with a backup set of phone numbers obtained from commercial phone number databases.
6. *Pickup of questionnaires*: On the following Sunday (for households with a weekday travel day) and Monday (for households with a weekend travel day), a different set of field staff goes out to the CCDs to collect the questionnaires.

   Two attempts are made to personally contact the household (since the effect of personal contact at this stage is not as strong as in the delivery stage). As soon as possible after collecting questionnaires from the household, the interviewer performs a quick edit check to identify whether the full number of travel diaries have been completed and handed over. If any travel diaries are missing, the interviewer returns to the household to enquire whether the diaries were inadvertently excluded from the package of completed questionnaires. If the household is not at home, but they have left their questionnaires out for collection,

then the questionnaires are collected and a 'Thank-you' postcard is left to indicate that the questionnaires have been collected.
If no household member is at home, and they have not left their questionnaires out for collection, the interviewer leaves a 'We-Missed-You' postcard together with a reply-paid envelope for the return of the questionnaires by mail to the survey office.

7. *Data coding*: A special program (*Speedit*) was developed by TUTI to enable data coding, including geocoding, and editing to take place in the field office as soon as questionnaires are returned. The program is readily adapted to new survey designs (especially if they are of the same general format as VISTA). The number of data-coding workstations can be modified to suit the survey workload, such that completed surveys are processed within a week of being received in the office. The geocoding module has six options: Home Address geocoding, Full Street Address geocoding (when the full address is known), Partial Street Address geocoding (when only part of the address is known, e.g. street without number), Landmark geocoding (e.g. a bank name), Cross-street geocoding and Town centroid geocoding (when the only detail is the suburb to which a person travelled); and has been designed to accept external data files of street names, suburbs, postcodes, cadastre file addresses, landmarks, cross-streets and town centroids; it can be easily modified for use in another geographic locality. All data output from the geocoding module can then be used in GIS programs such as MapInfo using the required map grid projection.
8. *Data editing*: The *Speedit* program performs both data entry and data editing at the same time. One of the advantages of this is that, because all geocoding is performed as the data is entered, it is possible to run detailed editing checks involving trip distances and speeds, since these can be calculated as soon as the geocodes for successive locations are found. Thus, in addition to the usual range checks and logic checks, these more complex trip editing routines can be run immediately, thus enabling the data coder to immediately correct problems with the data or flag the data for follow-up clarification calls (which are typically performed within days of the respondent filling out the survey forms). Data entry and editing in VISTA is typically completed within a week of the questionnaires arriving in the survey office at the end of a week of travel days.
9. *Identification of data problems*: Following the coding and editing of data from the questionnaires, the data are merged into an integrated file for each household. This file is then analysed looking for missing data and for data entries that fail the range checks or logic checks. These households, and the problem data, are earmarked for further investigation.
10. *Telephone clarification calls*: Those households in which problem data have been identified and not rectified, and for which a phone number is available, are then phoned to seek clarification about the problem data. The clarified information is then input into the data file. Approximately 15% of responding households are phoned during this process for VISTA.
11. *Reminder calls and letters*: About one week after the collection of completed questionnaires from households, those who have not yet responded are contacted

via phone to remind them to return the questionnaire if they have completed it. Those who are not contactable by phone are sent a reminder letter. However, unlike VATS, the non-respondents are not urged to complete the survey for a later travel day. They are only encouraged to send back the questionnaires if they have already completed them for the nominated travel day.

29.3.2. *Face to Face*

29.3.2.1. The Sydney experience The Sydney *Household Travel Survey* (HTS) is a large-scale, continuous survey of the travel patterns of residents of the Greater Metropolitan region of Sydney (Battellino & Peachman, 2003). Every year approximately 3500 households (or 10,000 individuals) are interviewed about their travel on a particular day, with interviews spread across the whole year. Information on all travel and activities undertaken by all household members over a given 24-h period is collected via face-to-face interview.

To maximise the statistical reliability of HTS results for a given year, each annual HTS estimate consists of three years of pooled HTS data weighted to the estimated resident population (ERP) from the Australian Bureau of Statistics for that year. This sample size enables reliable trip estimates at the statistical division (Sydney wide) level in wave 1, statistical sub-division (regions within Sydney) level after wave 2 and the statistical local area (SLA — local government area of which there are 50 in Sydney) level at the end of wave 3. Interviews are clustered within these SLAs to reduce the fieldwork costs. Each day of the year is represented with sampled households being given a nominated 'travel day' for which data will be collected. All members of the selected households are interviewed. The scope of the survey population is all residents of private dwellings in the survey area. Thus, visitors, tourists and residents of institutions are not included.

The sample is drawn on a quarterly basis by a process of area address listing; more details are provided by Battellino and Peachman (2003). Once the sample is selected, no replacement of dwelling or 'travel day' is allowed. Area address listing encounters the expected problems associated with this method, such as vacant dwellings and human error in recording addresses. Approach letters addressed to 'the householder' are sent before the interviewer attempts to make contact. The sample loss due to vacant dwellings, dwellings under construction/being demolished, or non-private dwellings have been around 5% and for incorrect addresses, less than 2%. There have been very few issues in relation to the sampling process, which have been introduced because of the continuous nature of the survey.

29.3.2.2. Questionnaire design The questionnaires used for the continuous survey were primarily the same as those used for the Sydney 1991 Home Interview Survey: a household questionnaire, collecting household, person and vehicle data for the household, a 24-h trip diary questionnaire for members of the household aged 15 years and over and a 24-h trip diary for members aged less than 15 years. This had the benefit of providing data which would be comparable with that collected in 1991 and to a large extent with 1981. Thus, for the first time, longer term trend data on travel patterns is available. Despite the fact that there is no real information on the intervening trends between 1981, 1991 and 1997, the ability to compare the data at these points in time has proved to be very valuable in providing advice for decision makers.

It is important when conducting the survey on a continuous basis that the questionnaire remains essentially the same to ensure the comparability of the data. However, there is room to progressively incorporate small improvements and updates. There is also the added benefit of being able to use the existing survey strategy to collect supplementary data on relevant issues as required, at marginal costs. There is also the potential to increase the sample size in any particular area, using the same survey strategy, to provide more statistically reliable data, for example in relation to the impact of a large infrastructure project, in a time frame shorter than would be provided by the standard sampling procedures.

29.3.2.3. Fieldwork control At the beginning of the continuous HTS, the fieldwork was contracted to a research company on an annual basis with a rollover of contract, if certain performance indicators were met, for a maximum of three years. As the HTS is a face-to-face interview, a sizable and stable team of interviewers, especially suited and trained in personal interviewing techniques, is required. The advantage of the continuous nature of the survey for interviewers is that it provides a reliable and consistent source of work and, given the interesting nature of the topic, many find it rewarding.

Notwithstanding, the HTS team is refreshed each year and few have stayed since the first wave; the team has some 30 interviewers, which dwindles to the low 20s during the wave and is refreshed with a new recruitment of up to 10 interviewers each year. It has been found that the composition of the team and the fieldwork managers/ supervisors can have a significant effect on response rates.

Because of the continuous nature of the survey, there is potential for interviewer fatigue and boredom, or over familiarisation, leading to laxity in the compliance with procedures. Much effort is put in by the fieldwork company and survey management staff to ensure that the fieldwork staff is made to feel part of the team. This is done by regular team meetings, every two months, where interviewers are consulted about issues arising in the field and suggestions sought for improvements, annual re-training and debriefing days. Fieldwork staff is also provided with information on the importance of the survey and use of the data. The performance of interviewers is also monitored by strict validation procedures that require that at least 10% of interviewer-completed workloads are validated by telephone. The validation must be done within one week of the receipt of the workloads and before the data is clerically checked, entered and processed. Where a significant proportion of completed interviews by an interviewer (i.e. 10% below the average for a reporting period of one month) do not have the household's telephone number or have incorrect telephone numbers, the survey consultant validates using the face-to-face interview method by making a single visit to each household in the affected workloads. Any interviewer suspected of not following field procedures and providing incorrect data is validated at a higher rate. The team leaders also conduct monitoring of interviewers, with a minimum of two supervised interviews undertaken per interviewer per year. The aim of the supervised interview is to check that all interviewers are conducting the interviews using correct procedures, are clearly explaining the survey to respondents and have a clear understanding of the survey questions. An extra-supervised interview is undertaken where the results of the initial supervised interview revealed that the interviewer's performance was not at an acceptable level.

There are strict control procedures for checking that information is collected only for the nominated travel day and that respondents are contacted according to certain rules around this day. Workloads are checked weekly to obtain immediate information on the level of response obtained by interviewers so that the survey consultant is able to act at once where an interviewer is having difficulty recruiting the respondents or specific issues are identified that require special attention. In addition to the wide range of data checks that are carried out on the data following data entry, response rates, trip rates, rate of sample loss and number of calls made to households are also monitored monthly on an interviewer basis. The percentage of persons staying at home on the designated travel day are also checked as a form of 'soft refusal' (see Madre, Axhausen, & Brög, 2007). This means that any indications of potential interviewer bias can be detected, and rectified, at an early stage.

Thus, although many interviewers are very familiar with the survey, there is a constant need for monitoring of interviewers' performance; project management staff cannot become complacent or assume that the continuous process will become self-managing. However, the necessity for continual vigilance of the fieldwork is not a deficiency, or a negative side-effect, of a continuous survey, but recognition of the fact that it offers an opportunity to rectify problems that a one-off survey does not.

For example some problems were discovered with regard to the adherence to procedures by a few interviewers in the second wave of the HTS. It was then possible to disregard these interviews and resample them in the third wave. This would not have been possible if it had been a one-off survey; these interviews would have been lost. The continuous nature of the survey also allows for the ongoing assessment of the effectiveness and practicality of fieldwork procedures and provides the opportunity to refine them over time. This, it is believed, has contributed to a continuation of the high quality of the data.

29.3.2.4. The Santiago de Chile experience The Santiago continuous mobility survey started in August 2001 with a first wave of 15,537 households (interviewed until April 2002), which should have been followed by roughly 5000 households every subsequent year. As it happened, after that first year the government decided to examine the results and the process was deferred for almost two years, to be re-commenced in April 2004 for another three years (the data collection stopped in May 2007). The government is currently studying the continuation of the process.

29.3.2.5. Questionnaire design and fieldwork control The main characteristics of the survey in this sense are very similar to those of the Sydney HTS. The design of the questionnaire was initially inspired by that of the VATS and Sydney HTS surveys, but the Santiago experience (and the evidence after a year-long, very thorough, pilot study of various methods) showed that, in the great majority of cases, face-to-face interviewing was the best approach to obtaining data from homes at an acceptable cost in a developing country such as Chile. Notwithstanding, there are cases where the personal interview method is simply not feasible; for example when household members travel too much and it is not possible to contact them (in spite of several attempts at different times in several days), or when the security personnel at a high-income condominium or building does not allow the interviewer to enter the premises. In these 'difficult' cases, the self-completion survey method with personal delivery and recovery of the survey form provided a solution to the problem. However, it was found that only people capable of following written instructions (and indeed, reading them) were able to complete the forms in Santiago. In practice, this meant that acceptable response rates were only possible in medium- to high-income households.

It was also found that self-completion may not be advantageous, cost wise, in a developing country. First, as the forms need to be attractive and easy to follow, this translated into much higher production costs than for personal interview material.[5] On the other hand, the sample must be over dimensioned (as response rates are clearly inferior in this method) and each household given a large number of forms (eight in Santiago) to make sure that there were enough for each household member. Finally, the cost of delivering and recovering the forms were also higher due to the over dimensioning of the sample.

In the first wave of the survey, self-completion forms were used in 3% of the households (490 cases) and only 12 had one or two missing items; this required special training of the interviewers (respondents filled the questionnaires when they 'remembered', so they had to be followed closely to make sure they jotted down all their trips in the memory joggers, etc.). However, and surprisingly, in the 2004–2007 wave, no self-completion forms were used; although the current administrator rightly claims that they were able to achieve higher response rates among high-income people than in the 2001–2002 wave, the fact remains that most probably they did not obtain information about the 'difficult' cases mentioned above.[6]

The field procedures have remained basically unchanged between the two periods and are almost identical to those described for the Sydney HTS. However, data validation and interviewer control have needed to be more stringent, as it appears that interviewers are not as meticulous as in Australia (although recruitment and

5. In the Santiago survey, each trip required two legal size (216 × 356 mm) pages printed in two colours, apart from the pages dedicated to presentation and instructions.
6. In fact, we have recently learned that high-income zones are now represented by (high-income) households with the lowest income values in that category (i.e. if the high-income group is, say, over US$ 1500/month, we understand that most households in this group have incomes just over this value).

training processes are similar); this behaviour could be, perhaps, attributable to the rather low wages received by interviewers (US$ 13.5 per household including transport) and other staff (e.g. the data validators in the field receive US$ 3.25 per household and the supervisors and digitisers US$ 500/month[7]). To improve on this, during the second wave of the survey, 20% of all households were revisited for validation purposes and a further 30% of the sample was re-contacted by phone.

29.3.2.6. Sample design The first wave-sampling frame utilised the registry of homes at the National Tax Office, and in the second wave the census addresses (the last was taken in 2001). At every period, for example three years in the second wave, the appropriate number of addresses (i.e. 15,000) were generated randomly in time and space using Sobol low discrepancy sequences[8] (Bratley, Fox, & Niederreiter, 1992) looking three years ahead (i.e. for a total sample of 15,000 households).

In this simple two-dimensional case, each sequence corresponds to a pairing (space and time) covering the study area in orderly form. The first column of the complete sequence would be the traditional random sample (in space) of a typical O–D survey (which is usually stratified by zone in practice). The second column is obtained by means of a linear projection between the initial and final dates of the data collection effort. In practice, it is necessary to order the pairings in terms of their 'spatial' component and associate them to a traditional spatial sample (spatially ordered[9]) so that they will be associated to an address in the sample; the pairings are then ordered by their 'time' element.

The following example illustrates the method. Assume that we need to generate a sample of 10 addresses in five zones (all zones have the same size[10]), to be surveyed (survey date) over two days (Saturday and Sunday). For this we first generate a two-dimensional Sobol sequence with 10 elements as in Table 29.1.

We then generate a sample of 10 addresses for the five zones, ordered by their zone number; in practice this is done as follows. First, a block is selected for each observation by randomly assigning to every block in a zone a probability of being chosen proportional to its number of households. Second, for each observation, a random number between 1 and the number of households in the block is chosen and this serves to choose the household in practice, that is the third in the block[11] in Table 29.2.

7. The minimum wage in Chile was US$ 300/month in 2007.

8. http://en.wikipedia.org/wiki/Low-discrepancy_sequence.

9. The ordering can be simplified using zonal codes that assist in finding their spatial location, as has been traditional in Santiago; in this way zones starting with the same number are closer.

10. In practice, not all zones are equal, and the number of cases to consider in each zone is the nearest integer value of the product of the number of households in the zone (from the census) and the sampling rate (i.e. 15,000 over the total number of households in Santiago).

11. In practice, a series of numbers is generated as for each household sampled; several replacement households are also selected in each case.

Table 29.1: Space and time Sobol numbers.

Number	Space	Time
1	0.5000	0.5000
2	0.7500	0.2500
3	0.2500	0.7500
4	0.3750	0.3750
5	0.8750	0.8750
6	0.6250	0.1250
7	0.1250	0.6250
8	0.1875	0.3125
9	0.6875	0.8125
10	0.9375	0.0625

Table 29.2: Associating zone and address to places.

Place	Zone	Address
1	1	1
2	1	2
3	2	1
4	2	2
5	3	1
6	3	2
7	4	1
8	4	2
9	5	1
10	5	2

Then ordering the Sobol table by its spatial component and integrating it with the sample of addresses, we get Table 29.3. And reordering the table in ascending order by time and distributing it according to date (i.e. 1: Saturday; 2: Sunday), we get the data shown in Table 29.4. Finally, eliminating redundant columns and ordering by Sobol number, we get Table 29.5 that satisfies the objectives of being random in both space and time.

A preliminary calculation, based on the spatial distribution of zones in the Santiago area (by the six macro-zones defined) and assuming that the survey would start on Thursday, 1 January 2004, and end on Sunday, 31 December 2006, produced the sample distribution by day and area in the three study years shown in Table 29.6 (note that the days between Thursday and Sunday have an additional week).

Table 29.3: Ordering the Sobol table by its spatial component.

Number	Space	Time	Place	Zone	Address
7	0.1250	0.6250	1	1	1
8	0.1875	0.3125	2	1	2
3	0.2500	0.7500	3	2	1
4	0.3750	0.3750	4	2	2
1	0.5000	0.5000	5	3	1
6	0.6250	0.1250	6	3	2
9	0.6875	0.8125	7	4	1
2	0.7500	0.2500	8	4	2
5	0.8750	0.8750	9	5	1
10	0.9375	0.0625	10	5	2

Table 29.4: Ordering by space and time.

Number	Space	Time	Place	Zone	Address	Date	Day
10	0.9375	0.0625	10	5	2	1	Saturday
6	0.6250	0.1250	6	3	2	1	Saturday
2	0.7500	0.2500	8	4	2	1	Saturday
8	0.1875	0.3125	2	1	2	1	Saturday
4	0.3750	0.3750	4	2	2	1	Saturday
1	0.5000	0.5000	5	3	1	2	Sunday
7	0.1250	0.6250	1	1	1	2	Sunday
3	0.2500	0.7500	3	2	1	2	Sunday
9	0.6875	0.8125	7	4	1	2	Sunday
5	0.8750	0.8750	9	5	1	2	Sunday

29.3.2.7. Data input, digitising and checking The rigorous procedures in place in the Santiago survey are very similar to those reported for the VISTA survey (indeed they started by mimicking the early procedures in the original VATS survey), so they will not be repeated here for lack of space. In particular, on-line checking software for data entry, which allows multiple staff to work at the same time, was designed so that the software also completes street and place names, easing the task of the digitisers and reducing errors substantially. However, in the second wave (and a little like the later experience in VATS), the strict, almost on-line, procedures for data entry and checking were relaxed and performed not every day or week, but every month and later every several months. This meant that finding out an interviewer who was performing imperfectly (e.g. much lower trip rates) took three or four months to detect.

This, in turn, led to severe data losses, replacement of personnel, re-training of interviewers, much higher costs and the need to apply, in some cases, non-trivial correction factors to non-compulsory trips.

Table 29.5: Ordering by Sobol number.

Number	Zone	Address	Day
1	3	1	Sunday
2	4	2	Saturday
3	2	1	Sunday
4	2	2	Saturday
5	5	1	Sunday
6	3	2	Saturday
7	1	1	Sunday
8	1	2	Saturday
9	4	1	Sunday
10	5	2	Saturday

Table 29.6: Sample distributions over three-year period.

Day	North	West	East	CBD	South	South-East	Total	%
Monday	289	464	346	107	478	453	2137	14.25
Tuesday	283	481	326	99	502	442	2133	14.22
Wednesday	292	444	342	105	507	447	2137	14.25
Thursday	295	465	356	107	475	450	2148	14.32
Friday	277	498	333	100	512	429	2149	14.33
Saturday	297	460	346	100	482	464	2149	14.33
Sunday	279	469	348	99	515	437	2147	14.31
Total	2012	3281	2397	717	3471	3122	15,000	100.00

29.3.3. Survey Content

The VATS survey attempted to obtain travel information from all members of the household, plus any visitors who stayed in the household on the night before the travel day. But while it is relatively easy to obtain demographic information about visitors, it is often difficult to obtain travel diary data for them, especially if they return to their own residence on the travel day. Also, asking for children under five years of age to complete the travel diary merely imposes an extra burden on the parents. This is especially the case when the young children's diaries are often identical to one of the parents, with whom they travel with during the travel day. For these reasons, the VISTA survey does not attempt to obtain travel diary data for visitors (but does ask for their demographics), and does not seek completion of travel diaries for children under five. Instead, it reconstructs them from data provided by other household members.

In Santiago, the scope of the mobility survey did not include all travellers in the urban area but only residents and visitors to households. People in hotels, other people in non-private dwellings (such as hospitals and military barracks) and travellers that pass through the area on the survey days were judged to be relatively few (in comparison to residents) and also too difficult to contact. However, trips are recorded for all household members, even if they are too young to do this independently; in this case the parents have to report their travel. Also, in the case of relatively young individuals and the old and infirm, a small amount of proxy reporting is accepted.

All the ongoing surveys mentioned contain questions on four essential topics: household data, persons, vehicles and trip stages for each person (in the VISTA survey only for those aged five and above). In most surveys, the client group inevitably wants to ask more questions than can reasonably be accommodated in a reasonably sized questionnaire. In general, the approach in these situations is to stay with a rigid maximum size of questionnaire (say, a folded A3 sheet for the household, person and vehicle forms; and a 16-page A4 booklet for each travel diary) and to force the required content to fit within these boundaries (with a minimum font size). This forces the client to think about what information they really need, while ensuring that the respondent burden is contained within what has been found to be acceptable in past surveys.

29.3.4. Sampling

In most Australian surveys, a clustered random sample of households within CCD has been used. VATS was a slight exception, in that it was not clustered, because of the nature of mail-out/mail-back surveys. The idea was to minimise the time and cost spent by interviewers in the field in moving between sampled households. In Santiago, the situation is different as the sample is truly random in time and space; the only concession to logistics being that the days of the week are aggregated into eight-week groups, where, for example a household with a date defined for a given Monday in the period is assigned to any of the eight possible Mondays in the period to optimise field work. This also ensures that the summer period (January and February) is completely covered.

Sample frames have varied over time. Earlier ones tended to use lists of addresses from utility companies (water supply, electricity connections, etc.), as these were the only comprehensive lists available for use in travel surveys. More recently, the widespread use of GIS programs has meant that many agencies have comprehensive address lists available in this format. For example several surveys conducted by TUTI in Melbourne in 2005 and in New Zealand in 2006 obtained lists of residential addresses from the local government agencies in whose areas the surveys were being conducted. This strategy, however, tends only to work well when the number of agencies covered is relatively small. When data has to be obtained from many agencies, the effort involved in standardising the content and format of the data sets can become significant. In Santiago, as mentioned above, the sampling frame was first the registry of homes in the Tax Office, but this was inevitably a couple of years out of date, leading to several losses due to changes (demolitions, building of new buildings, change to commercial use) which can be very rapid in developing nations. Furthermore, as this data corresponds to regularised (i.e. legal) dwellings housing actual taxpayers, we later discovered that it was biased towards higher incomes (poor people do not pay taxes nor have their dwellings regularised). In the second wave, the 2001 census information was used (which was released late in 2002); it had no income bias, but by 2006 it had also become relatively obsolete.

In recent times, the growing availability of commercially available GIS data sets of addresses and property boundaries has opened up new opportunities, which have particular relevance for ongoing travel surveys. For example in the Sunshine and Gold Coasts surveys in Queensland and in the VISTA survey, the sample has been chosen by using three GIS data sets (the Geocoded National Address File, an overlay of planning zones and a map of cadastral property boundaries) to choose a sample of residential addresses within the sampled CCD regions. This method is particularly appropriate for ongoing surveys, because such databases tend to be updated on a regular basis by the agencies developing them for commercial purposes (such as market research and location planning).

29.4. Challenges of the Current Practice

29.4.1. Response Rates

As in most areas and surveys, response rates in Santiago and Sydney have tended to decline with time in spite of the efforts spent on re-training interviewers and following strict fieldwork control rules. In the Santiago 2001–2002 wave, the response rate was 68% (Ampt & Ortúzar, 2004), but as roughly half of the non-response corresponded to no contacts, many of which were probably non-eligible, the real response rate was probably higher. The situation in Sydney is depicted in Figure 29.1;[12] full response corresponds to the case when household and all trip and person details from all members are obtained. Part response is when all household data is obtained, but full trip and person details are not obtained for one household member.

In Santiago, the final statistics for the 2004–2007 wave have not yet been released, but we have learned that due to cost reasons the strict procedures mentioned above were relaxed as follows. As each interviewer (only 10–16 for the whole city during the second wave) had to cover a very large area (Santiago has more than 1.5 million households and the study area is some 2000 km^2), they were given not one, but three sample addresses.

The idea was that they should try the first address in the first place, and if after four visits they could not get a response, try the second address, and so on; however, it is

12. We are grateful to Tim Raimond for having provided us with this information.

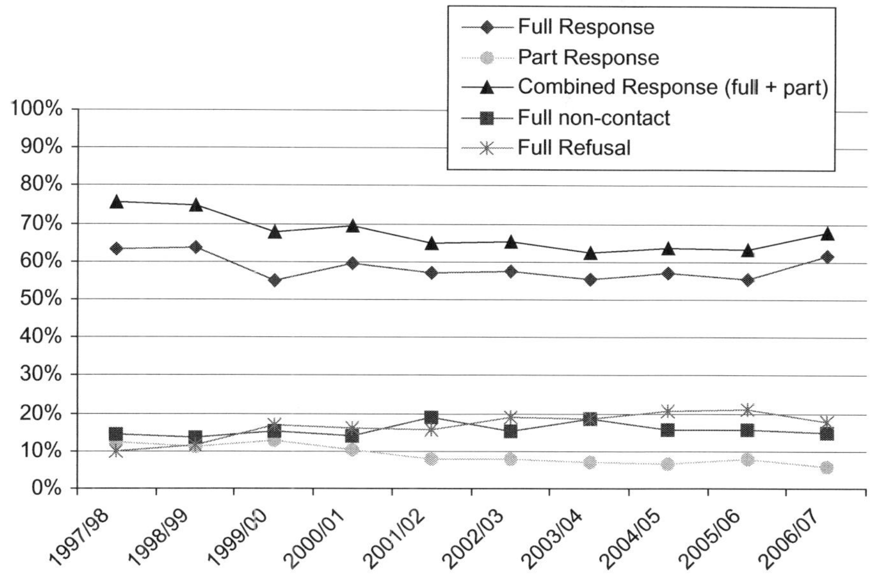

Figure 29.1: HTS response rates wave 1 (1997/1998) to wave 10 (2006/2007).

obvious that the interviewers, in general, used the addresses to their advantage (although the 20% validation checks should guarantee that at least they tried a couple of times before changing). The final result was that in the three years, the interviewers contacted (or, at least, were provided with) 34,669 addresses; of these, they were rejected or found nobody in 18,316 addresses (furthermore, 795 households had to be replaced because they had false data). So, as the final sample consisted of 15,558 complete households, we may deduce that the response rate dropped to 45%.

In the VATS survey, the response rate started out at around 60% in 1994, but had dropped to about 50% by 1999. This could have been due to a general decline in response rates in all surveys over that period, or to a problem with maintaining ongoing staff motivation over an extended period. After the major design and administration changes in 2000, the VATS response rate was reported to have fallen to about 25%, but lack of complete documentation of the survey in those years precludes a precise observation.

In all of the surveys using the current VISTA design, from the Perth survey in 2002 through to the Sunshine Coast survey in 2007, an average response rate of 55% has been obtained, ranging from a low of 49% to a high of 60%. The smaller cities tended to deliver higher response rates, with lower response rates in the larger cities.

29.4.2. Respondent Burden

Using an ongoing survey methodology, with one to two days travel, poses relatively low levels of respondent burden. The burden varies with the survey method, and self-completion is considered by some to be more difficult for the respondent for the following reasons:

- It takes longer for each individual to complete the forms because they are not as familiar with the questions as an interviewer would be.
- It is more difficult for people who are not used to completing forms (this can include people for whom the language of the survey form is a second language).

This can lead to under-reporting by certain socio-demographics and by people making certain trips (e.g. those who make many trips). (See discussion on respondent burden by Bricka, 2006.) Nonetheless, in the various Australian self-completion surveys from VATS through VISTA, the proportion of survey respondents not born in Australia (who might be thought to have more difficulties with self-completion surveys printed in English only) is virtually identical with the proportion of persons not born in Australia as identified in the census.

On the other hand, in contrast to face-to-face interviews, self-completion forms are able to be completed at any time of day or night, meaning that certain types of people find them much easier than a face-to-face approach. In face-to-face interviews, much of the burden of response is taken by the interviewer who reads the questions and encourages the respondent to continue to the end of the survey. Given that the methodology discussed is based on cross-sectional sampling, the considerable burden of panels where people are asked to contribute on a repeating basis is not an issue.

29.4.3. Weighting and Expansion

For standard cross-sectional travel surveys, there are often four types of weighting (and expansion): household weights, person weights, non-responding household weights and non-reported trip weights. The household weights may be obtained by comparing the survey results with the census results on the basis of household size (number of persons) and dwelling type within geographic regions. The person weights can be based on a comparison with the census on the basis of age and gender of persons within geographic regions.

Non-responding household weights are not used in some surveys (i.e. in VISTA and in the Santiago cases), and for self-completion surveys non-reported trip weights are calculated based on who completed the travel diary (self-reporting vs. proxy reporting) and when the travel diary was filled in (on the Travel Day compared to after the Travel Day). In Santiago, non-reported trip weights have been used for cases where there is a suspicion of laxity on the part of the interviewers (i.e. just prior to a full re-training exercise).

As always in the calculation of expansion weights, there is a trade-off between the desire for precision (lots of controlling variables and categories) and the limitations of zero and small counts in cells of the survey data cross-tabulations. Clearly, larger sample sizes can allow greater disaggregating in the expansion variables, and this may be improved as the sample size increases over many years of an ongoing survey. If census data is used for the control data set for expansion purposes, the timing of the travel survey and the census is important. Many transport agencies think it is a good idea to conduct a travel survey in the same year as the census (years ending in 1 and 6 in Australia; unfortunately the census is only every decade in Chile and most developing countries). However, this can pose several problems. Firstly, running a travel survey over the same period as the census can create severe respondent burden problems, with negative impacts on response rates for the travel survey (which, unlike the census, is not compulsory). Secondly, even if the travel survey is run in the same year as the census, it is quite normal to experience a significant delay before the release of the census data (in Australia, where census data release is relatively quick, the results from the 2006 census were not released for about 15 months; the same happens in countries such as Chile). Therefore, there can be a delay in being able to expand the travel survey data up to the census data. Thirdly, even though the new census data will eventually be available for weighting, there are many activities in the design phases of the travel survey (such as selection of stratification regions and drawing of the sample) that rely on census data. This will necessarily have to be done with data from the previous census, and this can pose a problem if census regions change between the old and the new census. This can mean that the sample is drawn with one set of geographic boundaries, whereas the expansion is then to be done with a different set of boundaries. If the census is to be used for expansion purposes, then it is probably best, for survey design purposes, to do the travel survey a year or so after the census, so that the new census data can be used for both sample design and expansion purposes.

With ongoing surveys there are two major questions to be addressed in weighting and expansion. Firstly, how many years of data can be aggregated for expansion purposes, and secondly, which year of census data should be used for this purpose. The answer to the first question will depend on how rapidly travel patterns are changing in the region, and on the sample size collected in each year. For example as noted earlier, the Sydney HTS pools three years of data to obtain reliable trip estimates at the zone level. Similar procedures were adopted with VATS, although for some analyses where the data was not expected to change significantly over time, all six years of the database (1994–1999) were pooled for analysis.

For a single year, or for a pool of several years of survey data, the question often arises as to which year of census data should be used for expansion purposes, but this may not matter all that much as the purpose of expansion is to transform the sample data so that it represents the travel behaviour of a much larger population. Therefore, if the underlying travel data does not change much over time (e.g. household trip rates by demographic characteristics), then expanding it to different census years merely changes the magnitude and distribution of the expanded result. Indeed, one could also expand the data to a hypothetical distribution of households corresponding to some future development that has not yet occurred.

In the Santiago case, researchers had to face the problems associated with a rapidly expanding city. The 2001 census was used to generate the sample, but by 2005, it was obvious that large parts of the study area, particularly to the north of the city, had acquired a significant middle- to high-income population with much higher car ownership than similar households in the more traditional quarters of the city. This meant that the car ownership figures in the survey were smaller and biased in their spatial distribution, in relation to independent data available; the same happened with other sources of independent data such as number of schoolchildren.

Furthermore, drastic changes occurred in relation to travel supply: four large urban highways were built after 2001 (and started operating at the end of 2005) and a large extension to the underground (one and a half lines serving massively populated new quarters) also started operating in 2006. This meant that the 'state of travel' in 2006 could only be adequately represented by the data for that year, but the typical sample bias related to household size distribution and socio-demographics were difficult to ascertain as the census data had become highly obsolete.

Correcting the problem of rapid, large-scale growth requires, among other things, 'listing' households in large areas where there were no households at the beginning of the project. Usually this is not only a data issue, but also needs additional funding. It also requires devising complex data expansion and re-weighting methods and this requires further research.

29.5. Uses of the Data

We have discussed the way data is used in general in the first section. The uses of the data depend to a large extent on the needs of the commissioning organisation. As we noted earlier, these needs are often for modelling of current and forecast travel patterns in cities and urban areas — always for cars, but with increasing frequency also for public transport, cycling and walking (especially in developing countries).

The data uses are not, however, confined to modelling and, to some extent, only lack of imagination can limit the uses of the data. In particular, the data are often used for planning and policy purposes. Examples that we know of include working out the influence of changing the size of free travel zones to school, the impact of all children within 1 km of a school walking or cycling and estimating differences in car-kilometre between those people who travel to work and those that work at home. There are also non-traditional uses of the data outside the transport field, such as petrol companies using it to optimise the location of their service stations along arterial roads and outdoor advertising agencies trying to tailor the content of roadside billboards to the demographics of passing motorists.

There are many advantages to encouraging the widespread, imaginative use of the data, since the more it is used, the more likely that continuing funding will be forthcoming. As an example, in the early days of the VATS survey, the main funding source was from users of the data — a good way to encourage innovative uses. We would note, however, that over the longer term this was not a viable way to fund the data collection process.

In the current political/social environment where there is a growing interest in climate change and its effects, it is likely that there will be new uses for the type of data discussed here. For example the type of vehicles people drive, on what roads, at what times of day and potentially at what speeds are likely to be important elements of information in policy and decision making of the future. As an example, this could give information on who/what type of trips/people are contributing most to congestion. And all of these data elements could potentially be included in ongoing travel surveys — particularly if new data collection mechanisms such as GPS devices can be used.

29.6. Ongoing Surveys and Climate Change

In this section we focus on how ongoing surveys are likely to need to change, both in the data they collect and the way it is collected, to reflect the 21st century focus on climate change.

29.6.1. Collecting Data Related to Climate Change

If, in an era in which climate change is likely to be the most significant issue to be addressed in transport planning, the travel survey data is to be used for measuring changes in energy use and greenhouse gas emissions, then new types of data might need to be collected, such as much more detailed information about household vehicles (i.e. make, model, year of manufacture, engine capacity and fuel type). Also, coupling the survey with a GPS device carried by respondents can allow information such as speed and route to be added to the already rich data collected; this has been tried in the Sydney HTS. Finally, as can be seen from the following example, some surveys have already included and used typical ongoing data for environmental modelling purposes.

A study conducted for Transport Canada by TUTI (Richardson & Seethaler, 2002) addressed the potential for developing an integrated vehicle use and fuel consumption survey, where previously these had been undertaken as separate surveys in Canada. As part of that project, it was demonstrated how relatively conventional travel survey data (VATS) could be used to estimate fuel consumption and greenhouse emissions across an urban area. Using data from the 1995 VATS survey, the analysis estimated greenhouse emissions from passenger car travel in the Melbourne region, taking account of the following factors:

- vehicle type used on trips (make, model, engine size);
- fuel consumption rates by vehicle type;
- emissions factors (kg CO_2 per litre fuel consumed);
- linking each vehicle to each trip;
- trip length;
- average trip speed;
- engine temperature (as a function of time since last trip and length of current trip);
- ambient temperature (as measured by meteorological records).

The emissions associated with each trip could then be analysed as a function of all the other characteristics recorded in the travel survey (such as trip purpose, trip location, trip timing and driver characteristics), to obtain an understanding of how and why greenhouse emissions were being created as a result of passenger vehicle usage.[13]

29.6.2. The Climate Change Impacts of Data Collection

A further issue that is likely to be raised in the near future is the actual impact of the survey collection on the environment. There are three types of emissions that must be included when estimating the carbon footprint from an activity:

- *Scope 1*: emissions created directly by people (e.g. fuel used in a private vehicle).
- *Scope 2*: indirect emissions from energy consumption (e.g. from fossil fuels burned to generate electricity).
- *Scope 3*: indirect emissions embodied in products or services consumed by an individual (e.g. paper consumed, flights taken).

Self-completion surveys require the use of large amounts of paper, which have embodied emissions. The emissions created by questionnaire delivery in mail-out surveys (i.e. VATS) are proportional to their volume and weight, as a fraction of total emissions generated by mail deliveries. However, VISTA has given much higher response rates as a result of personal drop-off and collection, which usually involves car travel, at least in Australia, the United States and Europe.

Face-to-face interviews have less paper involved, but almost always involve a larger amount of vehicle travel, even when samples are clustered. Further clustering could reduce vehicle travel but the need to obtain high response rates is likely to mean successive follow-ups to the same area even if the sample is clustered. Notwithstanding, in countries such as Chile the interviewers travel by public transport in most cases and some also use bicycles.

It could be argued that on-line methods will reduce climate change effects to the greatest extent, but this will need to be considered in the context of computer ownership and use, each of which has significant Scope 2 and Scope 3 emissions. As an example of the magnitude of likely emissions associated with field surveys, and the costs of offsetting these emissions, the current VISTA survey of 17,250 responding households in metropolitan Melbourne and regional Victoria has been estimated to require a total of approximately 300,000 km of field staff travel by car for the various components of the survey. This will generate about 85 tonnes of CO_2.

These emissions can, however, be offset for a total cost of about $ 1000 (e.g. www.treesmart.com.au), an amount that has been built into the budget for VISTA.

13. This study is similar in methodology to the Fuel Diary in the German Mobility Panel (Appendix 29.A.1); it has run since 1994 and covers vehicle use and fuel consumption for passenger transport in the whole of Germany.

This cost is not excessive, but emphasises the fact that greenhouse emissions can be alleviated at modest cost if everyone, including survey organisations, does their bit to reduce and then offset their own emissions.

29.7. Key Issues for Reflection and for the Future

This paper has covered a wide range of issues associated with large-scale, ongoing mobility surveys. Since it is a resource paper, we conclude by raising some key questions for thought:

- What are the best ways to deal with issues of retaining staff over time, and with keeping their levels of motivation and performance at a desired level?
- What types of questions will be needed in ongoing surveys in the future, given the likely emphasis on issues of climate change?
- When do ongoing surveys in urban areas give the most benefits?
- How can commissioning organisations expand the number of users of the data they collect?
- When might it be appropriate to use other methods (e.g. Internet) as part of ongoing data collection?
- Do you need to have equal sample sizes for each year, or are there advantages to taking a large sample every (say) four years and carrying out surveys of smaller samples in the intervening years?
- How can ongoing travel surveys overcome the over-surveying of the population, with consequent continuing falls in response rates?
- How do researchers tackle the integration weighting of successive 'sample years' when drastic or massive changes in transport supply are experienced in the city?

Acknowledgements

We are grateful to Martin Lee-Gosselin, Tim Raimond, Dirk Zumkeller and Johanna Zmud for their useful comments and support. As usual, any errors remaining in this paper are our responsibility alone. We also wish to thank FONDECYT, through Project 1050672, and the Millennium Institute on Complex Engineering Systems (Project P05-004F) for supporting the research activities of the second author.

References

Ampt, E. S., & Ortúzar, J. de D. (2004). On best practice in continuous large-scale mobility surveys. *Transport Reviews*, *24*, 337–363.

Battellino, H., & Peachman, J. (2003). The joys and tribulations of a continuous survey. In: P. Stopher & P. Jones (Eds), *Transport survey quality and innovation*. Amsterdam: Pergamon.

Bratley, P., Fox, B., & Niederreiter, H. (1992). Implementation and tests of low discrepancy sequences. *ACM Transactions on Modeling and Computer Simulation*, *2*, 195–213.

Bricka, S. (2006). Scheduling considerations in households travel surveys. In: P. Stopher & C. Stecher (Eds), *Travel survey methods: Quality and future directions*. Oxford: Elsevier.

Brög, W., & Meyburg, A. H. (1982). Consideration of non-response effects in large-scale mobility surveys. *Transportation Research Record*, *807*, 39–46.

Madre, J.-L., Axhausen, K. W., & Brög, W. (2007). Immobility in travel diary surveys. *Transportation*, *34*, 107–128.

Mistro, R. D., & Behrens, R. (2008). How variable is the variability in traffic? Can TDM succeed? *Proceedings of the 27th South African transport conference*, Pretoria, South Africa, 7–11 July.

Pas, E. (1988). Weekly travel activity behavior. *Transportation*, *15*, 88–109.

Richardson, A. J. (2003). Behavioural mechanisms of non-response in mailback travel surveys. *Transportation Research Record*, *1855*, 191–199.

Richardson, A. J., Ampt, E. S., & Meyburg, A. H. (1995). *Survey methods in transport planning*. Eucalyptus Press, Melbourne. Available at: http://www.transportsurveymethods.-com.au/

Richardson, A. J., & Seethaler, R. K. (2002). *A study of the potential of developing an integrated vehicle-use and fuel consumption survey*. TUTI Report no. 9-2002. The Urban Transport Institute, Taggerty.

Wermuth, M. J. (1985). Non-sampling errors due to non-response in written household travel surveys. In: E. S. Ampt, A. J. Richardson & W. Brög (Eds), *New survey methods in transport*. Utrecht: VNU Science Press.

Appendix 29.A.1. National data sources — not ongoing surveys

Country	Survey name	Year (duration)	Reporting period
Austria	Austrian National Travel Survey	1995	Daily mobility: one day Long-distance travel: two weeks
Belgium	Belgian National Travel Survey (MOBEL)	1998–1999	Daily mobility: one day Long-distance travel: four weeks
Denmark	Danish National Travel Survey	1992–2003 (continuous survey), 2006, 2007	Daily mobility: one day (evenly distributed throughout the year)
Finland	Finnish National Person Travel Survey	2004–2005	Daily mobility: one day Long-distance travel: one month (evenly distributed throughout the year)
France	French National Travel Survey	1993–1994	Daily mobility: one weekday, two weekend days Long-distance travel: three months Seasonality: eight waves of six weeks distributed throughout the year

Appendix 29.A.1. (*Continued*)

Country	Survey name	Year (duration)	Reporting period
Germany	Deutsches Mobilitätspanel (MOP)	Continuous since 1994	One week
Germany	Mobilität in Deutschland	2002	Daily mobility: one day Long-distance travel: three months
Germany	INVERMO	1999–2002	Eight weeks
Greece	Athens Metro Development Study	1996	Daily mobility: one day
Italy	ITALY NHTS	2004/2005	Daily mobility: one day Long-distance travel: seven days + six months for international journeys, reporting periods evenly distributed throughout the seasons
Italy	14. Censimento Generale della Popolazione e delle Abitazioni	2001	Daily communing: one day
The Netherlands	Onderzoek verplaatsingsgedrag (OVG)/ Mobiliteitsonderzoek Nederland (MON)	Ongoing since 1978	Daily mobility and long-distance travel: two days evenly distributed throughout the year
Norway	Norwegian National Person Travel Survey	2005	Daily mobility: one day Long-distance travel: one month (evenly distributed throughout the year)
Switzerland	Swiss Microcensus on Travel Behaviour	2005 (conducted every five years since 1974)	Daily mobility: one day Long-distance travel: two week excursions; Journeys with overnight stays: four months
Sweden	Swedish National Travel Survey (RES)	2005/2006	Daily mobility: one day Long-distance travel: one month for 100 km journeys, two months for 300 km
Spain	Spanish MOVILIA Survey	2000/2001–2002	Daily mobility: one weekday, one weekend day (October–December) Long-distance travel: 12 months
United Kingdom	Great Britain National Travel Survey	January 2002–January 2005	Daily mobility: seven days Long-distance travel: four weeks (evenly distributed throughout the year)

Chapter 30

Moving from Cross-Sectional to Continuous Surveying: Synthesis of a Workshop

Dirk Zumkeller and Peter Ottmann

30.1. Introduction

Transport policy and infrastructure development is embedded in an overall framework that addresses issues of economic growth, ecological and social sustainability, spatial accessibility and social access to opportunities. The further development of these elements results in often very expensive transport infrastructure that must be viewed from a long-term perspective and that is hardly reversible. Furthermore, these developments show considerable change over time with respect to a globalizing economy, climate change, ageing societies and the social divide. To further develop the transport infrastructure, it is thus very important to have an understanding of current as well as future travel demand. This should eventually result in a better understanding of processes, including the early identification of growing, declining and stagnating developments. While the understanding of current demand can often be captured by a single survey, the understanding of behavioural processes clearly requires more than isolated snapshots. The resulting question for the workshop was whether and how to switch from cross-sectional to continuous surveys.

There is no lack of travel surveys around the world. Still, many researchers and practitioners seem to reinvent the wheel rather than benefit from experiences in other countries. Fortunately, on the other side we see some very serious developments towards continuous surveys, ranging from small local experiments to national travel surveys. In order to generate a mosaic of different surveys all over the world, it was very helpful that the workshop members came from all over the world and were thus in a good situation to describe the worldwide state of the art.

Transport Survey Methods

ISBN: 978-1-84855-844-1

30.2. Definitions

In order to communicate on a sound scientific foundation, the workshop first tried to clarify the bewildering confusion of terms that complicates discussion in the field of ongoing surveys. Quite a few terms in the field of transport surveys are ambiguous. For example, it is far from clear if the terms 'ongoing', 'longitudinal' and 'continuous' are synonymous, or if each of them has a slightly different meaning. It seems there is indeed an ambiguity in these terms, although 'continuous' often refers more to the administrative aspects of a survey, while 'longitudinal' focuses more on the technical qualities of the data set. Meanings may vary between countries and scientific disciplines. A discussion of different terms and survey concepts can be found in Kish (1987). To decrease the confusions mentioned above, we recommend the following use of terms in the transport field:

- *Cross-sectional survey*: snapshot of one independent sample for a given time
- *Repeated cross-sections*: sequence of independent cross-sectional samples
- *Panel*: one sample, two or more observation periods
- *Pseudo panel*: repeated measurements of similar individuals at different times
- *Rotating panel*: exchange of panellists to keep sample reflective of the population
- *Continuous/longitudinal/ongoing survey*: ambiguous terms with non-specific distinction, thus all of them shall in the following refer to both repeated cross-sectional surveys and panels
- *Multiday*: continuous reporting period > one day
- *Multiperiod*: interrupted reports by same individuals at different times
- *Time series*: any indicator over time
- *Attrition*: any loss of units in a panel for any reason
- *Mortality*: special case of attrition — death for an individual; dissolution for a household; destruction for a vehicle
- *Fatigue*: reduced reporting accuracy over time
- *Conditioning*: adaption of behaviour or reporting to survey issues
- *Unit non-response*: non-response per unit
- *Item non-response*: non-response of any item (e.g. non-reported trips)
- *Non-mobility/non-Trippers*: (usually in reference to one day) a person who chooses not to leave home.

30.3. Experiences with Ongoing Surveys

To stimulate the discussion, the following papers were presented:

Resource paper:

- Ampt, Ortúzar and Richardson: Large Scale Ongoing Mobility Surveys: The State of the Practice

Contributed papers:

- Stopher, Kokelman, Greaves and Clifford: Sample Size Requirements for Multi-Day Travel Surveys: Some Findings
- Zumkeller, Chlond and Kagerbauer: Regional Panels against the Background of the German Mobility Panel: An Integrated Approach
- Ruiz, Timmermanns and Polak: Improving Continuous Surveys: Analysis of Attrition and Reported Immobility in the Madrid — Barcelona Corridor Panel Survey

Experiences in ongoing travel surveys have been collected in different parts of the world (see Bonnel & Armoogum, 2005 for an extensive overview). Ampt, Ortúzar, and Richardson (2009) focus on different cases in the Southern Hemisphere, mainly in Sydney, Victoria (both Australia) and Santiago de Chile. All three surveys cover regional areas rather than the whole country, and none are panels, i.e. the same households are not followed over different observation periods. These cases proved the administrative advantages of ongoing surveys, e.g. funding is easier to obtain for later waves of ongoing surveys than for first-time projects, and qualification of staff tends to increase over the years, although the Victoria example also showed difficulties in retaining a qualified survey team over a period of years.

Experience in the United States (e.g. with the Puget Sound Panel Survey and the National Household Travel Survey) also proves the administrative advantage of continuous surveys in terms of funding. Currently, the United States is conducting a survey with a sample size of 25,000 households nationwide. Regional add-ons can be bought, so the sample will be about 150,000 households in total, and runs throughout the year 2008. The concept to cover the whole year has already been applied successfully in the former period 2001–2002, which allowed accounting for seasonal differences; in the special case of 2001, the tragic events of 9/11 could be analysed in terms of their severe transport impact.

Canada, on the other hand, does not have a nationwide travel survey, but does have a patchwork of several ongoing regional surveys of long duration (see Chapleau, 1992). Unlike the metropolitan travel surveys in the United States, Canada has continued to use very large samples since the 1970s. For example, in Montreal's repeated cross-section survey, about 86,000 households were covered in 2008. Japan also conducts ongoing surveys exclusively on a regional level.

There is a vast variety of surveys in Europe: Ampt et al. (2009) provide a list of some of the most important ones. Germany has a combination of different approaches, both cross-sectional and ongoing: large cross-sectional surveys every few years with a sample size of about 50,000 persons; additionally a rotating panel with a sample size of about 1500 persons has been ongoing since 1994 in order to gather data on change over time (see Zumkeller, 2008). In addition there are several different surveys with regional focus, e.g. a panel approach in the Rhine-Main region (see Zumkeller, Chlond, & Kagerbauer, 2008) or samples in different cities. The German experience with the somewhat irregular NTS and the continuous panel approach demonstrates that it is easier to get funds and support for an ongoing survey than for intermittent cross-sections. In order to increase acceptability, surveys need to be embedded in an institutional context where expertise, funding and marketing is bundled; this is the case for the Swiss NTS (see BFS, 2007). Recruiting, non-response and attrition have extensively been analysed for the German Mobility Panel in a selectivity study (see Kuhnimhof, Chlond, & Zumkeller, 2006).

Examples for Africa and large parts of Asia (except for Japan) were not discussed in the workshop and seem not to be prominent in scientific publications, although it is likely that local authorities have conducted various surveys there as well. An important example seems to be South Africa, which undertook its first national face-to-face household travel survey covering almost 46,000 interviewed households in 2003 (see Department of Transport, 2005).

30.4. The Advantages and Downsides of Continuous Surveys and Panels

Most of the mentioned examples show administrative advantages of ongoing surveys over pure cross-sectional approaches: funding for further waves appears easier to obtain, qualification of the survey team increases over time and other synergies may lead to further decreases in marginal costs. Ongoing surveys also have methodological and quality advantages. In the special case of a panel, a decreased variance between the different observation periods allows for smaller sample sizes. Similarly, Stopher recommends extending the survey period from one day to several days, as intrapersonal variance can better be accounted for and thus the sample size can be reduced. While this suggestion has often been rejected for reasons of survey costs and respondent burden, new technologies such as GPS surveying may be a viable solution here (see Stopher, Kokelman, Greaves, & Clifford, 2008; Marchal et al., 2008).

Still, perhaps the most important advantage of ongoing surveys and particularly panel surveys is the possibility to understand change. While a cross-section is just a snapshot of the current situation, it does not address the question of how behaviour changes over time or after measures have been introduced. Generally, an observation of changes in aggregate demand can be attributed either to a change in the underlying population, in external factors or in changing behaviour of individuals. The former two can be analysed with different kinds of ongoing surveys, while the latter requires a panel approach. A possible example here would be public transport patronage: people may start as public transport users because they cannot afford a car, but some of them may remain faithful to public transport when their income increases. These people can only be identified with a panel.

None of the cases of changing behaviour could be observed in a single snapshot. Eventually, the observation of change may only be an illusion for methodological reasons, and these are mainly problems of ongoing surveys. Firstly, there may be a change in the survey between two waves, e.g. questions, or the means of questioning, have been changed. A GPS survey may report more short foot walks than a paper-and-pencil diary, although the underlying behaviour may still be the same. It is therefore of vital importance to be careful with changes in methodology within an ongoing survey, and inevitable innovations must be introduced in a staged process.

Furthermore, virtual or even actual change of behaviour in a panel survey can be caused by conditioning: a respondent may change his behaviour or at least the reporting of behaviour in the course of the panel participation due to participation. This effect is obviously worrisome, as changes caused by conditioning can be observed only within the participants of a panel, while the major part of the underlying population may actually not change. To solve this problem, a control group is required. In addition, a panel survey may only represent the population at the beginning of the survey, when the participants were recruited. If there is a change in the underlying population, e.g. due to migration into a country or to births and deaths, a panel may not represent the composition of a society after a few years. Finally, the respondent burden for the individual is higher in a panel survey, and attrition between waves can take place. The problems of attrition and recruiting in a panel survey are discussed by Ruiz, Timmermanns, and Polak (2008) based on the Madrid-Barcelona Corridor Panel.

A solution to these problems specific for panels is a rotating approach, in which participants are replaced after a few years. The decision of how many years respondents are asked to participate is arbitrary; the German Mobility Panel has had good experiences with a three years' tenure. If the rotating panel is supposed to represent the underlying population, newly recruited participants have to be chosen accordingly. It should be noted that in some cases it is actually desired to have a panel sample not representative of the actual (current) population, but instead of the population at the time when the panel started. In that case, drop-outs should be replaced with comparable persons.

30.5. Conclusion and Outlook

Surveys are pointless in isolation, but are a prerequisite for sound decision-making on infrastructure construction, or more generally for sound transportation and land-use policy. The important question for an efficient and useful survey design is, therefore: What does one want to measure? With the answer to this question, the right tool (ongoing vs. cross-sectional surveys) can be chosen as well as the required sample size. Ongoing surveys have important advantages over cross-sectional approaches: in administration, methodology and perhaps most notably in understanding changing behaviour. Nevertheless, it is not an 'either — or' choice; the most efficient design is probably a continuous survey with medium sample size, combined with a cross-section, with large sample sizes, conducted every few years.

The workshop provided interesting insights into the state of the art of ongoing surveys around the world. In addition, important methodological issues were addressed. While there is already extensive experience with continuous surveys in general, there seems to be less research on panel surveys. Further research is required on issues in ongoing surveys, such as how to deal with respondent burden and conditioning; how to best implement design changes in an ongoing survey; how to retain and train staff over time; what risks and opportunities are presented by new technologies for continuous surveys; and how best to deal with exceptional years, such as 2001 (due to the terrorist attack on 9/11) or as in 2006 in Santiago de Chile (when there were drastic changes to the transport infrastructure). Furthermore, an extensive comparative study on different ongoing approaches (pure panel, rotating panel, pseudo-panel, repeated cross-sections) is yet to be undertaken.

Acknowledgements

We would like to thank the resource paper and contributed paper authors, discussant Tim Raimond, and all other workshop participants for their valuable contributions. Workshop participants were Bastian Chlond, Germany; Itzhak Ditzian, Israel; Peter Endemann, Germany; Mark Freedman, United States; Marie-Odile Gascon, France; Martin Kagerbauer, Germany; Kees Maat, Netherlands; Jean-Loup Madre, France; Kim McDonough, Canada; Nancy McGuckin, United States; Andres Monzon de Caceres, Spain; Juan de Dios Ortuzar, Chile; Peter Ottmann, Germany; Tim Raimond (discussant), Australia; Sophie Roux, France; Tomas Ruiz, Spain; Felix Schiffner, Germany; Angelika Schulz, Germany; Peter Stopher, Australia; Orlando Strambi, Brazil; Claude Weis, Switzerland; Dirk Zumkeller, Germany.

References

Ampt, E. S., Ortúzar, J. D., & Richardson, A. J. (2009). Large-scale ongoing mobility surveys: The state of the practice. In: P. Bonnel, M. Lee-Gosselin, J. Zmud & J.-L. Madre (Eds), *Transport survey methods: Keeping up with a changing world.* Bingley, UK: Emerald.

BFS. (2007). *Mobilität in der Schweiz*. Neuchâtel: Bundesamt für Statistik BFS.

Bonnel, P., & Armoogum, J. (2005). National Transport Surveys: What can we learn from international comparison? *Proceedings of the European Transport Conference*, London.

Chapleau, R. (1992). La modélisation de la demande de transport urbain avec une approche totalement désagrégée. *Proceedings of the 6th Annual World Conference on Transportation Research* (*WCTR*), Lyon (Vol. 2, pp. 937–948).

Department of Transport. (2005). *The First South African National Household Travel Survey 2003*. Technical Report, Pretoria.

Kish, L. (1987). *Statistical design for research*. New York: Wiley.

Kuhnimhof, T., Chlond, B., & Zumkeller, D. (2006). Nonresponse, selectivity and data quality in travel surveys — Experiences from analyzing the recruitment for the German mobility panel. *Transport Research Record* (1972), 29–37.

Marchal, P., Roux, S., Yuan, S., Hubert, J.-P., Armoogum, J., Madre, J.-L., & Lee-Gosselin, M. (2008). A study of non-response in the GPS sub-sample of French National Travel Survey 2007–08. Presented at the 8th International Conference on Transport Survey Methods, Annecy, France.

Ruiz, T., Timmermanns, H., & Polak, J. (2008). Improving continuous surveys: Analysis of attrition and reported immobility in the Madrid — Barcelona Corridor Panel Survey.

Presented at the 8th International Conference on Transport Survey Methods, Annecy, France.

Stopher, P., Kokelman, K., Greaves, S., & Clifford, E. (2008). Sample size requirements for multi-day travel surveys: Some findings. Presented at the 8th International Conference on Transport Survey Methods, Annecy, France.

Zumkeller, D. (2008). The dynamics of change — 15 years German mobility panel. Paper presented at the TRB 88th Annual Meeting. Washington, DC.

Zumkeller, D., Chlond, B., & Kagerbauer, M. (2008). Regional panels against the background of the German mobility panel: An integrated approach. Presented at the 8th International Conference on Transport Survey Methods, Annecy, France.

Chapter 31

Moving Towards Continuous Collection of Large-Scale Mobility Surveys: Are There Compelling Reasons? A Discussant Response

Tim Raimond

31.1. Introduction

What is the incentive for a city or country to move from traditional cross-sectional mobility surveys, conducted every 5–10 years (periodical surveys), to continuous data collection? What barriers may prevent jurisdictions from moving in this direction? Inspired by the papers of the 2008 ISCTSC conference, this paper briefly argues that on the combined grounds of data needs, data quality and collection efficiency, continuous collection of cross-sectional or panel data offers significant advantages for large-scale mobility surveys over periodical data collection.

The workshop resource and contributed papers actually discuss three different aspects of how to improve the current state of practice around the world in large-scale mobility surveys. They are:

- moving from periodical to continuous collection of data
- moving from cross-sectional to panel data
- moving from single-day observations to multi-day observations of behaviour

In order to move from being good ideas to being adopted practice, each of these potential enhancements to existing practice must be tested against the needs of city, regional or national authorities who commission large-scale mobility surveys. In what is a relatively simple equation for those who commission surveys, the survey must be able to meet the data-collection *purpose*, with acceptable *quality*, and maximum cost *efficiency*.

Transport Survey Methods

ISBN: 978-1-84855-844-1

31.2. Fit for Purpose?

31.2.1. Continuous Data Collection

Large-scale mobility surveys, at city, regional or national level, are usually justified primarily for providing data to feed strategic travel models, which in turn are used to evaluate a multitude of planning, policy or infrastructure options. To meet the needs of modern tour- or activity-based models, *current* data on travel modes, purpose, time of day, origin, destination and cost are required from a representative sample of the population. In addition, there is a need to understand what factors influence travel behaviour. Thus, detailed socio-demographics must be matched to travel data.

A secondary, but important, requirement from large-scale mobility surveys is the monitoring of *change* in travel behaviour. This enables a better understanding of change processes to improve the modelling of responses to change, and provides an indicator of when strategic travel model parameters may need to be re-estimated in the light of fundamental behavioural shifts.

Either periodical or continuous large-scale mobility surveys are able to meet the first need for current, detailed data on travel behaviour. A continuous survey approach offers a better opportunity to monitor changes in behaviour, but given the relative stability in travel behaviour over time, regularly conducted periodical surveys should, in theory, also be able to monitor change, though in practice, comparability of data over time is a major challenge for periodical surveys.

In summary, the survey purpose of a large-scale mobility survey is probably not a sufficient condition for justifying continuous data collection — frequent periodical collection would suffice — though continuous collection would certainly be advantageous.

31.2.2. Panel Surveys

Zumkeller, Chlond, and Kagerbauer (2008) argue that monitoring of change could be further enhanced by using a panel rather than a repeated cross-sectional approach to surveying. The German Mobility Panel (and Regional Panels) has the benefits of a continuous collection; also, as a rotating panel, it allows the observation of individual change over time, whilst maintaining a representative sample of the population through panel refreshment. Especially in the European context of an increasing focus on understanding behavioural change, the rotating panel survey seems fit for purpose.

31.2.3. Multi-Day Collection

Stopher, Kockelman, Greaves, and Clifford (2008) show that there is considerable intra-personal variability in travel over the course of a week. This is not identified in traditional single-day surveys, perhaps leading to poorer modelling of travel behaviour. This improvement in the understanding of travel behaviour, which is a key purpose of large-scale mobility surveys, is certainly a strong argument for adopting a multi-day collection approach.

31.3. Data Quality?

31.3.1. Continuous Data Collection

It cannot be stated often enough that there is absolutely no point collecting data if you do not collect it accurately. Continuous surveys offer many opportunities to monitor, understand, control and improve data quality in ways that are not possible with periodical surveys. In particular, they offer continuity in team and approach that make comparison over time possible. The unquantifiable differences due to different teams and methods in periodical surveys make comparison over time highly problematic.

A continuous survey allows travel behaviour at different stages in the economic cycle, or the seasons, or any other changing external factor, to be monitored and better understood.

A perceived trade-off with continuous surveys is that they usually have a smaller annual sample size than periodical surveys. In order to increase the level of confidence in estimates from continuous surveys, multiple waves of data (often three subsequent waves) are pooled and weighted to produce an annual estimate. Given the relative stability over time in the way people are observed to make travel choices, this is probably a very minor trade-off when compared with the advantages.

Continuous surveys facilitate continuous improvement in all survey processes, with the outcome of improved data quality. For example:

- Project management: a continuous survey provides continuity for professional survey managers in the commissioning organisation. Staff knowledge and skills that are developed can be more readily retained, and improvements to existing methods considered, tested and implemented.
- Interviewers: a continuous survey provides an opportunity to maintain a well-trained set of interviewers over long periods of time. Usually a smaller pool of interviewers can be used than that required for large periodical surveys, so that only the best interviewers are retained. Recruitment and training programmes for interviewers can be continuously reviewed and improved.
- Monitoring: interviewer performance can be monitored, retraining undertaken, and if problems are found, re-sampling is possible.
- Response rates: all recent literature on large-scale mobility surveys states that response rates are declining (e.g. Stopher & Greaves, 2007). However, the process of continuous improvement possible in a continuous survey program may allow high response rates to stabilise, as has been the case in Sydney (Figure 31.1).

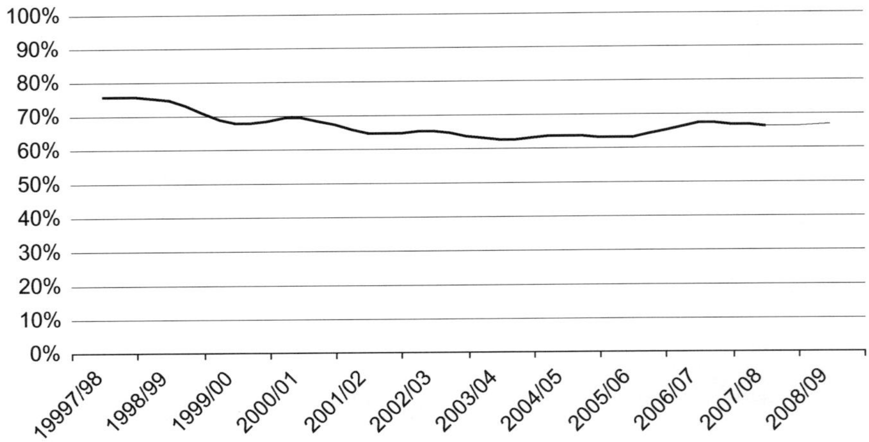

Figure 31.1: Response rate (full and part response) for the Sydney Household Travel Survey.

Continuity of team and approach, and therefore stability in response and data quality, may be the most important virtues of a continuous data-collection approach, and provide compelling reasons for moving to continuous data collection.

31.3.2. Panel Surveys

Panel surveys provide the opportunity to collect better quality data on the process of behavioural change by examining the behaviour of the same individuals in the population under different circumstances (Moons & Wets, 2006). This, in turn, should allow improved modelling of behavioural change. However, panel surveys also have significant drawbacks in terms of panel conditioning, attrition and representativeness of the population. These drawbacks provide sufficient reason to question the worth of moving wholly to a panel-based approach, especially in areas where there is continued strong growth and demand for infrastructure. Perhaps a hybrid approach similar to that adopted in Germany, where there is a larger-scale cross-sectional mobility survey, coupled with a panel with a smaller-sized sample, is the optimal solution. Where budget is an issue, perhaps qualitative approaches could substitute for a panel.

31.3.3. Multi-Day Collection

The primary argument of Stopher et al. (2008) is that multi-day surveys offer the opportunity to reduce sample sizes for the same quality of data. To date, the question of data quality has been difficult to resolve because of the attrition in response over multiple days. Stopher et al. (2008) argue that automated collection of multi-day travel information via a GPS overcomes this issue. However, it introduces a new set of quality concerns related to information derived from GPS devices: extensive data processing, imputation and sample loss due to people forgetting or not recharging the devices are ongoing issues with GPS-derived data.

An issue raised by Stopher et al. (2008) relevant to jurisdictions considering undertaking multi-day surveys is that a smaller sample size is not always desirable for the sake of external confidence in the survey results. Many decision-makers will not appreciate the statistical arguments as to why a smaller sample can give you just as much confidence. In these cases, maintaining existing sample sizes, but improving data quality with multi-day surveys, may be the prudent course for the future.

Data quality, being probably the most important factor affecting the methodology for large-scale mobility surveys, is influenced by a range of other factors discussed in passing in the workshop. One important factor is the fieldwork method. Many modern continuous surveys rely on face-to-face interviewing of respondents (e.g. Sydney, Santiago, London, New Zealand). Ruiz, Timmermans, and Polak (2008) note that 'personal contact is of central importance to increased participation'. Self-completion and phone techniques are severely limited by sample frame, response rates and proxy reporting issues, leading to serious questions about the quality of data collected by these methods.

31.4. Cost-Effectiveness?

31.4.1. Continuous Collection

There are many economies in conducting a mobility survey continuously. These include:

- once-only set up costs
- reduced interviewer training costs
- continuous monitoring of performance against efficiency benchmarks
- multi-year fieldwork contract

These efficiencies, along with the administrative advantages of staff continuity, and the survey becoming an accepted part of annual budgets after the first few years of operation, suggest that continuous collection is also justified on the basis of cost-efficiency.

31.4.2. Panel Surveys

Continuously collected panel surveys offer the same economies in collection efficiency as discussed above. However, panel surveys are administratively more complicated because of the need to attempt to follow households that may change circumstances, and to continually refresh the panel with an appropriate mix of households. These complications, and the added complexity of more demanding weighting requirements, may put some jurisdictions off this approach.

31.4.3. Multi-Day Collection

Stopher et al. (2008) document distinct cost-effectiveness advantages, in terms of reduced sample size, from this type of methodology. However, they focus on single variables in a small data set. The work needs to be extended to a mix of variables, such as those used in demand models, and larger studies, to see if the sample size reductions are maintained, and the reliability of the resulting data is acceptable.

31.5. A Case for Change?

Continuous data collection for large-scale mobility surveys offers compelling advantages in terms of improved data quality, and costs and administrative efficiency, with few if any trade-offs.

However, the choice of repeated cross-section versus panel, or a hybrid of the two, depends on survey purpose. More research is needed on the trade-off offered between the high response rate, continuous repeated cross-sectional data versus continuously collected panel data with the inevitable attrition after the first wave and conditioning of continuing panel members. What are the best methods to weight these respective approaches, and for a given level of effort, which will provide the more robust estimate of change?

The opportunity for greater efficiency and/or data quality through the use of multi-day surveys is also very close to mainstream reality, with GPS providing the chance to collect data from the same respondent over multiple days without the usual respondent fatigue issues. Again, more research is still required before either GPS, or the efficiency gains from multi-day approaches, can become completely accepted. In particular, it needs to be demonstrated that data quality issues can be adequately addressed, and that the needs of strategic travel models can be met with reduced sample sizes.

31.6. Barriers to Change

31.6.1. Continuous Surveys

There are some important barriers to moving from periodical to continuous data collection. A significant barrier, experienced acutely in the United States currently, is that updated data is required to feed new activity or tour-based models, and jurisdictions want this data immediately. Continuous surveys, with modest sample sizes and pooling of multiple waves, are perceived as too slow to provide updated estimates. This barrier is understandable, but there is only a one-off delay on start-up, then there is always up-to-date data. It is possible that the advent of new techniques that produce a richer data set more quickly, such as multi-day surveying, can be used to defray some of the perceived disadvantages of the slow start-up of continuous surveys.

A related barrier is a fear of the complexities of pooling data and producing annual estimates. In reality, however, the weighting principles for continuous surveys are little different from those used for periodical surveys.

Another barrier to change may have been a lack of evidence that the benefits of continuous surveying outweigh the costs. There is a growing weight of papers on the advantages of continuous collection (e.g. Battellino & Peachman, 2001) and the experiences of areas that have moved to continuous collections (e.g. Ampt, Ortúzar, & Richardson, 2009), as well as a clear growth in the use of continuous approaches at the city (e.g. London, TfL, 2007) and national (e.g. New Zealand, Ministry of Transport, 2008) level.

A further barrier to change is the fear of data not being compatible over time. Experience has shown that the transition from periodical to continuous surveys is best achieved in a conservative fashion, with similar survey instruments and methodologies applied in order to ensure the highest degree of data comparability over time.

31.6.2. Panel and Multi-Day Collections

Multi-day and panel surveys do present some technical barriers to easy adoption. In particular, the complexities of weighting these data sets may be a challenge for many jurisdictions. Unlike in the health field, the transport field does not have a strong history of appropriately weighting sample survey data to account for sample stratification, clustering, household or person interdependence, let alone multiple waves or days of data.

The increasing use of GPS and other mobile tracking may also face significant barriers in some countries because of civil liberty concerns, so it is important not to focus all our development efforts in this area.

Acknowledgements

Thanks to Dirk Zumkeller and John Peachman for their helpful comments and to the workshop participants for a stimulating discussion.

References

Ampt, E. S., Ortúzar, J. D., & Richardson, A. J. (2009). Large-scale ongoing mobility surveys: The state of practice. In: P. Bonnel, M. Lee-Gosselin, J. Zmud & J.-L. Madre (Eds), *Transport survey methods: Keeping up with a changing world.* Bingley, UK: Emerald.

Battellino, H., & Peachman, J. (2001). The joys and tribulations of a continuous survey. International Conference on Transport Survey Quality and Innovation, Kruger National Park, South Africa.

Moons, E., & Wets, G. (2006). Setting up a continuous panel for collecting travelling information: Discussion on methodological issues. MOLS2006 International Conference, University of Essex, U.K.

New Zealand Ministry of Transport. (2008). *Comparing travel modes: Household travel survey*, Wellington, January 2008.

Ruiz, T., Timmermans, H., & Polak, J. W. (2008). Improving continuous surveys: Analysis of attrition and reported immobility in the Madrid-Barcelona corridor panel. Presented at the 8th International Conference on Survey Methods in Transport, Annecy, France.

Stopher, P. R., & Greaves, S. P. (2007). Household travel surveys: Where are we going?. *Transportation Research Part A*, *41*, 367–381.

Stopher, P. R., Kockelman, K., Greaves, S. P., & Clifford, E. (2008). Sample size requirements for multi-day travel surveys: Some findings. Presented at the 8th International Conference on Survey Methods in Transport, Annecy, France.

Transport for London. (2007). Research briefing: London Travel Demand Survey Supplement, London, March 2007.

Zumkeller, D., Chlond, B., & Kagerbauer, M. (2008). Regional panels against a background of the German Mobility Panel. Presented at the 8th International Conference on Survey Methods in Transport, Annecy, France.

Chapter 32

Vehicle-Based Surveys: Towards More Accurate and Reliable Data Collection Methods

Dominika Kalinowska and Jean-Loup Madre

Abstract

Across Europe, on average more than 95% of all passenger cars and half of all light commercial vehicles are permanently available to a household. This includes both privately owned vehicles and company cars. The profiles of vehicle use can be specified as average annual distance driven per vehicle and for the fleet as a total, purpose of travel (trip destination), infrastructure use (urban, interurban or motorway road transport) and also fuel consumption together with data on CO_2 emissions. Indicators on vehicle use can be tracked in various ways:

- self-administered panels of households, which permit their vehicles to be followed for several years;
- national or local household travel surveys (with a seven-day trip diary);
- official vehicle inspection and vehicle registration files;
- 'vehicle surveys' based on vehicle registry data;
- traffic counts;
- data collected for road-charging purposes.

The paper will present a review of mainly vehicle-based survey methods used in France, Germany, Finland, the United Kingdom, the United States and Canada, describing existing sampling frames to their scope, advantages and limitations, as well as their costs. Issues addressed in this context will be further examined in terms of their methodological challenges as well as their purpose.

Transport Survey Methods

ISBN: 978-1-84855-844-1

The leading questions underlying this paper as well as the corresponding workshop are: why is it necessary to have data on passenger travel or transportation; and, looking at international experience, how good are vehicle-based surveys in delivering the required information? In discussing problems experienced in the different countries with data collection and evaluation methods, emphasis will be put on potential strategies for methodological and technological improvement and problem solving. One example is the potential use, benefits and constraints of new survey technologies presented by vehicle tracking techniques.

32.1. Introduction

Although vehicles are 'significant sampling units' (e.g. from registration or periodical inspection files), to address research and policy questions of transport planning authorities on environmental issues, energy consumption or fiscal policies, questionnaires on their ownership and use are filled either by households (for passenger transport) or by firms (for freight).

In this resource paper, the first point to raise is what research and policy questions (environmental issues, energy consumption, infrastructure building, etc.) can be addressed using vehicle-based surveys possibly also combined with administrative data. Responding to this question, two other aspects should be kept in mind:

- What is the particular advantage of vehicle-based survey data compared to other potential data sources of its kind regarding the objective of its implementation, for example the policy or research question to be addressed (Section 32.2)?
- Do there exist any national particularities within the international comparison concerning the variety of questions addressed with such data (Section 32.3)?

Finally, we will propose areas of research and development about longitudinal issues, accuracy of data and privacy rules (Section 32.4). A summary table is presented in the Appendix 32.A.1 of the paper to give a comparable overview of aspects discussed in the paper.

32.2. Why Do We Need Transportation Data?

Besides the general need to monitor trends in the volume of road traffic, well-defined political questions may need to be addressed on the basis of vehicle-survey-based data:

(a) *Local issues of infrastructure building, infrastructure investment and spatial planning*: What vehicles are passing through a node or a link of the road network? Why are they passing through? The main objective here is to quantify and describe traffic flows and to forecast how much they could change at different time horizons. Data providing the desired information should contain the 'vehicle population', composed of different categories observed at one point on the road.

(b) *Planning of public transport and intermodal transport*: Shifting passengers from individual motorised travel to public transport is an often-cited objective when strategies to reduce congestion, especially in large urban areas, are formulated. In the case of freight transport and intermodal transport, the issue often discussed is the possible advantages derived from a potential increase in rail transport.

(c) *Issues concerning external costs of vehicle use*: How should environmental and road congestion costs be quantified? Of particular interest in this context are total distance travelled and the corresponding fuel consumption. In addition, data on road use distribution according to the infrastructure network can provide valuable information. Data on vehicle mileage is used to assess emissions of greenhouse gases. Data on fuel consumption helps decision makers design fiscal frameworks for collection of taxes and charges, for example on fuel or road use, or for developing other regulative transport and environmental policy measures.

(d) *Road safety and accident prevention*: Data on vehicle use in terms of road travel distances together with their network patterns can be useful for the analysis of road safety issues, exposure to road accidents or vehicle safety. In addition, data on vehicle user characteristics could be useful.

(e) *General issues about long-term trends for vehicle ownership and use*: What is the life expectancy of cars? How much are they used for different purposes? The objective is to describe vehicles owned by the population living in a given area. Such information is an essential input for the calculation of data for vehicle mileage driven as well as for different components for the national accounts. For example in recent decades, the life expectancy of cars and total mileage covered by motorists has been steadily increasing. In 1985, one-third of the cars owned by French households had more than 80,000 km on the odometer; since 2000, the proportion is more than half, and in 2005, one in six cars had travelled more than 170,000 km. This upward trend has only been interrupted during the economic recoveries of the late 1980s (Madre & Lambert, 1989) and of the late 1990s. Vehicle survival laws therefore change over time (Gallez & Couturier, 2000). Of course, it is not possible to rely exclusively on survey results to obtain valid vehicle-use profiles (Madre, 2007), but it is critically important that simulations be based on observations.

A final issue is how to combine these data in order to portrait road traffic and its fuel consumption on a given territory (e.g. tables of road vehicle activity for national accounts, energy consumption and environment budget of trips in an urban area [Gallez, Hivert, & Polacchini, 1997; Quételard, Noppeet, & Hivert, 2003; Hivert, 2007]).

32.3. International Overview: Existing Challenges and Limitations

Independent of the survey method, data collection should meet the following criteria:

- It should cover the target field as long as possible, that is passenger cars, light commercial vehicles (LCVs), heavy goods vehicles (HGVs), etc.
- It should monitor vehicle use for as long as possible from when the vehicle was first put on the road to its destruction.
- It should provide the most accurate vehicle-use data available with current technology (e.g. satellite tracking), while still being able to measure the changes that have occurred since periods when measurements were less precise (only survey based).
- The data should address user needs and be accessible to the user community.
- Data sources should also refer to current time periods and correspond to comparability standards.
- In order to link different data sources, consistency is an important criteria.

In the following sections, we will briefly introduce some examples rather than an exhaustive overall survey of transport data collection methods for different types of data (traffic counts, surveys or administrative data) for different countries, emphasising existing shortcomings and pointing out innovative approaches and improvements.

32.3.1. France

32.3.1.1. Difficulty in getting a good database for sampling vehicles The removal of the annual road tax in 2001 for private individuals means that we no longer have an accurate count of the number of registered vehicles on French roads. For vehicles of less than 16 tonnes, this source provided a breakdown by Département of registration (NUTS level 3 area), engine rating and age (with 5, 20 and 25 years as the age group limits).

To carry out a survey of the uses of the vehicle fleet, it is necessary to possess a reliable sampling base. The central vehicle register (Fichier Central des Automobiles [FCA]) should provide this. However, only half of cars that are destroyed or sold abroad are declared to the Préfecture and dropped from the data file. Data are much more reliable in Germany, for example where registration papers are issued with a deposit (Section 32.3.2). The FCA was therefore becoming less and less reliable, at least in the case of old vehicles (most of the analyses from this were restricted to cars of less than 15 years of age). This is why the Economics Statistics and Forecasting Department at the French Ministry of Transport (SESP) has recently cross-checked the FCA with the vehicle inspection register (Metaphis, 2006).

In view of its poor management of vehicle mortality problems, the Central Vehicle Register has been little used as a sampling base for passenger car surveys; it is used every five years for LCVs and every year for buses and trucks. The yearly survey on buses has been stopped in 2007 because the sample was too small to obtain a good fit with yearly changes observed on a comprehensive basis (e.g. number of trips and vehicle-kilometres for urban transport) on which they could have been calibrated. The survey on HGVs is still ongoing and is harmonised by EUROSTAT.

Cars are used by all economic agents: companies, which often buy new vehicles (particularly car-hire companies); households, which buy new or second-hand cars from a company or another household as well as administrations, associations, etc.

When an exhaustive and reliable sampling base is not available, it is difficult to consider all the vehicle usage profiles in a single survey of vehicle users (households or firms). In France, the fullest coverage (more than 95% of the passenger cars on the road) is obtained by considering the vehicles, which are at the permanent disposal of a household (including company cars). It is also possible to observe some LCVs with these households surveys: in particular those that are owned by a household and which are used in a similar way as passenger car (Madre, 1992). Those that belong to one-person businesses are more difficult to observe; those corporate vehicles that remain on the firm's premises will not be covered.

As each of these categories accounts for approximately a third of all LCVs, a good percentage of them can be covered by sample-based household surveys in France.

32.3.1.2. Roadside surveys The origin–destination (O–D) surveys conducted in France cover a sample of 1000–2000 vehicles during one day, which is representative of yearly traffic. Vehicles are stopped at an urban or temporary traffic light; each driver is interviewed during the red light (45 s–1 min). Before 2000, vehicles were stopped by a police officer, but the police now consider this outside their main mission and only assure the safety of the interviewers. This type of survey is anonymous and not mandatory.

Survey questions are the origin and destination of the trip (at municipality level), trip purpose and frequency of trips made on the same route. For a truck, the weight and nature of load is asked. The interviewer notes the category of vehicle, the number of passengers (especially for a bus) and the NUTS level 3 zone (département) where the vehicle is registered; the latter data will be derived from the plate number by the European registration file in the future. If more data are needed, a mail-back questionnaire is distributed. Despite a low response rate by mail-back, a weighting procedure based on data collected by the interviewer allows the analysis of representative results. The sample scheme (location, period and duration of the survey) is an important issue. For each study it is necessary to survey approximately four one-way sites during 6–14 h. For a good extrapolation of the results, traffic counts are collected over at least two weeks.

These data are used for calibrating O–D matrices by adapting national traffic forecasts according to the characteristics of local traffic; for illustrating traffic counts, especially through the analysis of the main characteristics of peak hours traffic and for traffic forecasting on each lane of a modified network, by implementing traffic assignment procedures based on generalised cost (time and money).

These surveys are conducted either all around a conurbation (cordon survey) or on every road crossing a region, and more often on one route. Other types of survey exist (e.g. at parking areas on the motorways), but most of them are biased, with no serious possibility of re-weighting, and none of them give the same kind of results with the same accuracy.

32.3.1.3. Traffic counts In France, automated traffic counts are conducted annually on trunk roads and motorways. In addition, manual traffic counts are conducted every five years. The results obtained from the manual counts allow the split of the global road traffic figures obtained annually from automatic counts by vehicle category (motorcycles, passenger cars, LDV, etc.) and by origin, based on the registration plate (NUTS level 3 zone for French vehicles or country for foreign vehicles). These data are extrapolated from an adequate sample of counting points in terms of time and location. No large-scale number plate recognition has been undertaken, because of its cost, since it would be necessary to protect the video device to prevent it from being stolen or damaged.

32.3.2. Germany

In Germany, periodic usage patterns of motor vehicles are investigated based on several different data sources. One general advantage of the German pool of transportation data is the existence of a federal registry office, where relevant vehicle and vehicle holder data are stored on all the vehicles registered in the country. The registry not only contains information on vehicles currently in use but also on newly registered vehicles, reregistered vehicles and vehicles leaving the registry. Therefore, a broad database on the current vehicle fleet is available. The German register seems similar to the French FCA, except that in the French records, vehicles, which leave the fleet, are not well captured. Therefore, there exists only poor information on vehicles outflow from the registry.

32.3.2.1. Survey data For passenger cars, the most recent and most important survey data are the Mileage Survey 2002 (BASt, Bundesanstalt für Straßenwesen, 1996, 2005), the household mobility survey 'Mobility in Germany' 2002 (MiD, which is being updated in 2008) (MiD, 2002) and the German Mobility Panel (Mobilitaetspanel MOP, 2008), conducted annually. The German household panel survey collects data on everyday mobility behaviour in a travel diary covering one week in autumn each year. Also each year, during six weeks in spring, the MOP surveys the distance driven in private vehicles in Germany and their fuel consumption.

Another important data source on vehicle use for goods transportation purposes by HGVs available in Germany is the Survey of Road Freight Transport (Nationale Erhebung zum Gueterkraftverkehr deutscher Fahrzeuge or Güterkraftverkehrsstatistik) conducted each year by the Federal Bureau of Motor Vehicles and Drivers (Kraftfahrtbundesamt [KBA]). This survey corresponds to the harmonised pool of freight transport surveys by the EUROSTAT. The sampling units of the continuing survey are all tractive vehicles with at least 3.5 tonnes load capacity registered in the sampling register used for the survey, which is the German Central Vehicle Register (Zentrales Fahrzeugregister [ZFZR]) maintained by the KBA. Excluded from the survey are lorries with less than 3.5 tonnes load capacity, military vehicles, government vehicles and agricultural tractors. Information collected from the vehicle registry includes information for the stratification of the survey, such as type of vehicle and body type, owner group, region of registration, maximum permissible laden weight and load capacity, as well as information included into the survey output such as the date of first registration of the vehicle, maximum permissible laden weight, load capacity, engine power, number of axles, type of vehicle and body type, region of registration (Bundesland), owner group and exhaust emissions class.

Holders of the vehicles drawn from the register are obliged to respond to the survey. If the holder does not respond to the questionnaire, a reminder is sent 23 days after the date the questionnaire was due to be returned. A penalty procedure starts 23 days after the reminder if the questionnaire is still not received. The questionnaire mainly seeks information on the freight activity during the survey week: weight of goods handled, type of cargo transported on the journey, point of loading and unloading, and distance travelled on the journey, thus giving the tonne-kilometres of goods moved for each loaded journey. (Road freight transport methodology, 2005a, 2005b).

32.3.2.2. Traffic counts For information on road traffic figures differentiated by road category, annual automatic traffic counts, manual road traffic counts and periodical cross-border traffic counts are performed by the German Federal Highway Research Institute (BASt). For data collection by automatic traffic count, 600 counting stations on motorways and about 700 on other federal roads are in service. The manual road traffic counts are conducted every five years on 10,000 counting stations sited on motorways and other federal roads. In principle, the German motorway toll system, invented and designed for heavy trucks, could also provide reliable information on the total volume of traffic. However, the political accreditation of the system was tied to the adoption of strong data-use restrictions on the licensee, the Toll Collect corporation (www.toll-collect.de). So, the available data in Germany can only rely on road count stations for statistical purposes (Handbook on Statistics on Road Traffic: Methodology and experience, 2007).

A nationwide road traffic census (RTC 2005) was performed in 2005 as part of the usual five-year cycle to monitor traffic development and to determine the traffic volumes on the federal arterial roads, that is trunk roads and motorways. Traffic counts on the subordinate network (state and regional roads) were left optional to the German states, where experience from the RTC 2000 has shown that regional roads were evaluated in less detail and only in some of the states. In general, data from RTC cover not only passenger cars but also other vehicle categories including those used for freight transport. The report from RTC 2005 contains data for the average daily traffic volume (ADTV) and the annual distance travelled by vehicle type and road type for the whole federal area, as well as for each individual state.

32.3.2.3. Periodic inspection data In 2007, the Federal Ministry of Transport, Building and Urban Affairs conducted a pilot project, the main objective of which was to collect data on vehicles and their odometer readings from periodic vehicle inspections, and to apply these data as an empirical basis for the calculation of total annual volume of road traffic (i.e. vehicle-kilometres) for the various vehicle categories. However, the use of odometer readings to assess annual mileage driven by different vehicle categories has some limitations. Therefore, several assumptions need to be taken to extrapolate the data using different methodological approaches. One shortcoming of the odometer data is the incomplete information provided for fairly new automobiles (those less than three years old) as well as for old vehicles. In general, new vehicles are inspected for the first time after the second year of their utilisation. Therefore, no representative mileage readings can be observed in the sample. Those vehicles less than three years old that are found in the sample cannot be considered representative for their age category. New vehicles that are turned in for inspection before the obligatory due date tend to be owned by a commercial entity and therefore are used more intensively than the average. The unusually early inspections on these new vehicles are usually as a result of the vehicle being put up for sale or after an accident (Kalinowska, Kloas, & Kuhfeld, 2007).

32.3.3. Combination of Traffic Counts, Survey and Administrative Data in Finland

In Finland, collection of data on road traffic, that is vehicle-kilometres, is mainly based on road counts. In addition, as a trial, counter readings collected during vehicle inspections were employed to estimate figures on vehicle-kilometres. In Finland, 78,000 km of roadway is state-run public roads, 22,000 km is municipal streets and 350,000 km is private and forest roads. Public roads are classified into a main road network consisting of trunk roads (class I) and main connecting roads (class II), and lower road network consisting of regional and connecting roads. Public roads fall under the responsibility of the nine road districts of the Finnish Road Administration. The existing 432 municipalities are responsible for their corresponding street networks. Private road maintenance associations and local landowners are responsible for the construction and maintenance of private and forest roads. Except for a few major cities where regular street traffic censuses are conducted, extensive traffic censuses only cover public roads. Traffic counts covering private roads are performed occasionally and only at a few locations. For the overall traffic censuses, public roads are divided into 15,500 road sections, where traffic volumes are known to be homogeneous. The sample count rotates over a four-year period covering one quarter of the entire public road network every year. The mechanical sample count on each section is performed for one week long, twice during the census year: the first census week is in the June–August period and the second in September–October. The results obtained from the two census weeks determine the seasonal variation on each road section. Five types of seasonal variation are distinguished. Using a regression model together with the results on the seasonal variation and census results, average daily traffic is estimated separately for HGVs and vehicle combinations including trailers. In addition to average daily traffic and type of seasonal variation, average weekday traffic, summer average daily traffic, types of day-to-day and hourly variations of traffic and traffic volumes of the busiest, 100th busiest and 300th busiest hour of the year are estimated. In general the rotation interval of a traffic count is four years. It can be increased up to one year on road sections with rapidly growing traffic as well as reduced to eight years on road sections with relatively low average traffic volume (<150 vehicles per day). Data on main road traffic growth are obtained from 275 fixed traffic counting posts. Traffic developments on the lower road network have to be estimated based on sample censuses using general information on fuel sales and changes in automobile stock or land use.

Once the annual average daily traffic (AADT) has been estimated for an individual homogeneous road section, the corresponding value for the year total is obtained by multiplying the daily traffic by the length of the roads section and the number of days in the year. Estimates of vehicle-kilometres for streets, private roads and forest roads are based on old survey results, updated with information on traffic growth on public roads and streets. Estimates for the annual total of vehicle-kilometres are also obtained from the national passenger transport survey, conducted every six years, and the annually compiled Statistics on the Transport of Goods by HGVs in Finland. The last time a comprehensive transport survey, both for freight and passengers, was conducted in Finland was in 1987–1988. What followed were minor surveys. In 2007–2008, the next extensive transport survey was carried out in the Helsinki Metropolitan Area (HMA). It included several sub-surveys: a household interview with travel diary of around 20,000 persons, O–D surveys of both vehicles and public transport passengers and separate surveys for walking and cycling as well as for freight transport. The survey procedure was based on automatic number plate recognition (ANPR).

The use of kilometre counter readings obtained from vehicle inspections in estimating vehicle-kilometres was examined in Finland in the late 1990. The scope of the study was limited to passenger cars and vans; corresponding data on heavy vehicles and buses were collected with inquiries conducted among transport companies. Data on vehicle age and engine type (petrol, diesel) was added to the counter reading data. For passenger cars, further information such as engine capacity, total vehicle weight, as well as information on the municipality of registration, and information on the (private) owner of the vehicle (e.g. age, sex) were included into the sample. Around 260,000 km counter readings were available for 165,000 vans, and over 450,000 readings for around 200,000 passenger cars. Based on estimates on average mileage driven differentiated by the age and the engine type of the vehicle, annual vehicle mileages were estimated for the total stocks of passenger cars and vans.

As a result, all the different approaches used in Finland to estimate the annual totals of vehicle-kilometre volumes provided rather different results. Therefore, statistical figures on vehicle-kilometres for a base year had to be agreed by representatives from the relevant organisations. At the same time, more sophisticated econometric models were developed to improve the accuracy of the estimated results and to extrapolate estimated base year values for different time periods. In addition, data collection methods are being improved in order to obtain more precise data (Estimating vehicle kilometres in Finland, 2005).

32.3.4. A Combination of Automatic and Manual Traffic Counts in the United Kingdom

In the United Kingdom, information on traffic performance is obtained from both manual and automatic counts conducted on many different types of roads together with information on road lengths. Nevertheless, converting count data to information on total traffic volume is not a straightforward and easy exercise. The national road system is too extensive to allow an exhaustive collection of traffic data. Furthermore, the density and the mix of traffic by vehicle type vary in space as well as in time. Even same road categories may carry significantly different traffic volumes. Also, traffic flows can vary by time of day, by day of week and by month of year. Another kind of variation can be observed from comparing car and freight traffic levels. Car traffic flows may be fairly stable over the seven days of the week compared to goods traffic, with its lower performance during the weekends. All the aspects mentioned require a quite sophisticated sampling design, the collection of substantial volumes of data and complex computational procedures to estimate national traffic volumes.

For major roads the traffic is assessed by counts at a statistically random point on most links at regular intervals — once every three years in England and Wales and once every six years in Scotland. In total, about 5100 major road sites were counted in 2005. In addition to traffic count data, information is collected about the characteristics of each link. At each chosen point, trained enumerators count vehicles of each of 11 types (pedal cycles, two-wheeled motor vehicles, cars and taxis, buses and coaches, light vans and six separate categories of goods vehicle) for the 12 h from 7 a.m. to 7 p.m. Counts are conducted on weekdays, excluding days falling around or during public or school holidays. To minimise the effects of possible seasonal factors, counting is confined to the so-called 'neutral weeks' (in March, April, May, June, September and October).

Complete coverage of the minor road network is too extensive to be feasible. Minor road traffic estimates are therefore made by grouping them into road classes. A random sample of approximately 4500 sites across Great Britain (GB) is conducted each year, mostly including counts carried out in neutral weeks. Additionally, about 200 counts per year, the 'summer–winter counts', are carried out in non-neutral weeks and on weekends. Special attention is given to two-wheeled traffic, mainly because pedal cycles and motorcycles are not always accurately identified by automatic counters. The manual counts offer a complete coverage of major road sites and moderately good coverage of minor roads. Nevertheless, vehicle-type-related data on hourly traffic, traffic by night at weekends or during public holidays is rather sparse. To fill this gap, data from automatic counters is used. Around 160 sites exist in GB outside of London where traffic is monitored continuously, classified by 22 vehicle types, using automatic sensors. The numbers of detected vehicles sum up to hourly totals and are stored on-site until downloaded during the night to a computer in the department for transport (DfT) headquarters building. However, automatic counters do not deliver fully accurate results. Besides general problems with functioning, the equipment used cannot identify the vehicle type when the traffic is moving very slowly (5 mph or less) and fails to make a further distinction between similar vehicle types or to identify two-wheeled vehicle traffic.

The 56 automatic counters used in London are slightly different from those described. Being 'volumetric' classifiers, they only distinguish between short (up to 5.2 m) and long (greater than 5.2 m) vehicles. Manual counts of 24 h duration are conducted every three months to provide additional estimates of the breakdown of traffic by vehicle type and hour of the day (How the National Road Traffic Estimates are made, 2005).

32.3.5. Truck Surveys in the United States

The Vehicle Inventory and Use Survey (VIUS), carried out by the Census Bureau approximately every five years, provides a basis for estimating total nationwide truck-miles and tonne-miles of travel (by local, short, medium and long-distance intervals), but lacks any details on the geography of such moves, and contains very limited commodity detail.

The U.S. Commodity Flow Survey (CFS) (1993, 1997, 2002, 2007 pending) reports the true origin and destination (down to the Zip code level) of freight shipments by any combination or sequence of modes (truck, rail, water or air). However, this shipper establishment survey covers only some 60% of all commodities moved. It does not cover establishments involved in farming, forestry, fishing, construction and crude oil production; households; governments and foreign establishments; service businesses and most retail. The CFS also does not cover the first stage delivery of imported goods, since it does not survey non-U.S. establishments.

Other sources of truck activity data, such as FHWA's Highway Performance Monitoring System (HPMS) of state-supplied AADT truck counts, and the within and between state truck operating mileage data collected by the AAMVA's International Registration Program (IRP) offer additional truck movement information, but cannot be used to build statistically robust O–D movement patterns for anything below the state-to-state level of spatial resolution, and for limited truck size classes. Therefore, while some 70% of all tonnes are reported by the CFS to move all or in part by truck, there is currently no national dataset from which these truck movements can be generated, or from which national statistics can be drawn that show the working relationship between truck activity (e.g. vehicle-miles and tonne-miles driven by vehicle type) and the volumes and spatial patterns of the different commodities being traded. This makes it difficult to understand and quantify, among other things, the relationship between truck use, future highway infrastructure needs and the growth in commodity flows. The proposed cancellation of the VIUS enlarges this data gap and makes data fusion of existing sources even more problematic for state and regional, as well as federal, data users. In fact, the cancelled survey (VIUS) is the most comparable to the European harmonised truck survey, while the shipment-based CFS has a counterpart only in France (ECHO survey 1988 and 2004) with a more precise follow-up of the shipment involving the interview of each participant of the transport chain. Kockelman, Browne and Leonardi (2009) have discussed the respective advantages and drawbacks of truck-based versus shipment-based surveys for the analysis of long-distance freight transport.

32.3.6. Canadian Vehicle Survey

The Canadian Vehicle Survey (CVS) is voluntary and vehicle based. It combines a mail-in form and a computer-assisted telephone (CATI) survey. Since 1999, it has provided quarterly and annual estimates of road vehicle activity of vehicles registered in Canada. The estimation results are expressed in vehicle and passenger kilometres. Motorcycles, off-road vehicles (e.g. snowmobiles, dune buggies and amphibious vehicles), buses and special equipment (e.g. cranes, street cleaners, snowploughs and backhoes) are outside the scope of the survey. The estimates are provided by type of vehicle, by vehicle characteristics (e.g. fuel type, body type and age), by driver characteristics (e.g. age and gender) and by journey characteristics (e.g. purpose of trip, season, day of the week, time of day, road type and speed limit).

A quarterly sample of vehicles is drawn from vehicle registration records from the 13 jurisdiction vehicle registration lists (10 provincial and 3 territorial governments), which are created three months prior to the reference period. The CVS uses a two-stage sample design: firstly, the vehicle sample is selected from a stratified survey population, and secondly, a sample of consecutive days within the sample period is selected. The owners of the sampled vehicles are interviewed about vehicle usage. In the 10 provinces registered, owners of a sample of 20,000 vehicles are telephoned, interviewed to obtain background data and asked to complete a seven-day trip log. If respondents agree, a logbook is mailed to them. If respondents cannot be contacted by telephone, a trip log plus a short additional questionnaire are mailed out. There are three types of log, depending on the type of vehicle: a light log (for passenger vehicles), a bus log and a truck log for vehicles weighing more than 4500 kg. To increase the number of responses, respondents are contacted a second time either by telephone or by mail.

On the first or second day of the log, attempts are made to telephone each vehicle owner to answer any questions the respondents might have. Later an attempt is made to contact by telephone or by mail everyone who did not return a log. In the three northern territories, registered owners of 11,000 selected vehicles are sent postcards asking them to provide two odometer readings, one at the beginning of the quarter and another at the beginning of the next quarter, plus information about the vehicle status (owned, sold or scrapped) (Canadian Vehicle Survey, 2006).

32.4. Areas of Research and Development

Reports on practical country experience with established transportation data collection methods show that, in many cases, there is still some need for improvement. Significant advances already exist applied in the area of survey techniques, survey methods and data processing procedures. However, many of these improvements and technical innovations are still in the stage of development and are being applied in small-scale pilot projects to be further researched and optimised before entering broader applications. An international comparison and finally also an exchange of best practice experiences in this area of research and development serve as a learning platform for the participants. Best survey methods, sampling techniques as well as estimation approaches offer examples for further application. Often resulting from existing deficiencies, new ideas and improved technological solutions are elaborated. In the area of road traffic counts, innovative detection technologies enable the counting of vehicles by class, size and weight. Reliable results of well-differentiated traffic counts help to eliminate possible inaccuracies often occurring during the further processing of the survey data for estimation procedures. Estimates of relevant transportation data are therefore more consistent and give a better picture of the sampled universe. In addition, the emerging of new, for example satellite-based, technologies employed to track individual vehicles allow collection of data for a detailed analysis of underlying mobility and vehicle-use patterns in space and time. Advanced survey techniques of this kind allow researchers to obtain precise multi-day and multi-period information on a multitude of relevant variables. Thus, not only data on behavioural parameters in terms of distances travelled can be collected. The automatic data transmission from the vehicle can easily include additional information on the vehicle as well as the vehicle user, on land-use characteristics for travelled distances and purpose of the trip, travel speed, overall traffic situation, etc. Such data also offer a good basis for dynamic analyses. New satellite-based technologies (e.g. global positioning system [GPS] and, in the future, the global navigation satellite system being built by the EU and European Space Agency (ESA) GALILEO, as well as a system utilising cellular phones) also bring improvement in handling omitted or inaccurate data (e.g. rounded mileage or departure time). Another innovation recently introduced in some countries is the employment of electronic number plate recognition for O–D surveys of vehicles.

Independently from the survey technique and the data sample, it can be of great advantage if the survey data can be enhanced by linking with administrative data, for example from national vehicle or even population registries.

Existing estimation procedures are also further elaborated in order to validate or ameliorate the results. Finally, international exchange of research experiences and competencies in the field of transportation and mobility research are a practical way to ensure the comparability of national transportation and travel statistics. For trucks mainly used by companies, the survey harmonised by EUROSTAT is an excellent example of achieving data comparability (Council Regulation (EC), 1998).

32.4.1. Tracking Vehicles

In order to deal with issues of reliability and changing behaviour for a more sustainable mobility, we need to observe a given vehicle several times over as long a period as possible. The key challenge in this context is how to track a vehicle. Technical inspections should permit this, but in most cases the collected data is restricted to the mileage on the odometer and the tested items, which relate primarily to safety and pollution. In some cases, additional attributes of the inspected vehicle are included into the sample of odometer readings. But very often, due to data privacy protection laws, the unique identification of the vehicle is not permitted, making it impossible to survey it over its life cycle, so that a researcher cannot gather age- and period- specific information on vehicle use. In addition to what the French Department of Transport (SESP) has done recently to make the vehicle registration register (FCA) and the technical inspection registers comparable, in order to estimate the passenger car fleet as of 1 January 2006, it would be of undeniable value to be able to compare successive technical inspections for a sample of vehicles. 'The longitudinal data which could result from this, including the mileage on the odometer recorded during the technical inspections, would be very useful for analysing behaviour changes' (CNIS, 2006).

Household panels (Parc-Auto SOFRES, German Mobility Panel, etc.) allow us to follow the behaviour of households over three to four years. However, as the household may change cars, a particular vehicle is often tracked over a shorter time period. Moreover, it is not straightforward to identify the same vehicle in successive surveys: the household's main vehicle may become a second car if they buy a new one. Anyway, for an accurate measurement of annual change in car use (i.e. for national accounts), a panel study provides the most cost-efficient data source.

Another frequent problem in surveying vehicles within household mobility surveys is data entry errors, for example when the respondents are asked to copy vehicle matching keys such as the national vehicle model code commonly referred to in the official vehicle registration documents. Nevertheless, the combination of vehicle data and individual data on household members' mobility taken from a 10-day diary included into the survey makes it possible to estimate the vehicle's fuel consumption per 100 km (SECODIP panel). However, an excessively complex survey always presents a risk of excluding high-mileage drivers or generating high non-response rates.

When it is not possible to track individual vehicles, successive snapshots of the vehicle fleet can be taken from cross-sectional surveys. In this way, a 'pseudo-panel' (Deaton, 1985) can be constructed by matching vehicles or vehicle categories based on vehicle make or type, year of manufacturing, etc. The vehicle mileage survey conducted in Germany for the years 1993, 1998 and 2002 is one example of that kind of procedure. Pooling the samples for the different survey years and merging observed vehicles on the basis of a specific vehicle-type-manufacture key combination allows explaining the effects of different variables included into the study on the vehicle use collected in the sample. Main problem with this kind of mileage surveys, based on sampling vehicles from the national vehicle registry, is their relatively high cost, in particular when the sample is supposed to be representative for the entire vehicle population and in addition data on vehicle user and holder is taken into account. Therefore, this survey type is often conducted, for example as seldom as every five years. The disadvantage is that continuous behavioural changes between the survey periods remain unobserved.

Another approach to obtain data on travel times, traffic volumes or route choice behaviour of drivers is the application of innovative survey methods such as floating car data (FCD). The disadvantages of this method are high installation and data-transfer costs. One possible alternative is the application of plate recognition (ANPR). The great advantage of ANPR is the possibility to link recognised car registration code with official vehicle registry data, which in general contains a variety of additional data on the technical characteristics of the surveyed vehicle as well as on the vehicle holder. This kind of data extension allows for the investigation of a broad spectrum of research questions, even concerning environmental studies on the local concentration of exhaust gas pollution.

Another technology, application of which is still being explored and refined within the counting and classifying of vehicles, is the weighting of vehicles by means of loops and weigh-in-motion systems installed on the road surface to continuously collect vehicle size and weight data. To collect traffic volume data for planning and other purposes, in general vehicles are classified by axle spacing or length. Weigh-in-motion systems use weight data in addition to axle spacing and therefore adding accuracy to the traffic classification results. The disadvantage of this methodological extension is the relatively high cost of the method implementation, mainly due to the installation of the loops in the road, compared with mobile technologies (e.g. GPS or GSM). With regard to existing budget restrictions and resource constraints, the collection of sampled data instead of complete population data is discussed as possible alternatives. In this context, complex mathematical classification algorithms and sampling schemes need to be implemented to obtain consistent results on traffic classification in a cost-effective way.

32.4.2. How Can We Obtain a Fuller and More Accurate Description?

Conventional surveys are self-administered or involve memory. The description of the vehicles is often based on a facsimile of the vehicle registration papers (with the possibility of the copying and data entry errors mentioned above). This is supplemented with a few questions about the mileage on the odometer, the annual distance covered (difficult to estimate but unbiased on average [Hivert, 2001]), the purposes of vehicle use (yes/no, or a more difficult question requiring a breakdown of the distances covered) and the type of roads used (urban/interurban/motorway) during the period described in the survey wave (frequently one year). The design of this Parc-Auto panel survey has been adopted in Japan for a survey starting in autumn 2009.

A specific feature of the French National Travel Surveys — with a sample size of around 14,000 households in 1993 — is the trip diary, which aims to describe all trips made over seven days in a vehicle used by the household; in 2007–2008 it was allocated at random to any two- or four-wheel vehicle, with a higher probability for motorised two wheelers because of safety issues; in 1993–1994 it was allocated to four-wheel vehicle with a higher probability for secondary cars to analyse the potential market for electric vehicles; in 1981–1982 the diary was allocated to all four-wheel vehicles, with a high non-response rate in multi-car households. The estimation of distances is reliable as the kilometres on the odometer are recorded at each destination, but, in spite of the on-board clock, departure and arrival times are rarely given to within less than 5 min (Madre & Armoogum, 1997). During the interviews on daily or long-distance trips, there are practically no missing data concerning the locations visited (which are coded at municipality level), but the spontaneous estimation of the distances covered is difficult (underestimated by 25% within a municipality and by 5% up to 15 km outside one) and departure and arrival times are rounded, which makes the estimation of trip duration by subtraction very imprecise.

Surveys of heavy vehicle drivers are also an important data source for road safety analysis, especially for the measurement of work duration. Chrono-tachograph disks are an interesting data source, but a conventional time-use survey with additional questions on road safety is more appropriate; it is focussed on this specific population, which is not easy to identify (better through periodic driving licence inspection than in specific establishments). Such surveys have been conducted in France (Hamelin, 1987).

Of course, recording the main engine parameters as in France (Enquêtes EUREV sur l'Utilisation Réelle des Véhicules — real vehicle-use surveys) or in Canada provides a great wealth of data. But the high cost of the in-vehicle equipment limits these experiments to a few dozen vehicles for a few weeks.

Satellite tracking of travel (GPS now, GALILEO in the future) can describe the spatial and temporal structure of trips much more accurately (within 10 m and 1 s). Using the cigarette lighter in the vehicle as an electrical power supply, the first vehicle tracking experiments were performed in the 1990s: Lexington then Atlanta (Wolf, Loechl, Thompson, & Arce, 2003a; Wolf, Oliveira, & Thompson, 2003b), Paris (Flavigny, Hubert, & Madre, 1998) and London, California (Zmud & Wolf, 2003) and Sweden (www.rattfart.com).

Improvements in batteries mean it is now possible to track an individual for a day, and for one to two weeks if the device is recharged at night. We are testing this procedure on 1000 volunteers during the national travel survey 2007–2008 (similar to the GFK Group market research institutes for billboard watching in Italy and France). During their second visit, fieldworkers download the GPS traces onto their laptop computer and ask a few additional questions in order to make a rapid analysis of the traces (confirmation of and reason for the days without traces, travel modes and purposes during a recorded day, etc.). We should obtain a much more detailed description of travel, which overcomes the problems of rounding and the difficulty of obtaining precise addresses from respondents; it should allow identification of each road used, which is much more precise than an annual statement of the distances covered on urban networks/road/motorway.

The full-scale testing of the procedure within a conventional national travel survey should allow us to organise the transition between former and future travel data collection methods. Likewise, the car diary has been a stable feature of national transport surveys since 1981–1982, assisting the measurement of long-term changes. Apart from this diary, daily travel has been observed by means of an individual trip diary in 1981–1982, an interview on travel behaviour on the previous day and the previous weekend in 1993–1994, and on the previous day and the last Saturday or Sunday in 2007–2008.

All the conditions have therefore been met for us to observe vehicle use in the long term:

- by means of administrative data from vehicle inspections (biyearly electronic collection since 1992 of mileage on the odometer and details about tested items) and from the vehicle registration register (FCA), which can be cross-checked with vehicles inspection as shown by SESP;
- by means of innovative surveys using satellite tracking, which provide a precise and detailed description of travel over a period of about 10 days.

32.5. Concluding Remarks

To estimate relevant parameters on road travel, both for passengers and freight transport, or to assess how these can be affected by diverse exogenous factors, detailed information on the mobility unit of interest is required. Various research studies have shown that technical attributes of motorised vehicles serve as significant independent variables in explaining the use of a vehicle for purposes of passenger travel or freight transport (Kalinowska & Kuhfeld, 2006). Often, technical 'hard facts', especially when they come from a professionally administered database such as a vehicle registry, can be considered as more reliable than statements made by respondents during mobility surveys. At this point, surveying the vehicle as the primary survey unit reveals clear advantages and offers a very important, complementary source of data in travel and transport modelling research. Another situation arises when the vehicle user or vehicle holder, either a private person or a commercial entity, is asked in a survey situation to report mainly technical characteristics of his or her vehicle. Evaluations of a range of this kind of travel or transport surveys have shown that, for a significant number of respondents, giving advanced information on their vehicle is not a trivial exercise. At the same time, giving the respondents too much advice on how to find and copy certain information on his vehicle can easily overload the survey questionnaire. In the worst case, this can result in underreporting or a non-response. Data collected in such an indirect way is likely to contain different kind of errors or be incomplete, since it does not come directly from the survey population of the vehicles. There are also many countries that lack the extensive vehicle registries that otherwise could provide an excellent data source to supplement information collected on vehicles from the vehicle user or holder, and also to correct for possible errors reported by the survey respondents. Therefore, surveying the vehicle in the first place can help to avoid these kinds of shortcoming. Other types of errors associated with mobility surveys in which the respondents are asked to report on, for example the mileage they travelled with their car during a certain month are recall or measurement errors (misreading the odometer, rounding errors, etc.). On the other hand, accurate data, free of errors, allows for more precise analysis and as a result optimises the quality of our results, making them more consistent at the same time.

Another aspect, which gives some advantage to survey vehicles for obtaining parameters on vehicle use rather than surveying the holder or the user, is the simple fact that, for a majority of mobility surveys, the sample is drawn from a register of households or resident persons. As a consequence, vehicles covered in such surveys are, for the most part, private vehicles. Information on commercial vehicle ownership and use is then omitted.

If the sample is drawn from residential telephone directories, there exists a great probability that an important proportion of the surveyed universe will not enter the sample, since for several reasons they may not be included in the phone directory. The growing use of mobile telephones instead of fixed net telephones is only one reason for this kind of sampling error. A similar logic applies to the collection of traffic volume data, which is supposed to differentiate between mileages driven on the national territory versus by national residents. In principle, household mobility surveys are not designed to collect information on mileage driven by foreign vehicles on the national territory. In this case, conducting a vehicle-based survey to collect the relevant data is obvious.

Thus, numerous, well-documented deficiencies are known from experience in practical survey realisation. Depending on the underlying research objective, these factors are likely to motivate consideration of vehicles as the survey unit, in particular if the objective is to gather knowledge about road traffic volumes.

References

BASt, Bundesanstalt für Straßenwesen (1996). Inländerfahrleistung 1993. Berichte der Bundesanstalt für Straßenwesen, Heft M 61, Bergisch Gladbach.

BASt. Bundesanstalt für Straßenwesen (2005). Fahrleistungserhebung 2002. Berichte der Bundesanstalt für Straßenwesen, Heft V 120, Bergisch Gladbach.

Canadian Vehicle Survey 2006 (2006). Statistics Canada, Ottawa.

CNIS (2006). Inventaire des Besoins d'Information sur les Transports. Conseil National de l'Information Statistique.

Council Regulation (EC) (1998). No. 1172/98 on statistical returns in respect of the carriage of goods by road. *Official Journal of the European Communities*.

Deaton, A. (1985). Panel data from time series of cross-sections. *Journal of Econometrics*, *30*(1–2), 109–126.

Estimating vehicle kilometres in Finland. (2005). United Nations, Economic and Social Council, Economic Commission for Europe. Inland Transport Committee. Working Party on Transport Statistics, Workshop on statistics on the volume of road traffic (vehicle-kilometres). Transmitted by the Finnish Road Administration, Finland, (Copenhagen, 1–2 December 2005). TRANS/WP.6/AC.5/2005/6. Available at: http://www.unece.org/trans/main/wp6/transstatac5wdoc.html (02/21/2008)

Flavigny, P. -O., Hubert, J. -P., & Madre, J. -L. (1998). Analyse de trafic routier observé par GPS et comparaison avec d'autres sources statistiques. Rapport pour l'ADEME.

Gallez, C., & Couturier, N. (2000). Renouvellement du parc automobile et prospective transport-environnement. Rapport de convention ADEME-INRETS.

Gallez C., Hivert, L., & Polacchini, A. R. (1997). Environment energy budget of trips (EEBT): A new approach to assess the environmental impacts of urban mobility, Communication au 4e Colloque international *Transport et pollution de l'air*. Avignon. 9–13 juin 1997, 8 pages, pp. 326–334. Des actes publiés dans *International Journal of Vehicle Design*, *The Journal of Vehicle Engineering and Components*. Volume 20, Nos. 1–4, 1998, Inderscience Enterprises Ltd. GB, ISSN 0143-3369.

Hamelin, P. (1987). Lorry drivers' time habits in work and their involvement in traffic accidents. *Ergonomics*, *30*(9), 1323–1333.

Handbook on Statistics on Road Traffic: Methodology and experience (2007). United Nations, Economic Commission for Europe. Transport Division, WP.6, Task force on Road Traffic. Available at: http://www.unece.org/trans/main/wp6/transstatac5agenda.html (02/21/2008)

Hivert, L. (2001). Le parc automobile des ménages, étude en fin d'année 1999 à partir de la source "Parc Auto" SOFRES. Rapport de convention ADEME-INRETS.

Hivert, L. (2007). A trans-sectoral approach to estimate the incidence of urban sprawl on the greenhouse effect: Transport and housing CO_2 emissions analysed by residential location for households living in French large urban areas. *Communication (24 p.) to the 11th world conference on transport research WCTR 2007*, June 24–28, University of California, Berkeley, USA.

How the National Road Traffic Estimates are made (2005). United Nations, Economic and Social Council, Economic Commission for Europe, Inland Transport Committee. Working Party on Transport Statistics, Workshop on statistics on the volume of road traffic (vehicle-kilometres). Transmitted by the Government of the United Kingdom, (Copenhagen, 1–2 December 2005), TRANS/WP.6/AC.5/2005/14. Available at: http://www.unece.org/trans/main/wp6/transstatac5wdoc.html (02/21/2008)

Kalinowska, D., Kloas, J., & Kuhfeld, H. (2007). Hochrechnung der Kfz-Fahrleistungen auf der Basis der Ablesungen der Wegstreckenzähler bei Kfz-Hauptuntersuchungen. Forschungsprojekt im Auftrag des Bundesministeriums für Verkehr, Bau und Stadtentwicklung.

Kalinowska, D., & Kuhfeld, H. (2006). Motor vehicle use and travel behaviour in Germany — Determinants of car mileage. DIW Berlin Discussion Papers 602, Berlin.

Kockelman, K, Browne, M., & Leonardi, J. (2009). The acquisition of long distance freight data: Synthesis of a workshop.

Madre J.-L. (1992). Le parc automobile des ménages se diversifie. Colloque mi-parcours du PREDIT, Versailles.

Madre, J.-L. (2007). Observer l'utilisation des véhicules par les ménages: Vers des méthodes plus précises et plus fiables. *Ingénieurs de l'Automobile*, No. 788, Mai-Juin, pp. 40–42 (presentation at SIA Conference on January 18th 2007).

Madre, J.-L., & Armoogum, J. (1997). Accuracy of data and memory effects in home based surveys on travel behaviour. *Proceedings of the 76th annual meeting of Transportation Research Board*, Washington, DC.

Madre, J.-L. & Lambert, Th. (1989). Le vieillissement du parc automobile se ralentit depuis deux ans. Economie et Statistiques (INSEE) No. 22.

Metaphis, A. (2006). Bilan final des appariements entre le fichier central des automobiles (FCA) et les fichiers des contrôles techniques (UTAC) d'octobre 2003 à décembre 2005 et estimation du parc des voitures particulières (VP) au 1er janvier 2006. Note EI/SESP/ST1/0610003, Ministère de l'Equipement.

MiD. (2002). Survey Mobilität in Deutschland. Available at: http://www.mid2002.de/engl/index.htm (02/21/2008)

Mobilitaetspanel (MoP). German Household Panel Survey (mobility panel, MOP). Available at: http://mobilitaetspanel.ifv.uni-karlsruhe.de/ENGLISH/MOP_Frameset.htm (02/21/2008)

Quételard, B., Noppeet, J., & Hivert, L. (2003). Environment energy budgets of trips in Lille urban area. Communication au Colloque international TILT, 20 e anniversaire du GRRT, 2–4 décembre 2003, Lille, International Crashworthiness and Design Symposium 2003, proceedings tome 3.

Road freight transport methodology, Volume 1. (2005a). Methods and nomenclatures: Road freight transport methodology — Reference manual for the implementation of Council Regulation No 1172/98 on statistics on the carriage of goods by road, European Commission, Luxembourg: Office for Official Publications of the European Communities, 2005.

Road freight transport methodology, Volume 2. (2005b). Methods and nomenclatures: Road freight transport methodology—Methodologies used in surveys of road and freight transport in Member States and Candidate Countries, European Commission, Luxembourg: Office for Official Publications of the European Communities, 2005.

Toll-collect GmbH. Available at: www.toll-collect.de (02/21/2008)

Wolf, J., Loechl, M., Thompson, M., & Arce, C. (2003a). Trip rate analysis in GPS-enhanced personal travel surveys. Transport survey quality and innovation, Chapter 28, pp. 483–498. Elsevier, England.

Wolf, J., Oliveira, M., & Thompson, M. (2003b). The impact of trip underreporting on VMT and travel time estimates: Preliminary findings from the California Statewide Household Travel Survey GPS Study. *Transportation Research Record, 1854*, 189–198.

Zmud, J., & Wolf, J. (2003). Identifying the correlates of trip misreporting results from the California Statewide Household Travel Survey GPS Study. Paper presented at the 10th international conference on travel behavior research, Lucerne, August 2003.

Appendix 32.A.1

Data collection technique	Advantages	Limitations	Cost impact
Motorised vehicle surveys sampled on the basis of official vehicle register database	Make use of established administrative databases; no additional cost is required to generate vehicle information	Not available in every country	In general no additional costs are involved
	Contain relevant (technical) vehicle, vehicle holder and residential data	In some cases, data is only collected on inflows from car registrations; in these cases, data on vehicle outflows (mortalities, etc.) from the total fleet is missing	
	Contain geographical information on vehicle registration	Do not contain any data on transportation or mobility parameters, i.e. vehicle use	
	Can be linked to other official registry data, i.e. register of residents	Concerns exist as to data confidentiality requirements	
	Survey can be designed to collect data on vehicle movements (vehicle-kilometre) as well as on vehicle loads moved (tonne-kilometre)	When vehicle loads moved is the primary subject of the survey, certain service, refuelling and other trips may be unreported	
	Survey can be designated as voluntary or mandatory		

Appendix 32.A.1 (*Continued*)

Data collection technique	Advantages	Limitations	Cost impact
Motorised vehicle surveys sampled on the basis of vehicle inspection register	Data already collected for inspection purposes; no additional cost is required to generate vehicle information	Concerns exist as to data confidentiality	Data collection for one point in time is relatively inexpensive; however, combining data collected across several years for building a panel is likely to be much more costly
	Contain relevant (technical) vehicle, vehicle holder and residential data	Advanced estimation procedures are required to calculate annual mobility parameters on vehicle average or for fleet totals	
	Can be linked to other official registry data, i.e. register of residents or vehicle register data	Information on very recent and very old vehicles is not well covered by the data	
	Through collection of odometer readings, a database on vehicle use can be constructed	Odometer-based data can be manipulated or incorrect due to limited number of digits in some older odometers	
	When the inspected vehicle is identified as a unique survey unit observed over time through obligatory follow-up inspections, panel-like data can be obtained	Derived indicators on vehicle use do not contain any information of trip purpose, road type use, etc.	
	Available in many countries; applicable for international comparisons	In general no information on car user is available (eventually on car holder)	
		Give no information on foreign traffic on national roads	

Pull-over roadside vehicle surveys	Information on vehicle, vehicle origin (number plate), driver and travel as well as on transported load in case of heavy freight vehicles can be collected in a time-restricted, short-duration interview	When questionnaire data is collected during a red-light stopping phase, only a very limited number of questions can be asked
	Short questionnaire and face-to-face interview reduces the risk of non-response	
	Through distribution of mail-back questionnaires, additional information can be collected	Data is collected covering the entire day during two weeks in a year and is extrapolated to give representative annual traffic results; episodic instability of traffic may be underestimated
		If interviews are conducted at motorway service areas, data may contain an inherent bias since certain categories of individuals tend to stop more frequently at service areas than others (e.g. families with small children)
Quantitative vehicle-based surveys of vehicle drivers	Address a specific, often explorative topic, i.e. road safety or drivers' attitudes, habits and bchavioural characteristics	Complex survey design and data evaluation procedures required
	Account for complex and interdisciplinary approaches	Laborious data collection procedures required

Appendix 32.A.1 (*Continued*)

Data collection technique	Advantages	Limitations	Cost impact
	Collect information on generally observable attributes: socio-demographic and socio-economic characteristics of the vehicle driver or vehicle attributes Implications of results can be used for policy advisement purposes.	Due to elaborate sampling and survey realisation procedures, data can be collected only on a small scale, leading to a limited number of observations	
Automated road traffic counts	Distribution of classified road traffic over different road types	Inaccurate or missing information on vehicle classification	The investment cost for installing the electronic loop inside the road is rather high; operating cost for counting vehicles in categories automatically detected by the system is relatively low; the cost rises with more detailed categorisation of vehicle counts
	Possibility of recording the distribution of road traffic over exactly defined time periods, measuring of periodical (daily, weekly, hourly, etc.) or seasonal traffic intensity variations	Deriving information on total traffic volumes on different types of the national road network is problematic, mainly due to insufficient data coverage Accuracy of the data depends on traffic flow speed Technical devices can break down unexpectedly Data confidentiality concerns	

Automatic number plates recognition	Information on foreign road traffic on national road infrastructure		
	Distribution of classified road traffic by vehicle type over different road types	Need for large data storage and processing devices	
	Possibility of obtaining large amount of data (continuous survey)	Risk of violation or theft of the recognition device — high cost of protecting the device	
	Possibility of linking recorded vehicle data with official vehicle register data		
Manual road traffic counts	Allow adding information on type and residential origin of the vehicle	Involves a relatively laborious data collection process	
	Results can serve as extension of other surveys, e.g. automatic road counts	Data is mostly collected manually by individuals and is therefore likely to be error prone	
	Can better detect two-wheel traffic		
Periodic, cross-section national mobility surveys	Provide detailed information on mobility parameters if, e.g. O–D surveys or travel and activity diaries are included in the survey	Sophisticated estimation and data evaluation procedures needed to account for data deficiencies in terms of bias, non-response, etc.	Depending on the scope and the complexity of the survey design, implementation can be rather costly (one reason why national travel surveys are often conducted every several years rather than annually)
	Provide whole-country population representative data	Survey design, preparation and implementation procedures can be time and labour intensive	
	Individual or household context mobility patterns can be detected	Data confidentiality concerns need to be taken into account	

Appendix 32.A.1 (*Continued*)

Data collection technique	Advantages	Limitations	Cost impact
	Through advanced survey realisation procedures (sending out of memory joggers, CATI, PAPI), highly detailed data on various parameters can be collected — information on household and/or individual socio-demographics and socio-economy, mobility indicators such as trip purpose, trip chaining, travel mode, trip duration, distance travelled, etc.		
Mobility panels	Provide detailed information on mode-specific mobility parameters if, e.g. O–D surveys or travel and activity diaries are included into the survey Information on energy consumption for travel purposes (fuel consumption) can be included in the survey questionnaire If fuel consumption data is collected, it is possible to draw a comparison with other data sources on fuel consumption, often used for motor vehicle travel volume estimations Provide whole-country population representative data	Sophisticated estimation and data evaluation procedures needed to account for data deficiencies in terms of bias, non-response, etc.	Likely to be less costly than cross-sectional mobility surveys since the important recruitment cost is paid only once for several years

GPS/GALLILEO	Continuous vehicle movement recording technique	Recording and data transmission can be interrupted when the vehicle and therefore GPS device enter a tunnel	The implementation cost is likely to be lower than installing traffic-count loops in the road
	Timely and geographically stable data on travel activities can be obtained	Data transmission problems can occur in high density areas	
		Problems with keeping the transponder device in working order	
Electronic tolling systems	Depending on the charge collection and road use payment method, detailed data on the individual road user and the vehicle charged can be available	Main limitations are confidentiality concerns	High installation and start-up costs, in particular if free on-board units are distributed to the road users
	Individual road user and vehicle data can be linked and extended by other relevant data sources, e.g. technical vehicle data or residential location of the vehicle holder, etc.		

Chapter 33

Vehicle-Based Surveys: Synthesis of a Workshop

Klaas van Zyl, Dominika Kalinowska, Jean-Loup Madre and Bob Leore

33.1. Introduction

The purpose of the workshop was to review the state of the art of vehicle-based surveys in various countries, share information between the participants, identify problems and issues, and finally identify areas of future research.

The workshop was well attended by 14 people from a wide range of countries, including France, Germany, Finland, the United Kingdom, Canada, Brazil, Australia, and South Africa. A list of people attending the workshop, as well as those facilitating the workshop, is given at the end of this chapter.

The following papers were presented:

- Resource paper: "Vehicle-Based Surveys: Towards More Accurate and Reliable Data Collection Methods" (Kalinowska & Madre, 2009).
- Two contributed papers were presented, both on applications of automatic number plate recognition in Germany and Finland (Friedrich, Jehlicka, & Schlaich, 2008; Jarvi, 2008).

The resource paper mentioned above (Kalinowska & Madre, 2009) presents a review of mainly vehicle-based survey methods used in France, Germany, Finland, the United Kingdom, Bangladesh, the United States, and Canada. It describes existing sampling frames as to their scope, advantages, limitations, and costs. Problems and issues are further examined, and possible improvements are offered in terms of methodology and use of technology. One example describes the potential use, benefits, and constraints of new survey technologies presented by vehicle-tracking techniques.

Transport Survey Methods

ISBN: 978-1-84855-844-1

Some discussion was devoted to the scope of the workshop. It was agreed that the discussion be limited to surveys of road vehicles, specifically where the object of the survey is the vehicle, such as surveys of vehicle movements, characteristics of the vehicle population, features of the vehicle, and its use. However, it was noted that there is also a need to discuss in future conferences surveys of other modes of transport, such as train, aircraft, and ships.

The following process was followed:

- introduction of workshop members to each other, indicating their main areas of interest
- discussion of the scope of the workshop
- presentation of resource paper and discussion
- presentation of two contributed papers and discussion
- presentation by discussant, Bob Leore, to introduce main topics of discussion and give some background
- main discussion facilitated by the discussant
- conclusion confirming problems and issues, and identification of research needs

The discussion followed the following topics, and this workshop report also follows the same structure:

- national administrative data
- road link surveys
- new technologies
- research priorities and perspectives

Examples of survey methodologies in various countries are given to illustrate the survey problems and some solutions to these.

33.2. National Administrative Data

33.2.1. Vehicle Registration Data

Vehicle registration records are maintained by national transport authorities of nearly all countries. These records provide detailed information on the vehicle fleet, vehicle classes, vehicle license, and ownership information, among others.

The vehicle files can be used to sample vehicles for surveys on vehicle usage, such as kilometer driven per annum, fuel usage, and owner characteristics. In this way, the national transport authorities can monitor vehicle population size and growth, as well as average and total vehicle-kilometers driven per annum and factors affecting ownership and usage. By merging automatic number plate data with vehicle register data, the owners can be surveyed to obtain trip information. However, for confidentiality reasons, some countries do not allow access to vehicle registration data for this purpose. In Finland it is allowed, however (Jarvi, 2008).

33.2.2. Vehicle Inspection Programs

Many countries have vehicle inspection programs in order to ensure that vehicles are maintained in a road-worthy condition. These programs are a potential source of odometer readings, from which vehicle-kilometers driven can be deduced.

The frequency of inspection varies from country to country, from annual inspections in Great Britain, Canada, Australia, and Brazil to inspections every two years in France and Germany. The year of first inspection ranges from two to five years. Some countries, such as South Africa, require only vehicle road-worthiness tests with change in ownership.

Vehicle inspection data are largely an untapped source of vehicle information. In France, for example, inspection odometer data have existed in electronic format since 1992, but there is no evidence that it has been used for transportation research (Kalinowska & Madre, 2009).

The data can provide longitudinal information for the monitoring of trends in vehicle-kilometers over time and cross-sectional information on vehicle-kilometers by age, vehicle class, and engine size.

33.2.3. Problems and Issues

The following problems with national administrative data records are experienced:

- The files may contain transcription or other data quality errors.
- Scrapped vehicles may not be removed promptly.
- Registration files may not represent the on-road population properly owing to fraud or illegal activity by vehicle owners (e.g., owners who fail to register their vehicles).
- Confidentiality of information requires authority to restrict access to the files.
- Not all countries allow access for sampling. (Name and address of vehicle owners are treated as sensitive and are not shared with researchers).
- Foreign-registered vehicles are not covered.
- Odometer readings are not managed as a database for planning, causing poor quality in the monitoring of odometer readings, tabulation, storage, validation, and dissemination.

33.3. Road Link Surveys

33.3.1. Survey Methods

The most basic vehicle-based surveys are those carried out on the road by means of traffic counts or roadside interviews. These surveys are relatively cheap and easy to conduct, and they provide the basic input to transport planning and modeling studies.

Vehicle number plates recorded in road link surveys can also be identified in national administrative data to provide more information on the vehicle and its owner.

Workshop participants mentioned examples of traffic counting and roadside interview methods in their countries and problems experienced with these.

33.3.2. Traffic Counts

Traffic counts are done in all countries in some form, ranging from simple, short-term manual counts to permanent, automatic counting stations. Traffic counting instrumentation has become increasingly more sophisticated, allowing the measurement of a range of vehicle characteristics, such as volumes in short time intervals, vehicle class, number of axles, axle weight, speed, and following distance and times. Sensors run the gamut from single pneumatic tubes, capable of counting only the total volume, to sophisticated loops, piezoelectric cables, and weigh-in-motion (WIM) weight pads capable of measuring the full range of traffic characteristics.

Video cameras recording vehicle movements at intersections or at the roadside allow manual measurement afterward at the office, or by means of pattern-recognition equipment.

Road authorities need to estimate the annual average daily traffic (AADT) on their road networks, as well as trends in AADT and vehicle-kilometers, for road management and planning purposes. This is done by well-positioned permanent counting stations, supplemented by short-term seven-day counts on a random sample of the network covering one-third or one-quarter of the total network, so that the total network is covered every three to four years. Traffic counts can vary widely by time of the day, day of the week, week of the month, and month of the year. A sampling strategy that would capture this variation requires a well-planned and managed counting program.

Short-term counts ranging from 12 h to 24 h to multiple-day counts are normally done for local feasibility studies and planning purposes. These counts are most valuable when they can be related to a well-managed counting program by road authorities. For instance, conducting a roadside survey is obviously much easier during daylight periods than at night. However, neglecting night traffic can introduce significant biases: for example, dissymmetry between southbound and northbound volumes of traffic in the 1989 French Border Survey (Kalinowska & Madre, 2009).

Traffic-counting data formats and the way they are summarized can vary widely, which often provides a challenge to planners wishing to compare counts between different sources. For example, road and vehicle classification, level of data aggregation by time period, direction and road lane, and day of the week, month and year are some of the formats that can vary. Proper data storage and provision of easy access to the data by road authorities are equally important.

Road and traffic authorities often collect data for other purposes, such as law enforcement, overload control of heavy vehicles, traffic control, and toll collection.

These data are often not recorded and stored for planning purposes. Proper management of these non-planning data sources can provide a wealth of additional information to planners without much extra cost.

33.3.2.1. Problems and issues The following problems and issues were mentioned:

- poorly planned and managed road network surveys, due to inadequate funding and resources, or lack of network counting programs
- data recorded not always matching the needs of the planner
- lack of standardization in data formats (e.g., vehicle classification) leading to difficulty in comparing information between regions within a country and between countries, for example between European Union countries
- poor archiving of data for later use and difficulty in accessing data
- difficulty in accessing non-planning data sources (e.g., for law-enforcement, traffic control, etc.)

33.3.3. Roadside Interviews

Roadside interviews (RSIs) are sometimes done in conjunction with traffic counts to provide information about vehicle use. A sample of vehicles is pulled off the road at a safe and convenient location with the assistance of traffic police, and drivers are interviewed using a short survey. Typical information recorded includes vehicle occupancy, type of commodity carried (for goods movement), trip purpose, origin and destination (O–D) of trip, departure and arrival times, route, and driver characteristics. The sample characteristics are scaled up to the population by relating the sampled records to the total number of vehicles passing by the intercept location.

The same survey principles apply to these surveys compared to household surveys, which have been discussed in other workshops; however, RSIs suffer from a number of deficiencies. One objection is that owing to safety concerns it is not possible to select geographic locations randomly. Heavy truck roadside surveys conducted in Canada, for example, typically take place at dedicated truck inspection stations possessing lighting and traffic controls suitable for pulling trucks off the road in a safe, practical manner. Unfortunately, inspection stations are few in number and are usually located on major highways in rural areas. This may lead to surveying a mix of vehicles and trip characteristics that is different from the general population. An additional challenge is that roadside surveying is an imposition on a driver's work time (i.e., the survey is administered while he is conducting his trip) and may lead to non-response or rushed, unthoughtful answers. Another important concern is the high cost of conducting RSIs in the field and the low productivity attendant with conducting operations at temporary sites.

A variation in the RSI methodology is providing the driver with the survey form to complete afterward at his/her leisure. Although more vehicles can be sampled, a reduced response rate and typical problems of self-completion surveys come into play. A few characteristics collected by the surveyor for all vehicles (e.g., size of the vehicle and region where it is registered, number of occupants, and age and gender of the driver) can improve the weighting process in the case of low response rates.

RSIs are often done to provide information on vehicles crossing selected screen-lines, cordon counts around a selected area of interest, or cross-border movements. This can reduce the spatial bias as all points of entry/exit are covered.

Examples mentioned were:

- Germany: cordon surveys regularly conducted every two years in cities (Kalinowska & Madre, 2009)
- South Africa: toll road feasibility studies (van Zyl & Raza, 2006)
- Canada/U.S. border crossing (Transport Canada, National Roadside Survey, 1999 and 2006)
- Brazil: heavy vehicle O–D surveys on state roads (Kalinowska & Madre, 2009)
- Finland: global system for mobile communication (GSM) survey using cellular phone records (Jarvi, 2008)

33.3.3.1. Problems and issues The following problems and issues were mentioned:

- quality of data suffers from the challenging survey environment
- refusal of traffic police to assist with roadside interviews in some countries, for example, France, or difficulty in getting good cooperation from traffic police due to core duties, for example, South Africa
- inconvenience to drivers, causing them to be less cooperative or provide low-quality information
- data archiving and data accessibility problems, similar to those encountered with traffic-counting data

33.3.4. Mobility Surveys

As part of conventional household interviews, vehicle occupancy and usage are often covered. The typical problems and solutions to these surveys are covered in other workshops, and only a few aspects were discussed.

Vehicle users may be requested to record odometer readings to determine kilometers driven over a period of time, recall information on past trip making, or complete a trip-by-trip diary or a fuel diary.

Another format is to recruit willing participants to record vehicle usage, often coupled to global positioning system (GPS) recordings.

One problem is that non-household usage is not covered by household sample surveys, such as business travel and travel to foreign countries.

In France, for example, a dedicated panel survey is done on an annual basis, while light-duty vehicles are surveyed every five years (Kalinowska & Madre, 2009).

Other problems include:

- unwillingness of respondents to participate
- difficulty of respondents in recalling information, or errors in respondents' measurements of data such as kilometers traveled over the past month and fuel consumption

33.4. New Technologies

33.4.1. Survey Methods

During the last few decades, various new technologies have become available that can be used to obtain more accurate and detailed information on vehicle characteristics and movements. These are briefly discussed below.

33.4.1.1. Global positioning systems GPS is widely used for vehicle security purposes, fleet management, and onboard route navigation. Accurate coordinates are provided for the vehicle's position in real time, which allows capturing of the vehicle's route, speed, travel time, distance, and elevation. Galileo is the European equivalent to the U.S. GPS, but with improved accuracy. It should be in service by 2012 and will give global coverage (Kalinowska & Madre, 2009).

Problems include:

- Ownership of data by security and GPS companies, along with privacy issues, makes access to data difficult. Fitting of GPS to a sample of volunteers solves this problem, but the study may be restricted to a small sample owing to the high cost of acquiring and deploying the devices.
- Though massive amounts of data are collected, the GPS traces usually require considerable cleaning before they can be used for analytic purposes.
- Physical obstructions, urban canyons and tunnels, and tinted windows can block signals.
- After a period of inactivity, GPS receivers may require up to five minutes to acquire a satellite fix before they start logging data (cold start issue). This produces gaps in the trip traces.
- Most GPS receivers are not physically linked to the vehicle and thus do not provide engine performance data (e.g., RPM, fuel consumption).
- Trip purpose is not known, but may be inferred by overlaying the "bread-crumb trail" (i.e., GPS coordinates) on land-use maps.

33.4.1.2. Cell phone tracking A cell phone's position can be traced to the nearest cell phone tower zone and, in principle, can provide the same information as GPS, but with much less accuracy (Friedrich et al., 2008; Jarvi, 2008). Recently, though, cell phones with assisted-GPS technology have been introduced, providing the same precision as a dedicated GPS receiver. Cell phone tracking is attractive because the user base is much higher than that of GPS. As with GPS data, however, getting access to the data is a problem.

33.4.1.3. Onboard diagnostics (OBD-II) Every car and light truck, model year 1996 and newer, manufactured for the North American market must have an OBD-II data port (located inside the passenger compartment) with which one can monitor engine performance and emissions. Devices have been developed that plug into the OBD-II port and collect instantaneous data (i.e., second by second) on vehicle speed, fuel consumption, operating temperatures, distance, and emissions (Kalinowska & Madre, 2009).

Problems include:

- Fuel consumption for certain classes of vehicle is difficult to measure (e.g., diesel engines, engines that estimate fuel using the speed-density method).
- It can be difficult to install an OBD-II device in the vehicle, as the data port is usually located under the dashboard in the driver's side footwell.

33.4.1.4. Electronic number plates A few countries are investigating or planning the use of electronic number plates for improved security and law-enforcement purposes. These plates provide additional benefits in terms of automatic and accurate classified traffic counts and route and O–D information. Privacy of users is a concern, and access to individual data will be restricted by authorities. However, authorities can make summarized data available.

33.4.1.5. Electronic toll tags Electronic tags fitted inside windscreens of vehicles are used for automatic tolling. Data may be used for planning, but access to data is a problem. Coverage is a problem, as not all vehicle owners make use of toll tags unless made compulsory, such as in Germany for heavy vehicles.

33.4.1.6. Weigh-in-motion WIM instruments allow measurement of axle loads while the heavy vehicle is in motion. WIM's high capital and maintenance costs allow only limited network coverage. The road pavement also needs to be in good condition in order for data to be measured accurately.

33.5. Data Gaps and Research Priorities

33.5.1. Data Gaps

The following data gaps have been identified with respect to motor vehicle activity:

- dynamic fuel use data for purposes of environmental assessment, that is, idling, cold starts, and distribution of fuel use by speed class
- on-road fuel consumption versus fuel consumption based on lab tests

- use of nonconventional vehicle types for sustainable energy use and environmental reasons
  - hybrid technologies (those using two sources of energy, such as petrol/diesel combined with electric power)
  - biofuels, for example, fuel made from sugarcane, grain produce, and so on
- accurate location of accidents
- accurate routing of trips
- accurate heavy vehicle weights
- linking movements to trip purpose

33.5.2. Research Priorities and Perspectives

The following research priorities have been identified.

33.5.2.1. Fuel consumption and vehicle emissions A challenge is responding to new research questions of current political as well as scientific interest. One major international concern is high energy costs and the "peak oil" issue, as well as global warming. Accurate information on the development of fuel efficiency across car fleets is an important research area within the context of CO_2 and other greenhouse gas emissions for the development of regulative policy mechanisms to reduce these externalities.

Existing "onboard diagnostics" (OBD) devices allow the measurement of the exact fuel and emission cycles together with geographic positioning parameters and speed–time profiles as well as other additional vehicle-use and technical vehicle information.

33.5.2.2. Road safety There is a growing research interest in the changing demographic profile of vehicle users, especially aging and shrinking population, and related aspects dealing with vehicle and traffic safety. Of importance are accident statistics, such as exact spatial and time profiles, nature of accidents, causal factors, relationship with vehicle ownership and use, and so forth.

The use of GPS by traffic police, proper recording and management of accident data, and making this data available to planners will satisfy these research needs.

33.5.2.3. Road user charging Road user charging for road funding and congestion management is a policy mechanism that is used to varying degrees in different countries. In the United States and South Africa, tolling of intercity highways has been generally accepted. The impact of road user charging on vehicle use and ownership is an increasingly important research topic. Accurate information on trip generation, route choice, mode choice, vehicle occupancy, departure times, travel times, and costs by route is an important factor.

33.5.2.4. Utilizing full potential of new technologies Various new technologies were discussed in the workshop. The use of many of these technologies for vehicle-based surveys is still in the pilot stage. With the increasing difficulty in conducting conventional user interview surveys (e.g., via telephone or using paper-based trip diaries), the automatic, passive measurement of vehicle usage and vehicle characteristics is becoming more important. The use of new technologies can reduce the burden, time, and cost of surveys significantly.

List of Participants

The workshop participants were Markus Friedrich, Germany; Tuuli Jarvi, Finland; Prokop Jehlicka, Germany; Dominika Kalinowska, Germany; Bob Leore (discussant), Canada; Jean-Loup Madre, France; and Klaas van Zyl, South Africa.

References

Friedrich, M., Jehlicka, P., & Schlaich, J. (2008). Automatic number plate recognition for the observance of travel behaviour. Paper presented at the 8th International Conference on Transport Survey Methods, Annecy, France.

Jarvi, T. (2008). A pilot of using automatic number plate recognition for OD survey of vehicles. Paper presented at the 8th International Conference on Transport Survey Methods, Annecy, France.

Kalinowska, D., & Madre, J.-L. (2009). Vehicle-based surveys: Towards more accurate and reliable data collection methods. In: P. Bonnel, M. Lee-Gosselin, J. Zmud & J.-L. Madre (Eds), *Transport survey methods: Keeping up with a changing world*. Bingley, UK: Emerald.

van Zyl, N. J. W., & Raza, M. (2006). In search of the value of time: From South Africa to India. In: P. Stopher & C. Stecher (Eds), *Travel survey methods: Quality and future directions* (pp. 457–484). Oxford: Elsevier.

Chapter 34

Survey Mode Integration and Data Fusion: Methods and Challenges

Caroline Bayart, Patrick Bonnel and Catherine Morency

Abstract

Data fusion and the combination of multiple data sources have been part of travel survey processes for some time. In the current context, where technologies and information systems spread and become more and more diverse, the transportation community is getting more and more interested in the potential of data fusion processes to help gather more complete datasets and help give additional utility to available data sources. Research is looking for ways to enhance the available information by using both various data collection methods and data from various sources, surveys or observation systems. Survey response rates are decreasing over the world, and combining survey modes appears to be an interesting way to address this problem. Letting interviewees choose their survey mode allows increasing response rates, but survey mode could impact the data collected. This paper first discusses issues rising when combining survey modes within the same survey and presents a method to merge the data coming from different survey modes, in order to consolidate the database. Then, it defines and describes the data fusion process and discusses how it can be relevant for transportation analysis and modelling purposes. Benefiting from the availability of various datasets from the Greater Montréal Area and the Greater Lyon Area, some applications of data fusion are constructed and/or reproduced to illustrate and test some of the methods described in the literature.

Transport Survey Methods

ISBN: 978-1-84855-844-1

34.1. Introduction

Data fusion and the use of multiple databases have appeared in the literature on transport surveys for some time (Goulias, 2000). Their use, however, considerably predates this, but without direct reference to data fusion methodologies. One familiar example is when a weight is applied to a sampling unit on the basis of a comparison with population data or when a synthetic population is generated from aggregate tables that define the reference world. Moreover, population synthesis, which has been discussed for some time in the scientific literature (specifically in the framework of microsimulation), raises similar issues to data fusion (Miller, 1996; Stopher & Greaves, 2003; Voas & Williamson, 2000; Beckman, Baggerly, & McKay, 1996; Ton & Hensher, 2003).

Integrating a number of data sources in the transport analysis and decision-making process is becoming increasingly complex, primarily because of the rapid change in the state, form and quantity of the available data. In the current context, in which information systems and technologies can be deployed rapidly and are practically becoming standard tools, the transport community is moving towards a state that Olsen (1999) has described as data chaos: a growing quantity of data, and increasing reliance "*on the quality and comprehensiveness of disaggregate travel data to support modern analytical procedures*" (Greaves, 2006), the development of very demanding transport models and increasing difficulties (cost, response rates) in collecting conventional transport data. Although data fusion or integration does not yet seem to be the standard approach for managing data chaos of this type, it represents a very useful methodological approach which should be examined in depth and, above all, discussed and trialled. A number of professionals are experiencing difficulties in effectively managing and exploiting the new sources of data at their disposal, in particular the data which are generated on a continuous basis by automated systems. They are increasingly aware of the need to organise their data better, to preserve them and to use them in an integrated manner in order to make them more useful for everyday operational management and planning tasks. Data fusion methods and methods for integrating data acquisition techniques must adapt to these new types of data.

Moreover, an increasing number of surveys are confronted by the problem of mixing different methodologies in the same survey. This occurs when, in order to deal with a reduction in the response rate, an alternative mode is proposed to non-respondents, or when a choice of mode is made available in order to reduce respondent burden (Yun & Trumbo, 2000; Morris & Adler, 2003). Making contact with specific groups in a survey which targets a wide audience can require ad hoc protocols to avoid an inadequate response rate or biased responses (e.g. with illiterate individuals in a self-administered survey). The issues which then arise are the comparability of the data and the methodologies which can be used to separate, on the one hand, those differences in responses that are due to the methodology, from, on the other hand, real differences in the behaviour of each group.

The paper begins by setting out the reasons for increasing concerns about combining different survey modes and data fusion (Section 34.2). Analysis will then focus on the combination of data obtained using different methodologies in the same survey (Section 34.3). We shall then turn our attention to the combined use of data from different sources. The subsequent section (Section 34.4) begins with a presentation of the concepts and principles involved in data fusion, and continues with a description of the informational issues involved and the methods used. The authors then propose some points for discussion and suggest some possibilities for further research (Section 34.5).

34.2. The Rationale for Data Fusion and the Integration of Survey Methods

The increasing difficulty of obtaining survey data that are representative of the target population and the increasing complexity of the data that are required to feed ever more sophisticated models mean that it is generally no longer possible to collect all the data during a single survey or with a single methodology. At the same time, new technology is potentially able to provide a large quantity of data that partially answers some of the issues facing us today. Processing and combining these different sources of data is becoming extremely important as a means of increasing our knowledge of behaviours and how they are changing, as well as improving models.

34.2.1. The Representativeness of Transport Surveys

Response rates for conventional surveys are tending to fall (Atrostic & Burt, 1999). The increasing number of non-responses observed in travel surveys is explained by a number of factors which are unlikely to disappear in the future. The increasing number of surveys conducted in recent years, in particular for commercial ends, has reduced the public's level of acceptance. Households are increasingly acquiring equipment that restricts the intrusion of 'strangers' into their private lives (caller ID, answering machines, entry phones, etc.), which makes it more difficult to make contact with them and increases the cost of recruitment (Zmud, 2003). Weariness with surveys, associated with anxiety about revealing personal information, is tending to increase refusal rates. This propensity for non-response is eroding confidence that survey results are representative of the studied population (Cobanoglu, Warde, & Moreo, 2001). A large number of techniques are available to attempt to limit non-responses, for example giving prior warning and reducing respondent burden. In spite of the undeniable value of these techniques, non-response biases remain. It is primarily the desire to reduce these biases that explains the increased use of data weighting methods. They nevertheless always involve the assumption that non-respondents with certain socio-economic characteristics behave like respondents with the same characteristics. However, a considerable amount of research casts doubt on this hypothesis (Ampt, 1997; Richardson & Ampt, 1993; Richardson, 2000; Murakami, 2004). Combining survey modes therefore seems to be a possible solution insofar as the individuals who respond to one medium are not necessarily the same as those who respond to another (Bonnel & Le Nir, 1998; Bonnel, 2003; Bayart & Bonnel, 2008).

The reduction in the completeness of most official lists of residents is a major problem in many countries. This is obviously the case for telephone surveys because of the increasing number of individuals who only own cell phones and who are generally not listed in the telephone directories, or the growing number of individuals who do not wish their telephone number to appear in these directories. Once again, combining survey modes can provide a way of getting around the problem, as in the case of the German and Belgian national transport surveys in which the surveys are conducted by CATI for that part of the sample for which it is possible to obtain a telephone number and by post for the rest (Bonnel & Armoogum, 2005; Hubert & Toint, 2003).

34.2.2. Increased Awareness of the Complexity of Transport Phenomena

Growing concerns about critical issues such as sustainable development, social equity and public health problems also encourage the integration of databases from a variety of fields. There is also an increasing amount of multidisciplinary research that examines the relationships between travel behaviours and planning, as well as their impacts on health (pedestrian safety, exposure to road traffic, pollutant emissions, air quality, active transport and obesity [Sallis, Frank, & Saelens, 2004]). Likewise, the appraisal of transport policies from the standpoint of sustainable development requires a consideration of economic, environmental and social dimensions. It is very rare for the existing databases to permit a combined analysis of all these dimensions. The production of ad hoc data remains very costly and encounters major technical difficulties because of the cumbersome nature of the questionnaires generated by the investigation of multiple issues of this type. The fusion of existing data from various sources thus provides interesting possibilities for enriching the collected data and conducting cross analysis.

34.2.3. The Increasing Availability of Observation Systems

Technological developments, particularly in the areas of computer-aided operating systems and decision-making aid systems, are changing the way transport networks and the components of these networks are managed and operated. A number of transport authorities have installed systems for the temporal and spatial monitoring of their network: advance vehicle location and passenger counting systems for transit vehicles, smart cards to access transit services, GPS on taxis or adapted vehicles, online reservation and follow-up systems for members of car-sharing groups (Cassias & Kun, 2007). In addition to network-related data, collective data such as cell phone traces (tracking individuals — requiring coupling with network data in order to deduce the transport mode), Bluetooth markers (tracking portable devices or vehicles) or GPS traces of private vehicles may become available, if confidentiality concerns do not prevent their use. This data has the potential to provide information about the use of transport networks by the fusion of spatio-temporal traces and geographic layers (matching position on the ground to the network). One of many possible examples is provided by the Belgian firm BE MOBILE (in collaboration with the Belgian Transport Ministry and a radio station) which provides information on the state of networks based on journey time data from a fleet of trucks fitted with GPS tracking devices. Another possibility is to analyse the behaviour of users when additional information is collected. This may be done, for example during a survey conducted with PDAs fitted with GPS tracking systems: the traveller uses the PDA to validate and complete the information (trip purpose, mode used, etc.), which is automatically collected by the GPS tracking system. (See, e.g. Kochan, Bellemans, Janssens, & Wets, 2007; Pendyala, 1999; Stopher, Fitzgerald, & Xu, 2007 or Stopher, Jiang, & Fitzgerald, 2005.)

Other network operators also produce a large amount of data that can have transport applications (tracking the location of cell phones). This data is collected on a continuous basis, often without travellers being aware of it. The data can become very interesting for the transport community if effective data processing and analysis methods are developed. First and foremost, these tools frequently constitute a technical or administrative goal in their own right (limiting fraud by monitoring the rights of smart card cardholders, validating the distances travelled in the case of GPS tracking systems installed in car-sharing vehicles). However, the data produced are not generally very suitable for behavioural analysis or feeding models. Furthermore, as a result of data confidentiality constraints, it is quite unusual for the data to contain contextual information. This limits their potential use without input from other source to provide further confirmation of the value of data fusion methods.

34.3. The Mixing of Survey Modes or Methods

An increasing number of surveys are based on complex protocols which combine several modes or methodologies to increase the overall response rate, improve the rate of coverage of the target population or enhance the quality of the data that are produced (Couper, 2000; Gunn, 2002; Dillman et al., 2001). But proposing several data collection modes or methods carries a risk. The collection of information from different sources may provide results that lack comparability (Dillman & Christian, 2005). The danger when databases are merged is that a sample selection bias will be created that will compromise the accuracy of explanatory models of travel behaviours. This type of selection bias has received considerable coverage in the literature, from both theoretical and empirical standpoints (Winship & Mare, 1992), but as yet little attention has been paid to it with regard to transport surveys.

34.3.1. Sample Selection Bias

It has been known since the 1950s that estimating an equation on the basis of a subsample selected from the population may result in biases (Roy, 1951). However, the first econometric exploration of the consequences of such sample selection was Heckman's work in 1974. It was a female labour supply model, in which the researcher estimates the determinants of wage offers, but has only access to wage observations for those who work. Since people who choose to work are selected non-randomly from the whole population, estimating the determinants of wages from the subpopulation that chooses to work may introduce bias. Many papers have highlighted the importance of the selection bias in human and social science surveys (Maddala, 1986). Noteworthy examples are the model for migration in the United States analysed by Nakosteen and Zimmer (1980), or that for female employment rates analysed by Mroz (1987). In practice, the selection bias has two sources (Heckman, 1990). It results either from respondent self-selection or a selection decision by the study managers. When mixed survey modes are used, individuals choose to belong to one group or another or only respond if the proposed medium suits them. The responses are therefore not comparable, because the sample is no longer random and the presence of respondents is determined by external factors which may also affect the variable of interest in the studied model. It is highly likely that the socio-economic characteristics and the travel behaviours of the individuals who respond using the Internet are different from those of the individuals who respond to a face-to-face interview (Resource system group, 2002; Lozar Manfreda & Vehovar, 2002).

The Laboratoire d'Economie des Transports has conducted an Internet survey of non-respondents to the 2006 Lyon face-to-face household travel survey (Bayart & Bonnel, 2007); that is to say individuals who refused to allow an interviewer into their home or who could not be contacted during the first wave of interviews. The data from this survey highlight the problem of self-selection, with the respondents to the standard face-to-face survey and the non-respondents of the face-to-face survey who have responded to the Internet questionnaire. We shall illustrate this by proposing an explanatory model for travel. Let us consider an equation that permits an analysis of the effect of survey mode on an individual's average number of daily trips:

$$Y_i = \sum_k \beta_k X_{ki} + \alpha I_i + u_i \tag{34.1}$$

where Y_i is the average number of trips made by individual i (dependent variable), X_i a vector of explanatory variables, I_i a dummy variable that states whether the individual responded by Internet and u_i the error term. A question arises is whether the coefficient α measures the real impact on daily travel of the choice of responding to face-to-face interviews or on the Internet. The answer to this question is affirmative if individuals who decide to respond on the Internet would have reported the same number of trips if they had responded in the face-to-face situation. However, the variable I cannot be considered as exogenous in this model, as the contacted individuals chose whether or not to respond by face-to-face interaction. Respondent self-selection must be corrected during least-squares regression in order to obtain unbiased estimates of the coefficients. By using the two-stage estimation method developed by Heckman (1979), our explanatory model for travel can be formalised as follows for each individual *i*:

$$Y_{1i} = \sum_k \beta_{1k} X_{1ki} + u_{1i}, \quad \text{for Internet respondents} \tag{34.2}$$

$$Y_{2i} = \sum_k \beta_{2k} X_{2ki} + u_{2i}, \quad \text{for face-to-face respondents} \tag{34.3}$$

where Y_{1i} and Y_{2i} are the average number of trips made by individual *i*, X_{1i} and X_{2i} the two vectors of independent or explanatory variables for travel and u_{1i} and u_{2i} the two error terms which take into account of the unobserved forces possibly influencing results. We assume that u_1 and u_2 are normally distributed. The selection function that expresses the probability that an individual *i* will respond on the Internet can be written as follows:

$$I_i^* = \sum_m \delta_{mi} Z_{mi} + \varepsilon_i \tag{34.4}$$

$I_i = 1$, if and only if $I_i^* > 0$,

$I_i = 0$, if and only if $I_i^* \leq 0$,

where I^* is the selection variable, Z_i a set of variables that determines choice of the Internet and ε the error term which is assumed to be normally distributed. The observed travel for any individual *i* is defined as

$Y_i = Y_{1i}$, if and only if $I_i = 1$,

$Y_i = Y_{2i}$, if and only if $I_i = 0$.

It should be noted that only one of the parameters Y_{1i} and Y_{2i} is observed, depending on whether the individual chooses to respond in the face-to-face situation or on the Internet. In contrast to Eq (34.1), this model does not mean that the coefficients for the explanatory variables X_{1ki} and Z_{mi} are necessarily identical for the Internet respondents and the face-to-face respondents. Moreover, we assume that the error terms for the Eqs. (34.2)–(34.4) (u_1, u_2 and ε) obey a bivariate normal distribution with zero averages and correlations ρ_1 and ρ_2. Let us just consider all the individuals with a given (X_i, Z_i). Formally, the regression of Y_i on X_i in the truncated sample[1] is

$$E(Y_{1i}|I_i = 1) = E(Y_{1i}|X_{1ki}, Z_{mi}, I_i = 1) = \sum_k \beta_{1k} X_{1ki} + E(u_{1i}|Z_{mi}, I_i = 1) \tag{34.5}$$

$$E(Y_{2i}|I_i = 0) = E(Y_{2i}|X_{2ki}, Z_{mi}, I_i = 0) = \sum_k \beta_{2k} X_{2ki} + E(u_{2i}|Z_{mi}, I_i = 0) \tag{34.6}$$

1. A truncated sample is one where the observations are only made for certain individuals, who form a subset of the observed population (Tobin, 1958).

We have imposed normalization on the variance of ε_i. Let $\sigma_{\varepsilon i} = 1$.[2] Under the normality hypothesis, we can write[3] that $u_{1i} = \rho_1 \sigma_{u1i} \varepsilon_i$ and $u_{2i} = \rho_2 \sigma_{u2i} \varepsilon_i$. By replacement in Eqs. (34.5) and (34.6):

$$E(Y_{1i}|Ii = 1) = \sum_k \beta_{1k} X_{1ki} + \rho_1 \sigma_{u1i} E(\varepsilon_i|\varepsilon_i > -\sum_m \delta_m Z_{mi}) \tag{34.7}$$

$$E(Y_{2i}|Ii = 0) = \sum_k \beta_{2k} X_{2ki} + \rho_2 \sigma_{u2i} E(\varepsilon_i|\varepsilon_i \leq -\sum_m \delta_m Z_{mi}) \tag{34.8}$$

By definition, we have

$$E(\varepsilon_i|\varepsilon_i > -\sum_m \delta_m Z_{mi}) = \frac{\phi\left(-\sum_m \delta_m Z_{mi}\right)}{1 - \theta\left(-\sum_m \delta_m Z_{mi}\right)} \tag{34.9}$$

and

$$E(\varepsilon_i|\varepsilon_i \leq -\sum_m \delta_m Z_{mi} = -\frac{\phi\left(-\sum_m \delta_m Z_{mi}\right)}{\theta\left(-\sum_m \delta_m Z_{mi}\right)} \tag{34.10}$$

By replacement in expressions (34.7) and (34.8), we obtain[4]:

$$E(Y_{1i}|I_i = 1) = \sum_k \beta_{1k} X_{1ki} + \rho_1 \sigma_{u1i} \frac{\phi\left(\sum_m \delta_m Z_{mi}\right)}{\theta\left(\sum_m \delta_m Z_{mi}\right)} = \sum_k \beta_{1k} X_{1ki} + \rho_1 \sigma_{u1i} \lambda_{1i} \tag{34.11}$$

$$E(Y_{2i}|I_i = 0) = \sum_k \beta_{2k} X_{2ki} + \rho_2 \sigma_{u2i} \frac{-\phi\left(\sum_m \delta_m Z_{mi}\right)}{1 - \theta\left(\sum_m \delta_m Z_{mi}\right)} = \sum_k \beta_{2k} X_{2ki} + \rho_2 \sigma_{u2i} \lambda_{2i} \tag{34.12}$$

The functions ϕ and θ are, respectively, density and distribution functions of the normal distribution. The ratios λ_{1i} and λ_{2i} are known as the inverse Mills' ratio, λ is

2. In view of the nature of the data, only the sign of I_i^* can be observed and not its size, which means that it is not possible to estimate the variance of Eq. (34.4) (Cameron & Trivedi, 2005).
3. In fact, $\text{cov}(u, \varepsilon) = \rho \sigma_u \sigma_\varepsilon = \rho \sigma_u$ and $\text{cov}(\rho \sigma_u \varepsilon, \varepsilon) = \rho \sigma_u V(\varepsilon) = \rho \sigma_u$.
4. It should be remembered that $\phi(-\sum_m \delta_m Z_{mi}) = \phi(\sum_m \delta_m Z_{mi})$ and $1 - \theta(-\sum_m \delta_m Z_{mi}) = \theta(\sum_m \delta_m Z_{mi})$.

therefore a decreasing monotonic function of the probability that an observation will be selected.

It is now possible to estimate the travel functions without the sample selection bias, including the inverse Mills' ratio in expressions (34.2) and (34.3). Let the model be:

$$Y_{1i} = \sum_k \beta_{1k} X_{1ki} + \rho_1 \sigma_{u1i} \lambda_{1i} + e_{1i}, \quad \text{for Internet respondents} \tag{34.13}$$

$$Y_{2i} = \sum_k \beta_{2k} X_{2ki} + \rho_2 \sigma_{u2i} \lambda_{2i} + e_{2i}, \quad \text{for face-to-face respondents} \tag{34.14}$$

The selection bias therefore corresponds to a missing value bias. In fact, if expressions (34.2) and (34.3) are estimated by the ordinary least-squares method, two variables are omitted (respectively, $\phi(\sum_m \delta_m Z_{mi})/\theta(\sum_m \delta_m Z_{mi}) = \lambda_{1i}$ and $-\phi(\sum_m \delta_m Z_{mi})/1 - \theta(\sum_m \delta_m Z_{mi}) = \lambda_{2i}$), and we can anticipate that the model will be biased (the estimations of β_{1k} and β_{2k} will be non-convergent). Moreover, it is likely that the size, sign and significance of the coefficients differ when they are estimated using the two-stage method. The differences depend on the coefficients $\rho_{1i}\sigma_{u1i}$ and $\rho_{2i}\sigma_{u2i}$ and the coefficients of the relevant variables estimated in the selection model (Hoffman & Link, 1984). The parameters of the sample selection model can be estimated using the maximum likelihood method. However, Heckman's two-stage procedure (1979) is more often used. The first stage consists of estimating the selection equation using a probit model in order to obtain estimations of the δ_i values. For each selected observation, the model calculates the value of λ_{1i} or λ_{2i} (the inverse Mills' ratio). The second stage consists of estimating the parameters β_{1k} and $\rho_1\sigma_{u1i}$ by ordinary least-squares regression of Y_{1i} on X_{1ki} and λ_{1i} and the parameters β_{2k} and $\rho_2\sigma_{u2i}$ by ordinary least-squares regression of Y_{2i} on X_{2ki} and λ_{2i}.

The t-test is applied to identify any selection bias in the model ($\rho_1\sigma_{u1i}$ and $\rho_2\sigma_{u2i} = 0$). If it is not possible to reject the null hypothesis of an absence of selection bias in the model, we can apply the ordinary least-squares method to estimate the coefficients β_{1k} and β_{2k} directly (Berk, 1983). If the t-value is above the critical value, the effect of selection is significant and the two-stage method can provide unbiased estimates of β_{1k} and β_{2k}.

The interpretation of the coefficients $\rho_1\sigma_{u1i}$ and $\rho_2\sigma_{u2i}$ is complex, but interesting. The inverse Mills' ratios (λ_{1i} and λ_{2i}) are by definition positive. However, $\rho_1\sigma_{u1i}$ and $\rho_2\sigma_{u2i}$ may have either sign. If $\rho_1\sigma_{u1i}$ is negative, the individuals who choose to respond on the Internet report a lower average number of trips than individuals randomly selected in the sample. If $\rho_2\sigma_{u2i}$ is negative, the individuals who decide to respond to the face-to-face survey report a higher number of trips than randomly selected individuals. The opposite applies if $\rho_1\sigma_{u1i}$ and $\rho_2\sigma_{u2i}$ are positive.

Some criticisms have been made of the selection model, particularly with regard to the normality hypothesis (Lee, 1982). Alternative approaches, based on nonparametric estimators, are proposed in the literature. These would provide a means of dispensing with the hypothesis of normality. However, the results obtained differ little from those obtained with Heckman's parametric model (Greene, 2002), and the weaker hypotheses of the model generate less robust results (Winship & Mare, 1992). In addition, the difference between the parameter estimations obtained with the two-stage model and the ordinary least-squares method can be explained by a high degree of colinearity between the variables X_{ki} and Z_{mi}. The two-stage method is therefore a compromise between a selection bias and an error due to the introduction of a regressor, which is highly correlated with the explanatory variables in the model[5] (Stolzenberg & Relles, 1997).

34.3.2. Application to the Lyon Household Travel Survey

In our example, we have stated the hypothesis that the data collection mode has an impact on individuals' travel. The econometric model that is envisaged for this study is a sample selection model, the parameters of which we shall estimate with a two-stage procedure.

The first stage consists of estimating the survey medium 'choice' equation using a probit model. Before commencing the econometric analysis, we shall select variables which make it possible to distinguish the individuals who responded on the Internet. Two types of variables drew our attention in this context: the household's characteristics as regards to telecommunications (possession of an Internet connection at home or a cell phone and presence in the directory) and some socio-demographic characteristics (status, age, educational level, whether they stated annual income, number of persons in household and place of work). The coefficients estimated by the probit model are significant (Table 34.1). An examination of their sign gives a good idea of the importance of the various factors which influence choice of the Internet. Unsurprisingly, the possession of an Internet connection in the home or a cell phone and the absence of a subscriber from the directory increase the probability of responding to the survey by Internet. Likewise, the probability of responding online increases when individuals report their annual income, work in the city centre and have gone through higher education. Conversely, the probability of using the Web decreases with smaller household size and for individuals currently in school. This first stage also provides a selection correction factor for each individual, which is known as the inverse Mills' ratio.

The second stage consists of explaining the differences in travel behaviour using a specific model which includes a dependent variable (the average number of trips made by individuals), several independent or explanatory variables (the observed factors assumed to have an effect on the number of trips reported by individuals), the inverse Mills' ratio (a variable obtained in the first stage) and an error term which

5. In order to avoid a high degree of colinearity between X_{1i} and λ_{1i} and between X_{2i} and λ_{2i}, it is recommended that at least one explanatory variable in the selection equation should not be present in the equations of interest.

Table 34.1: Selection model (probit).

	Estimate	**Pr (>|z|)**
Sample size: 13,483		
Intercept	−2.408	<2e−16***
Age	0.0046	0.036*
Internet connection at home	0.424	7.0e−10***
Ownership of mobile phone	0.331	9.4e−06***
Telephone_ in directory	0.370	3.6e−11***
Telephone_no fixed line	0.317	0.00074***
Occupation_working unstated location	−0.859	0.0014**
Occupation_working in suburbs	−0.187	0.0031**
Occupation_not working	−0.291	0.00016***
Number of individuals in household	−0.098	1.3e−06***
Declared income	0.294	9.0e−07***
Educational level_ongoing	−0.387	0.0026**
Educational level_not higher	−0.358	5.2e−10***

Significance codes: 0 '***' 0.001 '**' 0.01 '*' 0.05 '.' 0.1 ' ' 1.

takes into account of the unobserved forces that could influence the measurement of results.

What we will do is estimate two models, one using the subsample of face-to-face respondents, and the other using the subsample of Internet respondents (another possibility is to include the variable that indicates the response mode in the model). The coefficients of the explanatory variables and the selection bias variable are estimated by a least-squares regression method (Table 34.2). We shall only consider here the explanatory variables with a direct impact on individuals' travel (gender, age group, residential location, number of individuals and children in the household, possession of a driving licence, number of cars in the household, the day individuals responded to the questionnaire and willingness to report their annual income). In each model, the coefficient of the selection bias variable is significant. Individuals are therefore subject to a selection bias, and this bias can be eliminated by using a two-stage estimation method. We thus obtain unbiased estimates.

The model that is restricted to the subsample that responded to the face-to-face survey provided interesting results, as all the variables are significant. Being a man and belonging to a large household has a negative impact on the propensity to travel. In contrast, living far from the centre, having cars available and having a driving licence increases the average daily number of trips. Travel also seems to be an increasing function of age up to 40 years of age (the square of age is also included in the model to take into account of the decrease in travel among individuals over the age of 40–50; this distribution better fits observed data) and of the number of children in the household. The Mills' coefficient has a negative sign: those who responded to the face-to-face survey would have reported a lower level of travel if they had responded to the Web survey.

Table 34.2: Travel model (OLS).

	Face to face		Web	
	Estimate	Pr (>|t|)	Estimate	Pr (>|t|)
Sample size: 13,155 (face to face) and 328 (Web)				
Intercept	2.56e + 00	<2e – 16***	4.68e + 00	0.0037**
Gender_male	– 1.44e – 01	0.0005***	– 7.22e – 01	0.0045**
Possession of driving licence	6.65e – 01	<2e – 16***	8.83e – 01	0.0821.
Number of cars per person ≥ 18 years	5.15e – 01	<2e – 16***	5.95e – 01	0.0999
Reference day Friday	1.93e – 01	0.0007***	– 5.00e – 01	0.0440*
No. of children in household	7.40e – 01	<2e – 16***	7.80e – 01	0.1102
(No. of children in household)^2	– 5.53e – 02	9.2e – 06***	– 4.36e – 01	0.0125*
No. of persons in household	– 2.18e – 01	<2e – 16***	4.87e – 01	0.0023**
Residential location/centre	1.07e – 05	0.0108*	– 6.24e – 05	0.0226*
Declared income	2.34e – 01	5.1e – 07***	2.68e – 01	0.4379
Age	5.59e – 02	<2e – 16***	– 4.09e – 02	0.3878
(Age)^2	– 6.97e – 04	<2e – 16***	3.72e – 04	0.4658
Mills	– 2.55e – 01	4.6e – 05***	– 1.11e + 00	0.0083**

Significance codes: 0 '***' 0.001 '**' 0.01 '*' 0.05 '.' 0.1 ' ' 1.

In the model that is restricted to the subsample of individuals who responded online, few of the coefficients achieved significance. This is explained in particular by the differences in the sizes of the two samples of respondents. A total of 13, 271 individuals responded to the face-to-face survey, compared with only 369 on the Web, that is a ratio of 1:36. The order of magnitude of the standard deviation of the estimated coefficients therefore varies in a ratio of 1:6 between the Web and face-to-face samples. Being male, taking Friday as the reference for trips and living far from the centre have a negative impact on travel (generally, women make more 'shopping' and 'accompanying' trips). However, the number of trips increases with household size. The Mills' coefficient has a negative sign: those who decided to answer on the Web would have reported higher daily trips if they had responded to the face-to-face survey.

34.4. The Data Fusion Process

34.4.1. Definition

Data fusion aims to increase the amount of information that can be obtained from a source using data which are available from another source (Gautier, 1999; Judson, 2006; Saporta, 2002). This process is also referred to as *data integration* or *merging.*

Statistical matching and *file grafting* are other commonly used terms that relate to the process of data fusion in which files are combined using linkages between similar records. The generic problem in data fusion is to make statistical inferences about the joint distribution of two sets of variables when no direct observation of this distribution is available. The relevant information exists separately in each of the two databases and some variables are available in both (common variable vector). The data fusion problem involves imputing a value for a variable, which is missing for all the observations.

Data fusion is performed when no database contains all the relevant data required for modelling. In fact, 'in order to obtain a more complete answer to their research questions, researchers frequently have to use more than one dataset. The type of analysis possible, however, is often constrained by the fact that each dataset possesses one desirable attribute while being deficient elsewhere" (Peters, Ekin, Leblanc, Clark, & Pickles, 2006).

34.4.1.1. Principles and conceptual framework The standard data fusion approach creates a set of combined data that contains the variables of interest (X, Y) as well as the common (or gateway) variables (Z). D'Orazio, Di Zio, and Scanu (2006) have identified two approaches to data fusion: the micro approach and the macro approach. The first involves the construction of a complete synthetic file that contains all the necessary data, even if some of them have not been observed directly. The second approach refers to the process by which the initial files are used to estimate the joint distribution functions (or similar parameters such as the correlations) of the variables of interest. In fact, the accuracy of the data fusion process is highly dependent on the ability to select a valid procedure for estimating the joint distribution between the different variables; as described by Polak (2006), 'the key to combining data is to be able to specify the joint probability distribution (PDF — probability distribution function) of all the parameters of interest' and, 'in most cases, the PDF is not easily available'.

34.4.1.2. Standard data fusion models Combining data from different sources can be straightforward when records can be matched without uncertainty, using a unique identifier such as the social security number or passport number. This approach is known as exact matching (record linkage). An example of record linkage can be found in Johansen (2005). However, exact record matching is generally not possible because the databases are not perfectly superimposed, or simply because the variables which could make it possible are not available. In such cases, statistical matching procedures are used to combine the data that is available in a number of sources (D'Orazio et al., 2006).

D'Ambrosio, Aria, and Siciliano (2007) have identified two types of data fusion model:

- *Explicit models*, that is using a model connecting the 'Y interest' variables with the 'Z gateway' variables in the donor file and *applying this model to the receptor file* where the Z variables are linked to the 'X interest' variables.

- *Implicit models*, that is finding for each individual in the receptor file one or more donor individuals that are as similar as possible, and then *transferring the values of the Y variables to the receptor individual*. In this approach, each record from one of the data sources is matched with a record from a second data source that generally does not represent the same unit but a similar or compatible unit. 'The matching process requires the definition of a distance function which permits the similarity of any pair of cases to be assessed in terms of a function D_{ij}, of the Z variables' (Rodgers-Willard, 1984). The most common method for identifying statistical twins (donor and recipient units that share similar common variables) is to apply nearest neighbour procedures (a measure of the similarity or distance between units observed in distinct datasets). In the framework of the MOBIDIC project (Desmet et al., 2007), the authors have applied a procedure which transferred a weighted average of the Y variables in the donor file, with the weighting depending on the degree of similarity between the observations. Aluja-Banet, Daunis-I-Estadella, and Pellicerc (2007) have presented GRAFT, a data fusion system based on the k-nearest neighbour (k-nn) imputation method.

The fusion of databases requires the identification of a LINK (gateway variable) which connects them; this compatibility may be spatial (same zone, nearest location, same node), temporal (same day, same time period, same year) or semantic (same or similar household or unit). Furthermore, the linkage between these datasets depends both on the resolution or granularity of the available data and the data to be produced.

34.4.1.3. Data fusion in the transport field: the opus project The OPUS project is without doubt the largest initiative that deals with data fusion in the transport field. OPUS stands for 'Optimizing the use of Partial Information in Urban and Regional Systems'; its principal goal is 'to contribute to improved decision making in the public and private sector within Europe by developing innovative statistical and database systems to enable the combination of data from disparate, cross-sectoral sources'. This project proposes a conceptual framework based on three principles: (1) the combination methods must be applied at a fairly high level of abstraction, (2) it should make use of existing theory as formulated in quantitative models and (3) it must use a statistical approach (insofar as possible and practical).

In the framework of this project, Chalasani and Axhausen (2005) have proposed a conceptual model for integrating transport survey data and spatial data based on Chen's entity-relationship approach (1976). The usefulness of object-oriented modelling has also been demonstrated in a transport context (see TOOM in Chapleau & Trépanier,1997 and Trépanier & Chapleau, 2001a).

34.4.2. Illustrations

There are many examples of the application of fusion methods in the areas of medical imaging, non-destructive testing (materials), remote sensing, signal coupling (Wei &

Lee, 2005), image or writing recognition (see in particular More & Ingman, 2008) or market studies.

However, the data fusion problem appears to be even more complex for the data that are available in the field of transport, since they generally need to be considered at a very high level of resolution and simultaneously in all the dimensions (spatial, temporal and semantic) as these dimensions are critical for the modelling of behaviour. Consequently, there are frequently one or more incompatibilities when data from different sources are combined and this will frequently reduce the precision of the data generated by fusion (due to the aggregation of data and/or the use of an imputation model). This may be because of partial overlaps or spaces between the zones, disparities between data observation periods (continuously collected data vs. transport surveys) or references (all residents vs. the users of a given service) or incompatibility between the coding classes (spatial analysis zones, definition of variables or age groups). This means data must be processed beforehand in order to make them compatible.

34.4.2.1. Illustration: the fusion of subsamples from transport surveys The customary way transport survey samples are used, that is the aggregation of all observations in an average weekday is little questioned. During these surveys, all the interviews are not conducted on the same day. For example during the most recent O–D survey in the Greater Montréal Region (2003), almost 100 different travel days were aggregated.

In this data fusion problem, Z represents a vector of attributes that were directly measured and Y_i represents a vector of travel behaviour variables (trip sequence, modal choice, mileage, etc.) for each type of day $i =$ Monday, Tuesday, …, Friday. The goal of data fusion is to construct an integrated database that contains the travel behaviour variables for the five weekdays for a set of individuals. The fusion process involves selecting attribute vectors from the five files to be combined, that is the common variables Z (gateway). Figure 34.1 shows the data fusion process, which resembles a missing data problem. In this case, the five survey subsamples are used to construct a synthetic file that contains, for all the individuals in the sample, a description of the travel choices made on five days. In addition, a preliminary classification is performed to simplify the process by checking for certain attributes of the individual's household (Y, HH), in particular residential location (zone of residence) and household type (size and car ownership). The imputation of average behaviours per type of day can be performed with a model that estimates the relationship between the demographic attributes of the individual ($Z(P)$), such as age, gender and principal occupation, or by coupling individuals on the basis of semantic proximity. From a statistical point of view, it is necessary during this process to ensure that each subsample is representative of the population.

34.4.2.2. Illustration: combining travel surveys (microdata) and census data (zonal) Considering migration datasets from Australia and the United Kingdom, Duke-Williams and Blake (1999) stated that: 'Due to the differences in the

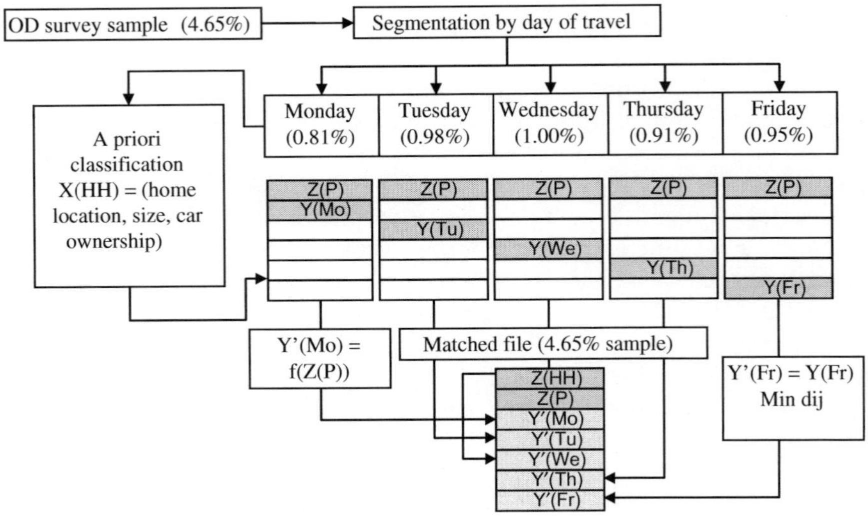

Figure 34.1: Fusion method for transport survey samples.

data a considerable amount of processing is required — dis-aggregation, estimation and re-aggregation — in order to achieve some degree of initial comparability".

There are few examples that explicitly involve the fusion of data from datasets that are in principle incompatible with regard to the granularity of the available data. Morency (2004) has presented a data fusion methodology for integrating microdata from the Montréal O–D surveys and the aggregated data from Canadian censuses in order to enrich the observations of variables which are not measured in transport surveys. This process is shown in Figure 34.2.

Briefly, this fusion process, which involves the weighting of all observations, involves (1) the spatial association of records on the basis of the residential location of each household and the area covered by each sector; (2) for each zone, the estimation of joint distributions for household size (four classes) and population segments (11); (3) the imputation of probabilities of having a certain attribute for each household and person in the O–D survey on the basis of type of household or people and factoring the weights in consequence.

34.4.2.3. Illustration: data fusion between cross-sectional data and continuous data The initial objective of smart card systems is to validate access to public transport networks. To do this, they collect, on a continuous basis, the various access transactions and by virtue of this fact constitute an impressive database on the spatio-temporal use of the different lines and routes. Figure 34.3 identifies the principal incompatibilities and provides some examples of possible objectives for fusion between these two datasets.

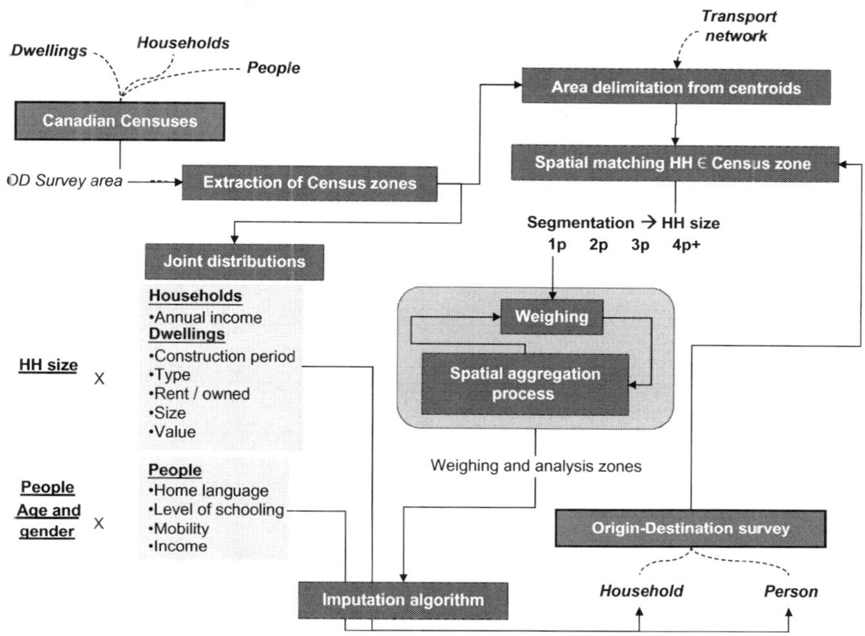

Figure 34.2: Data fusion method for aggregated census data and O–D survey microdata.

34.4.3. *Assessing the Quality of Data Fusion*

Assessing the quality and validity of the datasets outputted from a data fusion process is definitely one of the main challenges. According to Gautier (1999), the quality of data fusion depends primarily on two factors:

- the existence of common data (Z) in both data sources;
- the strong linkage between these common data (Z) and the variables to be transferred (Y).

Kiesl and Rassler (2006) have identified four levels of validity for data fusion:

1. *Preserving marginal distributions*: the first and lowest level of validity is attained if the marginal distribution (f′y = fy) and the joint distributions (f′y, z = fy, z) in the donor file are preserved in the fused file. This validation is performed using statistical tests such as the $\chi 2$-test or the t-test;
2. *Preserving correlation structures*: the second level of validity is achieved if, in addition, the correlation structure is preserved ($\mathrm{cov}'(X, Y, Z) = \mathrm{cov}(X, Y, Z)$);
3. *Preserving joint distributions*: the third level of validity is achieved if the joint distribution is preserved after data fusion (f′x, y, z = fx, y, z);

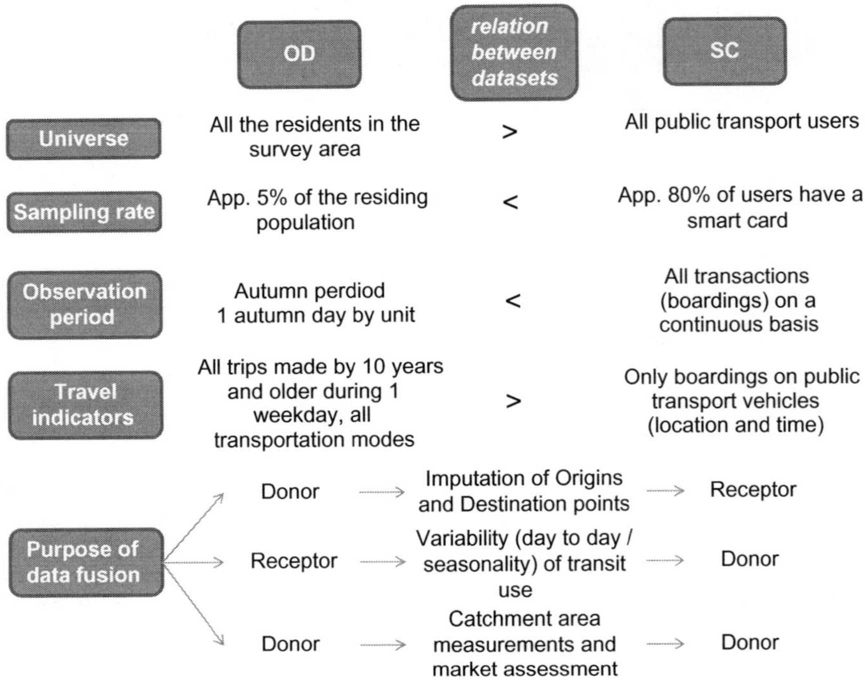

Figure 34.3: Comparison between O–D survey and smart card data for fusion purposes.

4. *Preserving individual values*: the fourth and final level of validity is achieved when the true values are reproduced ($y'_i = y_i$ for all the observations i in the receptor file).

In reality, the last level of validity cannot be attained (except in the very special, but very rare, case of exact matching). In fact, data fusion algorithms are able to verify the first or the first two levels of validity but they are unable, in practice, to verify the third and fourth levels.

Chesher and Nesheim (2006) have proposed an extensive review of the literature on the statistical properties of linked datasets. In this connection, More and Ingman (2008) have confirmed that 'The process of data fusion as such is not characterised by any quality characteristics, which makes it impossible to estimate the effectiveness and reliability of the fused data'. Although their work is concerned with the fusion of data signals, it is nevertheless relevant. These scholars emphasised the lack of indicators for evaluating the quality and reliability of data fusion processes and described a tool that provides a qualitative assessment of the results of the fusion. They propose an alternative approach for evaluating the process based on the concept of relevant information (identifying the data that should be produced by the fusion process and, above all, their level of resolution), the concept of information loss in the fusion process (ratio between the information that is reconstructed by fusion and the information obtained from measurements or surveys) and the concept of an indicator of information divergence, which is a concept taken from information theory.

34.5. Conclusion and Research Issues

This paper has documented the issues that relate to the combination of survey modes and methods and discussed issues relating to the combined exploitation of data from distinct sources. On the basis of the experiments and the concepts that have been discussed, some possibilities for research will now be described in order to feed debate and initiate new research.

34.5.1. Mixing Survey Modes and Methods

In view of how difficult it is to recruit respondents (increase in non-response rates, cost of interviews) and societal changes (the increasing number of surveys, discontinuation of fixed telephone lines, etc.) and technological innovations (Internet, GPS), it appears to be appropriate to combine different modes or methods in the same survey in order to improve the quality and statistical representativeness of the data. However, this combination process raises the problem of *data comparability*. Attention has been paid to this issue in certain disciplines, but it has not yet received much coverage in the field of transport. We propose that the two-stage model proposed by Heckman in 1974 should be adapted. Application to the Lyon household travel survey, in which two survey media were used, has revealed the presence of a *selection bias*. The traditional estimation of a travel equation using the ordinary least-squares method for the entire population therefore ignores an important factor, the determinants of survey mode choice. On the other hand, taking account of this selection bias provides a means of estimating unbiased relationships. It would be beneficial to develop this type of approach in the area of travel surveys in order, ultimately, to propose methods for combining samples obtained with different methodologies in order to increase the statistical representativeness of the surveys and reduce the non-response problem.

34.5.1.1. Directions for further research on the 2006 Lyon household travel survey

- Test whether the estimated coefficients are statistically different in the two groups (face-to-face and Internet respondents).
- Perform a sensitivity analysis to estimate the conditional value of the daily number of trips made by individuals with a variety of values for the independent variables.
- Estimate the difference in the travel behaviours of the two groups by evaluating the difference between the average number of trips that the Internet respondents would have reported in a face-to-face interview and the number of trips they

actually reported on the Internet. It would then be possible to allocate 'correcting coefficients' to the Internet respondents with a view to integrating this sample with the file of face-to-face respondents in order to correct the entire resulting file to make it representative of the population.
- Implement a multidimensional analysis to take into account of trips' characteristics (mode, travel purpose, etc.) in the two-stage model.

34.5.1.2. The fusion of distinct sources of data Data fusion raises many issues (Venigalla, 2004) that should be studied in greater detail in the specific framework of the modelling of travel behaviour. We propose the following classification for this:

Informational and methodological issues: The availability and compatibility of databases, storage and updating, on a continuous basis, of administrative and reference databases:

- Better identifying the reference world for each database and ensuring the continuous storage of this data.
- Homogenising and standardising variables (integration of conceptual systems, definition, meaning) and coding dictionaries (classes and georeferencing). Godbout and Grondin (2005) refer to differences in concepts between two sources and propose various ways of getting around this problem.
- Developing compatible spatial references: georeferencing (projection, coordinate systems), divisions (traffic analysis zone, census zones) and transport networks (see discussion with respect to the levels of resolution for networks in Trépanier & Chapleau, 2001b).
- Guaranteeing temporal compatibility by implementing an updating process (difference between the dates of transport surveys and censuses), data factoring or suitable filtering of observations (subset of data for observation systems such as smart cards).
- Nothing is static in transport! Several variables which affect daily travel behaviours are changing all the time: new roads are built, capacities change, the patterns and service times of public transport are altered, new housing units are constructed, etc. It is important, at any time, to be able to reconstruct, retrospectively, these attributes with precision.

Statistical issues:

- Identifying the most appropriate level of resolution (depending on the anticipated purposes of the merged file) and the most appropriate object (household, individual, trip).
- Selecting the common variables, which will serve as gateways. Coupling can be based on various types of compatibility:
  1. spatial compatibility: proximity between units, inclusion of a unit in a zone;
  2. temporal compatibility: observations made on the same day or a comparable day;
  3. semantic compatibility: proximity between attribute vectors.

- Identifying the most appropriate model for determining the joint distribution between the observed variables (Z) and the unobserved variables (Y). Chesher and Nesheim (2006) have pointed out that fusion only produces reliable data if the conditional independence property between the variables Y and Z given X is met.
- Examining the validity of using a third-party database in order to better identify the above correlation (existing database, performing a small-scale survey specifically to measure this correlation, etc.) and validating the hypothesis of independence.
- Validating the results of fusion by comparing the distributions of variables, as proposed by Kiesl and Rassler (2006).

Organizational and political issues:

- Opening current data collection processes to examination: if the data can be constructed using another dataset, why bother to do a survey?
- Ensuring that micro-databases are confidential (Sweeney, 2002).

Reference

Aluja-Banet, T., Daunis-I-Estadella, J., & Pellicerc, D. (2007). GRAFT, a complete system for data fusion. *Computational Statistics and Data Analysis*, *52*, 635–649.

Ampt, E. S. (1997). Response rates — Do they matter?. In: P. Bonnel, R. Chapleau, M. Lee-Gosselin & C. Raux (Eds), *Les enquêtes de déplacements urbains: Mesurer le présent, simuler le future* (pp. 115–125). Lyon: Collection Programme Rhône-Alpes Recherches en Sciences Humaines.

Atrostic, B. K., & Burt, G. (1999). Household non-reponse: What we have learned and a framework for the future. In: *Statistical policy working paper* (Vol. 28, pp. 153–180). Washington, DC: Federal Committee on Statistical methodology, Office of Management and Budget.

Bayart, C., & Bonnel, P. (2007). Web surveys: An opportunity to reduce non-response in traditional household travel survey. Paper presented at the 11th world conference on transportation research, 24–28 June, Berkeley, 19pp.

Bayart, C., & Bonnel, P. (2008). Enquête web auprès des non-répondants de l'enquête ménages déplacements de Lyon 2005–2006. Rapport pour le PREDIT, Laboratoire d'Economie des Transports, Lyon, 262pp.

Beckman, R. J., Baggerly, K. A., & McKay, M. D. (1996). Creating synthetic baseline populations. *Transportation Research A*, *30*, 415–429.

Berk, R. A. (1983). An introduction to sample selection bias in sociological data. *American Sociological Review*, *48*(3), 386–398.

Bonnel P. (2003). Postal, telephone and face-to-face surveys: How comparable are they?. In: P. R Stopher & P. M. Jones (Eds), *Transport survey quality and innovation* (pp. 215–237). Oxford: Pergamon.

Bonnel, P., & Armoogum, J. (2005). National transport surveys — What can we learn from international comparisons? Paper presented at the European Transport Conference, 3–5 October, Strasbourg, Actes sur CD ROM, 24pp.

Bonnel, P., & Le Nir, M. (1998). The quality of survey data: Telephone versus face-to-face interviews. *Transportation*, *25*(2), 147–167.

Cameron, A. C., & Trivedi, P. K. (2005). *Microeconometrics methods and applications*. New York: Cambridge University Press.

Cassias, I., & Kun, A. L. (2007). *Vehicle telematics: A literature review*. Technical Report ECE.P54.2007.9. Available at: http://www.project54.unh.edu/Reference/Download.pm/2691/Document.PDF, consulted February 2, 2008.

Chalasani, V. S., & Axhausen, K. W. (2005). Conceptual data model for the integrated travel survey and spatial data. In: R. Khan, R. Banks, R. Cornelius, S. Evans, & T. Manners (Eds), *Proceedings of ASC 2005: Maximising data value*, ASC, Chesham (pp. 123–135). Available at: http://www.ivt.ethz.ch/vpl/publications/reports/ab302.pdf, consulted January 10, 2008.

Chapleau, R., & Trépanier, M. (1997). *Transportation object-oriented modeling: An extension of the totally disaggregate approach*. Presented at the *8th international conference on travel behavior research*, Austin, TX, 16pp.

Chen, P. P. (1976). The entity-relationship model — toward a unified view of data. *ACM Transactions on Database Systems*, *1*(1), 9–36.

Chesher, A., & Nesheim, L. (2006). *Review of the literature on the statistical properties of linked datasets*. DTA Occasional Paper No.3, 54pp. Available at: http://www.berr.gov.uk/files/file24832.pdf, consulted February 26, 2008.

Cobanoglu, C., Warde, B., & Moreo, P. J. (2001). A comparison of mail, fax and web-based survey methods. *International Journal of Market Research*, *43*(4), 441–452.

Couper, M. P. (2000). Web surveys: A review of issues and approaches. *Public Opinion Quarterly*, *65*(2), 230–253.

D'Ambrosio, A., Aria, M., & Siciliano, R. (2007). Robust tree-based incremental imputation method for data fusion. Paper presented at 7th international symposium on intelligent data analysis, Ljubljana, Slovenia. Available at: http://velblod.videolectures.net/2007/ijs/ida07_ljubljana/dambrosio_antonio/ida07_ljubljana_dambrosio_antonio.pdf, consulted January 12, 2008.

Desmet, R., Gusbin, D., Hoornaert, B., Lambrecht, M., Mayeres, I., Paul, J.M., Poulain, M., Eggerickx, A.B., Sanderson, J. P., Toint, P., Cornélis, E., & Malchair, A. (2007). *Démographie, géographie et mobilité: perspectives à long terme et politiques pour un développement durable (MOBIDIC)*. Rapport final (SP1747), Politique scientifique fédérale, Bruxelles.

Dillman, D. A., & Christian, L. M. (2005). Survey mode as a source of instability in responses across surveys. *Field Methods*, *17*(1), 30–52.

Dillman D. A., Phelps, G., Tortora, R. D., Swift, K., Kohrell, J., & Berck, J. (2001). Response rate and measurement differences in mixed mode surveys using mail, telephone, interactive voice response and the Internet. *The American Association for Public Opinion Research (AAPOR)*, 56th Annual Conference.

D'Orazio, M., Di Zio, M., & Scanu, M. (2006). *Statistical matching: Theory and practice*. New York: Wiley.

Duke-Williams, O., & Blake, M. (1999). Database fusion for the comparative study of migration data. Paper presented at the 4th international conference on GeoComputation, July, Fredericksburg, VA, USA. Available at: http://www.geocomputation.org/1999/068/gc_068.htm, consulted January 12, 2008.

Gautier, J.-M. (1999). Mégabase de consommateurs, sondages et statistique. In: G. Brossier & A. M. Dussaix (Eds), *Enquêtes et sondages-méthodes, modèles, applications, nouvelles approaches* (pp. 73–80). Paris: Dunod.

Godbout, S., & Grondin, C. (2005). Getting around a conceptual difference between two data sources when producing estimates. *Statistics Canada international symposium series — Proceedings, symposium 2005: Methodological challenges for future information needs*. Available at: http://www.statcan.ca/english/freepub/11-522-XIE/2005001/9481.pdf, consulted January 10, 2008.

Goulias, K. (2000). Resource paper for the workshop on multi-method and multi-instrument surveys, surveys using multiple approaches. In: *Transport surveys: Raising the standard*, Transportation Research Circular EC-008, Washington, DC.

Greaves, S. (2006). Simulating household travel survey data. In: P. Stopher & C. Stecher (Eds), *Travel survey methods, quality and future directions* (pp. 625–650). Oxford: Elsevier.

Greene, W. H. (2002). *Econometric analysis*. Upper Saddle, NJ: Prentice Hall.

Gunn, H. (2002). Web-based surveys: Changing the survey process. *First Monday*, *7*(12).

Heckman, J. (1979). Sample selection bias as a specification error. *Econometrica*, *47*(1), 153–161.

Heckman, J. (1990). Varieties of selection bias. *The American Economic Review*, *80*(2), 313–318.

Hoffman, D., & Link, C. R. (1984). Selectivity bias in male wage equation: Black–white comparison. *The Review of Economics and Statistics*, *66*(2), 320–324.

Hubert, J.-P., & Toint, P. (2003). *La mobilité quotidienne des Belges*. Belgium: Presses Universitaires de Namur.

Johansen, H. (2005). Health studies using linked administrative hospital data. *Statistics Canada international symposium series – Proceedings, symposium 2005: Methodological challenges for future information needs*. Available at: http://www.statcan.ca/english/freepub/11-522-XIE/2005001/9482.pdf, consulted January 10, 2008.

Judson, D. H. (2006). *Comparing methods for data integration, presentation*. Research Methods Festival Programme. Available at: http://www.ccsr.ac.uk/methods/festival/programme/opus1/judson.ppt, consulted June 20, 2007.

Kiesl, H., & Rassler, S. (2006). *How valid can data fusion be*? IAB Discussion Paper no 15/2006. Available at: http://doku.iab.de/discussionpapers/2006/dp1506.pdf, consulted January 15, 2008.

Kochan, B., Bellemans, T., Janssens, D., & Wets, G. (2007). Dynamic activity-travel diary data collection using a GPS-enabled personal digital assistant. Conference publication, *Innovations in Travel Demand Modelling*. Available at: http://www.imob.uhasselt.be/index.php?page = 17&publication_action = download&publication_id = 833, consulted March 23, 2007.

Lee, L. F. (1982). Some approaches to the correction of selectivity bias. *The Review of Economic Studies*, *49*(3), 355–372.

Lozar Manfreda, K., & Vehovar, V. (2002). Do mail and web surveys provide same results?. In: A. Ferligoj & A. Mrvar (Eds), *Development in social science methodology* (pp. 149–169). Slovenia: Ljubljana.

Maddala, G. S. (1986). *Limited dependant and qualitative variables in econometrics*. London: Cambridge University Press.

Miller, E. J. (1996). Microsimulation and activity-based forecasting. In: *Summary, recommendations and compendium of papers, travel mode improvement program activity-based travel forecasting conference* (pp. 151–172), Washington, DC, June 1996. U.S. Department of Transportation.

More, K., & Ingman, D. (2008). Quality approach for multi-parametric data fusion. *NDT & E International*, *41*, 155–162.

Morency, C. (2004). *Contributions à la modélisation totalement désagrégée des interactions entre mobilité urbaine et dynamiques spatiales*. Thèse de Doctorat, Génie Civil, École Polytechnique de Montréal, 530pp.

Morris, J., & Adler, T. (2003). Mixed mode survey. In: P. R. Stopher & P. M. Jones (Eds), *Transport survey quality and innovation* (pp. 239–252). Oxford: Pergamon.

Mroz, T. A. (1987). The sensitivity of an empirical model of married women's hours of work to economic and statistical assumptions. *Econometrica*, *55*(4), 765–799.

Murakami, E. (2004). Travel survey methods and technologies resource paper. Paper presented at international conference on travel survey methods, August, Costa Rica.

Nakosteen, R. A., & Zimmer, M. A. (1980). Migration and income: The question of self-selection. *Southern Economic Journal*, *46*, 840–851.

Olsen, D. R. (1999). *Interacting in Chaos*. Interactions (pp. 42–54). ACM.

Pendyala, R. M. (1999). *Measuring day-to-day variability in travel behavior using GPS data*. Final Report, Federal Highway Administration Report no. DTFH61-99-P-00266, Tampa, FL.

Peters, S., Ekin, P., Leblanc, A., Clark, K., & Pickles, S. (2006). Grid enabled data fusion for calculating poverty measures. *Proceedings of the UK e-science all hands meeting*. Available at: http://www.allhands.org.uk/2006/proceedings/index.html, consulted January 10, 2008.

Polak, J. (2006). *OPUS: Optimising the use of partial information in urban and regional systems*. Presentation, Research Methods Festival Programme. Available at: http://www.ccsr.ac.uk/methods/festival/programme/opus1/polak.ppt, consulted June 20, 2007.

Resource System Group, INC. (2002). *Documentation for SBIR phase II final report: Computer-based intelligent travel survey system*. DTRS57-00-C-10030, Prepared for the FHWA, October.

Richardson, A. J. (2000). Behavioural mechanisms of non-response in mailback travel surveys. Paper presented at the 79th annual meeting of the transportation research board, Washington, DC.

Richardson, A. J., & Ampt, E. (1993). *The Victoria integrated travel, activities and land-use toolkit*. VITAL Working Paper VWP93/1, Transport Research Centre, University of Melbourne.

Rodgers-Willard, L. (1984). An evaluation of statistical matching. *Journal of Business and Economic Statistics*, *2*(1).

Roy, A. D. (1951). Some thoughts on the distribution of earnings. *Oxford Economic Papers*, *3*, 135–146.

Sallis, J. F., Frank, L. D., & Saelens, B. E. (2004). Active transportation and physical activity: Opportunities for collaboration on transportation and public health research. *Transportation Research Part A, Policy and Practice*, *38*(4), 249–268.

Saporta, G. (2002). Data fusion and data grafting. *Computational Statistics and Data Analysis*, *38*(4), 465–473.

Stolzenberg, R. M., & Relles, D. A. (1997). Tools for intuition about sample selection bias and its correction. *American Sociological Review*, *62*(3), 494–507.

Stopher, P. R., Fitzgerald, C., & Xu, M. (2007). Assessing the accuracy of the Sydney household travel survey with GPS. *Transportation*, *34*(6), 723–741.

Stopher, P. R., & Greaves, S. P. (2003). *Simulating household travel survey data in metropolitan areas*. Technical Report, Conducted for Louisiana Department of Transportation and Development Louisiana Transportation Research Center, 59pp. Available at: http://tmip.fhwa.dot.gov/clearinghouse/docs/household/household.pdf, consulted January 10, 2008.

Stopher P. R., Jiang, Q., & Fitzgerald, C. (2005). Processing GPS data from travel surveys. Transporting the Future: Transport in a changing environment, 28th Australian Transport Research Forum (ATRF), 28–30 September, Sydney.

Sweeney, L. (2002). Achieving k-anonymity privacy protection using generalization and suppression. *International Journal on Uncertainty, Fuzziness and Knowledge-based Systems*, *10*(5), 571–588.

Tobin, J. (1958). Estimation of relationship for limited dependent variables. *Econometrica*, *26*(1), 24–36.

Ton, T., & Hensher, D. (2003). Synthesising population data: The specification and generation of synthetic households. *TRESIS2.0, 9th world conference of transportation research*, Oxford, 22–27 July.

Trépanier, M., & Chapleau, R. (2001a). Analyse orientée-objet et totalement désagrégée des données d'enquêtes ménages origine-destination. *Revue Canadienne de Génie Civil*, *28*(1), 48–58.

Trépanier, M., & Chapleau, R. (2001b). Linking transit operational data to road network with a transportation object-oriented GIS. *Urban and Regional Information Systems Association Journal*, *13*(2), 23–27.

Venigalla, M. (2004). Household travel survey data fusion issues. Resource Paper, *National Household Travel Survey Conference: Understanding Our Nation's Travel*, Transportation Research Board, Washington, DC. Available at: http://www.trb.org/Conferences/NHTS/Workshop-DataFusion.pdf, consulted September 5, 2007.

Voas, D., & Williamson, P. (2000). An evaluation of the combinatorial optimisation approach to the creation of synthetic microdata. *International Journal of Population Geography*, *6*, 349–366.

Wei, C. H., & Lee, Y. (2005). Applying data fusion techniques to traveler information services in highway network. *Journal of the Eastern Asia Society for Transportation Studies*, *6*, 2457–2472.

Winship, C., & Mare, R. D. (1992). Models for sample selection bias. *Annual Review of Sociology*, *18*, 327–350.

Yun, G. M., & Trumbo, C. W. (2000). Comparative response to a survey executed by post, e-mail & web form. *Jornal of Computer-Mediated Communication*, *6*(1).

Zmud, J. (2003). Designing instruments to improve response: Keeping the horse before the cart. In: P. R. Stopher & P. M. Jones (Eds), *Transport survey quality and innovation* (pp. 89–108). Oxford: Pergamon.

Chapter 35

Best Practices in Data Fusion: Synthesis of a Workshop

John W. Polak and Eric Cornelis

35.1. Introduction

In recent years both researchers and policy makers have developed an increasing appetite for more comprehensive information on travel behaviour. For example, the gradual shift from trip-based to activity-based methods of modelling and analysis has created a demand for increased spatial, temporal, functional and semantic granularity in information on the characteristics of both in-home and out-of-home activity participation. The demand for better and more comprehensive data has spurred innovations in survey methods at several levels, from the design of individual survey instruments to the analysis of complex survey data. One important line of development has been a growing interest in formal methods of data fusion — that is, methods of combining data from different sources (e.g. data from different sample surveys, census records, operational data streams and data generated by ITS systems) in order to render more complete descriptions of relevant features of travel behaviour. The interest in data fusion reflects the influence of two interrelated factors. First, the recognition that, notwithstanding the continuous improvements being made in the design and administration of individual survey instruments, significant limits exist in the scope and quality of data that can be elicited via any single instrument. And second, the perception that there exist significant opportunities to enrich existing forms of data with new data sources arising from ITS technologies and systems. Both these arguments converge in the proposition that it is both increasingly necessary and increasingly desirable to have the capability to

Transport Survey Methods

ISBN: 978-1-84855-844-1

coherently combine information from different sources. Against this backdrop, the objectives of this workshop were threefold:

- To provide a brief overview of the state of the art and the state of practice in the field of transport data fusion (in terms of scope, methods and applications), with particular reference to the interests of the survey methods community.
- To identify key research challenges and opportunities for the future, taking into account existing and emerging data sources, real-world problems and the scope for multi- and inter-disciplinary working.
- To identify tangible ways to accelerate methodological innovation and adoption in practice through, for example, the prioritisation of research challenges, the better understanding of the barriers to successful applications of existing data fusion methods and the development of networks and communities of interest.

35.2. Workshop Organisation and Formal Papers

The workshop format comprised eight hours of presentation and discussion, spread over two days, with a heavy emphasis on discussion. Four papers were formally presented in the workshop: a resource paper by Bayart, Bonnel and Morency (with a formal response from Cornelis) plus three contributed papers by Nakamya, Moons and Wets; Yennamani and Srinivasan; and Kuhnimhof, Ottmann and Zumkeller. Each paper was discussed after presentation and key themes carried forward for further discussion in open session. A total of 15 participants took part in the workshop, representing a range of disciplinary and national backgrounds.

The resource paper by Bayart, Bonnel, and Morency (2009) provided an overview both of the motivations for data fusion in travel behaviour research and of some of the key conceptual issues and techniques that have been, or could be, deployed to develop effective data fusion approaches.

At the conceptual and methodological level, the authors recognised that a wide range of different statistical methods for data fusion are available both in the transport literature and in cognate fields in the economic, social and engineering sciences. They noted that a distinction is often drawn in the literature between, on the one hand, data fusion approaches that are concerned with linking together two or more data sets at the record level to produce a new synthetic data set (which they term a micro-approach) and, on the other hand, those that are concerned with estimating the parameters of the underlying joint distribution of all the variables of interest (which they term a macro-approach). They pointed out, however, that this distinction is more apparent than real, since any data fusion method must make assumptions regarding the underlying joint distribution of all variables; what distinguishes the micro- and macro-approaches is whether the assumptions are induced non-parametrically directly from the data or inferred parametrically from a model fitted to the data.

Whilst the majority of attention was focused on statistical issues, Bayart et al. (2009) made the important point that in practice, data fusion is not just a statistical problem — it also has significant institutional and organisational dimensions. For example, data can be fused only if they are accessible and sufficiently well documented to enable appropriate statistical techniques to be applied. This implies a requirement for organisational and institutional structures capable of handling questions of data confidentiality, security, storage, preservation, discovery and retrieval. These are, of course, generic issues, but they have a particular importance in the context of data fusion.

Following on from the general issues raised in the resource paper, the three contributed papers examined different facets of data fusion. The paper by Nakamya, Moons, and Wets (2008) illustrated the practical benefits of data fusion by demonstrating how combining traditional travel diary and time-use data sets using an iterative proportional fitting algorithm can lead to improvements in the estimation performance of linear regression models of daily person trip rates. Yennamani and Srinivasan (2008) also addressed the issue of dealing with travel diary and time-use data sets. Their paper, however, was concerned with characterising the implications of these different approaches to data collection for the estimation of mobility. Their results suggest that whilst the overall method effect is modest, in certain demographic segments it can be significant and of unpredictable sign. The paper by Kuhnimhof, Ottmann, and Zumkeller (2008) presented a method for imputing expenditures on car and public transport in a national travel survey data set, on the basis of information from a national income and expenditure survey. The method was found to work well in the case of car expenditure but less so for public transport.

35.3. Discussion

The formal papers generated a lively and wide-ranging discussion. The key issues to emerge from this discussion, in relation to the objectives of the workshop, may be summarised as follows.

35.3.1. State of the Practice and State of the Art

During the discussion, it rapidly became apparent that although the formal designation of interest in data fusion within the survey methods community is relatively recent, data fusion has been an active (although badly fragmented) area of work in traffic and transport modelling for many years. Some of the most notable and well-established examples highlighted during discussion include:

- the use of filtering (e.g. Kalman filter, extended Kalman filter), pattern-matching (e.g. K-nearest neighbour) and neural network techniques for the estimation of traffic state variables such as flow, travel time and queues, using data from

multiple types of traffic sensor including inductive loop detectors, automatic number plate recognition and GPS-equipped probe vehicles
- techniques such as entropy maximisation, information minimisation, generalised least squares and model-based Bayesian inference for the estimation and updating of OD matrices, based on historical OD data and contemporary traffic counts
- techniques such as iterative proportional fitting and statistical matching for the creation of synthetic populations to support the application of disaggregate discrete choice modelling techniques to large-scale (regional and national) prediction and forecasting
- techniques such as full information maximum likelihood (FIML) and model-based Bayesian inference to enable the use of multiple data types such as revealed preference data, stated preference data and attitudinal data in the estimation of discrete choice models of travel behaviour

In addition to these well-established areas of data fusion work, the workshop also identified a number of emerging areas of data fusion activity:

- the use of FIML and classical sample selection techniques to deal with data from multiple or mixed methods of survey administration (such as activity/travel diary data collected through a combination of personal interview, telephone interview and Internet-based interview)
- the use of statistical matching and imputation techniques to estimate complete out-of-home and in-home time-use and travel records from the combination of travel diary and time-use data
- the use of various pattern-matching techniques to extract activity/travel features (such as significant stops, mode changes, etc.) from GPS, GSM and related trace data, using conventional activity/travel diary data as kernels
- the use of FIML and Bayesian techniques to estimate population exposure to environmental hazards via the combination of data on pollution concentrations and individual travel/activity data

Whilst there is no doubt that the area of fusion is one that is attracting increasing attention, several workshop participants felt that the area still lacked a coherent identity, that the understanding of fundamental aspects of problem typology was incomplete and that the transmission of relevant methodology across sub-disciplinary boundaries within transport is hampered by inconsistencies in conceptual approach and terminology.

35.3.2. Key Research Challenges

The workshop considered research challenges under three headings: enabling research, methodological research and research infrastructure.

Enabling research was conceived of as research that, whilst not directly concerned with data fusion *per se*, would have a significant impact on the capability of the field to develop effective data fusion approaches in the future. Three key challenges were perceived to exist under the heading of enabling research:

- The development of better metadata standards and tools to enable data collection agencies to more fully characterise the complete survey sampling, survey administration and data pre-processing stages of the data lifecycle. Such metadata play a crucial role in enabling subsequent analysts, including those wishing to perform data fusion, to fully characterise the relevant features of the measurement process.
- The development of better techniques for the treatment of privacy, security and disclosure management. This was seen as crucial in order to avoid circumstances in which breaches of privacy or security lead to key data assets becoming "locked down." It was also argued that data fusion raises particular issues in terms of data disclosure, since for example, the potential always exists for two non-disclosive data sets to be combined in such a way as to lead to the inadvertent disclosure of sensitive information. It is important to examine carefully the trade-off between degrading the quality of scientific conclusions and protecting personal information (relevant to issues such as speeding, finding people and taxation).
- The development of a better understanding of how to make the business case for data fusion. This will entail, for example, developing better tools for quantifying the improvements in precision and accuracy that can be achieved through the appropriate fusion of data and the translation of these improvements into financial (e.g smaller sample size) and reputational (e.g. enhanced trust and confidence) goods. It will also entail developing a better understanding of some of the existing organisational and institutional barriers to the use of fused data, such as perceptions that such data are "made up" and hence invalid and the threat (real and imagined) posed to incumbent data collection and analysis processes ("… now that you can 'create' data, you don't need funding to collect data …").

Methodological research was seen in terms of addressing key weaknesses and gaps in existing and emerging data fusion methods. Three key challenges were perceived to exist:

- Validation was seen as a vital issue, and one having several important dimensions. At the most basic level, there is a need for better statistical tools to detect genuinely conflicting information that may not, in any useful sense, be fusible. A number of the workshop participants likened this to a model specification test, and although the analogy is attractive, the step from analogy to formal method remains to be taken.
- A second aspect of validity for which there was perceived to be a lack of readily available tools and methods was the assessment of the confidence that can be placed in fused data. With such a wide range of fusion methods available (e.g. from simple *ad hoc* voting methods to complex non-linear filters), it is far from clear which methods are most appropriate for particular problems. Moreover, many of the most commonly used methods of data fusion (e.g. deterministic

matching methods and mean imputation) do not inherently provide estimates of the statistical confidence of the resulting fused data. This was seen as a significant disadvantage for such methods.

- Closely related to the above observation was the perceived lack of generic methods and tools to fully understand how uncertainty in different data sources, and in the underlying structural or modelling assumptions used in the fusion process, are propagated into uncertainty in the fused data that results. Bayesian analysis offers one approach to this type of challenge, but, except in very simple cases, Bayesian approaches do not currently scale well, particularly in the context of the high dimensionality characteristic of much network-based transport data.

Finally, under the rubric of research infrastructure, the workshop identified two aspects of the general context in which data fusion work is undertaken that it believed merited further attention.

- As alluded to above, a number of workshop participants expressed concern regarding the lack of a consistent taxonomy of data fusion problems, imprecision in the use of concepts and the incomplete dissemination throughout the field of the full range of data fusion methods currently available. Although such concerns are difficult to avoid in a relatively new and immature area, there was nevertheless a strong view within the workshop that the survey methods community should seek to actively address these issues through focused capacity-building activities.
- It was felt that an important part of the process of capacity building in this area would be the development of a set of reference data sets and reference cases, covering a range of data fusion problem types, which could serve as benchmarks for those wishing to evaluate the performance of existing or newly developed approaches.

35.4. Conclusion

The formal treatment of data fusion problems in the travel survey methods community is still in its infancy. Yet it is already clear that the area both holds substantial promise in terms of benefits for the future and is characterised by a range of problems that will challenge current and future generations of researchers. We hope that this workshop has helped to highlight some of these issues and to stimulate the participants (and others) to address these challenges with zeal, rigour and creativity in the future.

List of Participants

The workshop participants were Caroline Bayart, France; Patrick Bonnel, France; Olivia Christophersen, United Kingdom; Eric Cornelis (discussant), Belgium;

Christine Fox, the United States; Dana Gruschwitz, Germany; Jean-Paul Hubert, France; Tobias Kuhnimhof, Germany; Elke Moons, Belgium; Catherine Morency, Canada; Marcela Munizaga, Chile; Pascal Pochet, France; John Polak, United Kingdom; Sivaramakrishnan Srinivasan, the United States; and Jarvi Tuuli, Finland.

References

Bayart, C., Bonnel, P., & Morency, C. (2009). Survey mode integration and data fusion: Methods and challenges. In: P. Bonnel, M. Lee-Gosselin, J. Zmud & J.-L. Madre (Eds), *Transport survey methods: Keeping up with a changing world.* Bingley, UK: Emerald.

Kuhnimhof, T., Ottmann, P., & Zumkeller, D. (2008). Adding value to your data: Analysis of travel expenses based on trip diary and enriched odometer reading data. Paper presented at the 8th International Conference on Transport Survey Methods, Annecy, France.

Nakamya, J., Moons, E., & Wets, G. (2008). Comparison between enriched travel data and the original survey data by means of a model based approach. Paper presented at the 8th International Conference on Transport Survey Methods, Annecy, France.

Yennamani, R., & Srinivasan, S. (2008). Disaggregate comparison of travel reported by two American surveys, the NHTS and the ATUS. Paper presented at the 8th International Conference on Transport Survey Methods, Annecy, France.

Chapter 36

Lessons from an Overview of National Transport Surveys, from Working Group 3 of COST 355: "Changing Behavior Toward a More Sustainable Transport System"

Jimmy Armoogum, Kay W. Axhausen and Jean-Loup Madre

Abstract

This chapter summarizes some of the results from the Working Group "National Travel Surveys" in COST Action 355. All 50 presentations could not be reviewed here; thus, we focus on three crucial topics:

- the periodicity of data collection, but also its longitudinal aspects (advantages of continuous surveying, repeated cross-sections vs. panel surveys, etc.),
- new technologies for improving the efficiency and accuracy of mobility surveys (computer-assisted telephone, Web-based, interviews, GPS, GSM, RDS, etc.),
- innovative approaches, exemplified by qualitative surveys combined with conventional quantitative ones, and by biographical approaches.

36.1. Introduction

Changes in behavior are necessary to reverse worrying long-term trends of growing mobility, with increasing oil consumption and greenhouse gases (GHG) emissions. Most data on mobility are collected through conventional instruments. The analysis of changes in behavior presupposes the comparability of these instruments over time, but also between countries and urban areas across Europe. Travel survey data are needed, both to portray existing situations and to help identify problems related to

Transport Survey Methods

ISBN: 978-1-84855-844-1

the operation of transport systems, as well as to estimate/validate those models which are essential for planning activities. In fact, the most typical (and more difficult) need is for data to allow calibrating the strategic transport-planning models used to forecast the effect of medium to long-term policies for project evaluation and environmental assessment.

A first issue is whether these surveys should be conducted from time to time (generally with increasing time intervals) or on a continuous basis (as in, the United Kingdom, the Netherlands, or Denmark). New technologies may have a large potential to improve survey data quality. Since all data needed for a comprehensive analysis of changes in behavior cannot be collected in the same survey, pairing different data sources (e.g., trip-based surveys with time use or family expenditure surveys) is an important issue.

We will focus our attention in this synthesis mainly on three topics:

- The periodicity of mobility surveys (Section 36.2). For the description and analysis of trends, as well as of changes in behavior, conventional travel surveys (covering, i.e., only one weekday in winter during school holiday periods when traffic flows are maximal) are not enough; for environmental issues, mobility has to be described for the whole year.
- New technologies (Section 36.3). The use of new technologies may decrease survey cost and also have a positive impact on data quality. For example, Web-based surveys (Section 36.3.1) might help to contact persons difficult to reach by conventional modes, making use of automated coding and editing of data. In addition, new devices such as mobile phone (GSM), GPS, and Galileo (Section 36.3.2) may help improve the accuracy of timing data (departure, arrival, trip duration) and of location data (origin, destination, trip distance, and route choice).
- Innovative methods (Section 36.4). New procedures such as qualitative approaches (Section 36.4.1) or biographies (Section 36.4.2) may help improve data quality and suggest original quantitative analyses of mobility surveys.

36.2. Periodicity of Mobility Surveys

In most countries and urban areas, personal travel surveys are conducted infrequently (e.g., every decade) and using evolving methodologies which often make the comparison difficult between subsequent surveys for the assessment of trends (e.g., when the American NPTS shifted from a trip-based to an activity-based approach, Stopher, 1992). These one-off surveys are subject to unpredictable events (e.g., strikes or extreme weather conditions) and do not allow a clear distinction between long-term trends and short-term events (economic boom or recession).

Although there are many advantages associated with collecting mobility data on a continuous basis, managing to do so in practice is not easy. We have considered advantages and drawbacks of the respective methodologies with respect to survey administration (e.g., training and fatigue of interviewers). Budgetary implications are also an important consideration in deciding to move to a continuous survey. Indeed, it seems easier to obtain funds when up and running (requesting similar budgets in successive years), than for one "big bang" survey. With a periodic survey there is an uneven requirement for funds, so large amounts have to be found at the time of the survey. If the political climate is not conducive to providing this funding, the survey could be delayed or not undertaken at all. It is acknowledged that a withdrawal of funding could occur at any time during a continuous survey, however, it is anticipated that if the survey is progressing successfully and producing relevant results, its prospects of continued financing will be enhanced. It is also expected that there might be significant economies of scale from undertaking the survey on a continuous basis. Even more importantly, there will be gains from the continuous learning, which will allow researchers to make near-continuous improvements to the survey design.

The issue of sampling is also an important question:

- A new sample may be created each year (Section 36.2.1), but drawn with a strong geographical basis (e.g., in the same postcode sector in Great Britain or the same block in Sydney or New Zealand) and temporal specification (a pre-allocated day is essential, contrary to what is done in France), or
- A true panel survey (Section 36.2.2) may be employed, which allows measuring changes at an individual level if the reporting period each year is long enough (seven days being better than two days); the drawbacks from selection bias and attrition are lower with a rotating panel (e.g., each household being withdrawn from the survey after a few years).

36.2.1. Subsequent Cross-Sections (A New Sample Each Year)

Most of the continuous mobility surveys currently underway are being conducted in Europe and in the southern hemisphere. Only a few of these are being conducted at the national level (those in Great Britain, Denmark, the Netherlands, and New Zealand). There are a large number of regional level surveys ongoing (those in Melbourne, Sydney, Perth, Santiago, Halle, Leipzig, Nürnberg, Burgenland/Niederösterreich, Wien, and Wiesbaden), probably because it is easier for a region to fund a survey when the budget is spread over many years than for a country to do. Regional continuous surveys have to be conducted in wide enough areas to enable analysis of urban sprawl, which is a major determinant of changes in travel behavior.

Specific variations in data due to unpredictable events can be isolated by comparing the survey period with other time periods. Surveying across the year allows controlling for seasonal effects (e.g., long-distance trips for holidays, bicycle use, walking, etc.). Moreover, subsequent years can be aggregated for the analysis of subgroups (e.g., regions). A critical deficiency of discontinuous surveys is that there is an unavoidable loss of staff and knowledge in the inter-survey period; these assets have to be reestablished for the next survey. This applies to both project management staff in the government organization conducting the survey, and staff in the market research organization conducting the fieldwork. According to British experience, the team has to remain the same at the sponsor level, especially if the field organization changes after a new tender; it is also worthwhile to highlight a recommendation from researchers in Sweden to avoid loss of motivation once the contract is obtained. However, in Denmark, the main reason for a substantial increase in the zero trip rate over time was traced directly to interviewer performance: an increasing number of interviews was conducted by a small number of interviewers who obtained much higher zero trip rates than other interviewers (Christensen, 2005).

36.2.2. Panel Surveys

The main advantage of panel surveys is their ability to measure the effects of any changes in external factors for individuals and households (Stopher et al., 2006; Stopher & Swann, 2008). In contrast, the repetition of cross-sections only allows the comparison of aggregate values or marginal distributions (net changes). The gross changes are captured through temporal and intrapersonal aspects. Thus, panel data facilitates the building of models that account for the dynamics of change. However, panel surveys also have some drawbacks:

- Panel conditioning: people may adapt their behavior in regard to the topic of the survey. That is a problem when attitudes or opinions are what is being surveyed, but is generally not critical for surveys about transport behavior and activities.
- Attrition, mortality, and fatigue effects: for panel surveys it is crucial to have volunteers respond for many waves; thus it is not easy to have a purely random sample.
- Selectivity phenomena: as a consequence of these biases it becomes likely that a multistage recruitment process and repetition within a panel certainly creates selectivity related to the characteristics of participating households.
- Refreshment of the sample: an ongoing panel can be kept representative. It makes sense to replace the dropouts by new households (rotating panel). A simple replacement of dropouts by households with the same socio-economic or demographic characteristics would be a solution, but it would be better to distinguish between "new units" (e.g., young individuals who have just left the household of their parents) and "rotating units" who replace dropouts.

The selective impacts on data quality have been studied for the German Mobility Panel. The middle class is overrepresented (good education, good income, middle-aged); there is a mobility interest bias: high dropout rate of non-trippers (particularly elderly, with permanent disability). Considering the heterogeneity between household members, surveying households counter-balances selectivity. From a larger perspective, one can conclude that balanced recruitment of different mobility styles is vital. It is important not to trade data quality for a high response rate, especially in the case of a panel survey, where the quality of data is crucial for the measurement of changes.

36.2.3. Measurement of Change and Survey Instruments

With regards to intrapersonal behavior variability and length of reporting period, Axhausen, Löchl, Schlich, Buhl, and Widmer (2007) have shown with the 2003 Thurgau six-week diary the strong intrapersonal variance in choices, modes used, and other aspects of travel behavior. These long-duration surveys also allow us to gain an understanding of the activity space of the travelers. Still, the analysis showed that the four interviewers employed in this survey had a substantial effect on the number of reported trips. Stopher, Clifford, and Montes (2008) have shown the benefits of using multiday data for modeling travel behavior.

It is obvious that the measurement of change is more accurate with a panel (Zumkeller, Madre, Chlond, & Armoogum, 2006). But it seems difficult to measure exactly the same indicators with these two (cross-section and panel) different survey instruments in a large-scale survey. In panel surveys, we should take care to avoid repetitive, burdensome questions. For example in the Netherlands as well as in the German urban areas where the New Kontiv design is used, it consists of:

- Asking respondent as little as possible (e.g., precoded items + an open answer);
- Allowing the respondent to choose the survey instrument (face-to-face, phone, or preferably mail-back), and thus obtain a lower nonresponse rate for households only accessible by mail;
- Proposing optional follow-up surveys to obtain additional data for specific subgroups (e.g., disabled people, children under six) or research topics (e.g., public transport, road accidents);
- Decentralizing the organization: everybody involved in the interview process has to know as much as possible about the survey.

The German statistical system for measuring mobility seems to be the most complete with: a cross-section survey every 6–13 years, an ongoing panel survey including regional add-on surveys measuring local mobility (Kuhnimhof, 2007) and the INVERMO a long-distance panel survey (Last, 2005).

A sample scheme chronologically and geographically balanced (even if not a panel) can improve the accuracy of time-series. It is important to choose a survey design giving a high and nondecreasing response rate (e.g., the New Kontiv) and a permanent and motivated staff is essential. In the future, new technologies (e.g., follow-up by GPS or GALILEO) could help surveying for longer periods, providing more accurate data on the spatial and temporal framework of mobility, with a relative low burden for interviewees.

36.3. Use of New Technologies to Gather Mobility Data

New technologies such as automatic satellite localization and mobile phone may improve the accuracy of the temporal detail (departure and arrival times, trip duration, etc.) and of the geographical detail (location of activities, origin and destination, trip distance, itinerary, etc.) of each trip. In addition, computer-assisted interview systems allow the detection of errors during the interview. Geo-coding encounters problems, on which there are experiences to share. Beside these improvements, the utilization of new technologies may reduce respondent burden and survey cost, which should have a major impact on data accuracy and quality.

36.3.1. Web-Based Surveys

The development of Web surveys has been very rapid in many respects. It is therefore important to analyze if Web-based surveys could be helpful for collecting travel data.

There are some obvious advantages to collecting data using a computer-assisted Web interview (CAWI), such as:

- Interactivity: this advantage is shared by all computer-assisted survey. Web-based surveys allow real-time entry of data, which improves the data consistency and quality.
- Availability of the interviewees for responding when they have time for this task.
- Able to reach individuals who do not respond to other survey modes.
- People who travel the most are often more difficult to contact by administered surveys, in particular because they are less often at home. Web-based surveys, like postal surveys, allow respondents to respond when they wish to.

Thus, the cost of this type of survey is low, because there is no need to have interviewers, no need to enter data. There are also some drawbacks:

- No sampling database is an exhaustive list of the whole population;
- More so than for other self-administered surveys, the response rates are low;
- Measurement errors may cause serious bias as definitions of mobility concepts are not so trivial;
- Technical problem such as unavailability of the server, browser incompatibilities (presentation of the survey on the screen), excessively long data loading times (high speed/low speed Internet), etc.

The use of the Web for transport surveys is likely to increase, in the same way as its use by the population. In view of the problems of coverage and Internet skills, its application as a single survey instrument is, for the time being, mainly restricted to the study of specific populations for which the construction of the sample can be controlled. Its use as a survey mode in combination with others is very promising in view of the steady rise in nonresponses for other survey modes in many countries. To reduce the bias from nonresponse, Bayart, Bonnel, and Morency (2009) have tested a Web-based survey in parallel with a household travel survey in the Lyon area. The idea is to provide a Web survey option to those households that refuse to respond through conventional instruments or that are not reachable after a certain number of attempts. The two main objectives of this research are to test the feasibility of a Web survey for nonrespondents and to compare mobility results of both survey modes.

The importance of the Web for marketing surveys is increasing considerably and seems particularly suitable for SP surveys involving controlled samples. Even in countries where access to the Internet remains quite limited, the Web appears to be quite useful for surveying a targeted population in the context of SP surveys (Hojman, Ortúzar, de, & Rizzi, 2004). The authors give the example of two Web-based surveys conducted in Chile to determine willingness to pay to reduce accident risks. These surveys produced results very consistent with those from other studies. However, beyond these applications, it is necessary to specify the domains and conditions in which Web-based surveys are applicable, both when it is the only survey mode used and when it is combined with other media.

36.3.2. GPS-Based Surveys

Data accuracy is a combination of sampling errors and non-sampling errors. Therefore it is not obvious how to compute confidence intervals due to the non-sampling errors such as nonresponse errors and measurement errors. Respondents are often unable to describe exhaustively their travel behavior and have a vague or even biased perception of main characteristics of their trips (for instance, the distance traveled). Interviewees are generally unable to describe their mobility with the degree of accuracy suggested in the questionnaire (e.g., in the 1993–1994 NTS, 1 min for departure and arrival time, 1 km for car annual mileage and daily mobility, even 100 m for trips under 2 km). For most analysis we do not need so much accuracy, but we have to be aware that rounding modifies variables' distributions. Summarizing the main findings obtained by comparing different instruments used in previous NTSs, it appears that:

- Time variables are less rounded when reported in diaries than when collected by interview.
- Fortunately, memory effects affect time (of departure or arrival) more than duration, which needs to be known more accurately, especially for modeling.
- The deterioration due to memory obviously increases when the facts reported have occurred a long time before the interview.
- Car diary surveys are more accurate than other types, probably because of the clock that is displayed on most car dashboards.

The measurement of trip distances is also an important issue. Controlled by the odometer, trip distance is well estimated through car diaries. If we compare trips by class of crow-flight distance between origin and destination, we notice a substantial underestimation of trip distance for trips with their origin and destination in the same municipality (about 25%); this underestimation is also observed for travel time, but this is of less importance. For longer trips (between municipalities within 15 km) the underestimation drops from 10% in the weekly stage diary of 1981–1982 to 5% in the 1993–1994 interviews. This improvement is probably due to the local maps, which were given to interviewers. On the other hand, long-distance trip length seems slightly overestimated.

Indeed in the field of travel behavior, since the mid-1990s, attention has focused on the potential of location-aware systems such as GPS (Global Positioning Systems), RDS (triangulation on FM radio stations), or GSM (Global System for Mobile communications). RDS is interesting for freight transport or long-distance travel, but does not provide accurate enough data for the analysis of daily mobility. Initially, the use of GPS was mostly limited to travel in private motor vehicles, because the power requirements of equipment in continuous use could easily be met with a connection to vehicle electrics, the problems of reception were minimized, and the linking of movement to ground features was simplified by staying on road networks. Nevertheless, even in the 1990s, some experiments used GPS to survey personal mobility in all modes of transport and off-road networks. These successful experiences in the United States, Canada, Japan, Australia, and Europe were conducted on relatively small samples, generally at a local/regional level. Dramatic technical improvements (smaller units, better precision, greater storage capacity, reduced power requirements) and decreases in prices allow GPS to be applied to large-scale surveys like National Travel Surveys. A first nationwide experience has been embedded in the 2007–2008 French National Travel Survey, which presented an opportunity to compare measurement tools before a larger use of new technology, while keeping the ability to measure long-term trends.

A GPS data-logging device allows the measurement of some details that are never given by respondents in conventional surveys:

- Description of very short trips, which are often forgotten;
- Route choice;
- Precise information on access/egress time and waiting time;
- The description of short trips made from an unusual place of residence (e.g., during holidays or long professional trips).

Moreover, the relatively low burden for the respondent (once she or he is trained) allows substantially extended survey duration: at least one week with GPS, compared to two days with the conventional questionnaire. The gain in accuracy is less because of cluster effect (travel patterns are quite similar on weekdays for the same person), but the gain is still important (Stopher, Kockleman, Greaves, & Clifford, 2008).

We should point out, that there are some drawbacks such as:

- Energy supply (battery last only about 15–20 h);
- Signal reception problems in tunnels or urban canyons;
- The interviewee may forget to take the GPS receiver with him (for some trips, some days, etc.); or
- Loan the device to another person;
- The interviewee may want to "play" with the device and therefore we record more trips than there should be; however, this may also happen in "conventional"

surveys, especially in face-to-face interviews in which some individuals may wish to appear socially well integrated and therefore may not describe what they think it is not acceptable and may invent some records.

GPS is certainly a promising technology for surveying travel behavior, because it provides much more accurate spatial and temporal data than conventional methods. But raw data are not directly usable, for the following reasons:

- Traces are not allocated to stage, trips, etc., yet;
- There are missing segments;
- There is information neither on transport means nor on trip purposes.

Thus, for post-processing these data, more or less sophisticated software packages have to be written depending on the accuracy needed by the users (e.g., much more spatial accuracy would be required to assess how many/which advertisement posters were viewed by the subject than would be needed for other types of travel surveys). In order to reduce the burden of interviewees, further research should focus on:

- Imputation of modes and transfer places from average speed and its variability, route, etc.;
- Imputation of purpose from destination location, arrival time, etc.; and
- Over the longer term, the question of missing data must be addressed by means such as automated reconstitution of continuous sequences in space as well as in time.

The comparability with data collected in the other countries is also important. EUROSTAT harmonizes several surveys in most EU member states (e.g., on time use or family expenditure), but nothing seems to be planned to similarly harmonize surveys on daily mobility. The generalization of GPS-based survey may introduce some data harmonization. But, in the near future it should be useful to have an assessment of such methods in term of feasibility at a large scale, the nonresponse due to GPS-based survey (Cortrino, 2009), and other sources of bias.

36.4. Innovative Survey Approach to Understanding Travel Behavior

36.4.1. Qualitative Approaches

The qualitative method has the advantage of being able to identify, from the speech and story of individuals' perception and experience, information that could not be captured through a closed-form questionnaire. This approach emphasizes the individual circumstances and highlights the complexity of the systems and of the actors, factors which are hardly observable by quantitative approaches. By highlighting the diversity of choices and trade-offs and the various related constraints, this approach makes it possible to understand individual choice in greater detail. The main drawbacks of such techniques are that their samples are generally not representative of the population (the sample size is very low compared with more quantitative approaches) and that it is burdensome (an interview may last for hours).

Qualitative approaches have their strengths and weaknesses, as well as the quantitative approaches, we must try to exploit and maximize the benefits of both. In the case of a mixed methodology, sample size can be reduced because it has no ambition of being absolutely representative or complete. Interpretation of results is made easier, while crossing qualitative and quantitative information.

If we have no *a priori* insights into the subject of study, it is clear that a qualitative phase is essential before drafting a quantitative survey. A return to the qualitative approach after a quantitative phase could clarify the meaning of the results and explain them. Therefore, qualitative approaches can be used to complement the quantitative approach at different levels:

- When we have to produce the questionnaire for a quantitative survey, the qualitative approach should help to design adequate questions and also to propose good response categories.
- When we are analyzing a quantitative survey, a qualitative approach should be used to enrich, providing insight and meaning to the results.

However, with the qualitative approach it is difficult to generalize the results; we can only show a typology of behavior. Complexity and diversity make it difficult to produce correlations: statements of individuals surveyed often complement and contradict each other at the same time. While the qualitative approach is clearly of interest in terms of thorough understanding of the behavior, it is not enough by itself to allow political decision-making. Indeed, political action based on a few instances would certainly be doomed to failure without an aggregate representation. Qualitative studies offer essential insights to understand subject behavior, but quantification is necessary for policy makers.

36.4.2. Biographical Approaches

Mobility trends over the long run have raised growing concerns about sustainability issues. The knowledge of mobility partly relies on household transport surveys. In France, such surveys have been conducted nationally four times since 1966, giving four cross-section points about the nation's travel behavior. But the derivation of change over time from comparing these four points lacks insight in two respects: the monitoring of individual change in behavior, and the narrative of a history of mobility. For both of these aspects, biographic surveys can bring new data to the field (Papon, Hubert, & Armoogum, 2007).

While historical sources describe the past of transport technology and the economic and social changes involved, asking people which mode they traveled in years as early as 1930 can record enough data from the subjects' memories to enable the description of a history of mobility, with the following details:

- Studying the development of motorization, including both two- and four-wheel vehicles, from 1920 onwards by social category and geographical area;
- Reconstituting modal share and mileage since 1940 for home to work travel, and since 1930 for home to school travel, or from 1940 for all purposes, by using existing household travel surveys conducted in 1966, 1974, 1982, and 1994 to adjust the fitting of the results, and stratifying by type of place, occupation, age, and gender, taking into account the structural turn-over of population: migrations, births, and deaths (utilizing differential survival laws by categories);
- Focusing on the transition of mobility from walking to car use that occurred in France during the 1950s (thus before any transport survey), especially its geographical, social, and generational diffusion;
- Understanding the stakes of this evolution for sustainable development during the 21st century, especially possible reductions in car use.

In France, this kind of survey has been conducted on other topics, such as "family, work and migration biography survey" by INED in 1981 (Riandey, 1985), "biographies and relatives" by INED in 2000–2001 (Lelièvre, Vivier, & Bonvalet, 2002) or "history of life" by INSEE in 2003. Lelièvre (1999) supervised a review of 14 previous biographical surveys.

Transport biographical surveys have been conducted in the United Kingdom (Pooley and Turnbull, 2000) and Switzerland (Axhausen, 2006; Beige, 2008). For the 2007–2008 French National Travel Survey, a new biographical section has been introduced. The survey is based on a chronological grid, where all events are recorded concerning:

- the place of residence at municipality level,
- the number of motor vehicles (two- and four-wheelers) available in the household,
- the main activity (school or work) and place of activity (municipality),
- the usual transport mode for commuting, or overall if no commuting occurred.

The main expected outputs of these additions are the following:

- a better understanding of personal travel behavior through the individual's personal history, making it possible to assess the likelihood of future changes with far greater appropriateness than the usual cross-sectional elasticity estimates; and
- a sketch of the general history of mobility in France from the 1930s on, with adjusted vehicle ownership, modal share, and mileage for commuting.

36.5. Conclusions

The transport sector is one of the major sources of global warming, from individual travel behavior (especially car dependency) and from freight delivery (mainly by trucks). Travel and freight surveys, as well as behavioral data collection, are essential to the formulation of transportation policies that will encourage more environment-friendly transport modes, and various data-collection methodologies have been proposed in recent decades. Within the four years of this COST action about 50 papers have been presented.

There have been many presentations of mobility surveys at the local and national level. Unfortunately these surveys are not harmonized, which complicates international comparisons. It is impossible up to now to study the impact of a given policy on mobility at the European level.

Transportation policies are becoming more sophisticated. Thus, more detailed data on attributes of travel behavior are required for travel demand analysis and modeling. For a description and analysis of trends, as well as of changes in behavior, conventional travel surveys (covering, i.e., only one weekday in winter out of school holiday periods, when traffic flows are maximal) are not enough. For environmental issues, mobility has to be described throughout the year (e.g., 24 h a day, seven days a week, and even possibly all seasons of the year, i.e., 365 days). Moreover travel surveys are burdensome (mainly due to the large number of items and the repetition of the same questions such as location, mode, purpose, etc.). Mobile communication technologies including GPS, GSM, and Radio Data System (RDS) have advanced rapidly and their prices are decreasing. They demonstrate great potential as survey instruments for tracking individual travel behavior as well as freight movement by surveying during longer period and providing more accurate data on the spatial and temporal framework of travel with a relative low burden for interviewees.

Hence we are at the turning point, where aiming at producing guidelines toward European harmonized travel surveys (either for passenger and freight) should take advantage of the development of new technologies. For a period when behaviors are changing, due to the rapid increase of fuel price and other factors, an important issue is continuous data collection. According to the keynote presentation by Pasi at the Annecy Conference, this is already the case for Continuing Survey of Road Goods Transport harmonized by EUROSTAT (but data on energy consumption do not exist in all countries, thus are not centralized). On daily mobility, continuous surveys are ongoing in a few countries (the Netherlands since 1978, the United Kingdom since 1988, the German Mobility Panel since 1994, Denmark and Sweden).

Since there does not seem to be a perspective in the European Framework Research Program (FP7) for implementing any harmonized surveys on transport, our approach is bottom-up: based on COST 355 experience (especially in WG3 on National Travel Surveys), a new COST Action has been launched: SHANTI (Survey Harmonization with New Technologies Improvement). It will allow researchers from different countries to harmonize their point of view by networking. Indeed, in addition to countries maintaining a continuous effort, several National Travel

Surveys have been conducted in 2008 (e.g., the United States, Spain, Germany, and France) or are planned for the near future (e.g., Luxembourg and Belgium).

Elaboration and analysis of surveys are funded in each country, as well as research projects (e.g., "Behavior and mobility within the week" in Belgium, six-week diaries in Switzerland and Germany about the rhythm of activities). It will also be an opportunity to tackle serious challenges at conceptual and technical levels (e.g., testing instruments for a comprehensive analysis of attitudes to social networking or the connection between participation to physical and virtual activities).

Acknowledgements

This research was funded by Action 355, "Changing Behavior Toward a More Sustainable Transport System," of the European Cooperation in Science and Technology (COST). The authors also acknowledge the valuable comments of two anonymous reviewers.

References

Axhausen, K. W. (2006). New survey items for a fuller description of traveler behaviour (biographies and social networks). TRB Travel Demand Forecasting Conference, Austin, May 2006.

Axhausen, K. W., Löchl, M., Schlich, R., Buhl, T., & Widmer, P. (2007). Fatigue in long-duration travel diaries. *Transportation*, *34*(2), 143–160.

Bayart, C., Bonnel, P., & Morency, C. (2009). Survey mode integration and data fusion: Methods and challenges. In: P. Bonnel, M. Lee-Gosselin, J. Zmud & J.-L. Madre (Eds), *Transport survey methods: Keeping up with a changing world*. Bingley, UK: Emerald.

Beige, S. (2008). Long-term and mid-term mobility decisions over the life course. Dissertation, Department Bau, Umwelt und Geomatik, ETH Zürich, Zürich.

Christensen, L. (2005). Possible explanations for an increasing share of no-trip respondents. In: P. Stopher & C. Stecher (Eds), *Travel survey methods: Quality and future directions*. Oxford: Elsevier.

Contrino, H. (2009). Nonresponse challenges in GPS-based surveys: Synthesis of a workshop. In: P. Bonnel, M. Lee-Gosselin, J. Zmud & J.-L. Madre (Eds), *Transport survey methods: Keeping up with a changing world*. Bingley, UK: Emerald.

Hojman, P., Ortúzar, J. de D., & Rizzi, L. (2004). Internet-based surveys to elicit the value of risk reductions. 7th International Conference on Survey Methods in Transport, Costa Rica, August 1–6, 2004.

Kuhnimhof, T. (2007). Regional longitudinal travel demand surveys in Germany — Profiting from synergies. COST 355 meeting in Madrid, May 10, 2007.

Last, J. (2005). The INVERMO survey design (composition from cross-sectional, longitudinal, panel and SP) and the data collected. COST 355 meeting in Berlin, November 25, 2005.

Lelièvre, E. (1999). Collecter des données de mobilité: des histoires migratoires aux biographies d'entourage. In: *Espace, Populations et Sociétés, La mobilité spatiale des populations*, no. 1999/2, pp. 195–205.

Lelièvre, E., Vivier, G., & Bonvalet, C. (2002). L'approche de la mobilité à partir des données de l'enquête Biographies et entourage. In: J. P. Levy & F. Dureau (Eds), *L'accès à la ville. Les mobilités spatiales en question*. Paris: L'Harmattan.

Papon, F., Hubert, J.-P., & Armoogum, J. (2007). *Biography and primary utility of travel: New issues in the measurement of social contexts in the next French National Travel Survey* (June 24–28). Berkeley, CA: WCTR.

Pooley, C. G., & Turnbull, J. (2000). Modal choice and modal change: The journey to work in Britain since 1890. *Journal of Transport Geography*, *8*, 11–24.

Riandey, B. (1985). L'enquête Biographie familiale, professionnelle et migratoire (INED 1981). Le bilan de la collecte. In: *Chaire Quételet 1983 — Migrations internes. Collecte des données et méthodes d'analyse* (pp. 117–149). Département de démographie, Université Catholique de Louvain, Belgium.

Stopher, P., Clifford, E., & Montes, M. (2008). Variability of travel over multiple days: Analysis of three panel waves. *Transportation Research Record*, *2054*, 56–63.

Stopher, P., Greaves, S. P., Fitzgerald, C., Lauer, N., Xu, M., & Perkins, A. (2006). A panel approach to evaluating voluntary travel behavior change programs — South Australia pilot survey. 85th Annual Meeting of the Transportation Research Board (TRB), Washington, DC, United States, 26th January, 2006.

Stopher, P., Kockleman, K., Greaves, S. P., & Clifford, E. (2008). Reducing burden and sample sizes in multiday household travel surveys. *Transportation Research Record*, *2064*, 12–18.

Stopher, P., & Swann, N. (2008). Six-wave odometer panel for evaluation of voluntary travel behavior change programs. *Transportation Research Record*, *2049*, 119–127.

Stopher, R. P. (1992). Use of an activity-based diary to collect household travel data. *Transportation*, *19*(2), 159–176.

Zumkeller, D., Madre, J.-L., Chlond, B., & Armoogum, J. (2006). Panel surveys. In: P. Stopher & C. Stecher (Eds), *Travel survey methods: Quality and future directions*. Oxford: Elsevier.

Subject Index